Energy and Civilization

Energy and Civilization

A History

Vaclav Smil

The MIT Press
Cambridge, Massachusetts
London, England

This book was set in Stone Sans and Stone Serif by Toppan Best-set Premedia Limited.
Printed on recycled paper and bound in the United States of America.

Library of Congress Cataloging-in-Publication Data

Names: Smil, Vaclav, author.
Title: Energy and civilization : a history / Vaclav Smil.
Other titles: Energy in world history
Description: Cambridge, MA : The MIT Press, [2017] | Revised edition of: Energy in
    world history / Vaclav Smil. 1994. | Includes bibliographical references and
    index.
Identifiers: LCCN 2016030742 | ISBN 9780262035774 (hardcover : alk. paper)
Subjects: LCSH: Power resources--History. | Power resources--Social aspects. |
    Technology and civilization. | Energy consumption--Social aspects.
Classification: LCC TJ163.5 .S623 2017 | DDC 333.7909--dc23 LC record available
    at https://lccn.loc.gov/2016030742

10 9 8 7 6 5 4 3 2 1

Contents

Preface and Acknowledgments

I finished writing *Energy in World History* in July 1993; the book came out in 1994, and it remained in print for two decades. Since 1994 energy studies have been through a period of great expansion, and I have added to it by publishing nine books dealing explicitly with energy matters and a dozen interdisciplinary books with significant energy components. Consequently, once I decided to revisit this fascinating topic it was obvious that superficial updating would not do. As a result, this is a substantially new book with a new title: the text is nearly 60% longer than the original, and there are 40% more images and more than twice as many references. Boxes sprinkled throughout the book contain some surprising calculations, as well as many detailed explanations of important topics and essential tables. I have also quoted from sources ranging from the Classics—Apuleius, Lucretius, Plutarch—to nineteenth- and twentieth-century observers such as Braudel, Eden, Orwell, and Senancour. Graphics were updated and created by Bounce Design in Winnipeg; two dozen archival photographs were secured from Corbis in Seattle by Ian Saunders and Anu Horsman. As is always the case with interdisciplinary studies of this kind, this book would not have been possible without the work of hundreds of historians, scientists, engineers, and economists.

Winnipeg, August 2016

1 Energy and Society

Energy is the only universal currency: one of its many forms must be transformed to get anything done. Universal manifestations of these transformations range from the enormous rotations of galaxies to thermo-nuclear reactions in stars. On Earth they range from the terra-forming forces of plate tectonics that part ocean floors and raise new mountain ranges to the cumulative erosive impacts of tiny raindrops (as the Romans knew, *gutta cavat lapidem non vi, sed saepe cadendo*—A drop of water hollows a stone not by force but by continually dripping). Life on Earth—despite decades of attempts to catch a meaningful extraterrestrial signal, still the only life in the universe we know of—would be impossible without the photosynthetic conversion of solar energy into phytomass (plant biomass). Humans depend on this transformation for their survival, and on many more energy flows for their civilized existence. As Richard Adams (1982, 27) put it,

> We can think thoughts wildly, but if we do not have the wherewithal to convert them into action, they will remain thoughts. ... History acts in unpredictable ways. Events in history, however, necessarily take on a structure or organization that must accord with their energetic components.

The evolution of human societies has resulted in larger populations, a growing complexity of social and productive arrangements, and a higher quality of life for a growing number of people. From a fundamental bio-physical perspective, both prehistoric human evolution and the course of history can be seen as the quest for controlling greater stores and flows of more concentrated and more versatile forms of energy and converting them, in more affordable ways at lower costs and with higher efficiencies, into heat, light, and motion. This tendency has been generalized by Alfred Lotka (1880–1949), an American mathematician, chemist, and statistician, in his law of maximum energy: "In every instance considered, natural selection will so operate as to increase the total mass of the organic system, to increase the rate of circulation of matter through the system, and to increase the total energy flux through the system so long as there is present an unutilized residue of matter and available energy" (Lotka 1922, 148).

The history of successive civilizations, the largest and most complex organisms in the biosphere, has followed this course. Human dependence on ever higher energy flows can be seen as an inevitable continuation of organismic evolution. Wilhelm Ostwald (1853–1932, recipient of the 1909 Nobel Prize in Chemistry for his work on catalysis) was the first scientist to expand explicitly "the second law of energetics to all and any action and in particular to the totality of human actions. ... All energies are not ready for this transformation, only certain forms which have been therefore given the name of the free energies. ... Free energy is therefore the capital consumed by all creatures of all kinds and by its conversion everything is done" (Ostwald 1912, 83). This led him to formulate his energetic imperative: "Vergeude keine Energie, verwerte sie"—"Do not waste any energy, make it useful" (Ostwald 1912, 85).

Three quotations illustrate how Ostwald's followers have been restating his conclusions and how some of them have made the link between energy and all human affairs even more deterministically explicit. In the early 1970s Howard Odum (1924–2002) offered a variation on Ostwald's key theme: "The availability of power sources determines the amount of work activity that can exist, and control of these power flows determines the power in man's affairs and in his relative influence on nature" (Odum 1971, 43). In the late 1980s Ronald Fox concluded a book on energy in evolution by writing that "a refinement in cultural mechanisms has occurred with every refinement of energy flux coupling" (Fox 1988, 166).

One does not have to be a scientist to make the link between energy supply and social advances. This is Eric Blair (George Orwell, 1903–1950), writing in 1937 in the second chapter of *The Road to Wigan Pier*, after his visit to an underground coal mine:

> Our civilization, pace Chesterton, is founded on coal, more completely than one realizes until one stops to think about it. The machines that keep us alive, and the machines that make machines, are all directly or indirectly dependent upon coal. In the metabolism of the Western world the coal-miner is second in importance only to the man who ploughs the soil. He is a sort of caryatid upon whose shoulders nearly everything that is not grimy is supported. For this reason the actual process by which coal is extracted is well worth watching, if you get the chance and are willing to take the trouble. (Orwell 1937, 18)

But restating that fundamental link (as Orwell did) and claiming that cultural refinements have taken place with every refinement of energy flux (as Fox does) are two different things. Orwell's conclusion is unexceptionable. Fox's phrasing is clearly a restatement of a deterministic view expressed two generations earlier by the anthropologist Leslie White (1900–1975), who called it the first important law of cultural development: "Other things being equal, the degree of cultural development varies directly as the amount of energy per capita per year harnessed and put to work" (White 1943, 346). While there can be no dispute either about Ostwald's fundamental formulation or about energy's all-encompassing effect on the structure and dynamics of evolving societies (pace Orwell), a deterministic linking of the level of energy use with *cultural* achievements is a highly arguable proposition. I examine that causality (or the lack of it) in the book's closing chapter.

The fundamental nature of the concept is not in doubt. As Robert Lindsay (1975, 2) put it,

> If we can find a single word to represent an idea which applies to every element in our existence in a way that makes us feel we have a genuine grasp of it, we have achieved something economical and powerful. This is what has happened with the idea expressed by the word energy. No other concept has so unified our understanding of experience.

But what is energy? Surprisingly, even Nobel Prize winners have great difficulty in giving a satisfactory answer to that seemingly simple question. In his famous *Lectures on Physics*, Richard Feynman (1918–1988) stressed that "it is important to realize that in physics today, we have no knowledge of what energy is. We do not have a picture that energy comes in little blobs of a definite amount" (Feynman 1988, 4–2).

What we do know is that all matter is energy at rest, that energy manifests itself in a multitude of ways, and that these distinct energy forms are linked by numerous conversions, many of them universal, ubiquitous, and incessant, others highly localized, infrequent, and ephemeral (fig. 1.1). The understanding of these stores, potentials, and transformations was rapidly expanded and systematized mostly during the nineteenth century, and this knowledge was perfected during the twentieth century when— a telling comment on the complexities of energy transformations—we understood how to release nuclear energy sooner (theoretically by the late 1930s, practically by 1943, when the first reactor began to operate) than we knew how photosynthesis works (its sequences were unraveled only during the 1950s).

TO \ FROM	ELECTRO-MAGNETIC	CHEMICAL	NUCLEAR	THERMAL	KINETIC	ELECTRICAL
ELECTRO-MAGNETIC		CHEMILUMINES-CENCE	NUCLEAR BOMBS	THERMAL RADIATION	ACCELERATING CHARGES	ELECTRO-MAGNETIC RADIATION
CHEMICAL	PHOTO-SYNTHESIS	CHEMICAL PROCESSING		BOILING	DISSOCIATION BY RADIOLYSIS	ELECTROLYSIS
NUCLEAR	GAMMA-NEUTRON REACTIONS					
THERMAL	SOLAR ABSORPTION	COMBUSTION	FISSION / FUSION	HEAT EXCHANGE	FRICTION	RESISTANCE HEATING
KINETIC	RADIOMETERS	METABOLISM	RADIOACTIVITY / NUCLEAR BOMBS	THERMAL EXPANSION / INTERNAL COMBUSTION	GEARS	ELECTRIC MOTORS
ELECTRICAL	SOLAR CELLS	FUEL CELLS / BATTERIES	NUCLEAR BATTERIES	THERMO-ELECTRICITY	ELECTRICITY GENERATORS	

Figure 1.1
Matrix of energy conversions. Where more possibilities exist, no more than two leading transformations are identified.

Flows, Stores, and Controls

All known forms of energy are critical for human existence, a reality precluding any rank ordering of their importance. Much in the course of history has been determined and circumscribed by both universal and planetary flows of energy and by their regional or local manifestations. The fundamental features of the universe are governed by gravitational energy, which orders countless galaxies and star systems. Gravity also keeps our planet orbiting at just the right distance from the Sun, and it holds a sufficiently massive atmosphere that makes Earth habitable (box 1.1).

As with all active stars, fusion powers the Sun, and the product of those thermonuclear reactions reaches Earth as electromagnetic (solar, radiant) energy. Its flux ranges over a broad spectrum of wavelengths, including visible light. About 30% of this enormous flow is reflected by clouds and surfaces, about 20% is absorbed by the atmosphere and clouds, and the remainder, roughly half of the total inflow, is absorbed by oceans and

Box 1.1
Gravity and the habitability of Earth

Extreme tolerances of carbon-based metabolism are determined by the freezing point of water, whose liquid form is required for the formation and reactions of organic molecules (the lower bound), and by temperatures and pressures that destabilize amino acids and break down proteins (the upper bound). Earth's continuously habitable zone—the range of orbital radius ensuring optimal conditions for a life-supporting planet—is very narrow (Perkins 2013). A recent calculation concluded that we are even closer to the limit than previously thought: Kopparapu and co-workers (2014) concluded that, given its atmospheric composition and pressure, Earth orbits at the inner edge of the habitable zone, just outside the radius where the runaway greenhouse effect would bring intolerably high temperatures.

About two billion years ago enough carbon dioxide (CO_2) was sequestered by the ocean and by archaea and algae to prevent this effect on Earth, but if the planet had been a mere 1% farther from the Sun, virtually all of its water would have been locked in glaciers. And even with temperatures within an optimum band the planet could not support highly diversified life without its unique atmosphere, dominated by nitrogen, enriched by oxygen from photosynthesis, and containing a number of important trace gases regulating surface temperature—but this thin gaseous envelope could not persist without the planet being sufficiently large to exert enough gravity to hold the atmosphere in place.

continents, gets converted to thermal energy, and is reradiated into space (Smil 2008a). The geothermal energy of Earth adds up to a much smaller heat flux: it results from the original gravitational accretion of the planetary mass and from the decay of radioactive matter, and it drives grand tectonic processes, which keep reordering oceans and continents and cause volcanic eruptions and earthquakes.

Only a tiny part of the incoming radiant energy, less than 0.05%, is transformed by photosynthesis into new stores of chemical energy in plants, providing the irreplaceable foundation for all higher life. Animate metabolism reorganizes nutrients into growing tissues and maintains bodily functions and constant temperature in all higher species. Digestion also generates the mechanical (kinetic) energy of working muscles. In their energy conversions, animals are inherently limited by the size of their bodies and by the availability of accessible nutrition. A fundamental distinguishing characteristic of our species has been the extension of these physical limits through a more efficient use of muscles and through the harnessing of energies outside our own bodies.

Unlocked by human intellect, these extrasomatic energies have been used for a growing variety of tasks, both as more powerful prime movers and as fuels whose combustion releases heat. The triggers of energy supplies depend on the flow of information and on an enormous variety of artifacts. These devices have ranged from such simple tools as hammerstones and levers to complex fuel-burning engines and reactors releasing the energy of nuclear fission. The basic evolutionary and historical sequence of these advances is easy to outline in broad qualitative terms. As with any nonphotosynthesizing organism, the most fundamental human energy need is for food. Foraging and scavenging by hominins were very similar to the food acquisition practices of their primate ancestors. Although some primates—as well as a few other mammals (including otters and elephants), some birds (ravens and parrots), and even some invertebrates (cephalopods)—have evolved a small repertory of rudimentary tool-using capabilities (Hansell 2005; Sanz, Call, and Boesch 2014; fig. 1.2), only hominins made toolmaking a distinguishing mark of their behavior.

Figure 1.2
Chimpanzee (*Pan troglodytes*) in Gabon using tools to crack nuts (Corbis).

Tools have given us a mechanical advantage in the acquisition of food, shelter, and clothing. The mastery of fire greatly extended our range of habitation and set us further apart from animals. New tools led to the harnessing of domesticated animals, the building of more complex muscle-powered machines, and the conversion of a tiny fraction of the kinetic energies of wind and water to useful power. These new prime movers multiplied the power under human command, but for a very long time their use was circumscribed by the nature and magnitude of the captured flows. Most obviously, this was the case with sails, ancient and effective tools whose capabilities were restricted for millennia by prevailing wind flows and persistent ocean currents. These grand flows steered the late fifteenth-century European transatlantic voyages to the Caribbean. They also prevented the Spaniards from discovering Hawaii, even though Spanish trading ships, the Manila Galleons (Galeón de Manila), sailed once or twice a year across the Pacific from Mexico (Acapulco) to the Philippines for 250 years between 1565 and 1815 (Schurz 1939).

Controlled combustion in fireplaces, stoves, and furnaces turned the chemical energy of plants into thermal energy. This heat has been used directly in households and in smelting metals, firing bricks, and processing and finishing countless products. The combustion of fossil fuels made all of these traditional direct uses of heat more widespread and more efficient. A number of fundamental inventions made it possible to convert thermal energy from the burning of fossil fuels to mechanical energy. This was done first in steam and internal combustion engines, then in gas turbines and rockets. We have been generating electricity by burning fossil fuels, as well by harnessing the kinetic energy of water, since 1882, and by fissioning a uranium isotope since 1956.

The combustion of fossil fuels and the generation of electricity created a new form of high-energy civilization whose expansion has now encompassed the whole planet and whose primary energy sources now include small but rapidly rising shares of new renewable sources, especially solar (harnessed by photovoltaic devices or in concentrating solar power plants) and wind (converted by large wind turbines). In turn, these advances have been predicated on a concatenation of other developments. To use a flow-model analogy, a combination of gates (valves) had to be set up and activated in proper sequence to enable the flow of human ingenuity.

The most notable gates required to release great energy potentials include requisite educational opportunities, predictable legal arrangements, transparent economic rules, the adequate availability of capital, and conditions conducive to basic research. Not surprisingly, it usually takes generations to allow much increased or qualitatively improved energy flows or to harness entirely new sources of energy on a significant scale. Timing, overall power, and the composition of the resulting energy flows are exceedingly difficult to predict, and during the earliest phases of such transitions it is impossible to appraise all eventual impacts that changing prime movers and fuel bases will have on farming, industries, transport, settlements, warfare, and Earth's environment. Quantitative accounts are essential to appreciate the constraints of our actions and the extent of our achievements, and they require knowledge of basic scientific concepts and measures.

Concepts and Measures

Several first principles underlie all energy conversions. Every form of energy can be turned into heat, or thermal energy. No energy is ever lost in any of these conversions. Conservation of energy, the first law of thermodynamics, is one of the most fundamental universal realities. But as we move along conversion chains, the potential for useful work steadily diminishes (box 1.2). This inexorable reality defines the second law of thermodynamics, and entropy is the measure associated with this loss of useful energy. While the energy content of the universe is constant, conversions of energies increase its entropy (decrease its utility). A basketful of grain or a barrelful of crude oil is a low-entropy store of energy, capable of much useful work once metabolized or burned, and it ends up as the random motion of slightly heated air molecules, an irreversible high-entropy state that represents an irretrievable loss of utility.

This unidirectional entropic dissipation leads to a loss of complexity and to greater disorder and homogeneity in any closed system. But all living organisms, whether the smallest bacteria or a global civilization, temporarily defy this trend by importing and metabolizing energy. This means that every living organism must be an open system, maintaining a continuous inflow and outflow of energy and matter. As long as they are alive, these systems cannot be in a state of chemical and thermodynamic equilibrium (Prigogine 1947, 1961; von Bertalanffy 1968; Haynie 2001). Their negentropy—their growth, renewal, and evolution—results in greater heterogeneity and increasing structural and systemic complexity. As with so many other scientific advances, a coherent understanding of these realities came only during the nineteenth century, when the rapidly evolving disciplines of physics, chemistry, and biology found a common concern in studying transformations of energy (Atwater and Langworthy

Box 1.2

Diminishing utility of converted energy

Any energy conversion illustrates the principle. If an American reader uses electric light to illuminate this page, the electromagnetic energy of that light is only a small part of the chemical energy contained in the lump of coal used to generate it (in 2015 coal was used to produce 33% of the electricity generated in the United States). At least 60% of coal's energy was lost as heat through a plant chimney and in cooling water, and if a reader uses an old incandescent light, then more than 95% of delivered electricity ends up as heat generated as the metal of the bulb's coiled filament resists the electric current. The light reaching the page either is absorbed by it or is reflected and absorbed by its surroundings and reradiated as heat. The initial low-entropy input of coal's chemical energy has been dissipated as diffused high-entropy heat that warmed the air above the station, along the wires, and around the light bulb and caused an imperceptible temperature increase above a page. No energy has been lost, but a highly useful form was degraded to the point of no practical utility.

1897; Cardwell 1971; Lindsay 1975; Müller 2007; Oliveira 2014; Varvoglis 2014).

These fundamental interests needed a codification of standard measurements. Two units became common for measuring *energy*: calorie, a metric unit, and the British thermal unit (Btu). Today's basic scientific unit of energy is the joule, named after an English physicist, James Prescott Joule (1818–1889), who published the first accurate calculation of the equivalence of work and heat (box 1.3). *Power* denotes the rate of energy flow. Its first standard unit, horsepower, was set by James Watt (1736–1819). He wanted to charge for his steam engines on a readily understandable basis, and so he chose the obvious comparison with the prime mover they were to replace, a harnessed horse commonly used to power a mill or a pump (fig. 1.3, box 1.3).

Another important rate is *energy density*, the amount of energy per unit mass of a resource (box 1.4). This value is of a critical importance for foodstuffs: even where abundant, low-energy-density foods could never become staples. For example, the pre-Hispanic inhabitants of the basin of Mexico always ate plenty of prickly pears, which were easy to gather from the many species of cacti belonging to the genus *Opuntia* (Sanders, Parsons, and Santley 1979). But, as with most fruits, pear's pulp is overwhelmingly (about

---

Box 1.3

Measuring energy and power

The official definition of a joule is the work accomplished when a force of one newton acts over a distance of one meter. Another option is to define a basic energy unit through heat requirements. One calorie is the amount of heat needed to raise the temperature of 1 cm^3 of water by 1°C. That is a tiny amount of energy: to do the same for 1 kg of water calls for a thousand times more energy, or one kilocalorie (for the complete list of multiplier prefixes, see "Basic Measures" in the Addenda). Given the equivalence of heat and work, all that is required to convert calories to joules is to remember that one calorie equals roughly 4.2 joules. The conversion is equally simple for the still common nonmetric English measure, the British thermal unit. One Btu contains roughly 1,000 J (1,055, to be exact). A good comparative yardstick is the average daily food need. For most moderately active adults it is 2–2.7 Mcal, or about 8–11 MJ, and 10 MJ could be supplied by eating 1 kg of whole wheat bread.

In 1782 James Watt calculated in his *Blotting and Calculation Book* that a mill horse works at a rate of 32,400 foot-pounds a minute—and the next year he rounded this to 33,000 foot-pounds (Dickinson 1939). He assumed an average walking speed of about 3 feet per second, but we do not know where he got his figure for an average pull of about 180 pounds. Some large animals were that powerful, but most horses in eighteenth-century Europe could not sustain the rate of one horsepower. Today's standard unit of power, one watt, is equal to the flow of one joule per second. One horsepower is equal to about 750 watts (745.699, to be exact). Consuming 8 MJ of food a day corresponds to a power rate of 90 W (8 MJ/24 h × 3,600 s), less than the rating of a standard light bulb (100 W). A double toaster needs 1,000 W, or 1 kW; small cars deliver around 50 kW; a large coal-fired or nuclear power plant produces electricity at the rate of 2 GW.

88%) water, with less than 10% carbohydrates, 2% protein, and 0.5% lipids, and has an energy density of just 1.7 MJ/kg (Feugang et al. 2006). This means that even a small woman surviving only on the carbohydrates of cactus pears (assuming, unrealistically, virtually no need for the other two macronutrients) would have to eat 5 kg of the fruit every day—but she could get the same amount of energy from only about 650 g of ground corn consumed as tortillas or tamales.

Power density is the rate at which energies are produced or consumed per unit of area and hence is a critical structural determinant of energy

Figure 1.3
Two horses turning a capstan geared to pumping well water in a mid-eighteenth-century French carpet manufactory (reproduced from the *Encyclopédie* [Diderot and d'Alembert 1769–1772]). An average horse of that period could not sustain a steady work rate of one horsepower. James Watt used an exaggerated rating to ensure customers' satisfaction with his horsepower-denominated steam engines installed to place harnessed animals.

Box 1.4
Energy densities of foodstuffs and fuels

Ranking	Examples	Energy density (MJ/kg)
Foodstuffs		
Very low	Vegetables, fruits	0.8–2.5
Low	Tubers, milk	2.5–5.0
Medium	Meats	5.0–12.0
High	Cereal and legume grains	12.0–15.0
Very high	Oils, animal fats	25.0–35.0
Fuels		
Very low	Peats, green wood, grasses	5.0–10.0
Low	Crop residues, air-dried wood	12.0–15.0
Medium	Dry wood	17.0–21.0
	Bituminous coals	18.0–25.0
High	Charcoal, anthracite	28.0–32.0
Very high	Crude oils	40.0–44.0

Sources: Specific energy densities for individual foodstuffs and fuels are listed in Merrill and Watt (1973), Jenkins (1993), and USDA (2011).

systems (Smil 2015b). For example, city size in all traditional societies depended on fuelwood and charcoal, and it was clearly limited by the inherently low power density of phytomass production (box 1.5, fig. 1.4). The power density of sustainable annual tree growth in temperate climates is at best equal to 2% of the power density of energy consumption for traditional urban heating, cooking, and manufactures. Consequently, cities had to draw on nearby areas at least 30 times their size for fuel supply. This reality restricted their growth even where other resources, such as food and water, were adequate.

Yet another rate, one that has assumed great importance with advancing industrialization, is the *efficiency of energy conversions*. This ratio of output/input describes the performance of energy converters, be they stoves, engines, or lights. While we cannot do anything about the entropic dissipation, we can improve the efficiency of conversions by lowering the amount of energy required to perform specific tasks (box 1.6). There are fundamental (thermodynamic, mechanical) constraints to these improvements, but we have pushed some processes close to the practical efficiency limits,

Box 1.5
Power densities of phytomass fuels

Photosynthesis converts less than 0.5% of incoming solar radiation into new phytomass. The best annual fuelwood productivities of traditional fast-growing species (poplars, eucalyptus, pines) were no more than 10 t/ha, and in drier regions the rates were between 5 and 10 t/ha (Smil 2015b). With the energy density of dry wood averaging 18 GJ/t, the harvest of 10 t/ha would translate into a power density of about 0.6 W/m^2: (10 t/ha × 18 GJ)/3.15 × 107 (seconds in one year) = ~5,708 W; 5,708 W/10,000 m^2 (ha) = ~0.6 W/m^2. A large eighteenth-century city would have required at least 20–30 W/m^2 of its built-up area for heating, cooking, and artisanal manufactures, and its fuelwood would have had to come from an area at least 30 and up to 50 times its size.

But cities required plenty of charcoal, the only preindustrial smokeless fuel preferred for indoor heating by all traditional civilizations, and charcoaling entailed further substantial energy loss. Even by the mid-eighteenth century the typical charcoal-to-wood ratio was still as high as 1:5, which means that in energy terms (with dry wood at 18 GJ/t and charcoal [virtually pure carbon] at 29 GJ/t), this conversion was only about 30% efficient (5 × 18/29 = 0.32), and the power density of wood harvests destined for charcoal production was only about 0.2 W/m^2. Consequently, large preindustrial cities located in a northern temperate climate and relying heavily on charcoal (China's Xi'an or Beijing would be good examples) would have required a wooded area at least 100 times their size to ensure a continuous supply of that fuel.

Figure 1.4
Charcoaling in early seventeenth-century England as depicted in John Evelyn's *Silva* (1607).

Box 1.6
Efficiency improvements and the Jevons paradox

Technical advances have brought many impressive efficiency gains, and the history of lighting offers one of the best examples (Nordhaus 1998; Fouquet and Pearson 2006). Candles convert just 0.01% of chemical energy in tallow or wax to light. Edison's light bulbs of the 1880s were roughly ten times as efficient. By 1900, coal-fired electricity-generating plants had efficiencies of just 10%; light bulbs turned no more than 1% of electricity into light, and hence about 0.1% of coal's chemical energy appeared as light (Smil 2005). The best combined-cycle gas turbine plants (using hot gas exiting a gas turbine to produce steam for a steam turbine) are now about 60% efficient, while fluorescent lights have efficiencies up to 15%, as do light-emitting diodes (USDOE 2013). This means that about 9% of energy in natural gas ends up as light, a 90-fold gain since the late 1880s. Such gains have saved capital and operation costs and lowered environmental impacts.

But in the past, the rise of conversion efficiency did not necessarily result in actual energy savings. In 1865 Stanley Jevons (1835–1882), an English economist, pointed out that the adoption of more efficient steam engines was accompanied by large increases in coal consumption and concluded, "It is wholly a confusion of ideas to suppose that the economical use of fuels is equivalent to a diminished consumption. The very contrary is the truth. As a rule, new modes of economy will lead to an increase of consumption according to a principle recognized in many parallel instances" (Jevons 1865, 140). This reality has been confirmed by many studies (Herring 2004, 2006; Polimeni et al. 2008), but in affluent countries, those whose high per capita energy use has approached, or already reached, saturation levels, the effect has been getting weaker. As a result, rebounds attributable to higher efficiency at the end-use level are often small and decrease over time, and specific economy-wide rebounds may be trivial, and even net positive (Goldstein, Martinez, and Roy 2011).

though in most instances, including such common energy converters as internal combustion engines and lights, there is still much room for further improvement.

When efficiencies are calculated for the production of foodstuffs (energy in food/energy in inputs to grow it), fuels, or electricity they are usually called *energy returns*. Net energy returns in every traditional agriculture relying solely on animate power had to be considerably greater than one: edible harvests had to contain more energy than the amount consumed as food and feed needed not only by people and animals producing those crops but also by their nonworking dependents. An insurmountable problem arises when we try to compare energy returns in traditional agricultures that were powered solely by animate energies (and hence involved only transformations of recently received solar radiation) with those in modern farming, which is subsidized directly (fuel for field operations) and indirectly (the energies needed to synthesize fertilizers and pesticides and to build farm machinery) and hence has, invariably, lower energy returns than traditional cropping (box 1.7).

Finally, *energy intensity* measures the cost of products, services, and even of aggregate economic output, in standard energy units—and of energy itself. Among the commonly used materials, aluminum and plastics are highly energy-intensive, while glass and paper are relatively cheap, and lumber (excluding its photosynthetic cost) is the least energy-intensive widely deployed material (box 1.8). The technical advances of the past two centuries have brought many substantial declines in energy intensities. Perhaps most notably, the coke-fueled smelting of pig iron in

---

Box 1.7

Comparison of energy returns in food production

Since the early 1970s, energy ratios have been used to illustrate the superiority of traditional farming and the low energy returns of modern agriculture. Such comparisons are misleading owing to a fundamental difference between the two ratios. Those for traditional farming are simply quotients of the food energy harvested in crops and the food and feed energy needed to produce those harvests by deploying human and animal labor. In contrast, in modern farming the denominator is composed overwhelmingly of nonrenewable fossil fuel inputs needed to power field machinery and to make machines and farm chemicals; labor inputs are negligible.

If the ratios were calculated merely as quotients of edible energy output to labor input, then modern systems, with their miniscule amount of human effort and with no draft animals, would look superior to any traditional practice. If the cost of producing a modern crop included all converted fossil fuels and electricity converted to a common denominator, then the energy returns in modern agriculture would be substantially below traditional returns. Such a calculation is possible because of the physical equivalence of energies. Both food and fuels can be expressed in identical units, but an obvious "apples and oranges" problem remains: there is no satisfactory way to compare, simply and directly, the energy returns of the two farming systems that depend on two fundamentally different kinds of energy inputs.

Box 1.8
Energy intensities of common materials

Material	Energy cost (MJ/kg)	Process
Aluminum	175–200	Metal from bauxite
Bricks	1–2	Fired from clay
Cement	2–5	From raw materials
Copper	90–100	From ore
Explosives	10–70	From raw materials
Glass	4–10	From raw materials
Gravel	<1	Excavated
Iron	12–20	From iron ore
Lumber	1–3	From standing timber
Paper	23–35	From standing timber
Plastics	60–120	From hydrocarbons
Plywood	3–7	From standing timber
Sand	<1	Excavated
Steel	20–25	From pig iron
Steel	10–12	From scrap metal
Stone	<1	Quarried

Source: Data from Smil (2014b).

large blast furnaces now requires less than 10% of energy per unit mass of hot metal than did the preindustrial charcoal-based production of pig iron (Smil 2016).

The energy cost of energy (often called EROI, energy return on investment, although EROEI, energy return on energy investment, would be more correct) is a revealing measure only if we compare values that have been calculated by identical methods using standard assumptions and clearly identified analytical boundaries. Modern high-energy societies have preferred to develop fossil fuel resources with the highest net energy returns, and that is a major reason why we have favored crude oil in general, and the rich Middle Eastern fields in particular; oil's high energy density, and hence easy transportability, are other obvious advantages (box 1.9).

Box 1.9

Energy returns on energy investment

Differences in the quality and accessibility of fossil fuels are enormous: thin underground seams of low-quality coal versus a thick layer of good bituminous coal that can be extracted in open-cast mines, or supergiant Middle Eastern hydrocarbons fields versus low-productivity wells that require constant pumping. As a result, specific EROEI values differ substantially—and they can change with the development of more efficient recovery techniques. The following ranges are only approximate indicators, illustrating differences among leading extraction and conversion methods (Smil 2008a; Murphy and Hall 2010). For coal production they range between 10 and 80, while for oil and gas they have ranged from 10 to far above 100; for large wind turbines in the windiest locations they may approach 20 but are mostly less than 10; for photovoltaic solar cells they are no higher than 2; and for modern biofuels (ethanol, biodiesel) they are at best only 1.5, but their production has often entailed an energy loss or no net again (an EROEI of just 0.9–1.0).

Complexities and Caveats

Using standard units to measure energy storages and flows is physically straightforward and scientifically impeccable—yet these reductions to a common denominator are also misleading. Above all, they cannot capture critical qualitative differences among various energies. Two kinds of coal may have an identical energy density, but one may burn very cleanly and leave behind only a small amount of ash, while the other may smoke heavily, emit a great deal of sulfur dioxide, and leave a large incombustible residue. An abundance of high-energy-density coal ideal for fueling steam engines (the often used adjective "smokeless" must be seen in relative terms) was clearly a major factor contributing to the British dominance of nineteenth-century maritime transport, as neither France nor Germany had large coal resources of comparable quality.

Abstract energy units cannot differentiate between edible and inedible biomass. Identical masses of wheat and dry wheat straw contain virtually the same amount of heat energy, but straw, composed mostly of cellulose, hemicellulose, and lignin, cannot be digested by humans, while wheat (comprised of about 70% complex starchy carbohydrates and up to 14% protein) is an excellent source of basic nutrients. They also hide the specific origin of food energy, a matter of great importance for proper nutrition.

Many high-energy foods contain no, or hardly any, protein and lipids, two nutrients required for normal body growth and maintenance, and they may not provide any essential micronutrients—vitamins and minerals.

There are other important qualities hidden by abstract measures. Access to energy stores is obviously a critical matter. Tree stem wood and branch wood have the same energy densities, but without good axes and saws, people in many preindustrial societies could only gather the latter fuel. That is still the norm in the poorest parts of Africa or Asia, where children and women gather woody phytomass; and its form, and hence its transportability, also matters because they have to carry wood (branch) loads home on their heads, often for considerable distances. Ease of use and conversion efficiency can be decisive in choosing a fuel. A house can be heated by wood, coal, fuel oil, or natural gas, but the best gas furnaces are now up to 97% efficient, hence far cheaper to operate than any other option.

Burning straw in simple stoves requires frequent stoking, while large wood pieces can be left burning unattended for hours. Unvented (or poorly vented, through a hole in the ceiling) indoor cooking with dry dung produces much more smoke than the burning of seasoned wood in a good stove, and indoor biomass combustion remains a major source of respiratory illnesses in many low-income countries (McGranahan and Murray 2003; Barnes 2014). And unless their origins are specified, densities or energy flows do not differentiate between renewable and fossil energies—yet this distinction is fundamental to understanding the nature and durability of a given energy system. Modern civilization has been created by the massive, and increasing, combustion of fossil fuels, but this practice is clearly limited by their crustal abundance, as well as by the environmental consequences of burning coals and hydrocarbons, and high-energy societies can ensure their survival only by an eventual transition to nonfossil sources.

Further difficulties arise when comparing the efficiencies of animate and inanimate energy conversions. In the latter case it is simply a ratio of fuel or electricity inputs and useful energy output, but in the former case daily food (or feed) intake should not be counted as an energy input of human or animal labor because most of that energy is required for basal metabolism—that is, to support the functioning of the body's vital organs and to maintain steady body temperature—and basal metabolism operates regardless of whether people or animals rest or work. Calculating the net energy cost is perhaps the most satisfactory solution (box 1.10).

But even in much simpler societies than ours a great deal of labor was always mental rather than physical—deciding how to approach a task,

---

Box 1.10
Calculating the net energy cost of human labor

There is no universally accepted way to express the energy cost of human labor, and calculating the net energy cost is perhaps the best choice: it is a person's energy consumption above the existential need that would have to be satisfied even if no work were done. This approach debits human labor with its actual incremental energy cost. Total energy expenditure is a product of basal (or resting) metabolic rate and physical activity level (TEE = BMR × PAL), and the incremental energy cost will obviously be the difference between TEE and BMR. The BMR of an adult man weighing 70 kg would be about 7.5 MJ/d, and for a 60 kg woman it would be about 5.5 MJ/day. If we assume that hard work will raise the daily energy requirement by about 30%, then the net energy cost would be about 2.2 MJ/day for men and 1.7 MJ/day for women, and hence I will use 2 MJ/day in all approximate calculations of net daily energy expenditures in foraging, traditional farming, and industrial work.

Daily food intake should not be counted as an energy input of labor: basal metabolism (to support vital organs, circulate the blood, and maintain a steady body temperature) operates regardless of whether we rest or work. Studies of muscle physiology, especially the work of Archibald V. Hill (1886–1977), recipient of the Nobel Prize in Physiology in 1922), made it possible to quantify the efficiency of muscular work (Hill 1922; Whipp and Wasserman 1969). The net efficiency of steady aerobic performances is about 20%, and this means that 2 MJ/day of metabolic energy attributable to a physical task would produce useful work equal to about 400 kJ/day. I will use this approximation in all relevant calculations. In contrast, Kander, Malanima, and Warde (2013) used total food intake rather than actual useful energy expenditure in their historical comparison of energy sources. They assumed an average annual food intake of 3.9 GJ/capita, unchanged between 1800 and 2008.

---

how to execute it with the limited power available, how to lower energy expenditures—and the metabolic cost of thinking, even very hard thinking, is very small compared to strenuous muscular exertion. On the other hand, mental development requires years of language acquisition, socialization, and learning by mentoring and the accumulation of experience, and as societies progressed, this learning process became more demanding and longer lasting through formal schooling and training, services that have come to require considerable indirect energy inputs to support requisite physical infrastructures and human expertise.

A circle is closed. I have noted the necessity of quantitative evaluations, but the real understanding of energy in history requires much more than reducing everything to numerical accounts in joules and watts and treating them as all-encompassing explanations. I will approach the challenge in both ways: I will note energy and power requirements and densities and point out improving efficiencies, but I will not ignore the many qualitative attributes that constrain or promote specific energy uses. And while the imperatives of energy needs and uses have left a powerful imprint on history, many details, sequences, and consequences of these fundamental evolutionary determinants can be explained only by referring to human motivations and preferences, and by acknowledging those surprising, and often seemingly inexplicable, choices that have shaped our civilization's history.

2 Energy in Prehistory

Understanding the origins of the genus *Homo* and filling in the details of its subsequent evolution is a never-ending quest as new findings push back many old markers and complicate the overall picture with the discovery of species that do not fit easily into an existing hierarchy (Trinkaus 2005; Reynolds and Gallagher 2012). In 2015 the oldest reliably dated hominin remains were those of *Ardipithecus ramidus* (4.4 million years ago, found in 1994) and *Australopithecus anamensis* (4.1–5.2 million years ago, found in 1967). A notable 2015 addition was *Australopithecus deyiremeda* (3.3–3.5 million years ago) from Ethiopia (Haile-Selassie et al. 2015). The sequence of younger hominins includes *Australopithecus afarensis* (unearthed in 1974 in Laetoli, Tanzania, and in Hadar, Ethiopia), *Homo habilis* (discovered in 1960 in Tanzania), and *Homo erectus* (beginning 1.8 million years ago, with many finds in Africa, Asia, and Europe extending to about 250,000 years ago).

Reanalysis of the first *Homo sapiens* bones—Richard Leakey's famous discoveries in Ethiopia starting in 1967—dated them to about 190,000 years ago (McDougall, Brown, and Fleagle 2005). Our direct ancestors thus spent their lives as simple foragers, and it was only about 10,000 years ago that the first small populations of our species began a sedentary existence based on the domestication of plants and animals. This means that for millions of years, the foraging strategies of hominins resembled those of their primate ancestors, but we now have isotopic evidence from East Africa that by about 3.5 million years ago hominin diets began to diverge from those of extant apes. Sponheimer and co-workers (2013) showed that after that time, several hominin taxa began to incorporate ^{13}C-enriched foods (produced by C_4 or crassulacean acid metabolism) in their diets and had a highly variable carbon isotope composition atypical of African mammals. Reliance on C_4 plants is thus of ancient origin, and in modern agriculture two C_4 cultivars, corn and sugar cane, have higher average yields than any other species grown for its grain or sugar content.

The first evolutionary departure that eventually led to our species was not a larger brain size or toolmaking but bipedalism, a structurally improbable yet immensely consequential adaptation whose beginnings can be traced as far back as about seven million years ago (Johanson 2006). Humans are the only mammals whose normal way of locomotion is walking upright (other primates do so only occasionally), and hence bipedalism can be seen as the critical breakthrough adaptation that made us eventually human. Yet bipedalism—essentially a sequence of arrested falls—is inherently unstable and clumsy: "Human walking is a risky business. Without split-second timing man would fall flat on his face; in fact, with each step he takes, he teeters on the edge of catastrophe" (Napier 1970, 165). And besides making us prone to musculoskeletal injuries, bipedalism also leads to age-related bone loss, osteopenia (lower than normal bone density), and osteoporosis (Latimer 2005).

Many answers have been offered to the obvious question of why, then, do it, and some of them, as Johanson (2006) summarily argues, appear quite unpersuasive. To appear taller in order to intimidate predators would have had no effect on wild dogs or cheetahs or hyenas, who are not intimidated by much larger mammalian species. To become upright just to look over tall grass would only have attracted predators; reaching for fruit on low-hanging branches could be done without surrendering rapid quadruped running; and the cooling of bodies could be achieved by resting in shade and foraging only during cooler mornings or evenings. Differences in overall energy expenditure may offer the best explanation (Lovejoy 1988). Hominins, much like other mammals, spend most of their energy in reproduction, feeding, and ensuring safety, and if bipedalism helped to do all of these, then it would have been adopted.

As Johanson (2006, 2) puts it, "Natural selection cannot *create* a behavior like bipedalism, but it can act to select the behavior once it has arisen." Viewed in a narrower sense, it is not clear that bipedalism offered sufficient biomechanical advantage to promote its selection just on the basis of the energy cost of walking (Richmond et al. 2001), though Sockol, Raichlen, and Pontzer (2007), after measuring energy expenditure in walking chimpanzees and adult humans, found that human walking costs about 75% less energy than both quadrupedal and bipedal walking in chimpanzees. The difference owes to the biomechanical differences in anatomy and gait, and above all to the more extended hip and longer hind limb in humans.

Bipedalism started a cascade of enormous evolutionary adjustments (Kingdon 2003; Meldrum and Hilton 2004). Upright walking liberated hominin arms for carrying weapons and for taking food to group sites instead of consuming it on the spot. But bipedalism was necessary to trigger hand dexterity and tool use. Hashimoto and co-workers (2013) concluded that adaptations underlying tool use evolved independently of those required for human bipedalism because in both humans and monkeys, each finger is represented separately in the primary sensorimotor cortex, just as the fingers are physically separated in the hand. This creates the ability to use each digit independently in the complex manipulations required for tool use. But without bipedalism it would be impossible to use the trunk for leverage in accelerating the hand during toolmaking and tool use. Bipedalism also freed the mouth and teeth to develop a more complex call system as the prerequisite of language (Aiello 1996). These developments required larger brains whose energy cost eventually reached three times the level for chimpanzees, accounting for up to one-sixth of the total basal metabolic rate (Foley and Lee 1991; Lewin 2004). The average encephalization quotient (actual/expected brain mass for body weight) is 2–3.5 for primates and early hominins, while for the humans is a bit higher than 6. Three million years ago *Australopithecus afarensis* had a brain volume of less than 500 cm^3; 1.5 million years ago the volume had doubled in *Homo erectus*, and then it increased by roughly 50% in *Homo sapiens* (Leonard, Snodgrass, and Robertson 2007).

A higher encephalization quotient was critical for the rise of social complexity (which raised the survival odds and set hominins apart from other mammals) and was closely related to changes in the quality of food consumed. The brain's specific energy need is roughly 16 times that of skeletal muscles, and the human brain claims 20–25% of resting metabolic energy, compared to 8–10% in other primates and just 3–5% in other mammals (Holliday 1986; Leonard et al. 2003). The only way to accommodate large brain size while maintaining the overall metabolic rate (the human resting metabolism is no higher than that of other mammals of similar mass) was to reduce the mass of other metabolically expensive tissues. Aiello and Wheeler (1995) argued that reducing the size of the gastrointestinal tract was the best option because the gut mass (unlike the mass of hearts or kidneys) can vary substantially, depending on the diet.

Fish and Lockwood (2003), Leonard, Snodgrass, and Robertson (2007), and Hublin and Richards (2009) confirmed that diet quality and brain mass have a significantly positive correlation in primates, and better hominin diets, including meat, supported larger brains, whose high energy need was partly offset by a reduced gastrointestinal tract (Braun et al. 2010). While extant nonhuman primates have more than 45% of their gut mass in the colon and only 14–29% in the small intestine, in humans those shares are reversed, with more than 56% in the small intestine and only 17–25% in the colon, a clear indication of adaptation to high-quality, energy-dense foods (meat, nuts) that can be digested in the small intestine. Increased meat consumption also helps to explain human gains in body mass and height, as well as smaller jaws and teeth (McHenry and Coffing 2000; Aiello and Wells 2002). But a higher meat intake could not change the energy basis of evolving hominins: to secure any food they had to rely only on their muscles and on simple stratagems while gathering, scavenging, hunting, and fishing.

Tracing the genesis of the first wooden tools (sticks and clubs) is impossible as only those artifacts that were preserved in anoxic environments, most commonly in bogs, were able to survive for extended periods. Disintegration is not a problem with the hard stones used to fashion simple tools, and new findings have been pushing back the date of the earliest verifiable hominin stone tools. For several decades the consensus dated the earliest stone tools to about 2.5 million years ago. Cobble-based, these relatively small and simple Oldowan hammerstones (cores with an edge), choppers, and flakes made it much easier to butcher animals and to break their bones (de la Torre 2011). But the latest findings at the Lomekwi site in West Turkana, Kenya, pushed the date of the oldest known stone toolmaking to about 3.3 million years ago (Harmand et al. 2015).

About 1.5 million years ago hominins started to quarry larger flakes to make bifacial hand axes, picks, and cleavers of Acheulean (1.2–0.1 million years ago) style. The chipping of a single core yielded sharp cutting edges less than 20 cm long, and these practices produced a large variety of special handheld stone tools (fig. 2.1). Wooden spears were essential for hunting larger animals. In 1948 a nearly complete spear found inside an elephant skeleton in Germany was dated to the last interglacial period (115,000–125,000 years ago), and in 1996 throwing spears found at a Schöningen open-cast lignite mine were dated to 400,000–380,000 years ago (Thieme 1997), and stone points were hafted to wooden spears beginning about 300,000 years ago.

But new discoveries in South Africa put the earliest date of making hafted multicomponent tools about 200,000 years earlier than previously reported: Wilkins and co-workers (2012) concluded that stone points from Kathu Pan, made about 500,000 years ago, functioned as spear tips. True long-range projectile weaponry evolved in Africa between 90,000 and

Figure 2.1
Acheulean stone tools, first made by *Homo ergaster,* were formed by the removal of stone flakes to create specialized cutting blades (Corbis).

70,000 years ago (Rhodes and Churchill 2009). Another recent South African discovery showed that a significant technical advance—the production of small bladelets (microliths), primarily from heat-treated stone, to be used in making composite tools—took place as early as 71,000 years ago (Brown et al. 2012). Larger composite tools became common only about 25,000 years ago (Europe's Gravettian period) with the production of ground and hafted adzes and axes, and with the more efficient flaking of flint, yielding many sharp-edged tools; harpoons, needles, saws, pottery, and items from woven fibers (clothes, nets, baskets) were also invented and adopted during that time.

Magdalenian techniques (between 17,000 and 12,000 years ago; the era is named after a rock shelter at La Madeleine in southern France where the tools were discovered) produced up to 12 m of microblade edges from a single stone, and experiments with their modern replicas (mounted on spears) show their hunting efficacy (Pétillon et al. 2011). A stone-tipped spear became an even more potent weapon after the invention of spear throwers during the late Paleolithic. A leveraged throw easily doubled the velocity of the weapon and reduced the necessity for a closer approach. Stone-tipped arrows carried these advantages further, with an added gain in accuracy.

We will never know the earliest dates for the controlled use of fire for warmth and cooking: in the open, any relevant evidence was removed by many subsequent events, and in occupied caves it was destroyed by generations of later use. The earliest date for a well-attested use of controlled fire has been receding: Goudsblom (1992) put it at about 250,000 years ago; and a dozen years later Goren-Inbar and co-workers (2004) pushed it as far back as 790,000 years ago, while the fossil record suggests that the consumption of some cooked food took place as early as 1.9 million years ago. But without any doubt, by the Upper Paleolithic—30,000 to 20,000 years ago, when *Homo sapiens sapiens* displaced Europe's Neanderthals—the use of fire was widespread (Bar-Yosef 2002; Karkanas et al. 2007).

Cooking has been always seen as an important component of human evolution, but Wrangham (2009) believes that it had a "monstrous" effect on our ancestors because it greatly expanded the range and quality of available food, and also because its adoption brought many physical changes (including smaller teeth and a less voluminous digestive tract) and behavioral adjustments (such as the need to defend stores of accumulated food, which promoted protective female-male bonds) that led eventually to complex socialization, sedentary lives and "self-domestication." All prehistoric cooking was done with open fires, with meat suspended above the flames, buried in hot embers, placed on hot rocks, encased in a tough skin, covered by clay, or put with hot stones into leather pouches filled with water. Owing to the variety of settings and methods, it is impossible to quote typical fuel conversion efficiencies. Experiments show that 2–10% of wood's energy ends up as useful heat for cooking, and plausible assumptions indicate annual wood consumption maxima of 100–150 kg/year/capita (box 2.1).

Box 2.1

Wood consumption in open-fire cooking of meat

Realistic assumptions for setting plausible maxima of wood consumption in open-fire meat cooking during the late Paleolithic are as follows (Smil 2013a): average daily food energy intake of 10 MJ/capita (adequate for adults, higher than the mean for entire populations), with meat being 80% (8 MJ) of the total food intake; a food energy density of animal carcasses of 8–10 MJ/kg (typical for mammoths, generally 5–6 MJ/kg for large ungulates); an average ambient temperature of 20°C in warm and a mean of 10°C in colder climates; cooked meat at 80°C (77°C suffices for well-done meat); a heat capacity of meat of about 3 kJ/kg°C; cooking efficiency of an open fire at just 5%; and an average energy density of air-dried wood of 15 MJ/kg. These assumptions imply an average daily per capita intake of nearly 1 kg of mammoth meat (and about 1.5 kg of large ungulate meat) and a daily need for about 4–6 MJ of wood. The annual total would be 1.5–2.2 GJ or 100–150 kg of (some fresh and some air-dried) wood. For 200,000 people who lived 20,000 years ago the global need would be 20,000–30,000 t, a negligible share (on the order of 10^{-8}) of the standing pre-agricultural woody phytomass.

In addition to warming and cooking, fire was also used as an engineering tool: modern humans were heat-treating stones to improve their flaking properties as early as 164,000 years ago (Brown et al. 2009). And Mellars (2006) suggested there is evidence for the controlled burning of vegetation in South Africa as early as 55,000 years ago. Woodland burning as a tool of environmental management by foragers during the early Holocene would have been done to aid hunting (by promoting the regrowth of forage to attract animals and improving visibility), to make human mobility easier, or to improve or synchronize the gathering of plant foods (Mason 2000).

The great spatial and temporal variability of the archaeological record precludes making any simple generalizations concerning the energy balances of prehistoric societies. Descriptions of first contacts with surviving foragers and their anthropological studies provide uncertain analogies: information on groups that survived in extreme environments long enough to be studied by modern scientific methods offers a limited insight into the lives of prehistoric foragers in more equable climates and more fertile areas. Moreover, many studied foraging societies were already affected by prolonged contact with pastoralists, farmers, or overseas migrants (Headland and Reid 1989; Fitzhugh and Habu 2002). But the absence of a typical foraging pattern does not preclude recognition of a number of biophysical imperatives governing energy flows and determining the behavior of gathering and hunting groups.

Foraging Societies

The most comprehensive collections of reliable evidence show that the average population densities of modern foraging populations—reflecting a variety of natural habitats and food acquisition skills and techniques—ranged over three orders of magnitude (Murdock 1967; Kelly 1983; Lee and Daily 1999; Marlowe 2005). The minima were less than a single person/ 100 km² to several hundred people/100 km², with the global mean of 25 people/100 km² for 340 studied cultures, too low to support more complex societies with increasing functional specialization and social stratification. The average densities of foragers were lower than the densities of similarly massive herbivorous mammals that were able to digest abundant cellulosic phytomass.

Whereas allometric equations predict about five 50 kg mammals/km², chimpanzee densities are between 1.3 and 2.4 animals/km², and the densities of hunter-gatherers surviving into the twentieth century were well below one person/km² in warm climates, only 0.24 in the Old World and 0.4 in the New World (Marlowe 2005; Smil 2013a). Population densities were significantly higher for groups combining the gathering of abundant plants with hunting (well-studied examples include groups in postglacial Europe and, more recently, in the basin of Mexico) and for coastal societies heavily dependent on aquatic species (with well-documented archaeological sites in the Baltic region and more recent anthropological studies in the Pacific Northwest).

Mollusk collecting, fishing, and near-shore hunting of sea mammals sustained the highest foraging densities and led to semipermanent, even permanent, settlements. The coastal villages of the Pacific Northwest, with their large houses and organized communal hunting of sea mammals, were exceptional in their sedentism. These large density variations were not a simple function of biospheric energy flows: they were not uniformly decreasing poleward and increasing equatorward (in proportion to higher photosynthetic productivity), or corresponding to the total mass of animals available for hunting. They were determined by ecosystemic variables, by a relative dependence on plant and animal foods, and by the use of seasonal storage. Much like nonhuman primates, all foragers were omnivorous, but killing larger animals was a major energetic challenge as it targeted a much smaller reservoir of edibles than plant gathering, a natural consequence of the diminishing energy transfer between trophic levels.

Herbivores consume only 1–2% of the net primary productivity in deciduous temperate forests and up to 50–60% in some tropical grasslands, with 5–10% most representative of terrestrial grazing (Smil 2013a). Generally less than 30% of the ingested phytomass is digested; most of it is respired, and in mammals and birds only 1–2% of it is converted into zoomass. As a result, the most commonly hunted herbivores embodied less than 1% of the energy initially stored in the phytomass of the ecosystems they inhabited. This reality explains why hunters preferred to kill animals that combined a relatively large adult body mass with high productivity and high territorial density: wild pigs (90 kg) and deer and antelopes (mostly 25–500 kg) were common targets.

Where such animals were relatively common, such as on tropical or temperate grasslands or in tropical woodlands, hunting was more rewarding, but, contrary to the common perception of an abundance of animal species, tropical forests were an inferior ecosystem to be exploited by hunting. Most tropical forest animals are small arboreal folivorous and fructivorous species (monkeys, birds) that are active and inaccessible in high tree canopies (many are also nocturnal), and hunting them yields low energy returns. Sillitoe (2002) found that both gathering and hunting in a tropical rain forest of the Papua New Guinea highlands to be costly, with foragers expending up to four times more energy on hunting than they obtained in food. Obviously, such a poor energy return would not allow hunting to be a primary means of food provision (the negative energy return could be explained only by the capture of animal protein), and some forms of shifting farming were required to provide enough food.

Bailey and co-workers (1989) concluded that there were no unambiguous ethnographic accounts of foragers who lived in tropical rain forests without some reliance on domesticated plants and animals. Bailey and Headland (1991) later changed that conclusion as archaeological evidence from Malaysia indicated that high densities of sago and pigs would allow exceptions. Similarly, gathering was often surprisingly unrewarding in the species-rich tropics, as well as in temperate forests. These ecosystems store most of the planet's phytomass, but they do so mostly in the dead tissues of tall tree trunks, whose cellulose and lignin humans cannot digest (Smil 2013a). Energy-rich fruits and seeds are a very small portion of total plant mass and are often inaccessible in high canopies; seeds are often protected by hard coats and need energy-intensive processing before consumption. Gathering in tropical forests also needed more searching: a great variety of species means there may be considerable distances between the trees or vines whose parts are ready for collection (fig. 2.2). The harvesting of Brazil nuts is a perfect example of these constraints (box 2.2).

In contrast to the often frustrating hunting in tropical and boreal forests, grasslands and open woodlands offered excellent opportunities for collecting and hunting. They store much less energy per unit area than a dense forest, but a higher share of it comes in the form of easily collectible and highly nutritious seeds and fruits, or as concentrated patches of large starchy roots and tubers. High energy density (as much as 25 MJ/kg) made nuts particular favorites, and some of them, such as acorns and chestnuts, were also easy to harvest. And unlike in forests, many animals grazing on grasslands can grow to very large sizes, often move in massive herds, and give excellent returns on energy invested in the hunt.

And hominins could secure meat on grasslands and woodlands even without any weapons, as scavengers, as unmatched runners, or as clever schemers. In light of the unimpressive physical endowment of early humans and the absence of effective weapons, it is most likely that our ancestors were initially much better scavengers than hunters (Blumenschine and

Figure 2.2
Tropical rain forests are rich in species but relatively poor in plants that would support larger foraging populations. This image shows canopies at La Fortuna, Costa Rica (Corbis).

Box 2.2

Harvesting Brazil nuts

Because of their high lipid content (66%), Brazil nuts contain about 27 MJ/kg (compared to about 15 MJ/kg for cereal grains), are about 14% protein, and are also a source of potassium, magnesium, calcium, phosphorus, and high levels of selenium (Nutrition Value 2015). Harvesting the nuts is both demanding and dangerous. *Bertholletia excelsa* grows up to 50 m, with individual trees widely scattered. Between 8 and 24 nuts are contained in heavy (up to 2 kg) capsules covered with a coconut-like hard endocarp. Nut foragers must time their harvest: too early, and the pods are still inaccessible in the canopies and gatherers must waste energy on another trip; too late, and agoutis (*Dasyprocta punctata*), which are large rodents and the only animals able to open the fallen pods, will eat the seeds right away or bury some of them in food caches (Haugaasen et al. 2010).

Cavallo 1992; Pobiner 2015). Large predators—lions, leopards, saber-toothed cats—often left behind partially eaten herbivore carcasses. This meat, or at least the nutritious bone marrow, could be reached by alert early humans before it was devoured by vultures, hyenas, and other scavengers. But Domínguez-Rodrigo (2002) has argued that scavenging would not provide enough meat and that only hunting could secure sufficient animal protein on grasslands. In any case, human bipedalism and ability to sweat better than any other mammal made it also possible to chase to exhaustion even the fastest herbivores (box 2.3).

Carrier (1984) believes that the outstanding rates of human heat dissipation provided a notable evolutionary advantage that served our ancestors well in appropriating a new niche, that of diurnal, hot-temperature predators. The human ability to sweat profusely and hence to work hard in hot environments was retained by populations that migrated to colder climates: there are no major differences in the density of eccrine glands among populations of different climate zones (Taylor 2006). People from middle and high latitudes can match the sweating rates of hot-climate natives after a short period of acclimatization.

But once adequate tools were invented and adopted, hunting with them was preferable to running down the prey, and Faith (2007) confirmed, after examining 51 assemblages from the Middle Stone Age and 98 from the Later Stone Age, that early African hunters were fully competent at killing large ungulate animals, including buffaloes. The energy imperatives of

Box 2.3

Running and heat dissipation by humans

All quadrupeds have optimum speeds for different gaits, such as walk, trot, and canter in horses. The energy cost of human running is relatively high compared to the running cost of similarly massive mammals, but unlike them, humans can uncouple that cost from running speed for common velocities between 2 and 6 m/s (Carrier 1984; Bramble and Lieberman 2004). Bipedalism and efficient heat dissipation explain this feat. Quadrupedal ventilation is limited to one breath per locomotor cycle. The thoracic bones and muscles must absorb the impact on the front limbs as the dorsoventral binding rhythmically compresses and expands the thorax, but human breath frequency can vary relative to stride frequency: humans can run at a variety of speeds, whereas optimal quadruped speed is structurally determined.

The extraordinary human ability to thermoregulate rests on very high rates of sweating. Horses lose water at an hourly rate of 100 g/m^2 of their skin, and camels lose up to 250 g/m^2, but people lose more than 500 g/m^2, with peak rates of more than 2 kg/hour (Torii 1995; Taylor and Machado-Moreira 2013). Perspiration rate translates to heat loss of 550–625 W, enough to regulate temperature even during extremely hard work. People can also drink less than they perspire, and make up for any temporary partial dehydration hours later. Running turned humans into diurnal, high-temperature predators that could chase animals to exhaustion (Heinrich 2001; Liebenberg 2006). Documented chases include those of the Tarahumara Indians of northern Mexico running down deer and of Paiutes and Navajos besting pronghorn antelopes. The Kalahari Basarwa could chase duikers, gemsbok, and, during the dry season, even zebras to exhaustion, as some Australian Aborigines did with kangaroos. Hunters running barefoot reduced their energy costs by about 4% (and had fewer acute ankle and chronic lower leg injuries) than modern runners with athletic shoes (Warburton 2001).

hunting large animals had also made an incalculable contribution to human socialization. Trinkaus (1987, 131–132) concluded that "most of the distinguishing human characteristics, such as bipedalism, manual dexterity, and elaborate technology, and marked encephalization can be viewed as having been promoted by the demands of an opportunistic foraging system."

Hunting's role in the evolution of human societies is self-evident. Individual success in hunting large animals with primitive weapons was unacceptably low, and viable hunting groups had to maintain minimum cooperative sizes in order to track wounded animals, butcher them, transport their meat, and then pool the gains. Communal hunting brought by far the greatest rewards, with well-planned and well-executed herding of animals into confined runs (using brush and stone drive lines, wooden fences, or ramps) and capturing them in prepared pens or natural traps, or—perhaps the simplest and most ingenious solution—stampeding them over cliffs (Frison 1987). Many large herbivores—mammoths, bison, deer, antelopes, mountain sheep—could be slaughtered in such ways, providing caches of frozen or processed (smoked, pemmican) meat.

Head-Smashed-In Buffalo Jump near Fort Mcleod, Alberta, a UNESCO World Heritage Site, is one of the more spectacular sites of this inventive hunting strategy, which was used at the site for about 5,700 years. "To start the hunt ... young men ... would entice the herd to follow them by imitating the bleating of a lost calf. As the buffalo moved closer to the drive lanes (long lines of stone cairns were built to help the hunters direct the buffalo to the cliff kill site), the hunters would circle behind and upwind of the herd and scare the animals by shouting and waving robes" and stampeded the herd over the cliff (UNESCO 2015a). The net energy return in animal protein and fat was high. Late Pleistocene hunters may have become so skillful that many students of the Quaternary era concluded that hunting was largely (even completely) responsible for a relatively rapid disappearance of the late Paleolithic megafauna, animals with a body mass greater than 50 kg (Martin 1958, 2005; Fiedel and Haynes 2004), but the verdict remains uncertain (box 2.4).

Box 2.4
Extinction of the Late Pleistocene megafauna

The persistent killing of slow-breeding animals (those with a single offspring born after a long gestation) could lead to their extinction. If we assume that Late Pleistocene foragers had a high daily food requirement of 10 MJ/capita, that they ate mostly meat, and that most of it (80%) came from megafauna, then their population of two million people would need nearly 2 Mt (fresh weight) of meat (Smil 2013a). If mammoths were the only hunted species, that would have required the annual killing of 250,000–400,000 animals. Megaherbivore hunting also targeted other large mammals (elephants, giant deer, bison, aurochs), and procuring 2 Mt of meat from a mixture of these species would have required an annual kill of some two million animals. A more likely explanation for the Late Pleistocene extinctions is a combination of natural (climate and vegetation change) and anthropogenic (hunting and fire) factors (Smil 2013a).

All pre-agricultural societies were omnivorous; they did not have the luxury of ignoring any available food resource. Although foragers ate a large variety of plant and animal species, usually only a few principal food-stuffs dominated their diets. A preference for seeds among gatherers was inevitable. Besides being rather easy to collect and store, seeds combine a high energy content with relatively high protein shares. Wild grass seeds have as much food energy as cultivated grains (wheat is at 15 MJ/kg), while nuts have an energy density up to about 80% higher (walnuts contain 27.4 MJ/kg).

All wild meat is an excellent source of protein, but most of it contains very little fat, and hence it has very low energy density—less than half that of grains for small, lean mammals. Not surprisingly, there was a widespread hunting preference for large and relatively fatty species. A single small mammoth provided as much edible energy as 50 reindeer, while a bison was easily equal to 20 deer (box 2.5). This is why our Neolithic ancestors were willing to ambush huge mammoths with their simple stone-tipped weapons, or why Indians on the North American plains, seeking fatty meat for preparing long-lasting pemmican, spent so much energy in pursuing bison.

But energy considerations alone cannot provide a full explanation of foraging behavior. If they were always dominant, then optimal foraging—whereby gatherers and hunters try to maximize their net energy gain by minimizing the time and effort spent in foraging—would have been their universal strategy (Bettinger 1991). Optimal foraging explains the preference for hunting large, fatty mammals or for collecting less nutritious plant parts that do not need processing rather than energy-dense nuts, which may be hard to crack. Many foragers undoubtedly behaved in ways that maximized their net energy return, but other existential imperatives often worked against such behavior. Among the most important ones were the availability of safe night shelters, the need to defend territories against competing groups, and the needs for reliable water sources and for vitamins and minerals. Food preferences and attitudes toward work were also important (box 2.6).

Our inability to reconstruct prehistoric energy balances has provoked some inadmissible generalizations. For some groups the total foraging effort was relatively low, only a few hours a day. This finding led to foragers being portrayed as "the original affluent society," living in a kind of material plenty filled with leisure and sleep (Sahlins 1972). Most notably, Dobe !Kung people of the Kalahari Desert in Botswana, living on wild plants and meat, were thought to provide an excellent window on the lives of

Box 2.5
Body masses, energy densities, and food energy content of hunted animals

Animals	Body mass (kg)	Energy density (MJ/kg)	Food energy per animal (MJ)
Whales	5,000–40,000	25–30	80,000–800,000
Large proboscids (elephants, mammoths)	500–4,000	10–12	2,500–24,000
Large bovids (aurochs, bison)	200–400	10–12	1,000–2,400
Large cervids (elk, reindeer)	100–200	5–6	250–600
Seals	50–150	15–18	500–1,800
Small bovids (deer, gazelle)	10–60	5–6	25–180
Large monkeys	3–10	5–6	5–30
Lagomorphs (hares, rabbits)	1–5	5–7	3–17

Notes: I assume that the edible portion is two-thirds of the body mass of whales and seals and half of the body mass of other animals. The average energy density for whales I calculated by assuming that 25% of their body mass is blubber.

Sources: Based on data in Sanders, Parsons, and Santley (1979), Sheehan (1985), and Medeiros and co-workers (2001).

prehistoric foragers, who allegedly led contented, healthy, and vigorous lives (Lee and DeVore 1968). This conclusion, based on very limited and dubious evidence, must be—and has been—challenged (Bird-David 1992; Kaplan 2000; Bogin 2011.

Simplistic theorizing about affluent foragers ignored both the reality of much of the hard and often dangerous work in foraging and the frequency with which environmental stresses and infectious diseases ravaged most foraging societies. Seasonal food shortages forced the eating of unpalatable plant tissues and led to weight loss, and often to devastating famines. They also resulted in high infant mortality (including infanticide) and promoted low fertility rates. And, not surprisingly, a reanalysis of energy expenditure and demographic data collected in the 1960s found that the nutritional

Box 2.6

Food preferences and attitudes toward work

Food preferences are convincingly illustrated by a comparison of two otherwise highly similar foraging groups. The !Kung Basarwa (in Botswana) owe
their notoriety in the anthropological literature to their dependence on abundant and highly nutritious mongongo nuts, which gave them the best energy
returns documented in food gathering. But the /Aise, another Basarwa group
with access to the nuts, did not eat them because to them, the nuts did not
taste good (Hitchcock and Ebert 1984). Similarly, coastal groups in southern
Australia achieved high energy densities through fishing, but across the
strait archaeological evidence shows scale fish refuse missing from middens
in Tasmania (Taylor 2007).

An excellent example of cultural realities at variance with simplistic
energy models is Lizot's (1977) comparison of two nearby groups of Yanomami Indians (northern Amazonia). The group surrounded by forest consumed less than half the amount of animal food energy and protein that its
neighbors did, living in an environment less well endowed with wild pigs,
tapirs, and monkeys and possessing the same hunting skills and tools.
His explanation: people of the first group were simply lazier, hunted infrequently, and, briefly, preferred to eat less well. "During one of the weeks ...
the men did not go hunting once, they had just collected their favorite hal-
lucinogen (*Anadenanthera peregrina*) and spent entire days taking drugs; the
women complained there was no meat, but the men turned a deaf ear" (Lizot
1977, 512).

This represents a common case of a major variation in energy provided
by hunting that bears no relationship either to resource availability (the
presence of animals) or to the energy cost of the hunt (given simple and
virtually identical weapons) but is solely a function of attitude toward work.
Another example of actions not conforming to energetic explanations comes
from an analysis of data on meat sharing among Tanzanian Hadza (Hawkes,
O'Connell, and Jones 2001). The best explanation for the widespread sharing
of meat of large animals is to reduce the risk inherent in big-game hunting—
but Hadza sharing was not motivated by the expectation of risk-reducing
reciprocity but was done primarily to enhance a hunter's status as a desirable
neighbor.

status and health of Dobe !Kung "were, at best, precarious and, at worst, indicative of a society in danger of extinction" (Bogin 2011, 349). As Froment (2001, 259) put it, "Coping with hazards and a heavy burden of diseases, hunter-gatherers do not live—and have never lived—in the Garden of Eden; they are not affluent, but poor, with limited needs and limited satisfaction."

Approximate calculations for a small number of twentieth-century foraging groups show the highest net energy returns for the gathering of some roots. As many as 30–40 units of food energy were acquired for every unit expended. In contrast, many hunting forays, above all those for smaller arboreal or ground mammals in tropical rain forests, had a net energy loss or bare equivalence (box 2.7). Typical gathering returns were 10–20-fold, similar to those of hunting large mammals. Prehistoric returns were undoubtedly much higher in many biomass-rich environments, allowing for a gradual increase in social complexity.

In fact, many foraging societies reached levels of complexity usually associated only with later agricultural societies. They had permanent settlements, high population densities, large-scale food storage, social stratification, elaborate rituals, and incipient crop cultivation. Upper Paleolithic mammoth hunters in the Moravian loess region had well-built stone houses, produced a variety of excellent tools, and could fire clay (Klima 1954). The social complexity of the Upper Paleolithic groups of southwestern France was promoted by the strong Atlantic influence, which resulted in fairly cool summers but also exceptionally mild winters, and extended the growing season and intensified the productivity of the continent's most

---

Box 2.7
Net energy returns in foraging

I use the method described in box 1.10 and assume smaller statures of prehistoric foragers (average adult weight of just 50 kg). That would have required about 6 MJ/day (about 250 kJ/h) for basal metabolism, and a minimum existential adult food energy need of about 8 MJ, or roughly 330 kJ/h. Plant collecting required mostly light to moderate labor, while hunting and fishing tasks ranged from light to heavy exertions. Typical foraging activities needed about four times the basal metabolic rate for men and five times that for women, or almost 900 kJ/h. Subtracting the basic existential need puts the net energy input in foraging at roughly 600 kJ/h. Energy output is simply the value of edible portions of collected plants or killed animals.

southerly open tundra or steppelike vegetation, which supported herbivorous herds larger than anywhere in periglacial Europe (Mellars 1985). The complexity of these Paleolithic cultures is best attested by their remarkable sculptures, carvings, and cave paintings (Grayson and Delpech 2002; French and Collins 2015) (fig. 2.3).

The highest productivities in complex foraging were associated with the exploitation of aquatic resources (Yesner 1980). Excavations of Mesolithic sites in southern Scandinavia showed that once the postglacial hunters had depleted the stocks of large herbivores they became hunters of porpoises and whales, fishers, and collectors of shellfish (Price 1991). They lived in larger, often permanent settlements that included cemeteries. Northwestern Pacific tribes dependent on fishing had settlements of several hundred people living in well-built wooden houses. Regular runs of salmon species guaranteed a reliable and easily exploitable resource that could be safely stored (smoked) to provide excellent nutrition. Thanks to its high fat content (about 15%), salmon has an energy density (9.1 MJ/kg) nearly three

Figure 2.3
Charcoal paintings of animals on a wall of the Chauvet Cave in southern France. These remarkable likenesses were dated to between 32,900 and 30,000 years ago (Corbis).

times that of cod (3.2 MJ/kg). The superior case of high population density dependent on maritime hunting is that of northwestern Alaskan Inuit, whose net energy returns in killing migrating baleen whales were more than 2,000-fold (Sheehan 1985) (box 2.8).

A food supply dependent on a few seasonal energy flows required extensive, and often elaborate, storage. Storage practices included caching in permafrost; drying and smoking of seafood, berries, and meats; storing of seeds and roots; preservation in oil; and the making of sausages, nut-meal cakes, and flours. Large-scale, long-term food storage changed foragers' attitudes toward time, work, and nature and helped stabilize populations at higher densities (Hayden 1981; Testart 1982; Fitzhugh and Habu 2002). The need to plan and budget time was perhaps the most important evolutionary benefit. This new mode of existence precluded frequent mobility and introduced a different way of subsistence based on surplus accumulation. The process was self-amplifying: the quest for the manipulation of an ever larger share of solar energy flows set the societies on the road toward higher complexity.

Box 2.8
Alaskan whalers

In less than four months of near-shore hunting of baleen whales, whose migration routes led along the Alaskan coast, men in umiak (boats with a driftwood or whalebone frame covered with sealskin and crewed by up to eight people) amassed food for settlements whose precontact population reached almost 2,600 people (Sheehan 1985; McCartney 1995). The largest adult baleen whales weigh up to 55 t, but even the most commonly landed immature two-year-old animals averaged nearly 12 t. The high energy density of blubber (about 36 MJ/kg) and muktuk (skin and blubber, which also has a vitamin C content comparable to that of grapefruit) resulted in a more than 2,000-fold energy gain in hunting.

Lower but still exceptionally high energy returns resulted from exploiting annual salmon runs by the coastal tribes of the Pacific Northwest: the density of fish returning upstream was often so high that fishers could simply scoop them into boats or onto the shore. These high energy returns supported large permanent settlements, social complexity, and artistic creativity (large wooden totems). Eventual limits on the population growth of these coastal settlements were imposed by the necessity to hunt other marine species and land game in order to secure raw materials for clothing, bedding, and hunting equipment.

Although our understanding of hominin evolution has increased impressively during the last two generations, one key area of uncertainty remains: contrary to all popular claims about the benefits of Paleolithic diets, we still cannot reconstruct the representative composition of pre-agricultural subsistence. This should not be a surprise (Henry, Brooks, and Piperno 2014). Readily degradable plant remains of food consumption very rarely survive for tens of thousands of years and almost never for millions, making it exceedingly difficult to quantify the share of plant foods in typical diets. Bones often survive, but their accumulations from animal predations must be carefully distinguished from hominin acquisition, and even then it is impossible to interpret how representative they were of particular diets.

As Pryor and co-workers (2013) note, the widely accepted image of European Upper Paleolithic hunter-gatherers as proficient hunters of large mammals inhabiting largely treeless landscapes stems from the poor preservation of plant remains at such ancient sites. Their study showed that the potential of such sites to provide macrofossil remains of plants consumed by humans has been underestimated, and that "the ability to exploit plant foods may have been a vital component in the successful colonisation of these cold European habitats" (Pryor et al. 2013, 971). And Henry, Brooks, and Piperno (2014) analyzed plant microremains— starch grains and phytoliths—left in dental calculus and on stone tools and concluded that both modern humans and their Neanderthal coevals consumed a similarly wide range of plant foods, including rhizomes and grass seeds.

Changes in body height and mass and in cranial features (gracilization of the mandible) are indirect indicators of prevailing diets and could have arisen from a variety of food mixtures. The findings of stone tools used to kill and butcher animals cannot be readily related to average per capita meat intake over extended periods of time, and hence only direct stable isotope evidence (ratios of $^{13}C/^{12}C$ and $^{15}N/^{14}N$) provides an accurate determination of long-term protein sources, their trophic levels, and their terrestrial and marine origins; distinguishes phytomass synthesized by the two principal pathways (C_3 and C_4) and heterotrophs feeding on those plants; and informs us about the basic makeup of the total diet. Even these studies cannot be translated into reliable patterns of average macronutrient (carbohydrates, proteins, lipids) intake, but isotope data indicate that during the Gravettian period in Europe, animal protein was the main source of dietary protein, with aquatic species contributing about 20% of the total, and even more at coastal sites (Hublin and Richards 2009).

Before leaving the forager energetics I should note that foraging retained an important role in all early agricultural societies. In Çatalhöyük, a large Neolithic agricultural settlement on the Konya Plain dated to about 7200 BCE, early farmers had diets dominated by grains and wild plants, but excavations also show the bones of hunted animals, ranging from large aurochs to foxes, badgers, and hares (Atalay and Hastorf 2006). And at Tell Abu Hureyra in northern Syria, hunting remained a critical source of food for 1,000 years after the beginning of plant domestication (Legge and Rowley-Conwy 1987). In predynastic Egypt (earlier than 3100 BCE), the cultivation of emmer wheat and barley was complemented by the hunting of waterfowl, antelopes, wild pigs, crocodiles, and elephants (Hartmann 1923; Janick 2002).

Origins of Agriculture

Why did some foragers start to farm? Why did these new practices diffuse so widely, and why did their adoption proceed at what, in evolutionary terms, is a fairly rapid rate? These challenging questions may be sidestepped by agreeing with Rindos (1984) that agriculture has no single cause but arose from a multitude of interdependent interactions. Or, as Bronson (1977, 44) put it, "What we are dealing with is a complex, multifaceted adaptive system, and in human adaptive systems ... single all-efficient 'causes' cannot exist." But many anthropologists, ecologists, and historians have been trying to find precisely such principal causes, and there are many publications surveying diverse explanatory theories about the origin of agriculture (Cohen 1977; Pryor 1983; Rindos 1984; White and Denham 2006; Gehlsen 2009; Price and Bar-Yosef 2011).

Overwhelming evidence for the evolutionary character of agricultural advances makes it possible to narrow down the possibilities. The most persuasive explanation of agricultural origins lies in the combination of population growth and environmental stress, in recognizing that the transition to permanent cropping was driven by both natural and social factors (Cohen 1977). Because the climate was too cold and CO_2 levels were too low during the late Paleolithic and because these conditions changed with the subsequent warming, Richerson, Boyd, and Bettinger (2001) have argued that agriculture was impossible during the Pleistocene but mandatory during the Holocene. This argument is strengthened by the fact that between 10,000 and 5,000 years before the present, cropping evolved independently in at least seven locations on three continents (Armelagos and Harper 2005).

Fundamentally, crop cultivation is an effort to ensure an adequate food supply, and hence agriculture's origins could be fully explained as yet another instance of an energy imperative. Diminishing returns from gathering and hunting led to the gradual extension of incipient cultivation present in many foraging societies. As already noted, foraging and cultivation coexisted in various shares of food output for very long periods. But no sensible explanation of agriculture's origins can ignore the many social advantages of farming. Sedentary crop cultivation was an efficient way for more people to stay together; it made it easier to have larger families, to accumulate material possessions, and to organize for both defense and offense.

Orme (1977) even concluded that food production may have been unimportant as an end in itself, but there is no doubt that both the genesis and diffusion of agriculture had critical social cofactors. Any simplistic energy-driven explanation of agricultural origins is also weakened by the fact that the net energy returns of early farming were often inferior to those of earlier or concurrent foraging activities. Compared to foraging, early farming usually required higher human energy inputs—but it could support higher population densities and provide a more reliable food supply. This explains why so many foraging societies had continuous interaction (and often much trade) with neighboring farming groups for thousands, or at least hundreds, of years before they adopted permanent farming (Headland and Reid 1989).

There was no single center of domestication from which cultivated plants and milk- and meat-producing animals spread, but in the Old World the most important region of agricultural origin was not, as previously thought, the southern Levant but rather the upper reaches of the Tigris and Euphrates rivers (Zeder 2011). This means that food production started along the margins, rather than in the core areas, of optimal zones. The botanical record from Chogha Golan in the foothills of the Iranian Zagros Mountains provides the most recent confirmation of this reality (Riehl, Zeidi, and Conard 2013): cultivation of wild barley (*Hordeum spontaneum*) began there about 11,500 years ago, later augmented by the cultivation of wild wheat and wild lentils.

In process terms, it is essential to stress that there are no clear thresholds or sharp divides between foraging and agriculture, as extended periods of managing wild plants and animals precede their true domestication, which is characterized by clearly identifiable morphological changes. And, contrary to earlier understandings, the domestication of plants and animals proceeded almost concurrently and yielded an effective arrangement fairly rapidly (Zeder 2011). The oldest approximate dates for the first appearance are about 11,500–10,000 years before the present for the plant species emmer (*Triticum dicoccum*) and einkorn wheat (*Triticum monococcum*) and barley (*Hordeum vulgare*) in the Middle East (fig. 2.4), 10,000 years for China's millets (*Setaria italica*), 7,000 years for rice (*Oryza sativa*), 10,000 years for Mexican squash (*Cucurbita* species), and 9,000 years for corn (*Zea mays*) and 7,000 years for Andean potatoes, *Solanum tuberosum* (Price and Bar-Yosef 2011). The earliest animal domestications go back to 10,500–9,000 years ago, starting with goats and sheep, followed by cattle and pigs.

The two main explanations of Europe's Neolithic transition to farming have been through indigenous action animated by imitation (cultural diffusion) or driven by dispersing populations (demic diffusion). Radiocarbon dating of material from early Neolithic sites by Pinhasi, Fort, and Ammerman (2005) yielded results consistent with the prediction of demic diffusion, radiating most likely from the northern Levant and the Mesopotamian area and proceeding northwestward at an average pace of 0.6–1.1 km/year. This conclusion is supported by comparisons of mitochondrial DNA sequences from late European hunter-gatherer skeletons with those from

(a) (b) (c)

Figure 2.4

The earliest domesticated cereals. a–c. Emmer wheat (*Triticum dicoccum*), einkorn wheat (*Triticum monococcum*), and barley (*Hordeum vulgare*) were the foundation of the origins of agriculture in the Middle East (Corbis).

early farmers and from modern Europeans: they show persuasively that the first farmers were not the descendants of local foragers but had immigrated at the onset of the Neolithic (Bramanti et al. 2009).

Early agriculture often took the form of shifting cultivation (Allan 1965; Spencer 1966; Clark and Haswell 1970; Watters 1971; Grigg 1974; Okigbo 1984; Bose 1991; Cairns 2015). This practice alternated usually short (1–3 years) periods of cultivation with fairly long (a decade or more) fallow spells. Despite many differences (determined by ecosystems, climates, and dominant crops) there were many similarities, most of them clearly driven by efforts to minimize energy expenditures. The cycle started with the clearing of natural vegetation, and its slashing or burning was often sufficient to prepare the surface for planting. To minimize the walking distance, fields or gardens were opened as close to the settlements as possible, and clearing of the secondary growth was the preferred option: for example, Rappaport (1968) found that only one out of 381 Tsembaga (New Guinea) gardens was cleared in virgin forest. Some plots had to be fenced to prevent damage by animals: in that case, tree felling for fences needed the highest labor inputs. Plant nitrogen was largely lost in combustion, but mineral nutrients enriched the soil.

Men did the heavy work (in the absence of good tools, vegetation was simply burned; some trees had to be cut down for fencing), while women's labor was dominated by weeding and harvesting, and, because of their relatively high yields, cereals and tubers were the staples (Rappaport 1968). In all warmer regions there was much interplanting, particularly in intensively cultivated gardens; intercropping; and staggered harvesting. Shifting agriculture was important on all continents except Australia. In South America its ancient practice (mostly between 500 BCE and 1000 CE) left its marks throughout the Amazon basin in the form of *terra preta*, up to 2 m deep dark soils containing charred wood and crop residues, human waste, and bones (Glaser 2007; Junqueira, Shepard, and Clement 2010). In North America it reached as far north as Canada, where the Hurons cultivated corn and beans in long (35–60 years) rotation cycles and supported 10–20 people/ha (Heidenreich 1971).

In areas of low population density and abundant land availability, the practice was a fitting part of the evolutionary sequence from foraging to permanent cropping. A diminishing supply of land, environmental degradation, and increasing pressures for more intensive cropping have been steadily reducing its importance. Net energy returns have varied widely. Tsembaga horticulture in the highlands of New Guinea yielded approximately a 16-fold energy return (Rappaport 1968). Another New Guinea study found returns no higher than 6- to 10-fold (Norgan et al. 1974)—but the corn harvest of Kekchi Maya (Guatemala) brought at least a 30-fold energy return (Carter 1969). Most net returns were 11–15 for small grains and 20–40 for most root crops, bananas, and also for corn, while maxima were close to 70 for some roots and legumes (box 2.9). Feeding one person required mostly 2–10 ha of land to be cleared periodically, with the actually cultivated area ranging from just 0.1 to 1 ha/person. Even moderately productive shifting agriculture supported population densities an order of magnitude higher than the best foraging.

Where scarcity of precipitation, or its long seasonal absence, made cropping unrewarding or impossible, nomadic pastoralism has been an effective alternative (Irons and Dyson-Hudson 1972; Galaty and Salzman 1981; Evangelou 1984; Khazanov 2001; Salzman 2004). Managed grazing has been the energetic foundation of scores of Old World societies, and though some of them remained poor and isolated, others were among the most feared long-distance interveners in history: Xiongnu were in conflict with the early Chinese dynasties for hundreds of years, and the Mongol invasion of 1241 reached as far west as today's Poland and Hungary.

Animal husbandry is a form of prey conservation, a strategy of deferred harvests whose opportunity costs are greater for larger animals, especially

---

Box 2.9

Energy costs and population densities in shifting cultivation

Net energy cost is used to calculate returns of shifting farming. I assume that an average labor input requires 700 kJ/h. Outputs are edible harvests uncorrected for storage losses and seed needs.

Populations	Main crops	Energy inputs (hours)	Energy returns	Population densities (people/ha)
Southeast Asia	Tubers	2,000–2,500	15–20	0.6
Southeast Asia	Rice	2,800–3,200	15–20	0.5
West Africa	Millet	800–1,200	10–20	0.3–0.4
Mesoamerica	Corn	600–1,000	25–40	0.3–0.4
North America	Corn	600–800	25–30	0.2–0.3

Sources: Calculated from data in Conklin (1957), Allan (1965), Rappaport (1968), Carter (1969), Clark and Haswell (1970), Heidenreich (1971), Thrupp and co-workers (1997), and Coomes, Girmard, and Burt (2000).

bovids (Alvard and Kuznar 2001). Larger animals are preferred, but higher growth rates favor sheep and goats. Animals can convert grasses into milk, meat, and blood with remarkably low inputs of human energy (fig. 2.5). Pastoralists' labor was confined to herding the animals, guarding them against predators, watering them, helping with deliveries, milking regularly, and butchering infrequently, and sometimes building temporary enclosures. The sustainable population densities of such societies were no higher than those of foraging groups (box 2.10).

For millennia, nomadic grazing dominated parts of Europe and the Middle East, and large regions of Africa and Asia. In all of these places it sometimes blended into mixtures of seminomadic agropastoralism, especially in parts of Africa with a significant component of foraging. Often hemmed in by more productive farmers, and commonly dependent on barter with settled societies, some of these nomads had little impact beyond their confined worlds. But many groups exercised great influence on the Old World's history through their repeated invasions and temporary conquests of agricultural societies (Grousset 1938; Khazanov 2001). Some pure pastoralists and agropastoralists survive even today—above all in Central Asia and in Sahelian and eastern Africa—but it has been an increasingly marginal existence.

Figure 2.5
Maasai herder with his cattle (Corbis).

Box 2.10

Nomadic pastoralists

Helland (1980) illustrated low labor requirements in pastoral societies by pointing out the large numbers of major livestock species managed by a single herder in the East African setting: up to 100 camels, 200 cattle, and 400 sheep and goats. Khazanov (1984) lists similarly large figures for Asian pastoralists: two mounted shepherds for 2,000 sheep in Mongolia, an adult shepherd and a boy to tend 400–800 cattle in Turkmenia. The appeal of low labor needs was one of the key reasons for the reluctance of many pastoralists to abandon their peregrinations and to become settled farmers. As a result, many nomadic societies existed for generations as neighbors of sedentary farmers and abandoned their herds only because of devastating drought or substantial loss of available pastures.

Minimum per capita counts for pastoral subsistence were 5–6 heads of cattle, 2.5 3 camels, or 25–30 goats or sheep. Much higher cattle ownership among traditional Maasai (13–16 heads/capita) is explained by the minimum requirements for blood harvesting, done by piercing a tightened jugular vein and drawing 2–4 L every 5–6 weeks. During the periods of drought a herd of 80 cattle was needed to provide blood for a family of 5–6 people, or 13–16 animals/capita (Evangelou 1984). In all cases, nomadic population densities were low compared with those of settled farmers, in East Africa mostly between 0.8–2.2 people/km^2, and 0.03–0.14 heads/ha (Helland 1980; Homewood 2008).

3   Traditional Farming

While the transition from foraging to farming cannot be explained solely by energetic imperatives, the evolution of agriculture can be seen as a continuing effort to raise land productivity (to increase digestible energy yield) in order to accommodate larger populations. Even within that narrowed framework, important nonenergy considerations (such as an adequate supply of micronutrients, vitamins, and minerals) should not be forgotten, but because of the overwhelmingly vegetarian diets of all traditional peasant societies it is not a distorting simplification to focus on the output of digestible energy produced in staple crops in general and grains in particular.

Only grains combine fairly high yields—initially only about 500 kg/ha; eventually, in the most intensive traditional agricultures, more than 2 t/ha—with high shares of easily digestible carbohydrates and a moderately high level of proteins (some, above all corn, also have a significant amount of lipids). Their energy density at maturity (15–16 MJ/kg) is roughly five times that of fresh tubers, and their moisture content when air-dried is low enough to allow long-term storage (in vessels by households, on a large scale in granaries). Staple grains also mature fast enough—traditional varieties mature in 100–150 days—to permit increased food productivity through annual rotations with other crops (mainly oilseeds and grain legumes) or by double-cropping of cereals.

Boserup (1965, 1976) conceptualized the link between food energy and the evolution of peasant societies as a matter of choices. Once a particular agricultural system reaches the limits of its productivity, people can decide to migrate, to stay and stabilize their numbers, to stay and let their numbers decline—or to adopt a more productive way of farming. The last option may not be necessarily more appealing or more probable than the other solutions, and its adoption is often postponed or chosen only reluctantly because such a shift almost invariably requires higher energy inputs—in most cases of both human and animal labor. Increased productivity will support larger populations by cultivating the same (or even smaller) areas, but the net energy return of intensified cropping may not increase and may actually decline.

Reluctance to expand permanently cultivated land (a choice that entailed higher energy inputs, beginning with the clearing of primeval forests, the draining of swamps, or the building of terraced fields) led to much delayed reclamation of marginal lands. The villages of Carolingian Europe were overpopulated and their grain supplies were chronically inadequate, but only in parts of Germany and Flanders were new fields created in less easily cultivable areas (Duby 1968). Medieval Europe saw waves of German peasants moving from densely populated western regions to open up new farmland in wooded or grassland areas of Bohemia, Poland, Romania, and Russia that were seen as undesirable by nearby cultivators. Similarly, China began its colonization of the fertile but cold Northeast (Manchuria) only during the eighteenth century, and even now cultivation on outlying Indonesian islands is of low intensity compared to the high productivities on densely populated Java. And everywhere, it took millennia to shift from regular, extensive fallowing to annual cropping and then to multicropping.

Despite many differences in agronomic practices and in cultivated crops, all traditional agricultures shared the same energetic foundation. They were powered by the photosynthetic conversion of solar radiation, producing food for people, feed for animals, recycled wastes for the replenishment of soil fertility, and fuels for smelting the metals needed to make simple farm tools. Consequently, traditional farming was, in principle, fully renewable. In reality, it often led to the depletion of accumulated energy stocks, above all in its pioneering stages when new farmland was created through the widespread clearance of primeval forests. In any case, the whole enterprise relied on virtually immediate conversions of solar energy flows (with typical delays ranging from just the several months required to harvest crops to several decades before cutting down mature trees).

Yet even where cropping replaced natural grasslands (entailing a much smaller loss of stored phytomass), this renewability was no guarantee of sustainability. Poor agronomic practices lowered soil fertility or caused excessive erosion or desertification, resulting in reduced yields and even the abandonment of cultivation. In most regions traditional farming progressed from extensive to intensive cultivation: its prime movers—human and animal muscles—remained unchanged for millennia, but cropping practices, cultivated varieties, and the organization of labor were greatly transformed. Thus both constancy and change mark the history of traditional farming.

The advancing intensification of farming sustained higher population densities, but it also demanded higher energy expenditures, not only for direct farming activities but also for such critical supportive measures as the digging of wells, the building of irrigation canals, roads, and food storage structures, and the terracing of fields. In turn, these improvements required more energy to make a larger variety of better tools and simple machines, powered by domestic animals or by water and wind. More intensive cultivation relied on animal labor at least for plowing, usually by far the most energy-demanding field task. The Americas were notable exceptions: neither Mesoamerican cultivators nor Inca potato and corn growers had any draft animals. Keeping domestic animals required more intensive cropping to produce feeds. Animals were also used extensively for many other field tasks, as well as in grain threshing and milling, and were indispensable for the land-borne distribution of food. Their stabling, feeding, and breeding and the production of harnesses, shoes, and implements introduced new complexity and skills.

But not all steps toward intensive farming were as energy-intensive as multicropping, which put a recurrent strain on available labor during planting and harvesting periods; increased the reliance on more powerful draft animals, which led to a requirement for more land to produce their feed; or had to be supported through the construction and maintenance of irrigation canals, labor that entailed repeated heavy exertion. To use a mechanical analogy, some changes that made it possible to harness a higher share of the available photosynthetic potential involved the opening of critical nonenergy gates (valves) that were either throttling the existing flows or virtually preventing their conversion into digestible phytomass.

The availability of nitrogen, the key plant macronutrient, is perhaps the most important example of this effect, and the rotation of nitrogen-fixing leguminous crops with cereals and tubers increased overall food output while also bringing important agroecosystemic benefits. Similarly, advances in the design of irrigation devices and the adoption of new cultivars and new crop varieties helped boost productivities and annual harvests. In turn, intensified farming brought not only energetic benefits (more food and feed) but also contributed to the advancement of preindustrial civilizations as it demanded long-range planning, long-term investment, and improved labor organization even as it promoted wider social and economic integration.

Of course, not every form of cropping intensification required centralized organization and oversight. The digging of short, shallow irrigation canals or wells or the building of a few terraces or raised fields originated repeatedly with individual peasant families or villages. But the increasing scale of such activities eventually demanded hierarchical coordination and supralocal management. And the need for more powerful energy sources to process larger amounts of grain and oilseeds for growing cities was an important stimulus for the development of the first important substitutes for human and animal muscles, the use of water and wind flows for grain milling and oilseed pressing. Millennia of farming evolution resulted in a wide range of operating modes and productivities within the constraints of shared agronomic practices and common energetic imperatives.

Principal commonalities included basic field and postharvest operations, a widespread dominance of cereals in cropping, and sequences of production cycles that were determined largely by environmental conditions. Four major steps toward the intensification of traditional farming were a more efficient use of animal labor, advances in irrigation, increasing fertilization, and crop rotation and multicropping. Despite many environmental and technical constraints, traditional agricultures could support population densities that were orders of magnitude higher than those of all but a few foraging societies. Relatively early in their existence they began creating an energy surplus that allowed initially small but significant numbers of adults to engage in an expanding range of nonfarming activities, which eventually led to highly diversified and stratified preindustrial societies. Productive limits on traditional agricultures were removed only by rising inputs of fossil fuels, an energy subsidy that reduced agricultural labor to only a small fraction of the total labor force and enabled the rise of modern high-energy urban societies.

Commonalities and Peculiarities

The requirements of crop growing imposed a general pattern on the sequence of field work. Cultivation of identical crops led to the invention or adoption of very similar agronomic practices, tools, and simple machines. Some of these innovations came early, diffused rapidly, and then remained largely unchanged for millennia. Other inventions remained restricted to their regions of origin for very long time but, once diffused, underwent rapid improvement. The sickle and flail are in the first category, the iron moldboard plow and seed drill in the second. Tools and simple machines made field operations easier (thus providing a mechanical advantage) and faster, raised productivity, and enabled fewer people to grow more food, and the resulting energy surplus could be invested in structures and actions: without sickle and plow there would be no cathedrals—or no European voyages of discovery. I will first provide brief surveys of field operations, tools, and simple machines and then describe the dominance of cereal grains and the peculiarities of cropping cycles.

Field Work

A great deal of traditional farming required heavy work, but such spells were often followed by extended periods of less demanding activities or seasonal rest, an existential pattern quite different from the nearly constant high mobility of foraging. The shift from foraging to farming left a clear physical record in our bones. Examination of skeletal remains from nearly 2,000 individuals in Europe whose lives spanned 33,000 years, from the Upper Paleolithic to the twentieth century, revealed a decrease in the bending strength of leg bones as the population shifted to an increasingly sedentary lifestyle (Ruff et al. 2015). This process was complete by about two millennia ago, and there has been no further decline in leg bone strength since then, even as food production has become more mechanized, an observation confirming that the shift from foraging to farming, from mobility to sedentism, was a truly epochal divide in human evolution.

Environmental imperatives dictated the timing of field work in traditional farming, a requirement stressed in *De agri cultura*, the oldest extant compendium of farming advice, written by Marcus Cato during the second century BCE: "See that you carry out all farm operations betimes, for this is the way with farming: if you are late in doing one thing you will be late in doing everything." Seeding was done for millennia by hand, but all other field tasks required tools, whose assortment increased with time, and although there were some early designs of farm machines, such implements began to diffuse only during the early modern era (1500–1800).

Reviews of traditional farm tools, implements, and machines are available in books dealing with the history of agriculture in specific regions or countries that are cited later in this chapter, and in greater detail in more specialized volumes by White (1967) for the Roman world, Fussell (1952) and Morgan (1984) for Britain, Lerche (1994) for Denmark, Ardrey (1894) for the United States, and Bray (1984) for China. I have used all of these sources in describing all important implements and key cultivation practices and advances in the following pages; only animal harnesses will be dealt with in the section on traditional draft power.

In all of the Old World's high cultures the sequence started with plowing. In the words of a classic Chinese treatise, "No king or ruler of a state could dispense with it." Its indispensability is also reflected in ancient writing. Both the Sumerian cuneiform records and the Egyptian glyphs have pictograms for plows (Jensen 1969). Plowing prepares the ground for seeding much more thoroughly than hoeing: it breaks up the compacted soil, uproots weeds, and provides loosened, well-aerated ground in which seedlings can germinate and thrive. The first primitive scratch plows (ards), commonly used after 4000 BCE in Mesopotamia, were just pointed wooden sticks with a handle.

Later most of them were tipped with metal, but for centuries they remained symmetrical (depositing soil on both sides) and light. Such simple plows, which merely opened up a shallow furrow for seeds and left cut weeds on the surface, were the mainstay of both Greek and Roman farming (*aratrum* in Latin). They were used in large parts of the Middle East, Africa, and Asia until the twentieth century. In the poorest places they were, in extremis, pulled by people. Only in lighter, sandier soils would such an effort be speedier than hoeing (Bray 1984). The addition of a moldboard was by far the most important improvement. A moldboard guides the plowed-up soil to one side, turns it (partially or totally) over, buries the cut weeds, and cleans the furrow bottom for the next turn. A moldboard also makes it possible to till a field in one operation rather than by cross-plowing it, as required with ards. The first moldboards were just straight pieces of wood, but before the first century BCE the Han Chinese had introduced curved metal plates joined to the plowshare (fig. 3.1).

Heavy medieval European plows had a wooden moldboard and a coulter that cut an edge into the soil ahead of the wrought-iron share. During the second half of the eighteenth century Western plows still retained their heavy wooden wheels but carried well-curved iron moldboards (fig. 3.1). These moldboard plows became common in Europe and North America only with the availability of inexpensive steel, produced first by the Bessemer process during the 1860s, and soon afterward in much larger quantities in open-hearth furnaces (Smil 2016) (fig. 3.1). In most soils plowing leaves behind relatively large clods, which must be broken before seeding. Hoeing will work, but it is too slow and too laborious. That is why harrows have been used by all old plow cultures. Their development led from primitive brush harrows to a variety of wooden or metal frames to which were fastened wooden pegs or metal teeth or disks. Inverted harrows or rollers were often used to further smooth the surface.

After plowing, harrowing, and leveling, the ground was ready to be seeded. Although seed drills were used in Mesopotamia as early as 1300 BCE, and sowing plows were used by the Han Chinese, broadcast seeding by hand—wasteful and resulting in uneven germination—remained common in Europe until the nineteenth century. Simple drills that dropped

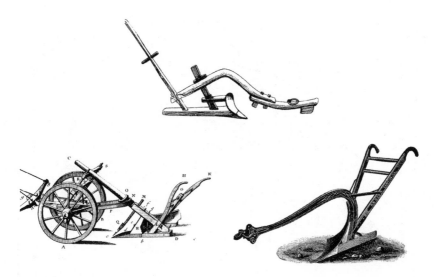

Figure 3.1
Evolution of curved moldboard plows. Traditional Chinese plow (top) had small but smoothly curving moldboard made from nonbrittle cast iron. Heavy European medieval plow, attached to a forecarriage (bottom left), had a pointed coulter in front of the share to cut the roots. The efficient American beam plow of the mid-nineteenth-century (bottom right) had its share and moldboard fused into a smoothly curving steel shape. *Sources:* Hopfen (1969), Diderot and D'Alembert (1769–1772) and Ardrey (1894).

seeds through a tube from a bin attached to a plow started to spread, first in northern Italy, during the late sixteenth century. Before long, many further innovations turned them into complex seeding machines. The intercultivation of growing crops was done largely by hoeing. Manures and other organic wastes were brought to fields on carts, in wooden cisterns, or in buckets carried at the end of shoulder beams, a common practice in East Asia. Then the wastes were pitchforked, poured, or ladled onto the field.

Sickles were the first harvesting tools to replace the short sharp stone cutters used by many foraging societies, and large scythes with cutting edges up to 1.5 m long are documented from Roman Gaul (Tresemer 1996; Fairlie 2011). Sickles have serrated (the oldest designs) or smooth edges and semicircular, straight, or slightly curved blades. Cutting with sickles was slow, and scythes, equipped with cradles for grain reaping, were preferred for harvesting larger areas (fig. 3.2). But sickle harvesting caused lower grain losses from ear shattering than broad sweeps with scythes, and the practice

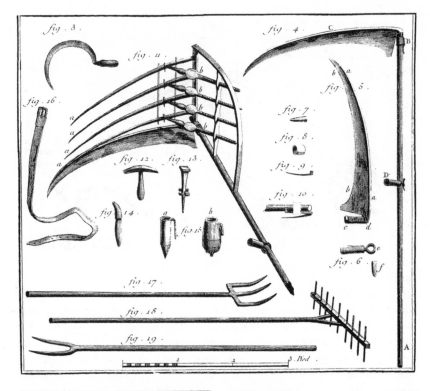

Figure 3.2
Sickle and scythes pictured in the French *Encyclopédie* (Diderot and d'Alembert 1769–1772). The simple scythe on the right was used for grass mowing, the cradled one for cereal harvesting. Also shown are tools for hammering (straightening) and sharpening the scythes, as well as a rake and pitchforks. The bottom illustrations show nineteenth-century American grain harvesting by sickle and cradled scythe.

was retained in Asia for harvesting easily shattered rice. Mechanical reapers came to American and European grain fields only in the early part of the nineteenth century (Aldrich 2002). Harvests were brought home as sheaves carried on heads, in panniers hung on shoulder beams or the sides of animals, and in wheelbarrows, carts, or wagons pushed or pulled by people or draft animals.

A considerable amount of energy went into crop processing. Grain spread on a threshing floor was beaten with sticks or flails; sheaves were hit against grates or pulled across special combs. Animals were used to tread on the spread grain or to pull heavy sleds or rollers over it. Before the adoption of crank-turned fans, winnowing, the separation of chaff and dirt from grain, was done manually with baskets and sieves. Tedious manual labor was also needed for grain milling before animals and water and windmills were used to mechanize the task. Oil was extracted from seeds by manual or animal-operated presses, as was the sweet juice from cane.

The Dominance of Grains

Although all traditional agricultures grew a variety of grain, oil, fiber, and feed crops, the described sequence of common field tasks was performed most often in cultivating cereals. Besides plowing, the dominance of cereal grains in annual cropping was certainly the other most obvious common trait of all Old World agricultures. Plowless Mesoamerican societies shared this trait with their reliance on corn, and even the Incas were only a partial exception: at high elevations and on steep mountain slopes they planted a wide range of potato varieties, but they also cultivated corn in lower altitudes, and quinoa grain on the high-lying Andean altiplano (Machiavello1991). That cultivation relied on the *chaki taklla*, a foot plow consisting of a wooden pole with a curved and sharp point and a crossbar to be pushed by foot to make a furrow.

Many other cereals were of only local or regional importance, among them Incan quinoa, recently in vogue in vegan Western diets, but the main genera of staple grains had gradually diffused worldwide from their areas of origin. Wheat spread from the Near East, rice from Southeast Asia, corn from Mesoamerica, and millets from China (Vavilov 1951; Harlan 1975; Nesbitt and Prance 2005; Murphy 2007). The importance of grains owed to a combination of evolutionary adjustments and energetic imperatives. Foraging societies gathered a wide variety of plants, and, depending on the exploited ecosystem, either tubers or seeds provided most of their food energy. In settled societies the tuber option as staple food was restricted.

The water content of freshly harvested tubers is too high for long-term storage in the absence of effective temperature and humidity controls. Even if this challenge could be overcome, their bulkiness requires large storage volumes to tide densely settled populations of more northerly latitudes (or in higher altitudes) over winter months. High-altitude Andean societies solved the problem by preserving potatoes as *chuño*. This dehydrated food-stuff, produced by Quechua and Aymara by an alternating process of freez-ing, trampling, and drying, can be stored for months, even years (Woolfe 1987). Tubers are low in protein (typically about one-fifth of that in cereals: some durum wheats have up to 13% protein, white potatoes just 2%). Legumes have twice as much protein as cereals (peas about 20%, beans and lentils between 18% and 26%) and soybeans more than three times as much (35–38%, with some cultivars reaching 40% by weight). But average yields of leguminous crops are a fraction of staple grain harvests: average yields for U.S. cereals were 2.5 t/ha in 1960 and 7.3 t/ha in 2013, with ana-logical pulse yields at 1.4 t/ha and 2.5 t/ha (FAO 2015a).

Dependence on cereal grains is thus a matter of clear energy advantages. Their primacy stems from the combination of fairly high yields, good nutri-tional value (high in filling carbohydrates, moderately rich in proteins), a relatively high energy density at maturity (roughly five times higher than for tubers), and a low moisture content suitable for long-term storage (in well-ventilated storage they do not spoil when grains contain less than 14.5% water). The dominance of a particular species is largely a matter of environmental circumstances (above all the length of the vegetation period, the presence of suitable soils, and the availability of adequate water) and taste preferences. In terms of total energy content, all cereals are remarkably similar: differences between mature seeds of various grains are mostly less than 10% (box 3.1).

The bulk of cereals' food energy is in carbohydrates, present mostly as highly digestible polysaccharides (starches). Rising share of starches in human diet resulted in a remarkable dietary adaptation of the first domes-ticated animal: genetic mutations increased starch digestion in dogs rela-tive to the carnivorous diet of wolves, a crucial step in domesticating the species (Axelsson et al. 2013). The protein content of cereals has a wider range, from less than 10% for many rice cultivars to 13% for hard summer wheat and up to 16% for quinoa. Proteins have the same energy density as carbohydrates (17 MJ/kg), but their primary role in human nutrition is not as a source of energy but as a provider of nine essential amino acids whose ingestion is essential to build and repair body tissues (WHO 2002). We

Box 3.1
Energy density and carbohydrate and protein content of principal grains

Cereal grains	Energy content (MJ/kg)	Carbohydrates (%)	Protein (%)
Wheat	13.5–13.9	70–75	9–13
Rice	14.8–15.0	76–78	7–8
Corn	14.7–14.8	73–75	9–10
Barley	13.8–14,2	73–75	9–11
Millet	13.5–13.9	72–75	9–10
Rye	13.3–13.9	72–75	9–11

Sources: Ranges compiled from USDA (2011) and Nutrition Value (2015).

cannot synthesize body proteins without consuming these essential amino acids in plant and animal foods.

All animal foods and all mushrooms supply perfect proteins (with adequate proportions of all nine essential amino acids), but all four leading staple grains (wheat, rice, corn, millet) and other important cereals (barley, oats, rye) are deficient in lysine, while tubers and most legumes are short of methionine and cysteine. Complete protein can be supplied even with the strictest vegetarian diets by combining foodstuffs with particular amino acid deficiencies. All traditional agricultural societies subsisting on largely vegetarian diets dominated by cereal grains have found independently (and obviously in the absence of any biochemical knowledge: amino acids and their role in nutrition were discovered only in the nineteenth century) a simple solution of this fundamental deficiency by including grains and legumes in mixed diets.

In China, soybeans (one of the few important food plants with complete protein), beans, peas, and peanuts have supplemented northern millets and wheat and southern rice. In India, protein from dal (a generic Hindi term for pulses, including lentils, peas, and chickpeas) has always enriched diets based on wheat and rice. In Europe the most common legume-cereal combinations relied on peas and beans and on wheat, barley, oats, and rye. In West Africa, peanuts and cowpeas were consumed alongside millets. And in the New World, corn and beans were not only eaten together in a variety of dishes, they were also commonly intercropped, grown together in alternating rows in the same field.

This means that even purely vegetarian diets could provide adequate protein intakes. At the same time, nearly all traditional societies valued meat highly, and where its consumption was proscribed they resorted either to dairy products (India) or fish (Japan) to consume high-quality animal protein. Two proteins in wheat are unique, not nutritionally but because of their physical (viscoelastic) properties. Monomeric gluten proteins (gliadin) are viscous; polymeric gluten proteins (glutenin) are elastic. When combined with water they form a gluten complex that is sufficiently elastic to allow a leavened dough to rise, yet strong enough to retain carbon dioxide bubbles formed during the yeast fermentation (Veraverbeke and Delcour 2002).

Without these wheat proteins there would be no leavened bread, the basic food of Western civilization. Yeast was never a problem: wild (naturally occurring) *Saccharomyces cerevisiae* is present on the skins of many fruits and berries, and many strains have been domesticated, resulting in changes in gene expressions and colony morphology (Kuthan et al. 2003). The dominance of cereals in traditional diets makes the energy balances of grain production the most revealing indicators of agricultural productivity. Data on typical agricultural labor requirements and their energy costs are available for a large variety of individual field and farmyard tasks (box 3.2).

But this level of detail is not necessary for calculating approximate energy balances. Using a representative average of typical net energy costs in traditional farming works quite well. The typical energy needs of moderate activities are about 4.5 times the basal metabolic rate for men, and five times for women, or 1 and 1.35 MJ/h (FAO 2004). Subtracting the respective basic existential needs results in net labor energy costs of 670 and 940 kJ/h. The simple mean is roughly 800 kJ/h, and I will use it as the net food energy cost of an average hour of labor in traditional agriculture. Similarly, gross grain output is calculated by multiplying the harvested mass by appropriate energy equivalents (typically by 15 GJ/t for grain with less than 15% moisture that can be stored).

The ratio of these two measures indicates the gross energy return, and hence the productivity, of these critical farming tasks. Net energy returns—after subtracting seed requirements and milling and storage losses—were substantially lower. Farmers had to set aside a portion of every harvest for the next year's seeding. The combination of low yields and high seed waste in hand broadcasting could mean that as much one-third or even one-half of medieval grain crops had to be set aside. With increasing harvests these shares fell gradually to less than 15%. Some grains are eaten whole, but

Box 3.2
Labor and energy requirements in traditional farming

Tasks	People/Animals	Hours per hectare	Energy cost
Hoeing			M–H
General	1/—	100–120	M–H
Wet soil	1/—	150–180	H
Plowing			M–H
Wooden plow	1/1	30–50	H
Wooden plow	1/2	20–30	H
Steel plow	1/2	10–15	M
Harrowing	1 /2	3–10	M
Sowing			L–M
Broadcasting	1/—	2–4	M
Seed drills	1/2	3–4	L
Weeding	1/—	150–300	M–H
Harvesting			M–H
Sickle (wheat)	1/—	30–55	H
Sickle (rice)	1/—	90–110	H
Cradle	1/—	8–25	H
Binding sheaves	1/—	8–12	M–H
Shocking	1/—	2–3	H
Reaper	1/2	1–3	M
Binder	1/3	1–2	M
Combine	4/20	2	M
Threshing			L–H
Treading	1/4	10–30	L
Flailing	1/—	30–100	H
Threshers	7/8	6–8	M

Notes: Light work (L) consumes less than 20 kJ of food energy per minute for an average adult man. Moderate (M) exertions range up to 30 kJ/min, and heavy (H) ones up to 40 kJ/min. Analogous rates for women are about 30% lower.

Sources: Ranges were compiled and calculated from data in Bailey (1908), Rogin (1931), Buck (1937), Shen (1951), and Esmay and Hall (1968). Energy cost indicators were estimated from metabolic studies reviewed in Durnin and Passmore (1967).

Box 3.3
Cereal milling

Whole grain flours incorporate the complete kernel, but white wheat flour is made only of the gain's endosperm (about 83% of total weight), with bran (about 14%) and germ (about 2.5%) separated for other uses (Wheat Foods Council 2015). The production of white rice entails even higher milling loss. The husk layer makes up 20% of rice grain mass; its removal produces brown rice. The bran layer makes up 8–10% of the grain, and different degrees of its removal produce more or less polished (white) rice that contains only 70–72% of the grain's initial weight (IRRI 2015).

Japanese testimonies of food shortages refer to people forced to eat brown rice, and when matters got worse brown rice mixed with barley, and ultimately just barley (Smil and Kobayashi 2011). Corn milling removes the tip cap, bran coat, and germ, leaving the endosperm, about 83% of the kernel. Corn flour for making tortillas and tamales, *masa harina,* is produced by nixtamalization, or the wet milling of kernels soaked in lime solution (Sierra-Macías et al. 2010; Feast and Phrase 2015). This loosens the hulls from the kernels, softens the kernels by dissolving hemicellulose, reduces the presence of mycotoxins, and enhances the bioavailability of niacin (vitamin B_3).

before actual food preparation (cooking or baking) most cereals are milled first, and in the process lose a significant share of the whole grain's mass (box 3.3).

Storage losses on traditional farms—to fungi and insect infestations and to rodents able to access bins or jars—would commonly reduce the edible grain total by anywhere between a few percent to more than 10%. As already noted, grain with less than 15% moisture can be stored for long periods of time; higher moistures, especially when combined with higher temperatures, provide perfect conditions for seed germination and for the growth of insects and fungi. In addition, improperly stored grain can be consumed by rodents. Even as recently as the mid-eighteenth century a combination of seeding requirements and storage losses could have reduced the gross energy gain of European grains by around 25%.

Cropping Cycles
The commonalities of annual crop cycles and the dominance of cereal cultivation obscured an astonishing variety of local and regional peculiarities. Some of them were of distinctly cultural origins, but most developed as responses and adaptations to different environments. Most notably,

environmental conditions determined the choice of leading crops, and hence the makeup of typical diets. They also molded the rhythm of annual farming cycles, which determined the management of agricultural labor. Wheat was able to spread from the Middle East to all continents because it does well in many climates (in semideserts as well as in rainy temperate zones, and it is the leading food cultivar in the temperate zone between 30 and 60°N) and elevations (from sea level to as high as 3,000 m above sea level) and on many soils, as long as they are well drained (Heyne 1987; Sharma 2012).

In contrast, rice is originally a semiaquatic plant of tropical lowlands and grows in fields flooded with water until just before the harvest (Smith and Anilkumar 2012). Its cultivation has also spread far beyond the original South Asian core, but the best yields have always been in rainy tropical and subtropical regions (Mak 2010). Constructing and maintaining ridged wet fields, germinating seeds in nurseries, transplanting seedlings, and providing subsidiary irrigation add up to much higher labor requirements than for wheat cultivation. Unlike wheat, corn yields best harvests in regions with warm and rainy growing seasons, but it, too, prefers well-drained soils (Sprague and Dudley 1988). Potatoes grow best where summers are cool and rains abundant.

Annual farming cycles were governed by water availability in both arid subtropics and in monsoonal regions, and by the length of growing seasons in temperate climate. In Egypt the Nile's floods determined the annual cycle of cultivation until the adoption of widespread perennial irrigation in the second half of the nineteenth century. Sowing began as soon as the water receded (usually in November), and no field work could be done between the end of June, when the waters started to rise, and the end of October, when they rapidly receded, with harvests 150–185 days later (Hassan 1984; Janick 2002). This pattern prevailed largely unchanged until the nineteenth century.

In monsoonal Asia, rice cultivation had to rely on summer precipitation, usually abundant but often delayed. For example, in intensive Chinese cropping, seedlings of early rice were transplanted from nurseries to fields in April. The first crop's harvest in July was followed immediately by the transplanting of the late rice, which was harvested in the late fall and, in turn, followed by a winter crop. Double-cropping in temperate zones worked under much less pressure. In Western Europe the overwintering crops planted in the fall were harvested five to seven months later. They were followed by spring-seeded crops, which matured in four to five months. In cold northern regions the ground would thaw by April but the planting of a single annual crop had to wait until late May, when the danger of killing frosts would have receded, and a crop could have only about three months to mature before the return of killing frosts.

A climatically dictated rhythm of cultivation put highly fluctuating demands on the mobilization and management of human and animal labor. Regions with a single annual crop had long months of winter idleness, so characteristic of grain farming in northern Europe and on the North American plains. Taking care of domestic animals was, of course, a yearlong task, but it still left much free time, and some of it was spent in domestic craft work, in repairing farming equipment, or in construction. Many days during the shorter winter in North China were devoted to maintaining and extending irrigation works.

Spring plowing and seeding called for a few weeks of hard work, followed by a few months of an easier routine (though the weeding of rice fields could be arduous). Harvest was the most taxing time, and the fall plowing could extend over a much longer period. Where less extreme climates allowed the planting of a winter crop—in Western Europe, on the North China Plain, in most of the eastern North America—there were two to three months between harvesting the summer crop and putting in the winter one. In contrast, in countries with much less evenly distributed precipitation, and especially in monsoonal Asia, there were only limited slots for performing field tasks. Timeliness in planting and harvesting was especially critical. Even a week's delay beyond the optimum planting period could cause a substantial yield reduction. An early grain harvest may need labor-intensive drying because of its high moisture, a delayed one may cause losses because of the shattering of overripe ears.

Before the introduction of reapers and binders, manual grain harvesting was the most time-consuming task, taking three to four times longer than plowing, and it put clear limits on the maximum area manageable by a single family. When a crop had to be harvested fast in order to plant the next one, the labor demand soared. An old Chinese proverb captured this need: "When both the millet and the wheat are yellow ripe, even the spinning girls have to come out to help." Buck (1937) quantified this requirement in his comprehensive studies of China's traditional farming: planting and harvesting (between March and September) in China's double-cropping area used virtually all (average of 94–98%) available labor. In parts of India the two peak summer months required more than 110% or even 120% of actually available labor, and a similar situation prevailed in other parts of monsoonal Asia (Clark and Haswell 1970). This common energetic bottleneck could be overcome only if all families worked arduously long hours—or by relying on migrant labor.

The use of animal labor, reserved in many agricultures just for the most demanding field tasks, was even more uneven. For example, the maximum work periods for South China's water buffaloes were two months of planting, harrowing, and grading in the early spring, six weeks of summer harvest, and a month of field preparation (again, plowing and harrowing) for a winter crop, altogether about 130–140 days, or less than 40% of the year (Cockrill 1974). In single-cropping regimes of northern Europe draft horses would do only 60–80 days of strenuous field work during the fall and spring plowing and the summer harvesting, but most of them were used extensively for transport. A typical working day ranged from just five hours for oxen in many African locations to more than ten hours for water buffaloes in Asian rice fields and for horses during European or North American grain harvests.

Routes to Intensification

No quest for higher yields could succeed without three essential advances. The first one was a partial replacement of human work by animal labor. In rice farming this eliminated usually only the most exhaustive human work as tedious hoeing was replaced by deep plowing using water buffalos. In dryland farming animal labor replaced human labor and sped up considerably many field as well as farmyard tasks, freeing people to pursue other productive activities or to work shorter hours. This prime mover shift did more than make the work quicker and easier; it also improved its quality, whether in plowing, seeding, or threshing. Second, irrigation and fertilization moderated, if not altogether removed, the two key constraints on crop productivity, shortages of water and nutrients. Third, growing a greater variety of crops, either by multicropping or in rotations, made traditional cultivation both more resilient and more productive.

Two Chinese peasant sayings convey the importance of removing the two constraints and diversifying the output: "Whether there is a harvest depends on water; how big it is depends on fertilizer" and "Plant millet after millet and you will end up weeping." The use of draft animals was a fundamental energetic advance with implications beyond field cultivation and harvesting. Draft animals became indispensable for fertilization, both as the source of nutrients in manures and as the prime mover distributing them to crops. In many places they also energized irrigation. More powerful prime movers and better water and nutrient supply also brought more multicropping and crop rotations. In turn, these advances could support large numbers of more powerful animals as the three intensification paths were linked by mutually reinforcing feedback loops.

Draft Animals

Domestication has resulted in many working breeds with distinctive characteristics, with weights spanning an order of magnitude, from just over 100 kg for small donkeys to just over 1,000 kg for the heaviest draft horses. Indian bullocks weighed less than 400 kg; Italian Romagnola or Chianina cattle were easily twice as heavy (Bartosiewicz et al. 1997; Lenstra and Bradley 1999). Most horses in Asia and parts of Europe were just ponies, less than 14 hands tall and weighing no more than an Asian ox. A hand, a traditional English measure, is four inches (10.16 cm), and the animal's height was measured from the ground to its withers, the ridge between the shoulder blades below its neck and head. Roman horses were 11–13 hands. The heaviest European breeds of the early modern era—Belgian Brabançons, French Boulonnais and Percherons, Scottish Clydesdales, English Suffolks and Shires, German Rheinlanders, Russia's Heavy Draft—approached and even topped 17 hands and weighed close to, or even just above, 1,000 kg (Silver 1976; Oklahoma State University 2015). Water buffaloes can range from just 250 kg to 700 kg (Cockrill 1974; Borghese 2005).

Traditional agricultures used animals for a variety of field and farmyard tasks, but plowing was undoubtedly the activity where they made the greatest difference (Leser 1931). In general, the tractive force of working animals is roughly proportional to their weight, and other variables determining their actual performance include the animal's sex, age, health, and experience, the efficiency of the harness, and soil and terrain conditions. As all of these variables can vary rather widely, it is preferable to summarize the useful power of common working species in terms of typical ranges (Hopfen 1969; Cockrill 1974; Goe and Dowell 1980). A typical draft is 15% of animal's body weight but for horses it is up to 35% during brief exertions (about 2 kW) and even more during a few seconds of supreme effort (Collins and Caine 1926). The combination of large mass and relatively high speed makes horses the best draft animals, but most horses could not work steadily at the rate of one horsepower (745 W), and usually delivered between 500 and 850 W (box 3.4, fig. 3.3).

Actual draft requirements varied widely with the task (the extremes of heavy and light work might be deep plowing and harrowing) and with soil type (demanding in heavy clay soils, much easier in sandy soils). Shallow plowing (with a single plowshare) and grass mowing needed a sustained draft of 80–120 kg, deep plowing required drafts of 120–170 kg, and a 200-kg pull was needed for a mechanical grain reaper and binder. Even an average horse pair can do all of these tasks, but a pair of oxen was inadequate

Box 3.4
Typical weights, drafts, working speeds, and power of domestic animals

Animals	Weights (kg) Common range	Large sizes	Typical draft (kg)	Usual speed (m/s)	Power (W)
Horses	350–700	800–1000	50–80	0.9–1.1	500–850
Mules	350–500	500–600	50–60	0.9–1.0	500–600
Oxen	350–700	800–950	40–70	0.6–0.8	250–550
Cows	200–400	500–600	20–40	0.6–0.7	100–300
Buffaloes	300–600	600–700	30–60	0.8–0.9	250–550
Donkeys	200–300	300–350	15–30	0.6–0.7	100–200

Note: Power values are rounded to the nearest 50 W.
Sources: Based on Hopfen (1969), Rouse (1970), Cockrill (1974), and Goe and Dowell (1980).

for deep plowing or harvesting with a reaper. At the same time, mechanical imperatives favored smaller animals: everything else being equal, their line of pull is more parallel to the direction of traction, resulting in higher efficiency, and in plowing, a lower pull line also reduces the uplift on the plow, making it easier for a plowman to guide it. Lighter animals are also often more agile, and they may compensate for their lower weight with tenacity and endurance.

Draft potential could be translated into effective performance only with practical harnesses (Lefebvre des Noëttes 1924; Haudricourt and Delamarre 1955; Needham 1965; Spruytte 1983; Weller 1999; Gans 2004). Traction must be transferred to the point of work—whether plowshare or reaper's edge—by a gear that allows its efficient transmission and also enables human control of the animal's movements. Such designs may look simple, but they took a long time to emerge. Cattle, the first working animals, were harnessed by yokes, straight or curved wooden bars fastened to the animal's horns or neck.

The oldest Mesopotamian harness (best used with strong, short-necked animals, and later common in Spain and Latin America) was the double head yoke, fixed either at the front or the back of the head (fig. 3.4). That was a primitive harness: merely a long wooden beam whose throat fastenings may choke the animal in heavier labor and whose traction angle is too large. Moreover, to avoid excessive choking of one ox or cow, the animals

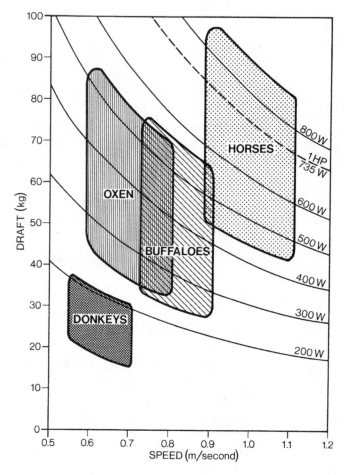

Figure 3.3
Comparisons of animal draft power showing the clear superiority of horses. Plotted
from data in Hopfen (1969), Rouse (1970), and Cockrill (1974).

must be of identical height, and a pair must be harnessed even when a
single animal would do for lighter work. A more comfortable single head
yoke was used in several parts of Europe (the eastern Baltic region, southwestern Germany). The single neck yoke, connected to two shafts or to
traces and a swingletree, was common throughout East Asia, and also in
Central Europe (fig. 3.4). Africa, the Middle East, and South Asia favored a
double neck yoke.

Horses are the most powerful draft animals. Unlike cattle, whose body
mass is almost equally divided between the front and the rear, horses' fronts

Figure 3.4
The head yoke was the first, and rather inefficient, harness for working oxen. The neck yoke became the dominant way of harnessing cattle throughout the Old World. Adapted from Hopfen (1963) and from a late Ming dynasty (1637) illustration.

are notably heavier than their rears (ratio of about 3:2), and so the pulling animal can take a better advantage of inertial motion than cattle (Smythe 1967). Except in heavy, wet soils, horses can work in fields steadily at speeds of around 1 m/s, easily 30–50% faster than oxen. Maximum two-hour pulls for paired heavy horses can be almost twice as high as for the best cattle pairs. The largest horses can thus work for short spells at rates surpassing 2 kW, or more than three standard horsepower. But humpback cattle are superior in the tropics thanks to their efficient heat regulation; they are also less susceptible to tick infestation. And water buffaloes thrive in wet tropics and convert roughage feed more efficiently than cattle and can graze on aquatic plants while fully submerged.

The oldest existing images of working horses do not show them laboring in fields but rather pulling light ceremonial or attack carriages. During most of antiquity, draft horses were harnessed by a dorsal yoke (Weller 1999). The dorsal yoke was a forked device, wooden or metal, that was placed on the animal directly behind its withers and held in place by a chest strap that ran across the animal's breast and was fastened on both sides of the yoke and by a surcingle (a strap running over the back and under the belly). Inaccurate reconstruction of the Roman harness by Lefebvre des Noëttes (1924) led to a mistaken but for decades a widely accepted conclusion that this was a very inefficient arrangement because it choked the animal, as the breast collar tended to ride up (box 3.5).

The breastband harness, introduced in China no later than the early Han dynasty, had its point of traction too far away from the animal's most powerful pectoral muscles (fig. 3.5). Nonetheless, the design spread across Eurasia, reaching Italy as early as the fifth century, most likely with migrating Ostrogoths, and northern Europe some 300 years later. But it took another Chinese invention to turn horses into superior working animals. The collar harness was first used in China perhaps as early as the first century BCE as a soft support for the hard yoke; only gradually was it transformed into a single component. By the fifth century CE its simple variant can be seen on the Donghuang frescoes. Philological evidence suggests that by the ninth century it had reached Europe, where it was in general use within about three centuries, and the design remained largely unchanged until horses were replaced by machines more than 700 years later. It is still used on a diminishing number of China's working horses.

The standard collar harness consists of a single oval wooden (later also metal) frame (hame), lined for a comfortable fit onto horse's shoulders, often with a separate collar pad underneath. Draft traces are connected to the hame just above the horse's shoulder blades (fig. 3.6). The horse's

---

Box 3.5

Comparing harnesses and draft power

For decades many writings repeated the claim that the ancient throat-and-girth harness was not suitable for heavy field tasks because of its excessively high point of traction and its choking effect created by the throat strap. This conclusion was based on actual experiments with a reconstructed harness done in 1910 by a French officer, Richard Lefebvre des Noëttes (1856–1936), and published in 1924 in his book, *La Force Motrice à travers les Âges*. These findings were accepted not only by many classicists but also by three leading twentieth-century historians of technical advances, Joseph Needham (1965), Lynn White (1978), and Jean Gimpel (1997).

But they were based on a mistaken reconstruction: experiments done by Jean Spruytte during the 1970s with a properly reconstructed dorsal yoke harness (placed directly behind the withers and fastened by chest straps) did not result in any choking, and the harness performed well when two horses were pulling a load of nearly 1 t (Spruytte 1977). This disproved the idea that "the classical cultures were 'blocked' by a defective system of harnessing animals" (Raepsaet 2008, 581). But in his tests, Spruytte used a light nineteenth-century carriage (much lighter than a Roman wagon), and hence, even if the difference in horse size is ignored, his tests did not fully replicate the conditions common two millennia ago. In any case, because of the weight limit (500 kg) put on the horse-drawn wagons by the Theodosian Code (439 CE), "It seems certain that the Romans were aware of the distress caused in horses by pulling heavy loads" (Gans 2004, 179).

---

movements are controlled by the bridle, a metal bit in its mouth attached to reins and braced by a headstall. The collar harness provided a desirably low traction angle and allowed a heavy exertion through deployment of the animal's powerful breast and shoulder muscles. The collar harness also allowed for effective teaming of animals in a single or double file for exceptionally heavy labor.

An efficient harness was not the only precondition of a horse's superior performance, and hence its introduction did not launch an agricultural revolution (Gans 2004). Hard-working horses were fed grain, which required a planting cycle, and needed relatively expensive harnessing and shoeing, while weaker and slower-working oxen could be fed with just straw and chaff and harnessed cheaply. Horseshoes are narrow U-shaped metal plates fitting the hoof's rim and fastened with nails driven into the insensitive hoof wall (fig. 3.6). Their use prevented excessive wear of the animal's soft

Figure 3.5
The breastband harness, reproduced here from the *Encylopédie* (Diderot and D'Alembert 1769–1772), remained in use for lighter duty until the twentieth century.

hooves and improved its traction and endurance. This was especially important in the cool and wet climate of Western and northern Europe. The Greeks did not have them; they encased hooves in leather sandals filled with straw. The Romans knew of horseshoes (but their *soleae ferreae* were fastened by clips and laces), but nailed shoes became common only after the ninth century.

Whippletrees (swingletrees), attached to traces, interconnected, and then fastened to field implements, equalized strain resulting from uneven pulling. They made it easier to lead the animals and allowed the harnessing of an even or odd number of horses. Horses also have better endurance (working 8–10 hours a day compared to 4–6 hours for cattle) and they live longer, and while both oxen and horses started working at 3–4 years of age, oxen lasted usually just for 8–10 years, while horses carried on commonly for 15–20 years. Finally, a horse's leg anatomy gives the animal a unique advantage by virtually eliminating the energy costs of standing. The horse has a very powerful suspensory ligament running down the back of the cannon bone and a pair of tendons (superficial and deep digital flexors) that can "lock" the limb without engaging muscles. This allows

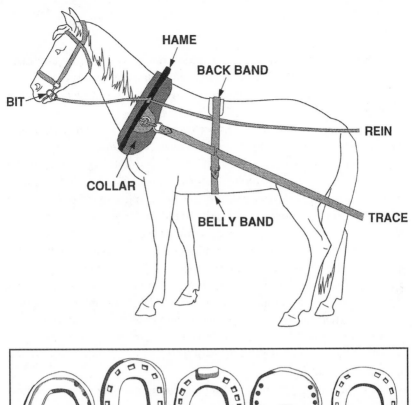

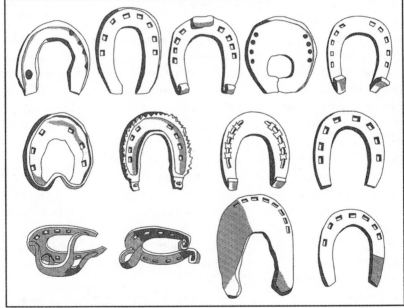

Figure 3.6
Components of a typical late nineteenth-century collar harness (based on Telleen 1977 and Villiers 1976)—and a variety of mid-eighteenth-century horseshoes (Diderot and d'Alembert 1769–1772). The shapes (starting on the left) are, respectively, typical English, Spanish, German, Turkish, and French horseshoes.

the animals to rest, even to doze, while standing, with hardly any metabolic cost, and to spend little energy while grazing (Smythe 1967). All
other mammals need about 10% more energy when standing as compared
to lying down.

Even smaller and poorly harnessed animals made a great difference
(Esmay and Hall 1968; Rogin 1931; Slicher van Bath 1963). A peasant working with a hoe would need at least 100 hours, and in heavy soils up to 200
hours, to prepare a hectare of land for planting cereals. Even with a simple
ox-drawn wooden plow that task could be done in just over 30 hours. Hoe-
dependent farming could never have attained the scale of cultivation made
possible by animal-drawn plowing. Besides speeding up plowing and harvesting, animal labor also made it possible to lift large volumes of irrigation
water from deeper wells. Animals were used to operate such food-processing machines as mills, grinders, and presses at rates far surpassing human
capabilities. Relief from long hours of tiresome labor was no less important
than the higher output rates, but more animal work required more cultivated land to grow feed crops. This was easily done in North America and
in parts of Europe, where the upkeep of horses at times claimed up to one-
third of all agricultural area.

Not surprisingly, in China and other densely populated Asian nations,
cattle were the preferred draft animals. Being ruminants, they could be
maintained solely on roughage from straw and from grazing. And when
working they do not have to be fed much grain either: concentrate feed can
come largely from such crop-processing residues as brans and oil cakes. I
estimated that in China's traditional farming the cultivation of feed for
draft animals claimed only about 5% of the annually harvested area. In
India, fodder crops also accounted traditionally for about 5% of all cultivated land but most of this feed went to milking animals, and some of it
ended up nourishing sacred cows (Harris 1966; Heston 1971). Feed for the
working bullocks probably claimed less than 3% of all farmland. In the
most densely populated parts of the Indian subcontinent cattle survived on
a combination of roadside and canal-bank grazing and on the feeding of
crop by-products, ranging from rice straw and mustard oil cakes to chopped
banana leaves (Odend'hal 1972).

Indian or Chinese draft animals were clear energetic bargains. Many of
them did not compete for crop harvests with people at all, while some pre-
empted at most an area of cultivated land, which would grow food for one
person a year. But their useful annual labor was the equivalent of three
to five peasants working 300 days a year. An average nineteenth-century
European or American horse could not give such a high relative return, but

Box 3.6
Energy cost, efficiency, and performance of a draft horse

A mature 500-kg horse needs about 70 MJ of digestible energy per day to maintain its weight (Subcommittee on Horse Nutrition 1978). If its feed is high in grains, this may imply just 80 MJ of gross energy intake; if it is mostly less digestible hay then it may rise to 100 MJ. Depending on the task, feed requirements during the working periods were 1.5–1.9 times the maintenance need. Brody(1945) found a 500-kg Percheron working at a rate of about 500 W expending about 10 MJ/h. With 6 hours of work and 18 hours of rest (at 3.75 MJ/h), this adds up to about 125 MJ/day.

Not surprisingly, traditional feeding recommendations concur: at the beginning of the twentieth century American farmers were advised to feed their working horses 4.5 kg of oats and 4.5 kg of hay a day (Bailey 1908), which translates to about 120 M J/day. With an average power of 500 W, a horse would do about 11 MJ of useful work during six hours, and while an average male human would contribute less than 2 MJ, though he could not maintain steady exertion above 80 W and managed only brief peaks above 150 W, a horse could work steadily at 500 W and have brief peak pulls in excess of 1 kW, an effort that would require the exertions of a dozen men.

it, too, was an energetic boon (box 3.6). Its annual useful labor was an equivalent of about six working farmers, and the land used for its feeding (including all the nonworking animals) could have grown food for about six people. Even if the nineteenth-century draft horse were to be seen merely as a substitute for tedious human labor, it would have earned its keep, but strong, well-fed horses could perform tasks beyond human capacity and endurance.

Horses could drag logs and pull out stumps when humans converted forests to cropland, break up rich prairie soils by deep plowing, or pull heavy machinery. There were, of course, additional energy costs of animal labor beyond maintaining a breeding herd and providing adequate feeding for field labor; these additional energy costs appeared above all in the making of harnesses and shoes and the stabling of the animals. But there were also additional benefits derived from the recycled manure and from milk, meat, and leather. Manure recycling has been important in all intensive traditional agricultures as the source of scarce nutrients and organic matter. In largely vegetarian societies, meat (including horsemeat in parts of Europe) and milk were valuable sources of perfect protein. Leather was used in making a large number of tools essential in farming and in traditional manufactures. And, of course, the animals were self-reproducing.

Irrigation

The water demand of crops depends on many environmental, agronomic, and genetic variables, but the total seasonal need is commonly about 1,000 times the mass of the harvested grain. Up to 1,500 t of water are needed to grow 1 t of wheat, and at least 900 t must be supplied for every tonne of rice. About 600 t will suffice for a tonne of corn, a more water-efficient C_4 crop and the staple grain with the highest water use efficiency (Doorenbos et al. 1979; Bos 2009). This means that for wheat yields of between 1 and 2 t/ha, the total water needs during the four months of the growing season were 15–30 cm. In contrast, annual precipitation in the arid and semiarid regions of the Middle East ranged from a mere trace to no more than 25 cm.

Cropping in such locations thus required irrigation as soon as the fields were established beyond the reach of seasonal floods, which saturated valley soils and allowed for the maturation of one harvest or as soon as the growing population required the planting of a second crop during the low-water season. Irrigation was also necessary to cope with seasonal water shortages. These are especially pronounced in the most northerly reaches of monsoonal Asia, in Punjab, and on the North China Plain. And, of course, rice growing required its own arrangements for flooding and draining the fields.

Gravity-fed irrigation—using canals, ponds, tanks, or dams—requires no water lifting and has the lowest energy cost. But in river valleys with minimal stream gradients and on large cultivated plains, it has always been necessary to lift large volumes of surface or underground water. Many lifts were only across a low embankment, but often they had to surmount steep stream banks or come from deep wells. Unavoidable inefficiencies, aggravated by the rough finishing of moving parts and often by the absence of lubricants, prolonged the task. Irrigation powered by human muscles represented a large labor burden even in societies where tedious work was the norm. Much ingenuity went into designing mechanical devices using animal power or water flow to ease that work—as well as to make higher lifts possible.

An impressive variety of mechanical devices was invented to lift water for irrigation (Ewbank 1870; Molenaar 1956; Oleson 1984, 2008; Mays 2010). The simplest ones—tightly woven or lined shovel-like scoops, baskets, or buckets—were used to raise water less than 1 m. A scoop or a bucket suspended by a rope from a tripod was slightly more effective. Both these devices were used in East Asia and the Middle East, but the oldest water-lifting method in widespread use was the counterpoise lift, commonly known as the Arabic *shaduf*. Recognizable first on a Babylonian cylinder seal of 2000 BCE and widely used in ancient Egypt, it reached China by about 500 BCE and eventually spread all over the Old World. A shaduf, basically a long wooden pole pivoted as a lever from a crossbar or a pole, was easily made and repaired (fig. 3.7).

Its bucket dipper was suspended from the longer arm and counterpoised by a large stone or a ball of dry mud. Its effective lift was usually 1–3 m, but serial deployment of the devices in two to four successive levels was common in the Middle East. A single man could raise about 3 m^3/h to the height of 2–2.5 m. Pulling on a rope could be very tiresome, but cranking an Archimedean screw (Roman *cochlea*, Arabic *tanbur*) to rotate a wooden helix inside a cylinder was even more demanding and allowed only low lifts (25–50 cm). Paddle wheels were commonly used in Asia. Chinese water ladders (dragon backbone machines, *long gu che*) operated as square-pallet wooden chain pumps, with a series of small boards passing over sprocket wheels and forming an endless chain to lift water through a wooden trough (fig. 3.8). The driving sprocket was inserted into a

UNE CHADOUF.

Figure 3.7
Nineteenth-century engraving of Egyptian peasants using a shaduf (Corbis).

Figure 3.8
China's ancient "dragon backbone machine" was powered by peasants leaning on a
pole and treading a spoked axle. Adapted from a late Ming dynasty illustration.

horizontal pole trodden by two or more men who supported themselves by leaning on a pole. Some ladders were operated by hand cranks or by animals walking in a circle.

All of the following devices were always powered either by animals or by running water. The rope and bucket lift, common in India (*monte* or *charsa*), was powered by one or two pairs of oxen walking down an incline while lifting a leather bag fastened to a long rope. An endless chain of clay pots on two loops of rope carried upside down below a wooden drum to fill at the lower end and to discharge into a flume at the top was used already by the Greeks. The arrangement is best known by its Arabic name, *saqiya*, and the device spread throughout the Mediterranean. When powered by a single blindfolded animal walking in a circular path, it raised water from wells usually less than 10 m deep at a rate not surpassing 8 m³/h. An improved Egyptian version, the *zawafa*, delivered water at a higher rate (up to 12 m³/h from a 6 m deep well).

Noria, yet another device widely used both in Muslim countries and in China (*hung che*), had clay pots, bamboo tubes, or metal buckets fastened to the rim of a single wheel. The wheels could be driven through right-angle gears by circle-walking animals or by a water current when equipped with paddles. The need to lift the buckets another wheel radius above the level of the receiving trough was a source of considerable inefficiency, which was eliminated by the Egyptian *tabliya*. This improved device, powered by oxen, included a double-sided all-metal wheel that scooped up water at the outer edge and discharged it at the center into a side trough. Comparisons of typical power requirements, lifts, and hourly outputs of traditional water-raising devices make the limits of human performance obvious (box 3.7, fig. 3.9).

The energy cost of human-powered irrigation was extraordinarily high. A worker could cut a hectare of wheat with a cradle scythe in eight hours, but he would need three months (8 h/day) to lift half of its water requirement just 1 m from an adjoining canal or stream. Large differences in crop response to watering make any generalizations regarding the energy return of traditional irrigation impossible. Not only are there substantial variations among crops, but yield response also depends on the timing of water availability (peanuts are fairly insensitive to a temporary water deficit, corn is rather vulnerable). A realistic example shows that energy returns could be easily 10-fold or higher (box 3.8).

In contrast, for some Inca projects, net energy returns had to be low. Gravity irrigation required no water lifting, but carving long and wide canals (main lines up to 10–20 m wide) out of rocks with simple tools was

Box 3.7
Power requirements, lifts, and capacities of traditional water-lifting devices

Devices	People/ Animals	Lift (m)	Capacity (m^3/h)	Work (kJ)	Input (kJ)	Efficiency (%)
Scoops	2/—	0.6	5	30	440	7
Suspended scoops	2/—	1	8	80	440	18
Shaduf	1/—	2.5	3	75	220	34
Archimedean screw	2/—	0.7	15	100	440	23
Paddle wheel	1/—	0.5	12	60	220	27
Water ladder	2/—	0.7	9	60	440	14
Rope and bucket	3/4	9	17	1500	5,690	26
Saqiya	1/2	6	8	470	2,740	17
Zawafa	1/2	6	12	710	2,740	26
Noria high lift	1/2	9	9	790	2,740	29
Noria low lift	1/1	1.5	22	325	1,480	22
Tabliya	1/1	2.5	12	295	1,480	20

Note: Energy costs were calculated by assuming an average power input of 60 W for people and 350 W for draft animals.
Sources: Compiled and calculated from data in Molenaar (1956), Forbes (1965), Needham and co-workers (1965), and Mays (2010).

labor-intensive. The main arterial canal between Parcoy and Picuy ran for 700 km to irrigate pastures and fields (Murra 1980), and the conquering Spaniards were astonished to see well-built canals carrying water to a small group of corn fields. All major irrigation projects required careful planning and execution in order to maintain appropriate gradients, and large numbers of workers had to be mobilized. The pay-off—a food energy output from irrigated crops surpassing the huge investment of labor—was obviously postponed for many years, even for decades. Only well-established central governments able to shift resources among different parts of its realm could undertake such programs of public construction. In most cases, water management for higher crop yields has involved field irrigation, but some agricultures intensified their cropping through an opposite process.

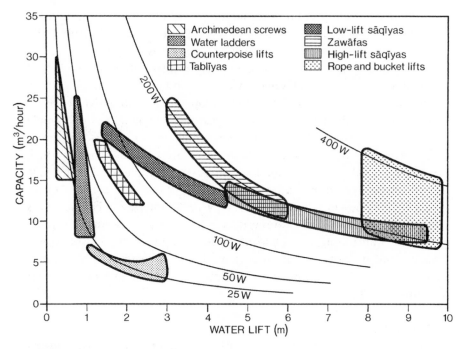

Figure 3.9
Comparison of lifts, volumes, and power requirements of preindustrial water-raising devices and machines. Plotted from data in Molenaar (1956), Forbes (1965), and Needham and colleagues (1965).

Box 3.8
Energy returns of wheat irrigation

A single specific calculation demonstrates the considerable energy returns of traditional irrigation. Field studies indicated that winter wheat yield declines by half when a 20% shortfall in annual water supply is concentrated during the critical flowering period (Doorenbos et al. 1979). A good late Qing dynasty harvest of 1.5 t/ha would be thus reduced by about 150 kg on a typical small field of 0.2 ha. Making up the deficit of 10 cm of rain by irrigation would require 200 t of water—but as ridge-and-furrow irrigation was only 50% efficient (owing to soil seepage and evaporation), the actual supply from a canal had to be twice as large. Lifting 400 t of water less than 1 m by a ladder operated by two treading peasants would take about 80 hours and require about 65 MJ of additional food energy, while the increased wheat yield would contain (after subtracting about 10% for seed and storage losses) about 2 GJ of digestible energy. Consequently, water ladder treading would return about 30 times more food energy than its food cost.

In many regions continuous farming would have been unthinkable without draining away the excessive water. Emperor Yu (2205–2198 BCE), one of the seven great pre-Confucian sages, owes his place in Chinese history largely to his master plan and heroic organization of a prolonged quest to drain away flood waters (Wu 1982). And the Mayas and successive inhabitants of the basin of Mexico adopted more intensive forms of crop cultivation that required water management ranging from simple terracing and spring irrigation to elaborate drainage systems and the extensive construction of raised fields (Sanders, Parsons, and Santley 1979; Flannery 1982; Mays and Gorokhovich 2010). A unique kind of drainage agriculture evolved over a period of many centuries in a part of China's Guangdong province (Ruddle and Zhong 1988). Intensively cropped dikes separate ponds stocked with several species of carp. Recycling of organic wastes—human, pig, and silkworm excreta, crop residues, grasses, weeds, and pond deposits—sustained high yields of mulberries for silkworms, sugar cane, rice, and numerous vegetables and fruits, as well as a high output of fish.

Fertilization

Atmospheric CO_2 and water supply carbon and hydrogen, the two elements forming the bulk of plant tissues as new carbohydrates. But other elements are absolutely necessary for photosynthesis. Depending on the amount needed, they are classified as macronutrients or micronutrients. The latter are more numerous, including above all iron, copper, sulfur, silica and calcium. There are only three macronutrients: nitrogen, phosphorus, and potassium (N, P, and K). Nitrogen is by far the most important one: it is present in all enzymes and proteins and is the element to be most likely in short supply in continuously cultivated soils (Smil 2001; Barker and Pilbeam 2007). Harvest of 1 t/ha of wheat (a typical French or U.S. harvest around 1800) removes (in grain and straw) about 1 kg each of calcium and magnesium (Ca and Mg), 2.5 kg of sulfur (S), 4 kg of potassium, 4.5 kg of phosphorus, and 20 kg of nitrogen (Laloux et al. 1980).

Rain, dust, weathering, and the recycling of crop residues would, in most cases, replenish withdrawals of phosphorus, potassium, and micronutrients, but continued cropping without fertilization would create nitrogen deficits, and because nitrogen's availability largely determines the grain size and protein content, these shortages caused stunted growth, poor yields, and poor nutritional quality. Traditional farming could replace nitrogen in only three ways: by directly recycling unwanted crop residues, that is, by plowing in part of straws or stalks that were not removed from the fields for feed, fuel, and other household uses; by applying a variety of organic matter, above all by spreading (usually composted) animal and human urine and feces and other organic wastes; and by cultivating leguminous crops to enhance nitrogen soil content for a subsequent nonleguminous crop (Smil 2001; Berklian 2008).

Cereal straws were a major potential source of nitrogen, but their direct recycling was limited. Unlike modern short-stalked plants, the traditional cultivated varieties yielded much more straw than grain, with straw-to-grain ratios commonly at 2:1. Plowing in so much plant mass could have been a great strain on many animals—but such a situation almost never arose. Only a small fraction of crop residues was returned directly to the soil because they were needed as animal feed and bedding (and only then recycled in manures), as household fuel, and as a raw material for construction and manufacturing. But in wooded regions straws and stalks were often simply burned in the field, with a virtually complete nitrogen loss.

The recycling of urine and excrement was perfected over centuries in Europe and East Asia. In Chinese cities, high shares of human waste (70–80%) were recycled. Similarly, by the 1650s virtually all of Edo's (today's Tokyo) human wastes were recycled (Tanaka 1998). But the usefulness of this practice is limited by the availability of such wastes and their low nutrient content, and the practice entails much repetitive, heavy labor. Even before storage and handling losses, the annual yield of human wastes averaged only about 3.3 kg N/capita (Smil 1983). The collection, storage, and delivery of these wastes from cities to the surrounding countryside created large-scale malodorous industries, which even in Europe persisted for most of the nineteenth century before canalization was completed. Barles (2007) estimated that by 1869, Paris was generating annually about 4.2 Mt N, about 40% from horse manure and about 25% from human wastes; by the late nineteenth century about half of the city's excreta was collected and industrially processed to make ammonium sulfate (Barles and Lestel 2007).

The recycling of much more copious animal wastes—which involved cleaning of stalls and sties, liquid fermentation or composting of mixed wastes before field applications, and the transfer of wastes to fields—was even more time-consuming. And because most manures have only about 0.5% N, and preapplication and field losses of the nutrient had commonly added up to 60% of the initial content, massive applications of organic wastes were required to produce higher yields. In the eighteenth century Flanders fields received on the average 10 t/ha, and some up to 40 t/ha, of manure, night soil, oil cakes, and ash, and typical rates in prerevolutionary France were about 20 t/ha (Slicher van Bath 1963; Chorley 1981). Similarly, detailed accounts for the 1920s China show nationwide average above 10 t/ha, and small farms in China's southwest averaging almost 30 t/ha (Buck 1937).

Every conceivable organic waste was used as a fertilizer in traditional farming. Cato's *De agri cultura* lists pigeon, goat, sheep, cattle, "and all other dung," as well as composts made of straw, lupines, chaff, bean stalks, husks, and ilex and oak leaves, and the Romans knew that rotating grain crops with legumes (they relied on lupines, beans, and vetch) would raise the yields. The Asian use of organic wastes was even more eclectic, ranging from wastes relatively rich in nitrogen (oilseed cakes, fish waste) to canal and river mud with a mere trace of the nutrient. As cities grew, food wastes, above all plant remains, created a new source of recyclable matter.

The organic fertilizer with the highest nitrogen content (around 15% for the best deposits) is guano, droppings of seabirds preserved in the dry climate of islands along the Peruvian coast. The conquering Spaniards were impressed with its use by the Incas (Murra 1980). Imports to the United States began in 1824, to England in 1840; the 1850s saw their rapid rise, but by 1872 the exports of the richest deposits, from Peru's Chincha Islands, had ended (Smil 2001). Afterward Chilean nitrates became the world's most important source of traded nitrogen as the agricultures of industrializing countries began to subsidize their harvests with inputs of fuels, metals, machines, and inorganic fertilizers, a process described in some detail in chapter 5.

Actual field applications varied widely with shares of recoverable manure (very high with penned animals, negligible with grazing animals), attitudes toward the handling human wastes (ranging from proscriptions to routine recycling), and the intensity of cropping. Any theoretical estimates of nitrogen in recycled wastes are far removed from its eventual contribution. This is because of very high losses (mainly through ammonia volatilization and leaching into groundwater) between voiding, collection, composting, application, and eventual nitrogen uptake by crops (Smil 2001). These losses, commonly of more than two-thirds of the initial nitrogen, further increased the need to apply enormous quantities of organic wastes. Consequently, in all intensive traditional agricultures, large shares of farm labor had to be devoted to the unappealing and heavy tasks of collecting, fermenting, transporting, and applying organic wastes.

Green manuring was effectively used in Europe since ancient Greek and Roman times and has been also widely employed in East Asia. The practice has relied mainly on nitrogen-fixing leguminous crops, initially on vetches (*Astragalus*, *Vicia*) and clovers (*Trifolium*, *Melilotus*), later also on alfalfa (*Mecdicago sativa*). These plants can fix as much as 100–300 kg N/ha per year and, when rotated with other crops (usually planted as a winter crop in milder climates), will add, in the three to four months before they are plowed in, 30–60 kg N to the soil, to be tapped by a subsequent cereal or oil crop and raise its yield.

Higher population density favors the planting of yet another food crop during the winter months. This practice inevitably decreases the total nitrogen availability and affects the yield. In the short run it will be energetically more advantageous, producing additional carbohydrates and oils. In the long run the provision of adequate nitrogen is of such importance that intensive agricultures cannot do without the nitrogen-fixing legumes and must plant them instead in edible varieties. This desirable practice, repeated every year or as part of longer crop-rotation sequences, represents perhaps the most admirable energetic optimization in traditional farming. Not surprisingly, it formed the core of all intensive agricultural systems relying on complex crop rotations, but it was only between 1750 and 1880 when standard rotations, including legume cover crops (exemplified by Norfolk's four-year succession of wheat, turnips, barley, and clover), were widely adopted in Europe and at least tripled the rate of symbiotic nitrogen fixation and secure rising yields of nonleguminous crops (Campbell and Overton 1993).

Chorley (1981, 92) recognized that this change was truly epochal, justifying his label of Agricultural Revolution:

> Although advance was broad-fronted, the outcome of many small changes, there was one big change of overriding importance: the generalization of leguminous crops and the consequent increase in the nitrogen supply. Is it not fanciful to suggest that this neglected innovation was of comparable significance to steam power in the economic development of Europe in the period of industrialization?

Wrigley (2002) illustrated the resulting achievements of English agriculture by contrasting its performance in 1300 and 1800, and Muldrew (2011) documented how the post-1650 changes led to an increasingly varied and nutritious diet and how these improvements in the worker's diet promoted higher productivity, steady employment, and rising affluence.

Crop Diversity

Modern farming is dominated by monocultures, annual plantings of the same crop, reflecting the regional specialization of commercial agriculture.

But repeated plantings of the same species have high energy and environmental costs. They need fertilizers to replace the removed nutrients, as well as chemicals to control pests, which thrive on vast uniform plantings. Monocultures of row crops such as corn, where much of the soil remains exposed to rain before the closing of canopies, promote heavy erosion when planted on sloping lands. And the constant cultivation of rice in flooded soils deprived of oxygen degrades the quality of the soils.

Long experience taught many ancient cultivators the perils of monocultures. In contrast, the rotation of cereals and leguminous crops either replenishes the soil nitrogen or at least eases the drain on its soil reserves. The cultivation of a variety of grain, tuber, oil, and fiber crops lowers the risk of total harvest failure, discourages the establishment of persistent pests, reduces erosion, and maintains better soil properties (Lowrance et al. 1984; USDA 2014). Crop rotations can be chosen to fit climatic and soil conditions and to satisfy specific dietary preferences; they are highly desirable from an agronomic point of view, but where more than one crop is grown per year (multicropping) they obviously require more labor. In places with dry spells, irrigation will be needed, and for intensive multicropping, with three or even four different species grown every year in the same field, substantial fertilization will be also needed. Where two or more crops are grown in the same field at the same time (intercropping), labor demands may be even higher. The fundamental reward of multicropping is its ability to support larger populations from the same amount of cultivated land.

The traditional variety of crops and their rotation schemes was enormous. For example, Buck's second survey of Chinese farming counted an astonishing 547 different cropping systems in 168 localities (Buck 1937). But several key commonalities are clear. None is more remarkable than the already noted nearly global practice of linking leguminous grains with cereals. Besides their contribution to soil fertility and protein supply, some legumes, most notably soybeans and peanuts, also yield good edible oils that were always welcome in traditional diets. Oilseed cakes, the compact material remaining after oil is expressed, became either high-protein feeds for domestic animals or excellent organic fertilizers.

The second commonality has also been noted already: rotations of green manures with food crops had an important place in every intensive traditional agriculture. A third commonality, rotation of crops, reflected the desire to produce fibers along with staple carbohydrates (grains, tubers) and oil crops. Consequently, traditional Chinese cropping included numerous schemes of rotating wheat, rice, and barley with soybeans, peanuts, or sesame and with cotton and jute. Besides staple cereals (wheat, rye, barley, oats) and legumes (peas, lentils, beans) European peasants cultivated flax and hemp for fiber. Mayan crops included the three staples of the New World farming—corn, beans, and squash—but also tubers (sweet potato, manioc, jicama) and agave and cotton for fiber (Atwood 2009).

Persistence and Innovation

The inertia of traditional farming practices was, in many instances, unmistakable even after several millennia: sowing dryland grains by hand broadcasting and the backbreaking labor of transplanting rice seedlings to wet fields; harnessing slow-moving oxen and guiding simple wooden plows; hand harvesting with sickles or scythes and threshing by flails or by using animals. But this apparent constancy of recurrent processes was also hiding numerous, though almost always very gradual, changes. These innovations ranged from the spread of better agronomic techniques to the adoption of new crops.

Crop diffusion had a profound effect by introducing new carbohydrate staples (corn, potatoes) and new micronutrient-rich vegetables and fruits. Some diffusions were relatively slow and followed more than one route. For example, cucumber (*Cucumis sativus*) was introduced to Europe through two independent diffusions, first (before the rise of Islam) overland from Persia (to eastern and northern Europe), then by a maritime route into Andalusia (Paris, Daunay, and Janick 2012). Undoubtedly the most consequential diffusion of new crops followed the European conquest of the Americas, which brought the worldwide adoption of potatoes, corn, tomatoes, and peppers and the pantropical cultivation of pineapples, papayas, vanilla, and cacao trees (Foster and Cordell 1992; Reader 2008). Perhaps the best way to appreciate agricultural evolution is to take a long look at the four most persistent traditional farming arrangements, and then at the rapid advances of North American preindustrial agriculture.

Historically the first one is the Middle Eastern agriculture, exemplified by Egyptian practices. There the natural limitations (a restricted availability of arable land and the virtual absence of precipitation) and an extraordinary environmental bounty (the Nile's annual floods bringing predictable water and nutrient supply) combined to produce a highly productive agriculture already during early dynastic times. At the beginning of the twentieth century, after a long period of stagnation, Egyptian peasants still produced some of the highest outputs achievable in solar farming (unsubsidized by any inputs of fossil energies).

Traditional Chinese farming is illustrative of East Asia's admirably productive cropping. These practices supported the world's largest culturally cohesive populations, and they survived surprisingly intact until the 1950s. This persistence made it possible to study them by modern scientific methods and to produce some reliable quantifications of their performance. Complex Mesoamerican societies depended on a unique and highly productive cropping done without plowing and without draft animals. European agriculture evolved from simple Mediterranean beginnings to rapid advances during the eighteenth and nineteenth centuries. The transfer of traditional European farming techniques to North America, and the unprecedented rate of agricultural innovation in the United States during the nineteenth century, created the world's most efficient traditional farming arrangement.

Ancient Egypt
Predynastic Egyptian agriculture, traceable from shortly before 5000 BCE, coexisted with a great deal of hunting (of antelopes and pigs, as well as crocodiles and elephants), fowling (geese, ducks), fishing (especially easy in flooded shallows), and plant gathering (herbs, roots). Emmer wheat and two-row barley were the first cereals, and sheep (*Ovis aries*) were the first domesticated animals. October and November seeding followed the receding waters of the Nile, crop weeding was rare, and the harvests came after five to six months. Calculations based on archeological records indicate that predynastic Egyptian farming could feed perhaps as many as 2.6 people/ha of cultivated land, but a more likely long-term mean was about half that rate.

Egyptian farming has always prospered because of irrigation, but in both the Old Kingdom (2705–2205 BCE) and the New Kingdom (1550–1070 BCE), this involved relatively simple manipulation of annual flood waters. This was done by building higher and stronger levees, blocking off drainage channels, and subdividing the flood basins (Butzer 1984; Mays 2010). Unlike in Mesopotamia or the Indus valley, perennial canal irrigation was not an option: the Nile's very small gradient (1:12,000) made radial canalization impossible, and its first limited use was in the Faiyum depression during Ptolemaic times (after 330 BCE).

Similarly, the absence of effective water-lifting devices greatly limited the dynastic irrigation of higher-lying farmland. Counterpoise lifts, used since the Amarna period of the fourteenth century BCE, were suitable only for irrigating small gardenlike plots. The animal-drawn saqiya, needed for continuous high-capacity lifts, was adopted only during Ptolemaic times.

Consequently, there was no dynastic cultivation of summer crops, just more extensive winter cropping. Wheat and barley were the dominant grains, harvesting was done with wooden sickles set with short notched or serrated flint blades, and the straw was cut high above the ground, sometimes just below the heads. This practice, also common in medieval Europe, made for easy harvesting, easier transport of the crop to the threshing floor, and cleaner threshing. In Egypt's dry climate the standing straw could be cut later as needed for use in weaving, brickmaking, or as cooking fuel, and the stubble was grazed by domestic animals.

Paintings from Egyptian tombs bring these scenes alive. Scenes from the tomb of Unsou show peasants hoeing, casting seed, harvesting with sickles, and carrying cut grain in panniers to be threshed by oxen (fig. 3.10). Inscriptions from Paheri's tomb express eloquently the energetic constraints and realities of the time (James 1984). An overseer prods the laborers: "Buck up, move your feet, the water is coming and reaches the bundles." Their reply— "The sun is hot! May the sun be given the price of barley in fish!"—sums up perfectly both their weariness and their awareness that grain destroyed by flood may be compensated by fish.

Figure 3.10
Scenes of Egyptian farming activities from the eighteenth dynasty (New Kingdom) tomb of Unsou in East Thebes (Corbis).

And the boy whipping the oxen tries to cheer the working animals: "Thresh for yourselves, thresh for yourselves. ... Chaff to eat for yourselves, and barley for your masters. Don't let your hearts grow weary! It is cool." Besides chaff, oxen were fed barley and wheat straw and grazed on wild grasses of the floodplain and on the cultivated vetches. As the cultivation intensified, cattle were seasonally driven to graze in the delta marshes. For plowing, oxen were harnessed by double head yokes, clods were broken by wooden hoes and mallets, and the scattered seed was trampled into the ground by sheep. Records from the Old Kingdom indicate not only large numbers of oxen but also substantial cow, donkey, sheep, and goat herds.

Butzer's (1976) reconstruction of Egypt's demographic history has the Nile valley's population density rising from 1.3 persons/ha of arable land in 2500 BCE to 1.8 people/ha in 1250 BCE and 2.4 people/ha by the time of Rome's destruction of Carthage (149–146 BCE). During Roman rule, Egypt's total cultivated land was about 2.7 Mha, with about 60% of it in the Nile Delta. This land could produce about 1.5 times as much food as was needed for its nearly five million people. The surplus was a matter of great importance for the prosperity of the expanding Roman Empire: Egypt was its largest grain supplier (Rickman 1980; Erdkamp 2005). Afterward Egyptian farming declined and stagnated.

Even as recently as the second decade of the nineteenth century the country cultivated only half as much land as it did during Roman rule. But because of higher yields this land supported about as many people as the area twice as large fed, at home and abroad, nearly two millennia before. Productivity rose rapidly only with the spread of perennial irrigation after 1843, when the first Nile barrages provided adequate water heads to feed canal networks. The national multicropping index rose from just 1.1 during the 1830s to 1.4 by 1900, and during the 1920s it surpassed 1.5 (Waterbury 1979). Farming was still powered by animate energies but, already helped by inorganic fertilizers, fellahin were feeding six people from every hectare of cultivated land.

China

Imperial China also underwent long periods of turmoil and stagnation, but its traditional farming was considerably more innovative than Egyptian agriculture (Ho 1975; Bray 1984; Lardy 1983; Li 2007). As elsewhere, the early stages of Chinese farming were not at all intensive. Before the third century BCE there was no large-scale irrigation and little or no double-cropping or crop rotation. Dryland millet in the north and rain-fed rice in the lower Yangzi basin were the dominant crops. Pigs were the oldest domesticated animals—the earliest evidence for them is about 8,000 years before present (Jing and Flad 2002)—and also by far the most abundant, but clear evidence of manuring emerges only after 400 BCE.

By the time Egypt was supplying the Roman Empire with surplus grain (during the Han dynasty, 206 BCE–220 CE), the Chinese had developed several tools and practices that Europe and the Middle East adopted only centuries, or even more than a millennium, later. These advances included, above all, iron moldboard plows, collar harness for horses, seed drills, and rotary winnowing fans. All of these innovations came into widespread use during the early Han dynasty (206 BCE–9 CE). Perhaps the most important was the widespread adoption of the cast-iron moldboard plow.

Made from nonbrittle metal (whose casting was perfected by the third century BCE), these mass-produced plows extended the possibilities of cultivation while easing the heavy work. Although heavier than wooden plows, they created much less friction and could be pulled by a single animal even in water-logged clay soils. Multitube planting drills reduced seed waste associated with hand broadcasting, and crank-operated winnowing fans greatly shortened the time needed to clean threshed grain. Efficient collar harness for horses did not make that much of a difference in field work because throughout the poor North, the less demanding oxen remained a more affordable choice than horses (whose feeding required good forage or grain), and only water buffaloes, harnessed by neck yokes, could be used in the southern wet fields.

No other dynastic period can compare with the Han in terms of fundamental changes in farming (Xu and Dull 1980). Subsequent advances were slow, and after the fourteenth century CE rural techniques were nearly stagnant. Slightly more than half of the increased grain output between the Ming dynasty (1368–1644) and the early period of the Qing dynasty (1644–1911) came from the expansion of cultivated area (Perkins 1969). Higher labor inputs—above all, more irrigation and fertilization—accounted for most of the remaining increase. Better seeds and new crops, most notably corn, made some regional difference.

Undoubtedly the most important, and lasting, contribution to intensified cropping in China was the design, construction, and maintenance of extensive irrigation systems (fig. 3.11). The antiquity of these schemes is best shown by the fact that nearly half of all projects operating by the year 1900 had been completed before the year 1500 (Perkins 1969). The origins of perhaps the most famous one, Sichuan's Dujiangyan, which still waters fields growing food for several tens of million people, go back to the third

Figure 3.11
Small area of extensive longji (dragon's back) rice terraces north of Guilin in Guangxi
whose origins go back to Yuan Dynasty (1271–1368). *Source:* https://en.wikipedia
.org/wiki/Longsheng_Rice_Terrace#/media.

century BCE (UNESCO 2015b). Min Jiang's bed was cut at the river's
entrance to the plain at Guanxian, and the stream was then repeatedly
subdivided through the building of rocky arrowheads in the midflow.

Water was diverted into branch canals and its flow was regulated by
dikes and dams. Baskets of woven bamboo filled with rocks were the main
building ingredient. Dredging and dike repairs during low-water seasons
have kept the irrigation system working for more than 2,000 years. The
construction and unceasing maintenance of such irrigation projects (as
well as the building and dredging of lengthy ship canals) required long-
range planning, the massive mobilization of labor, and major capital investment. None of these requirements could be met without an effective central
authority. There was clearly a synergistic relationship between China's
impressive large-scale water projects and the rise, perfection, and perpetuation of the country's hierarchical bureaucracies.

Human-powered water lifting was tedious and time-consuming, and its
energy costs were rather high—but so were the rewards of higher yields.
When irrigation supplied additional water to crops during their critical

growth periods its net food direct energy return (excluding the cost of building and maintaining irrigation canals) was easily around 30 (see box 3.8). Making up water shortfalls during less sensitive growth periods could still return around 20 times as much food energy in the additional harvest as the food consumed by peasants while treading water ladders.

In China's rice-growing areas applications of animal manure and night soil commonly averaged 10 t/ha during the late nineteenth and early twentieth centuries. Huge amounts of organic wastes were collected in cities and towns and moved to the countryside, creating a large waste-handling and transportation industry (box 3.9). This high intensity of Chinese manuring was admired by Western travelers, who were curiously unaware of how closely it was matched by earlier European experience (King 1927). But no culture surpassed the highest known applications of organic wastes that supported intensive farming in the dike-and-pond region of Guangdong province in South China, between 50 and 270 t of pig and human excrements per hectare (Ruddle and Zhong 1988).

Composting and regular applications of many other organic wastes—ranging from silkworm pupae to canal and pond mud, and from waterweeds to oilseed cakes—further increased the burden of collecting, fermenting, and distributing. Not surprisingly, at least 10% of all labor in

---

Box 3.9

Nitrogen content of organic materials recycled in China

Scale of traditional recycling (and hence the energies devoted to gathering, handling and applying the waste biomass) had to be so large because the organic materials applied to field or plowed in (as green manures) had very low nitrogen content: human and animal wastes are largely water, as are green manures; only oil cakes (residues after pressing edible oils) have relatively high nitrogen content. For comparison, urea, the leading modern synthetic fertilizer, contains 46% of nitrogen.

Materials	Nitrogen content (% of nitrogen in fresh weight)
Pig manure	0.5–0.6
Night soil (human waste)	0.5–0.6
Green manures (vetches and beans)	0.5–0.3
Oil cakes (soybean, peanut, rapeseed)	4.5–7.0
River and lake silt	0.1–0.2

Sources: Ranges are from data assembled in Smil (1983, 2001) and cited in a variety of historical and modern Chinese sources.

Chinese traditional farming was devoted to managing fertilizers, and on the North China Plain heavy fertilization of wheat and barley was commonly the most time-consuming part of the human labor (close to one-fifth) as well as of the animal draft power (about one-third) devoted to those crops. But this investment was very rewarding: its net energy return was commonly over 50 (box 3.10).

Overall food energy returns in traditional Chinese cropping were not so high even during the period of its peak performance in the early decades of the twentieth century. The main reason was the minimal mechanization of cropping, which meant the continuing dominance of human labor. A wealth of quantitative information on virtually all aspects of traditional Chinese farming of the 1920s and 1930s (Buck 1930, 1937) makes it possible to describe the system in much detail and to prepare accurate energy accounts. Most fields were very small (only about 0.4 ha) and were just five to ten minutes away from a farmhouse. Nearly half of the farmland was irrigated, and a quarter was terraced.

More than 90% of the cropped land was planted in grains, less than 5% with sweet potatoes, 2% with fibers, and 1% with vegetables. Only about one-third of all northern farms had at least one ox, and less than a third of southern holdings owned a water buffalo. Crops required most of the draft labor (90% for rice, 70% for wheat), but, except for plowing and harrowing, Chinese field work relied almost exclusively on human labor. Both oxen and water buffaloes were fed hardly any grain and hence energy returns can

---

Box 3.10
Net energy return of fertilization

A good late Qing dynasty winter wheat harvest of about 1.5 t/ha required just over 300 hours of human and about 250 hours of animal labor. Fertilization took, respectively, 17% and 40% of these totals. I assume, conservatively, that 10 t of fertilizer applied per hectare contained only 0.5% of nitrogen (Smil 2001). After inevitable leaching and volatilization losses, only half of it would be actually available to the crop. Each kilogram of nitrogen will result in an additional production of about 10 kg of grain. Compared to an unfertilized crop, there is a yield increment of at least 250 kg of grain. No more than 3–4% of this grain was used as animal feed. After milling, the grain yielded at least 200 kg of flour, or about 2.8 GJ of food energy, compared to an investment of about 40 MJ of additional food for human labor. The net energy return of fertilization was thus around 70, an impressive benefit/cost ratio.

be calculated just by considering human labor. Unirrigated northern wheat yielded usually no more than 1 t/ha, its production required more than 600 hours of labor, and it returned between 25 and 30 units of food energy in unmilled grain for every unit of food energy needed for field work and crop processing.

Local and regional rice yields were rather high already during the Ming dynasty, and the national average was around 2.5 t/ha during the early decades of the twentieth century, second only to Japan. About 2,000 hours of labor were needed to produce such harvests, giving gross energy returns of 20- to 25-fold. Gross energy returns for corn were up to 40, but cornmeal was never a favored food in China. For grain legumes (soybeans, peas, beans), the returns were rarely higher than 15, and commonly just around 10. That was also the return for plant oils pressed from rapeseed, peanuts, or sesame seeds. Grains supplied about 90% of all food energy, and meat consumption in peasant families was negligible (usually only on festive occasions). But these monotonous vegetarian diets eventually supported high population densities.

Ancient Chinese population densities could not have been much different from the Egyptian means, ranging from just around 1 person/ha in the poorest northern regions to well over 2 people/ha in southern rice areas, and there were also large intraregional differences, with the Northeast out of bounds for Chinese immigration during the first two centuries of Manchu-ruled Qing dynasty, and with very low densities in parts of the mountainous South. The gradual intensification of cropping combined with simple diets did eventually support much higher rates. An approximate reconstruction for the Ming (1368–1644) and Qing (1644–1911) dynasties starts with about 2.8 people/ha of farmland in 1400 and rises to 4.8 people/ha by 1600 (Perkins 1969). A slight decline during the prosperous Qianlong reign (1736–1796) was the result of an expanded population opening up new farmland. The increase in population densities resumed during the nineteenth century, and at its close the average rate was above 5 people/ha, higher than the contemporary mean for Java and at least 40% above the mean for India (fig. 3.12).

Buck's (1937) surveys for the early 1930s indicate a national average of at least 5.5 people/ha of cultivated land. This was nearly as much as in contemporary Egypt, but in Egypt all land was irrigated, and inorganic fertilizers were already used. In contrast, China's nationwide crop productivity average was depressed by northern dryland farming. The southern rice region surpassed the rate of 5 people/ha already by 1800, and large parts of it supported more than 7 people/ha by the late 1920s. In comparison to

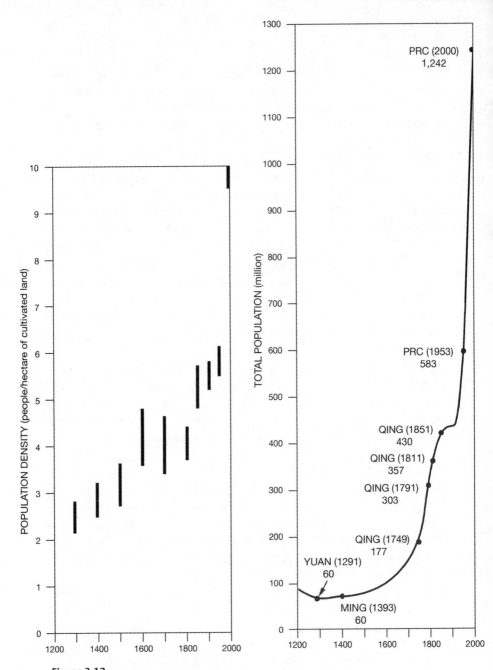

Figure 3.12

China's population density in long-term perspective. A substantial expansion of cultivated area during the Qing dynasty was soon overwhelmed by the country's continuing population growth. Density bars indicate the uncertainty of historical estimates. Based on data in Perkins (1969) and Smil (2004).

dryland wheat cropping, net energy returns were invariably lower in wet rice farming but that was more than counterbalanced by greater yields per hectare: double-cropping of rice and wheat in the most fertile areas could feed 12–15 people/ha.

Mesoamerican Cultures

Without any draft animals (and hence without any plowing), agricultures of the classical Mesoamerican civilizations differed substantially from those of the Old World cultures. But they, too, evolved more intensive cropping methods that could support impressively high population densities, and they domesticated several food crops that are now cultivated worldwide, above all corn, peppers (*Capsicum annuum*), and tomatoes (*Solanum lycopersicum*). The most important nonfood crop originating in Mesoamerica is cotton (*Gossypium barbadense*). Molecular analyses indicate the Yucatán Peninsula as the site of cotton's original domestication, while the gene pool of modern cotton cultivars has its origins in southern Mexico and Guatemala (Wendel et al. 1999).

The tropical Mayan lowlands and a much drier, high-lying basin of Mexico were the regions of greatest achievements. Although there was a great deal of interaction between the people of these regions, and although they shared corn as their staple crop, their history was substantially different. The reasons for the decline of the first culture remain contentious (Haug et al. 2003; Demarest 2004), while the other one was destroyed by the Spanish invasion (Leon 1998). Mayan society was developing gradually for a long time before the beginning of the Classic period, about 300 CE. The region encompassing parts of today's Mexico (Yucatán), Guatemala, and Belize sustained a complex civilization until about 1000 CE. Then, in one of the most enigmatic turns in world history, the classic Mayan society disintegrated and its population declined from about three million during the eighth century CE to just around 100,000 by the time of the Spanish conquest (Turner 1990).

Agricultural malpractice—excessive erosion and a breakdown in water management—has been suggested as one of the reasons for the Mayan collapse (Gill 2000). In the early stages of their development Mayans were shifting cultivators, but they gradually turned to intensive forms of cropping (Turner 1990). Upland Mayas built extensive rock-wall terraces that conserved water and prevented heavy erosion on continuously cultivated slopes. Lowland Mayas built some impressive canal networks and raised fields above the floodplain level to prevent seasonal inundation. Ancient Mayan elevated ridged fields, some dating from as early as 1400 BCE, are still discernible on modern aerial photographs. Their clear identification and dating during the 1970s disproved the long-prevailing notion of Mayas being limited to shifting agriculture (Harrison and Turner 1978).

The basin of Mexico had a succession of complex cultures, starting with the Teotihuacanos (100 BCE–850 CE), followed by the Toltecs (960–1168) and, since the early fourteenth century, by the Aztecs (Tenochtitlan was founded in 1325). There was a long transition from plant gathering and deer hunting to settled farming. Intensification of cropping through water regulation started early in the Teotihuacan era, and gradually evolved to such a degree that by the time of the Spanish conquest, at least one-third of the region's population depended on water management for its food (Sanders, Parsons, and Santley 1979).

Permanent canal irrigation around Teotihuacan was able to support about 100,000 people, but the most intensive cultivation in Mesoamerica relied on *chinampas* (Parsons 1976). These rectangular fields were raised up to between 1.5 and 1.8 m above the shallow waters of the Texcoco, Xalco, and Xochimilco lakes. Excavated mud, crop residues, grasses, and water weeds were used in their construction. Their rich alluvial soils were cropped continuously, or with only a few months of rest, and their edges were reinforced with trees. Chinampas turned unproductive swamps into high-yielding fields and gardens, and solved the problem of soil waterlogging. Accessibility by boats made for easy transportation of harvests to city markets. Chinampa cultivation provided an outstanding return on the invested labor, and the high benefit/cost ratio explains the frequency of the practice, which started as early as 100 BCE and reached its peak during the last decades of the Aztec rule (box 3.11).

Box 3.11
Raised fields in the basin of Mexico

A chinampa could yield up to four times as much as nonirrigated land. An excellent corn harvest of 3 t/ha would produce, after subtracting about one-tenth for seed and waste, about 30 GJ more food energy than a dryland plot. Fields were raised at least 1.5 m above the water level, so 1 ha of chinampa required a build-up of some 15,000 m^3 of lake silt and mud. A man working five to six hours a day would emplace no more than 2.5 m^3. Raising 1 ha of fields thus required about 6,000 man-days of labor. With a labor energy cost of 900 kJ/h, the task called for about 30 GJ of additional food energy—an amount gained in increased harvest in just a single year.

At the time of the Spanish conquest the Texcoco, Chaco, and Xochi-milco lakes had about 12,000 ha of chinampa fields (Sanders, Parsons, and Santley 1979). Their construction required at least 70 million man-days of labor. The average peasant had to spend no less than 200 days a year to grow enough food for his own family, so he could not work more than about 100 days on large hydraulic projects. As a good portion of this time had to be devoted to the maintenance of existing embankments and canals, seasonal labor of at least 60 and up to 120 peasants was needed to add 1 ha of a new chinampa. The means were different—but the pre-Hispanic basin of Mexico was clearly as much a hydraulic civilization as Ming China, its great Asian contemporary. Long-term, well-planned, centrally coordinated effort and an enormous expenditure of human labor were the key ingredients of its agricultural success.

Irrigated corn is an inherently higher-yielding crop than wheat, and population densities supported by the best Mesoamerican farming were very high. A hectare of a high-yielding chinampa could feed as many as 13–16 people deriving 80% of their food energy from the grain. Naturally, averages for the whole basin of Mexico were considerably lower, ranging from less than 3 people/ha in fringe areas to about 8 people/ha in well-drained soils with permanent irrigation (Sanders, Parsons, and Santley 1979). The basin's pre-Conquest (1519) population of around one million people, using all cultivable land in the valley, had an average density of about 4 people/ha. Nearly identical densities were supported by raised-field cultivation of potatoes in the wetlands around Lake Titicaca, the core area of the Incas, between today's Peru and Bolivia (Denevan 1982; Erickson 1988).

Europe
In Europe, much as in China, periods of relatively steady improvements alternated with stagnation in productivity, and major regional peacetime famines persisted until the nineteenth century. But until the seventeenth century European farming was generally inferior to Chinese accomplishments, always belatedly adopting innovations coming from the east. Greek farming, of which we know little, was certainly not as impressive as its contemporary Middle Eastern counterparts. The Romans had gradually evolved a moderately complex agriculture whose descriptions survive in the works of Cato (*De agri cultura*), Varro (*Rerum rusticarum libri III*), Columella (*De re rustica*), and Palladius (*Opus agriculturae*). These writings were often reprinted—perhaps their best single-volume collection with commentaries and notes was published by Gesner (1735)—and they exerted

significant influence up to the seventeenth century (White 1970; Fussell 1972; Brunner 1995).

Unlike densely populated core regions of China, where the shortage of grazing land and a high population density precluded extensive ownership of livestock, European farming always had a strong component of animal husbandry. Roman mixed farming included rotations of cereal and legume crops, composting, and the plowing in of legumes as green manure. Recycling of all possible organic wastes was often intensive, ranging from the highly valued pigeon excreta to oil cakes. Repeated liming (chalk or marl applications) of fields was done to reduce soil acidity. At least one-third of fields were in fallow.

Oxen, often shod, were the principal draft animals. Plows were wooden, sowing was done by hand, and harvesting was done with sickles. A mechanical Gallic reaper, described by Pliny and pictured on a few surviving reliefs, was of limited use. Threshing was done by treading animals or with flails, and yields were low and highly variable. Reconstruction of inputs into Roman wheat farming during the first centuries of the Common Era comes up with between 180 and 250 hours of human labor (and around 200 hours of animal labor) to produce typical harvests of merely close to 0.5 t/ha. Even so, gross food energy returns, ranging mostly between 30 and 40, were fairly high (box 3.12).

The productivity of European farming changed very slowly during the millennium between the demise of the Western Roman Empire and the beginnings of the great European expansion. In the early thirteenth century, wheat production proceeded by largely unchanged means, and could not support population densities higher than the predynastic Egyptian average. But the Middle Ages were definitely not a period devoid of important technical innovation (Seebohm 1927; Lizerand 1942; Slicher van Bath 1963; Duby 1968, 1998; Fussell 1972; Grigg 1992; Astill and Langdon 1997; Olsson and Svensson 2011). One of the most important changes was the adoption of the collar harness for draft horses.

Largely because of this improved harness horses started to replace oxen as the principal draft animals in every richer region of the continent. But the transition was very slow, and it commonly took centuries to accomplish. In better-off regions of Europe it occurred between the eleventh century, when horseshoeing and the collar harness became the norm, and the sixteenth century. Well-documented progress in England shows that horses made up only 5% of all demesne draft animals at the time of the Domesday count (1086), but about 35% on peasant holdings (Langdon 1986). By 1300 these shares had risen, respectively, to 20% and 45%, and after a period of

---

Box 3.12
Labor requirements of European wheat harvests, 200–1800

	Hours of labor (people/animals)/ha of wheat		
Tasks	Italy, 200	England, 1200	Netherlands, 1800
Plowing			
Oxen	37/74	25/150	
Horses			15/30
Harrowing	8/16	7/14	5/10
Sowing			
Broadcasting	4/—	4/—	
Seed drill			3/6
Manuring			40/60
Harvesting			
Sickle	50/—	50/—	
Cradle			24/—
Hauling	15/30	10/20	7/14
Threshing			
Treading (oxen)	30/60		
Flailing		30/—	33/—
Winnowing	25/—	25/—	30/—
Measuring, sacking	8/—	7/—	10/—

Sources: Calculations are based on information in Baars (1973), Seebohm (1927), White (1970), Stanhill (1976), and Langdon (1986).

---

stagnation, horses became the majority of draft animals only by the end of sixteenth century.

The relative richness of English data also illustrates the complexity of this transition. For long time horses were simply substituted for oxen as pacesetters in mixed teams. Their adoption had a clear regional pattern (East Anglia was far ahead of the rest of England), and smallholders were much more progressive in using them on their farms. Differences in prevailing soil types (clays favored oxen), the availability of feed (extensive pastures favored oxen), and access to markets to buy good working animals and to sell the meat (proximity to towns favored horses) combined to produce a complex outcome. Other countervailing factors included conservatism and resistance to change, and striving for lower operating costs and pioneering spirit. The transition was further hindered by poorly designed plows and by the weakness of most medieval horses.

The combination of wide wooden soles, heavy wooden wheels, and large wooden moldboards resulted in enormous friction. In wet soils it was not uncommon to use four to six animals, whether oxen or horses, to overcome this resistance. Despite its relative inefficiency, the combination of flat moldboard plows and larger animal teams (increasingly including horses) was essential in extending the land under cultivation. By dividing fields into raised lands and sunken furrows, moldboard plowing created a pattern of effective artificial drainage. Although certainly much less spectacular than chinampas, this form of controlling excessive field water had far more widespread spatial and historical repercussions. Moldboard plowing opened up the extensive water-logged plains of northern Europe to the cultivation of wheat and barley, crops native to dry Middle Eastern environments.

By the late Middle Ages the frontier of German settlement marked the easternmost extent of moldboard plowing. The technique encompassed European flatlands between the North Sea and the Urals only by the nineteenth century, and it was absent until that time also throughout most of the Balkans. Clearly, its use was both a revolutionary change ensuring agronomic advances in northwestern and Central Europe and in the Baltics and a key ingredient of the continuing agricultural prosperity of cool and wet lowlands. Heavy draft horses, common on European farms and roads during the nineteenth century, were a product of many centuries of breeding (Villiers 1976), but the progress was slow, and medieval horses were hardly larger than their Roman predecessors (Langdon 1986). Even during the late Middle Ages most of animals were no taller than 13–14 hands, and draft horse power started to rise appreciably only after animals measuring 16–17 hands and weighing close to 1 t became more common in Western Europe during the seventeenth century (fig. 3.13).

This explains the medieval English complaints that horses were useless in heavy clay soils. In contrast, the heavy draft horses of the nineteenth century were outstanding on wet land, in heavy soils, and on uneven ground. During the nineteenth century a pair of good horses easily did 25–30% more field work in a day than a team of four oxen. Three intensification gains resulted from this speed: a more frequent cultivation of existing fields (especially the plowing of fallow land to kill weeds), extension of cropping on a new land, and the freeing of labor for other field or farmyard activities. And in most European regions crop rotations could easily provide enough concentrate feed to make the maintenance of a two-horse team

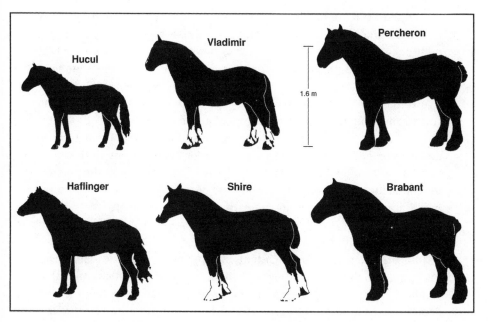

Figure 3.13
European draft horses ranged from small, ponylike animals less than 12 hands (1.2 m) tall to tall (more than 16 hands), heavy beasts (weighing around 1 t). Animal silhouettes, based on Silver (1976), have been scaled to size.

cheaper than the upkeep of four oxen. The combination of the slow rate of transition from oxen to horses, major regional fluctuations of farm output, and the persistence of very low staple grain harvests makes it impossible to show a steadily improving productivity attributable to a growing number of draft horses.

Their superiority became obvious only when the more powerful animals made up the majority of the draft stock, and when they started to work as a part of much more intensified farming during the seventeenth and eighteenth centuries. In road transport their advantages were recognized much earlier. Working horses also posed a major energy supply challenge. The heavy work possible with efficient harnessing and horseshoes had to be energized by better feed than just the roughages (grasses or straws) sufficient for working cattle. Powerful working horses needed concentrates, cereal or legume grains. Consequently, farmers had to intensify their cropping to provide for both their families and their animals, and intensive agriculture was born in regions where population densities were still too low to generate it without the need for animal feed.

The abundance of historical price figures makes it possible to reconstruct long-term production trends for a number of countries (Abel 1962). Naturally, there were substantial regional differences but large-scale cyclical fluctuations are unmistakable. Times of relative prosperity (most notably 1150–1300, the sixteenth century, and 1750–1800) were marked by extensive conversions of wetlands and forests to fields. They also spurred the colonization of remote areas, and brought a greater variety of foods to supplement bread, the ubiquitous staple. Periods of major economic declines and wars brought famines, large population losses and extensive abandonment of fields and villages (Centre des Recherches Historiques 1965; Beresford and Hurst 1971). Epidemics and wars brought great population losses in the fourteenth century. In the early decades of the fifteenth century Europe had about one-third less population than in 1300, and Germany lost about two-fifths of its peasants between 1618 and 1648.

Insecurity remained the persistent attribute of European peasant farming right up to the end of the eighteenth century, and peasantry's miserable condition was still evident even in richer parts of Europe during the first decades of the nineteenth century. Cobbett (1824, 111), traveling through France in 1823, was astounded to see "women spreading dung with their hands!" and noted that the farming implements used in French fields "seem to be about the same as … used in England a great many years, perhaps a century, ago." But soon afterward a fairly intensive cultivation finally became the norm in most of Atlantic Europe.

Its hallmarks were the gradual abandonment of fallowing and general adoption of several standard crop rotations. The cultivation of potatoes became widespread after 1770, livestock production was expanded, and heavier manuring became regular. In eighteenth-century Flanders annual applications of manure, night soil, oil cakes, and ash easily averaged 10 t/ha (Slicher van Bath 1963). The Netherlands emerged as the leader in crop productivity. Around 1800 Dutch farms cultivated wheat as the principal food crop, as well as barley, oats, rye, beans, peas, potatoes, rapeseed, clover, and green fodder; less than 10% of farmland was fallow, and there was a close integration with livestock production (Baars 1973).

The hours of labor required to cultivate a hectare of Dutch wheat changed little compared to medieval or Roman practices, and, as powerful horses replaced oxen, the hours of draft labor actually declined, but better cultivars and intensive fertilization resulted in the best wheat yields about four times higher than in the Middle Ages. As a result, the net energy return of the early nineteenth-century Dutch farming was more than 160-fold compared to less than 40-fold gain in medieval English wheat cultivation and a less than 25-fold return in Roman grain farming in Italy around 200 CE (box 3.13).

Farming intensification continued in most European countries after the recovery from an overproduction-induced depression in the early nineteenth century. Two German examples illustrate these changes (Abel 1962). In 1800 about a quarter of German fields were fallowed, but the share was less than 10% by 1883. Average annual per capita meat consumption was less than 20 kg before 1820, but it was almost 50 kg by the end of the century. Earlier three-crop rotations were replaced by a variety of four-crop sequences. In a popular Norfolk cycle, wheat was followed by turnips, barley, and clover, and six-crop rotations were also spreading. Applications of calcium sulfate, and of marl or lime to correct excessive soil acidity, became common in better-off areas.

The adoption of better-designed implements also accelerated during the nineteenth century, and it was accompanied by increasing numbers of draft livestock: between 1815 and 1913 the total of horses, oxen, and donkeys (in horse equivalents) rose by 15% in the UK, 27% in the Netherlands, and by 57% in Germany (Kander and Warde 2011). By 1850 yields were rising in every important farming region as the rapidly intensifying agriculture was able to supply food for growing urban populations. After centuries of fluctuation, population densities in the most intensively cultivated regions of the continent—the Netherlands, parts of Germany, France, and England—reached 7–10 people/ha of arable land by the year 1900. These levels already

---

Box 3.13

Energy costs and returns of European wheat harvests, 200–1800

	Energy costs and returns of wheat farming		
	Italy, 200	England, 1200	Netherlands, 1800
Hours of labor	177	158	167
Energy cost (MJ)	142	126	134
Grain yield (t/ha)	0.4	0.5	2.0
Food yield (GJ)	3.3	4.9	22.2
Net energy return	23	39	166
Hours of animal work	180	184	120

Sources: Calculations are based on information in Seebohm (1927), White (1970), Baars (1973), Stanhill (1976), Langdon (1986), and Wrigley (2006).

reflected considerable energy support received indirectly through machinery and fertilizers produced with coal. European farming of the late nineteenth century became a hybrid energy system: still critically dependent on animate prime movers but increasingly benefiting from many inputs of fossil energy.

North America

The history of postrevolutionary American farming is remarkable because of the fairly high, and accelerating, rate of innovation. These changes resulted in the world's most labor-efficient crop cultivation by the end of the nineteenth century (Ardrey 1894; Rogin 1931; Schlebecker 1975; Cochrane 1993; Hart 2004; Mundlak 2005). During the last decades of the eighteenth century, cropping in the northeastern states, and even more so in southern states, lagged behind the European advances. Wooden plows with wrought-iron shares and wooden moldboards shoddily covered with metal pieces caused high friction, heavy clogging, and strain on the yoked oxen. Sowing was done by hand, harvesting by sickle, and threshing by flailing, though in the South primitively by animal treading.

All of this changed rapidly during the new century. Changes in plowing came first (Ardrey 1894; Rogin 1931). Charles Neubold introduced a cast-iron plow in 1797; Jethro Wood's patents (1814 and 1819) made its interchangeable version practicable; and by the early 1830s improved cast-iron plows began to be replaced by steel plows. The first one was made from saw-blade steel by John Lane in 1833, and the production was commercialized by John Deere, whose original (1843) advertisement for moldboard made of wrought iron promised that the metal, ground smooth, would "scour perfectly bright in any soil, and will not choke in the foulest of ground" (Magee 2005).

And the production of inexpensive steel in Bessemer converters made moldboards readily available: Lane introduced his layered steel plow in 1868. Two- and three-wheel riding plows also became common during the 1860s (fig. 3.14). Gang plows, with up to ten shares, drawn by as many as a dozen horses, were used before the end of the century to open up new farmland in the northern Plains states and the Canadian Prairie provinces of Manitoba, Saskatchewan, and Alberta. Massive, steel moldboard plows made it possible to cut the heavily sodded grasslands and to open up North America's vast plains for grain cropping.

Advances in plowing were matched by other innovations. Seed drills and horse-powered threshing machines were widely used by 1850. The first mechanical grain reapers were patented in England between 1799 and

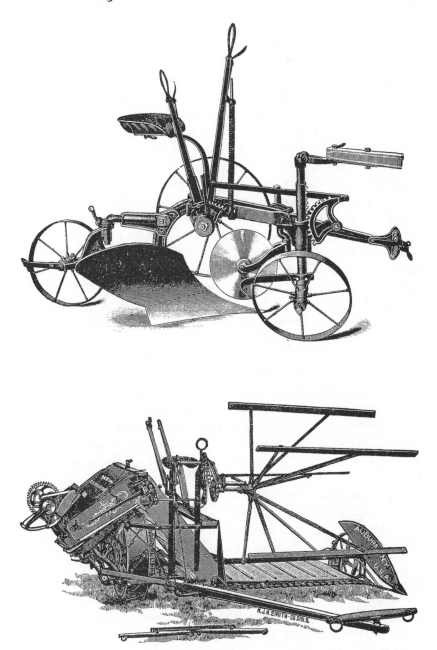

Figure 3.14
The three-wheeled steel riding plow (made by Deere & Co. in Moline, Illinois, during the 1880s) and the twine grain harvester (made during the last decades of the nineteenth century in Auburn, New York). These two innovations opened up the American plains for large-scale grain cropping. Reproduced from Ardrey (1894).

1822, and two American inventors, Cyrus McCormick and Obed Hussey, built on this foundation to develop practical mass-produced machines starting in the 1830s (Greeno 1912; Aldrich 2002). They started to sell heavily during the 1850s, and 250,000 of them were in use by the end of the Civil War. The first harvester was patented in 1858 by C. W. and W. W. Marsh, with two men binding the cut grain, and the first successful twine knotter was introduced by John Appleby in 1878.

This invention was the last ingredient needed for a fully mechanical grain harvester that discharged tied grain sheaves ready for stooking (fig. 3.14). The rapid diffusion of these machines before the end of the nineteenth century together with gang plowing made it possible to open up huge expanses of grasslands not only in North America but also in Argentina and Australia. But the performance of the best twine-binding harvester was soon surpassed with the introduction of the first horse-drawn combines, marketed by California's Stockton Works during the 1880s. Housers, the company's standard combines after 1886, cut two-thirds of California's wheat by 1900, when more than 500 machines were working in the state's fields (Cornways 2015).

The largest ones needed up to 40 horses and could harvest a hectare of wheat in less than 40 minutes—but they tested the limits of animal-powered machinery because harnessing and guiding up to 40 horses was an enormous challenge. But their deployment is the best illustration of the labor shift that took place in traditional American agriculture during the nineteenth century. At its beginning, a farmer (80 W) working in a field was aided by about 800 W of draft power (two oxen); by its end, a farmer combining his Californian wheat field had at his disposal 18,000 W (a team of 30 horses) as he became a controller of energy flows and ceased to be an indispensable energizer of farm work.

In 1800 New England farmers (seeding by hand, with ox-drawn wooden plows and brush harrows, sickles, and flails) needed 150–170 hours of labor to produce their wheat harvest. By 1900 in California, horse-drawn gang-plowing, spring-tooth harrowing, and combine harvesting could produce the same amount of wheat in less than nine hours (box 3.14). In 1800 New England farmers needed more than seven minutes to produce a kilogram of wheat, but less than half a minute was needed in California's Central Valley in 1900, roughly a 20-fold labor productivity gain in a century.

In terms of net energy expenditures, these differences were slightly larger: most of the longer labor hours in 1800 were spent in much heavier labor, with walking plows, scythes, flails, than in later decades, and seeding and storage losses declined appreciably. Compared to 1800, every unit of

Box 3.14
Labor requirements (human/animal) in hours/hectare and the energy cost of American wheat, 1800–1900

Tasks	1800	1850	1875	1900
Plowing				
Wooden plow	20/40			
Cast-iron plow		15/30		
Steel plow			8/24	
Steel gang plow				3/30
Harrowing				
Brush harrow	7/14			
Tooth harrow		5/10	5/15	1/4
Seeding				
Broadcasting	3/—			
Drilling		3/6	3/9	1/2
Harvesting				
Sickle	49/—			
Cradle		25/—		
Binder			11/6	
Combine				3/17
Hauling	10/10	8/8	5/5	2/10
Threshing				
Flailing	33/—			
Threshers		10/10	8/8	
Winnowing	40/—			
Hours of labor	162	66	40	9
Energy cost (MJ)	145	56	32	7
Gross food energy return	129	335	586	2680
Net food energy return	90	270	500	2400
Labor productivity (min/kg of grain)	7.2	2.9	1.8	0.4

The first representative case (1800) is a typical New England cultivation on which two oxen and one to four men powered all the tasks. The second sequence (1850) traces inputs in the horse-powered midcentury farming in Ohio. The third one (1875) shows further advances in Illinois, and the last account (1900) reviews the most productive form of horse-powered U.S. wheat growing in California. Figures in the table are total hours (men/animals) spent per hectare of wheat growing. Because the yields of American wheat did not show any upward trend during the nineteenth century, I assume a constant yield of 20 bushels per acre, or 1,350 kg/ha (18.75 GJ/ha). Accounts are based largely on performance rates compiled by Rogin (1931).

food energy needed for farm labor produced, on the average, about 25 times more edible energy in wheat grain in 1900. Naturally, these huge advances were only partially due to much higher efficiencies resulting from better machinery. The other principal reason for the rapidly rising energy returns of human labor was the substitution of horse power for human muscles. American inventors produced a vast range of efficient implements and machines, but they had only limited success in displacing draft animals as farming prime movers.

Threshing was the only major operation in which horses were gradually replaced by steam engines. America's rapidly expanding agriculture had to rely on growing stocks of horses and mules. These were generally powerful, large and well-fed animals—and their energy costs were surprisingly high. By 1900 they needed about 50% more energy in their feed than New England oxen in 1800, and they required not just hay or straw but also oats or corn. Growing this feed grain reduced the production of crops for humans, and it is possible to quantify these costs fairly accurately (USDA 1959). During the first two decades of the twentieth century the numbers of American horses and mules stayed around 25 million. Growing enough feed for their maintenance and work required about one quarter of America's cultivated land (box 3.15). This huge claim was possible only because of America's plentiful farmland. In 1910 the country had almost 1.5 ha/capita, twice as much as in 1990 and about ten times as much as contemporary China.

During the last decades of the nineteenth century it was not only a combination of clever designs and plentiful horse power that made American farming so productive. During the 1880s American coal consumption surpassed wood combustion, and crude oil started to gain in importance. The production and distribution of tools, implements and machines, and shipments of agricultural products became dependent on inputs of coal and oil. American farmers ceased to be just skilled managers of renewable solar flows: their outputs were subsidized by fossil fuels.

The Limits of Traditional Farming

The enormous socioeconomic contrasts between life during the Qin dynasty, China's first period of unification (221–207 BCE), and the closing decades of the Qing Empire (1644–1911), or between the Roman Gallia Celtica and prerevolutionary France, make us forget the immutability of prime movers and the constancy of basic cultivation practices across millennia of preindustrial history. Populations supported solely by the

Box 3.15
Feeding America's draft horses

In 1910 America had 24.2 million farm horses and mules (and only 1,000 small tractors); in 1918 the draft animal herd peaked at 26.7 million and the number of tractors rose to 85,000 (USBC 1975). With an average daily need of 4 kg of grain for working animals and 2 kg of concentrate feed for the rest (Bailey 1908), the annual feed requirements were roughly 30 Mt of oats and corn. With grain yields of about 1.5 t/ha, this would have required planting at least 20 Mha to feed grains. To supply roughage, working horses needed at least 4 kg/day of hay, while the rest could be maintained with about 2.5 kg/day, requiring an annual total of roughly 30 Mt of hay. With average hay yields of about 3 t/ha, at least 10 Mha of hay had to be harvested. Land devoted to horse feed had to be no less than about 30 Mha, compared to around 125 Mha of annually harvested land, which means that America's farm horse herd (working and nonworking animals) required almost 25% of the country's cultivated land. The USDA's (1959) calculation came up with a nearly identical total of 29.1 Mha.

exertion of people and animals and by the recycling of organic wastes and the planting of legumes had increased with more efficient use of animate power and with greater intensity of cropping practices.

In such highly productive regions as northwestern Europe, central Japan, and the coastal provinces of China, yields approached the limits imposed by the maximum rates of available energy and nutrient flows by the end of the nineteenth century. At the same time, preindustrial agricultures brought only very limited improvements in average harvests. They also provided no more than basic subsistence diets for most of the people even in good years, and could not prevent a great deal of chronic malnutrition and recurrent famines. Productive arrangements that were durable, resilient, and adaptable were also fragile, vulnerable, and inadequate to meet the rising needs.

Achievements

Advances in traditional farming were slow, and the adoption of new methods did not mean a general disappearance of old practices. Fallow fields, scythes, and inefficiently harnessed oxen did not disappear in late nineteenth-century Europe when annual cropping, grain harvesters, and good horse teams became common. The only way to reduce human labor in a system whose field tasks were powered solely by animate power was through the more widespread use of draft animals. This shift required not only better harnessing, feeding, and breeding but also innovation in the design of field tools and machines for specific operations replacing human labor.

Slow at first, these advances accelerated during the eighteenth century. Comparisons of wheat cultivation are perhaps the best markers of this progress. During the early decades of the eighteenth century farming a hectare of wheat in Europe and North America took almost 200 hours, or nearly as much time as during the high Middle Ages. By 1800 the U.S. mean had fallen below 150 hours, by 1850 below 100 hours. By 1900 it was less than 40 hours, and the most productive practices (California gang-plowing and combining) needed less than nine hours to accomplish the task (fig. 3.15).

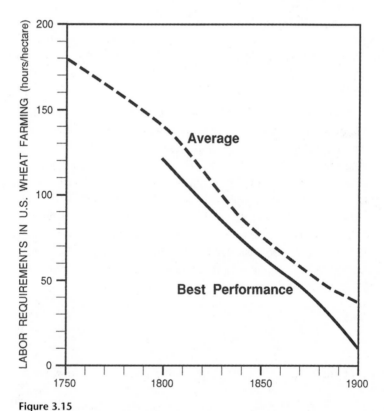

Figure 3.15
Improving efficiency of American wheat cultivation during the nineteenth century can be charted fairly accurately on the basis of data compiled by Rogin (1931) and the USDA (1959).

The gradual intensification of traditional farming achieved by the substitution of human labor by animal draft increased productivity, but for a long time it had a barely discernible effect on average yields. Scarcity and inaccuracy of available information make all long-term assessments difficult, but it is clear that stagnation and marginal gains were the norm in both Europe and Asia. Until the early decades of the nineteenth century we have no reliable national or regional averages. Most of the older figures in European sources are given as the relative returns of planted seed, usually in volume, rather than in mass terms. As those seeds were smaller than today's highly bred varieties, conversions to mass are uncertain. Moreover, even the best monastic or estate records have frequent gaps, and virtually all of them show wide year-to-year fluctuations. During the Middle Ages weather extremes could even bring yields too low to produce enough seed for the next planting.

The best estimates indicate that returns in the early Middle Ages were just twofold for wheat. The best long-term reconstructions of a national trend for the past seven centuries are available for England (Bennett 1935; Stanhill 1976; Clark 1991; Brunt 1999). In the thirteenth century, English wheat seed returns ranged mostly between three and four, with recorded maxima up to 5.8. This translates to a mean of just above 500 kg/ha. Careful analyses of all available English evidence show that doubling of this very low yield was irreversibly achieved only some five centuries later. English wheat yields remained at nearly medieval levels until about 1600, but afterward they increased steadily.

The countrywide mean for 1500 was doubled before 1800 and tripled by 1900, largely as a result of extensive land drainage and the widespread adoption of crop rotation and intensive manuring (fig. 3.16). By 1900 British agriculture was already greatly benefiting from much improved machinery and even more from the rapid advances of the nation's economy fueled by rising coal combustion. The effect of these fossil energy subsidies is also clearly discernible in the case of Dutch yields; in contrast, French wheat yields showed only a much milder upward trend even during the nineteenth century, and there was actually a decline for the efficiently but extensively grown American crop (fig. 3.16). Using the best available yield averages, an hour of medieval labor produced no more than 3–4 kg of grain. By 1800 the average rates were around 10 kg. A century later they were close to 40 kg, and the best performances were much above 100 kg.

Energy returns increased a bit faster as an average hour of late nineteenth-century field work called for less physical exertion than the

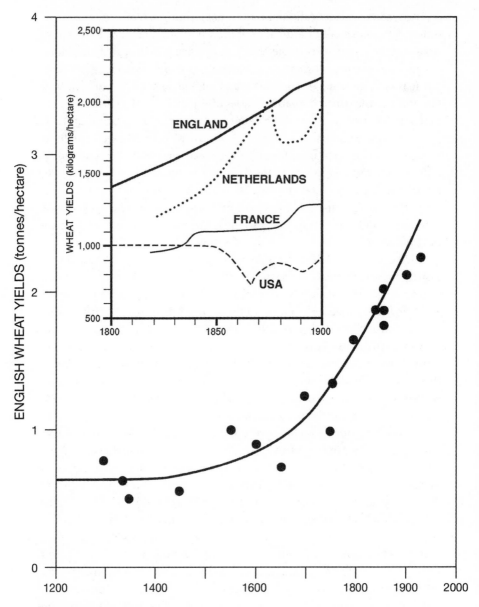

Figure 3.16
English wheat yields show a long period of stagnation followed by a post-1600 take-off. Yield gains during the nineteenth century were even more impressive in the Netherlands, but marginal in France, while in the United States the westward expansion of wheat farming to the drier interior had actually resulted in declining average yields. Plotted from data in USDA (1955), USBC (1975), Stanhill (1976), Clark (1991), and Palgrave Macmillan (2013).

typical medieval efforts: hand plowing with a heavy wooden moldboard plow pulled by oxen required much greater effort than riding on a steel plow pulled by a team of powerful horses. A complete late Roman or early medieval wheat-growing sequence would yield a roughly 40-fold net energy gain in the harvested grain. At the beginning of the nineteenth century good Western European harvests returned about 200 times more energy in wheat than they spent on its production. By the century's end the ratio was commonly above 500, and the best returns were above 2,500.

Net energy gains (after subtracting seed requirements and storage losses) were necessarily lower, no more than 25 for usual medieval harvests, 80–120 by the beginning of the nineteenth century, and typically 400–500 by its end. But these soaring labor productivities were bought with the growing deployment of draft power, and hence with substantial energy investment in animal feeding. In the Roman case every unit of useful power available from human work was supplemented by about eight units of animal-labor capacity. In early nineteenth-century Europe the typical ratio of human/ animal power capacity rose to around 1:15, but on the most productive American farms it was well above 1:100 during the 1890s. Human labor became a negligible source of mechanical energy, and the value of farmers' work shifted mostly to management and control, tasks of low-power needs but high-output rewards.

The energy costs of draft power grew even faster. A pair of Roman oxen subsisting on roughages did not need any grain to perform its field tasks, and hence its use did not lower peasant's potential grain supply. A medium-sized pair of early nineteenth-century European horses would consume almost 2 t of feed grain a year, about nine times the total of food grain eaten per capita. During the 1890s a dozen powerful American horses needed some 18 t of oats and corn per year, about 80 times the total of food grain eaten by their master. Only a few land-rich countries could provide so much feed. Feeding 12 horses would have required about 15 ha of farm-land. An average U.S. farm had almost 60 ha of land in 1900, but only one-third of it was cropland. Clearly, even in the United States, only large grain growers could afford to keep a dozen or more working animals; the 1900 average was only three horses per farm (USBC 1975).

Not every traditional society could intensify its farming by relying on higher inputs of animal labor. Cropping intensification based on more elaborate cultivation of a limited amount of arable land became the norm in the rice-growing Asia. The most notable examples of this development were Japan, parts of China and Vietnam, and Java, the most densely settled island of the Indonesian archipelago. This approach, aptly called by Geertz

(1963) agricultural involution, rested on the high yield potential of irrigated rice and on heavy energy investment embodied over decades and centuries in the construction and maintenance of irrigation systems, wet fields, and terraces.

While cropping intensification in dryland farming can easily lead to environmental degradation (above all to soil erosion and nutrient loss), paddy agroecosystems are much more resilient. Their assiduous cultivation is an enormous absorber of human labor. The process starts with the careful leveling of fields and the sprouting of seedlings in nurseries, and involves such micromanagement techniques as carefully spaced planting, weeding by hand, and harvesting of individual plants. Once established, this introversive tendency is difficult to break. The process supports progressively higher population densities but it leads eventually to extreme impoverishment. Labor productivity first stagnates, then starts declining as larger populations rely on increasingly marginal diets. Many regions of China showed clear signs of agricultural involution during the Ming and Qing dynasties.

After the conflicts of the first half of the twentieth century, Maoist policies based on mass rural labor in communal farming perpetuated the involution until the late 1970s. At that time 800 million peasants still represented more than 80% of China's total population, and they continued to subsist on barely adequate, although more equitably distributed, rations. Only Deng Xiaoping's abolition of communes and de facto privatization of farming during the early 1980s radically reversed the trend. A number of Asian rice-growing countries followed the involution spiral even after 1950. In contrast, Japan broke the trend with the Meiji Restoration in 1868. Between the early 1870s and the 1940s its total population grew 2.2 times. This rate was matched by increases in average rice yields, while the rural population declined by half, to just 40% of the total (Taeuber 1958).

Despite their fundamental differences, the two grand patterns of farming intensification—one based on substitution of human labor by animal draft, the other on maximization of peasant labor inputs—pushed agricultural production in the same direction, toward slowly increasing population densities. This process was essential for releasing a growing share of labor for nonagricultural work, a trend that led to occupational specialization, larger size of settlements, and the emergence and growing complexity of urban civilizations.

These changes can be reconstructed only in approximate terms. Past population totals are highly uncertain even for societies with a long tradition of relatively comprehensive counts (Whitmore et al. 1990). But it is much more difficult to find any reliable data for cultivated land, and even more so for the shares of cultivable land that were actually planted to annual or permanent crops. Consequently, it is impossible to present reliable trends in population densities. What can be done with confidence is to contrast the minima characteristic of early agricultures with a number of typical later performances (derived from written records), and then with the best achievements of the most intensive preindustrial ways of farming (well documented by modern research).

The average rates for all ancient civilizations appear to start around 1 person/ha of arable land. Only after many centuries of slow advances did this rate double. In Egypt this doubling took about 2,000 years, and it appears that a very similar amount of time was needed both in China and in Europe (fig. 3.17). By 1900 the best national averages were about 5 people/ha of cultivated land, and the best regional achievements peaked at more than twice that level (and during the twentieth century they grew much faster: by the year 2000 there were nearly 25 people/ha in Egypt, 12

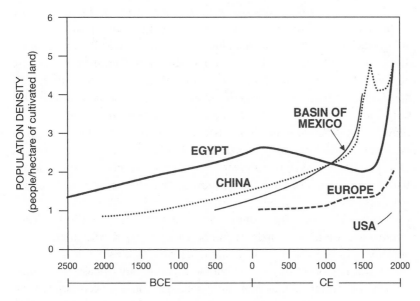

Figure 3.17
Approximate long-term trends of population densities per hectare of cultivated land in Egypt, China, the basin of Mexico, and Europe, 2500 BCE–1900 CE. Calculations are based on estimates and data in Perkins (1969), Mitchell (1975), Butzer (1976), Waterbury (1979), Richards (1990), and Whitmore and co-workers (1990).

in China, and 3 in Europe). But comparisons of population densities should also take into account the nutritional adequacy and variety of diets.

Nutrition

Numbers tracing the population densities of preindustrial societies reveal little about the adequacy and quality of typical diets. Calculating the average food requirements of a traditional society cannot be done with a reasonable certainty: too many assumptions are required to fill in the missing information. Production estimates must also rely on cumulative assumptions, and actual consumption was also affected by high, and highly variable, postharvest losses. Perhaps the only acceptable generalization, based on documentary and anthropometric evidence, is that there is no clear upward trend in per capita food supply across the millennia of traditional farming. Some early agricultural societies were in certain aspects relatively better off, or at least no worse off, than their successors. For example, Ellison's (1981) reconstruction of ancient Mesopotamian ration lists indicates that daily energy supplies between 3000 and 2400 BCE were about 20% above the early twentieth-century mean for the same region.

Calculations based on the Han dynasty records show that during the fourth century BCE in the state of Wei a typical peasant was expected to provide each of his five family members with nearly half a kilogram of grain a day (Yates 1990). That total is identical to the North Chinese mean during the 1950s, before the introduction of pumped irrigation and synthetic fertilizers (Smil 1981a). More reliable figures for early modern Europe also show some notable declines of staple food consumption even in cities enjoying privileged food deliveries. For example, the yearly per capita supply of grain in the city of Rome fell from 290 kg during the late sixteenth century to about 200 kg by the year 1700, and the average per capita availability of meat also declined, from almost 40 kg to just around 30 kg (Revel 1979).

And in most cases, the more recent diets were also less diverse. They contained less animal protein than the earlier intakes with higher consumption of wild animals, birds, and aquatic species. This qualitative decline was not offset by a more equitable availability of basic foodstuffs: major consumption inequalities, both regional and socioeconomic, were common by the end of the eighteenth century and persisted until the nineteenth century. Large shares, and recurrently even majorities, of people in all traditional farming society had to live with food supplies that were below the level necessary for a healthy and vigorous life.

In his 1797 survey of the state of the English poor, Frederick Morton Eden found that even in the richer southern part of the country staple foods were just dry bread and cheese, and in the household of a Leicester-shire worker there was

seldom any butter, but occasionally a little cheese and sometimes meat on Sunday. ... Bread, however, is the chief support of the family, but at present they do not have enough, and his children are almost naked and half starved. (Eden 1797, 227)

Reconstruction of food intakes by poor English and Welsh rural laborers found that between 1787 and 1796 they averaged just 8.3 kg of meat a year (Clark, Huberman, and Lindert 1995), and meat consumption by the poorer half of the English population was barely above 10 kg by the 1860s (Fogel 1991). And in eastern Prussia a third of the rural population could not afford enough bread as late as 1847 (Abel 1962).

Even during fairly prosperous times the typical diets—supplying more than adequate nutrition in terms of total energy and basic nutrients—were highly monotonous and not very palatable. In large parts of Europe bread (mostly dark, in northern regions with little or no wheat flour), coarse grains (oats, barley, buckwheat), turnips, cabbage, and later potatoes were the everyday staples. They were often combined in thin soups and stews, with evening meals indistinguishable from breakfasts and midday food. Typical rural Asian diets were, if anything, even more dominated by a few cereals. In premodern China millets, wheat, rice, and corn supplied more than four-fifths of all food energy. The Indian situation was almost identical.

Seasonally abundant vegetables and fruits regularly enlivened this monotony. Asian favorites included cabbages, radishes, onions, garlic, and ginger and pears, peaches, and oranges. Cabbages and onions were also among European mainstays, besides turnips and carrots; apples, pears, plums, and grapes brought the largest fruit harvests. The most important Mesoamerican species were tomato, chayote, chili, papaya, and avocado. Typical Asian rural diets were always overwhelmingly vegetarian, as were those of the Mesoamerican societies, which, dogs aside, never had any sizable domesticated animal. But some parts of Europe enjoyed relatively high meat intakes during prosperous periods. Still, typical diets had only occasional small servings of meat. Animal proteins were eaten largely in dairy products. Roasts, stews, beer, cakes, and wine were common only on such festive occasions as religious holidays, weddings, or guild banquets (Smil 2013d).

Even when everyday diets supplied enough energy and protein there could be frequent deficiencies of vitamins and minerals. Mesopotamian diets, with their staple of high-yielding barley, were short in vitamins A and

C: ancient inscriptions carry references to blindness and a scurvylike disease (Ellison 1981). During subsequent millennia these two deficiencies were common in most extratropical societies. Very low meat intakes caused chronic iron deficiencies where leafy green vegetables were rarely eaten. Rice-dominated diets had major calcium deficits, especially for child growth: in southern China the mean was less than half of the recommended daily intake (Buck 1937). Monotonous and inadequate diets and widespread malnutrition remain a norm today in many poor countries where population densities have surpassed the limits sustainable by even the most intensive traditional farming.

Limits

Despite the slow progress in yields and labor productivities, traditional farming was an enormous evolutionary success. There would be no complex cultures without high population densities supported by permanent cropping. Even an ordinary staple grain harvest could feed, on the average, ten times as many people as the same area used by shifting farmers. But there were clear limits to population densities achievable with traditional farming. Moreover, average food supply was rarely much above the existential minimum, and seasonal hunger and recurrent famines weakened even the societies with rather low population densities, good soils, and relatively good farming techniques.

Energy supply was the most common limit to the process of substituting animal draft for human labor. The production of concentrate feed needed for working animals could not be allowed to compromise adequate harvests of food grains. Even in land-rich agricultures with extraordinary feed production capacities the substitution trend could not have continued much beyond the American achievements of the late nineteenth century. Heavy gang plows and combines took animal-drawn cultivation to its practical limit. Besides the burden of feeding large numbers of animals used for relatively short periods of field work, much labor had to go into stabling, cleaning, and shoeing the animals. Harnessing and guiding large horse teams were also logistical challenges. There was a clear need for a much more powerful prime mover—and it was soon introduced in the form of internal combustion engines.

The limits to population density in societies undergoing agricultural involution were reached by the ability to subsist on gradually diminishing per capita returns of human labor. These gains became eventually limited by the maximum possibilities of nitrogen recycling. The most intensive application of traditional nitrogen sources—from recycling of organic wastes and planting of green manures—provided enough of the nutrient to support 12–15 people/ha of cultivated land. The production of manures could not be increased beyond the limit set by the availability of animal feed. In intensively cultivated regions they can come only from crop and food-processing residues. Moreover, heavy manuring and the use of human wastes is quite taxing in terms of the strenuous, repetitive labor needed to collect, transport, and distribute the organic matter.

The only universally available, effective alternative was to rotate the crops demanding fertilization with leguminous plants. Yet this is also a limited solution. Frequent planting of leguminous green manures would maintain high soil fertility—but it would inevitably lower the average annual output of staple cereals. Leguminous grains can be cultivated largely without an external nitrogen supply, but the two classes of foodstuffs are interchangeable only as far as their gross food energy content is concerned. Legumes are high in protein—but they are also difficult to digest and often have low palatability. Moreover, they cannot be used to bake bread or, with a few exceptions, to make noodles. Unfailingly, as soon as societies get richer one of the most notable nutritional shifts is their declining consumption of legumes.

Regardless of the historical period, environmental setting, or prevailing mode of cropping and intensification, no traditional agriculture could consistently produce enough food to eliminate extensive malnutrition. All of them were vulnerable to major famines, and even the societies practicing the most intensive cultivation were not immune to recurrent catastrophes, with droughts and floods being the most common natural triggers. In China in the 1920s peasants recalled an average of three crop failures within their lifetime that were serious enough to cause famines (Buck 1937).

These famines lasted on average about ten months, and forced a quarter of the affected population to eat bark and grasses. Nearly one-seventh of all people left their villages in search of food. A similar pattern would be found in most Asian and African societies. Some famines were so exceptionally devastating that they remained in collective memory for generations and led to major social, economic, and agronomic changes. Notable examples of such events are the frost- and drought-induced failures of corn harvests in the basin of Mexico between 1450 and 1454 (Davies 1987), the famous collapse of *Phytophthora*-infested Irish potato crops between 1845 and 1852 (Donnelly 2005), and the great Indian drought-induced famine of 1876–1879 (Seavoy 1986; Davis 2001).

Why could not preindustrial societies insulate themselves better from the recurrence of such drastic food shortages? They could—either by extending cultivated land, or by intensifying their cropping, or by doing both—and they were constantly trying to do so. But in an overwhelming majority of cases these moves were undertaken reluctantly, and they were commonly postponed for so long that natural disasters were repeatedly translated into major famines. There is a clear energetic reason for this tardiness. Both the extended and the intensified cultivation required a higher investment of energy. Even in societies that could afford a larger number of draft animals, most of those additional energy inputs had to come from longer hours and harder exertion of human labor.

Moreover, intensified food production often had a lower energy benefit/ cost ratio than its less intensive predecessors. Not surprisingly, traditional cultivators tried to postpone these greater labor burdens and lower relative returns. Usually they expanded or intensified their cropping only when forced to provide basic needs for gradually increasing populations. In the long run this reluctant expansion and intensification could support substantially larger populations—but per capita food availabilities, and the quality of average diets, had hardly changed centuries or even millennia later.

This reluctance to expand and intensify was repeatedly manifested by a great persistence of less energy-intensive agricultural practices. There was usually a very long transition from shifting to permanent cropping, and peasants were reluctant to expand farming into new areas or to intensify their cropping. When gradual population increases could not be sustained by local or regional production they were primarily accommodated by extending the cultivated area rather than by intensifying the cropping of existing land. Consequently, it took centuries, even several millennia, to adopt annual cropping instead of extensive and prolonged fallowing.

There is no shortage of historical examples illustrating each of these reluctant steps. Shifting cultivation in forest environments offered basic subsistence and meager material possessions—but for many societies it remained the preferred way of life even after many generations of contact with permanent farmers. Sharp contrasts between the alluvial farmers and mountain dwellers could be seen well into the twentieth century in southern Chinese provinces, throughout Southeast Asia, and in many parts of Latin America and sub-Saharan Africa, but the practice was surprisingly persistent even in Europe.

In the Île-de-France, the fertile region surrounding Paris, shifting cultivation (with fields abandoned after just two harvests) was still common in the early twelfth century (Duby 1968). And on the margins of the continent, in northern Russia and Finland, it was still practiced during the

Figure 3.18
Slash-and-burn cultivation in late nineteenth-century Europe. In this photograph, taken by I. K. Inha in 1892 in Eno, Finland, women clear-burn slopeland before it is plowed and planted with a grain or root crop.

nineteenth century and in some places well into the twentieth century (Darby 1956; Tvengsberg 1995) (fig. 3.18). Reluctance to expand cultivation is best seen in an aversion of lowland peasants to colonize the nearby marginal mountain or wetland soils. The villages of Carolingian Europe were overpopulated and their grain supplies constantly insufficient—but except in parts of Germany and Flanders there was little effort to create new fields beyond the most easily cultivable lands (Duby 1968). Later European history is replete with waves of German migrations from densely populated western regions. Armed with superior moldboard plows, they opened up farmlands in areas considered inferior by local peasants in Bohemia, Poland, Romania, and Russia—and set the stage for nationalist conflicts for centuries to come.

The expansion of cropping required additional labor to open up new farmland—but in most instances this one-time energy investment was a fraction of the additional inputs needed for multicropping, manuring, terracing, irrigation, ditching, or field raising in intensive agriculture. And so, even in relatively densely populated regions of Asia and Europe, it took millennia to move gradually from extensive fallowing to annual cropping and multicropping. In China, every dynasty adopted in its early years a policy of extending the cultivated land as the primary means of feeding a growing population (Perdue 1987). In Europe the fallowing of 35–50% of land was still common as recently as the beginning of the seventeenth century. And the more intensive triennial system coexisted in England with the biennial one since the twelfth century—and it generally prevailed only during the eighteenth century (Titow 1969).

Not surprisingly, the transition from shifting to permanent farming and its subsequent intensification happened generally first in areas with poorer soils, limited arable land, high aridity, or uneven precipitation. Environmental stress and high population density certainly do not explain every instance and timing of cropping intensification, but a strong relationship is unmistakable. An excellent early example comes from the archaeological findings in northwestern Europe. There is clear evidence that the transition from the Neolithic to the Bronze Age started first in areas with limited arable land, in today's Switzerland and Britain (Howell 1987).

An abundance of potential farmland in the core area of the Seine-Oise-Marne culture simply led to further expansion of extensive cropping rather than to its intensification and the attendant centralization. Archaeological evidence also indicates that intensification in Yucatan Maya started first in the environments that were either more marginal (drier) or more fertile (hence more densely settled) than the average locations (Harrison and Turner 1978). The historical record is in obvious accord: intensification usually proceeded first either in stressed environments (arid and semiarid climates, poor soils) or in densely inhabited regions.

For example, Hunan province, with good alluvial soils and usually abundant precipitation, is now by far the largest rice producer in China. In the early fifteenth century—more than a millennium after the dry and erosion-prone valley of the Wei He (the site of Xi'an, China's ancient dynastic capital) turned to intensive farming—it was still a sparsely peopled frontier. And farmers in densely settled Flanders were one to two centuries ahead of most of their German or French counterparts in reclaiming its wetlands and heavily fertilizing its soils (Abel 1962). These realities could be generalized as a fundamental preference of peasant societies to minimize the labor needed to secure basic food supply and essential possessions. Cultural differences aside, nearly all traditional peasants behaved as gamblers. They tried to stay on the slim margins of food surplus too long, betting that the weather would help to produce another fair harvest next year. But given the low staple grain yields and relatively high seed/harvest ratios they lost repeatedly, and often catastrophically.

Seavoy (1986) labeled this behavior—acquiring minimum levels of food safety and material welfare with the least expenditure of physical labor—a subsistence compromise. He also saw high birth rates as the other key strategy to reduce per capita labor exertion. The energy cost of pregnancy and of bringing up another child is negligible compared to its labor contribution, which can start at a very early age. According to Seavoy (1986, 20), "Having many children (an average of four to six) and transferring labor to them at the earliest possible age is highly rational behavior in peasant societies, where the good life is equated with minimal labor expenditures, not with the possession of abundant material goods."

But Seavoy's insistence on a universal peasant preference for practicing indolence as a primary social value is unacceptable. Similarly, Clark (1987), apparently unaware of Seavoy's hypothesis, attempted to explain the substantial difference between the early nineteenth-century agricultural productivity in the United States and Britain, on one side, and Central and Eastern Europe on the other, almost solely by faster work rates in the two English-speaking nations. Such sweeping generalizations ignore the influence of many other critical factors. Environmental conditions—soil quality, the amount and reliability of precipitation, the per capita availability of land, fertilizer, and food, the capability to support draft animals—have always made much difference. So did socioeconomic peculiarities (land tenure, corvée, taxation, tenancy, ownership of animals, and access to capital) and technical innovation (better agronomic methods, animal breeds, plows, and cultivation and harvesting implements).

Komlos (1988) considered some of these factors in his persuasive refutation of Clark's exaggerations. Undoubtedly, many cultures did put a low social value on the physical labor of cultivation, and there were important differences in work rates among traditional agricultures. But these realities arose from a complex combination of social and environmental factors, not merely from a simplistic distinction between indolent subsistence peasants devoid of any motivation to accumulate material wealth, on one side, and hard-working farmers driven by accumulation of commercial wealth on the other.

A much less contentious generalization concerning the physical work is that it was spread as much as possible. In practice this meant transferring much of it to women and children, generally persons of lower status in peasant societies. Women were responsible for a high share of field and household tasks in virtually every traditional society. And because even pregnancy and lactation did not have to be much of a burden in terms of additional food, and because the children often started working as early as four to five years of age, large families were the least energy-intensive way to minimize adult labor and to secure food in old age when infirmities set in.

In those traditional agricultures that were powered overwhelmingly by human labor it was clearly rational to minimize individual workloads by having large families. At the same time, this strategy made it much more difficult to increase average per capita food availability, and to avoid recurrent famines. Those traditional agricultures in which draft animals performed most, or virtually all, of the heavy tasks weakened the link between human labor and crop productivity—but could function only when a significant share of land (and harvests) was set aside to feed the working animals.

Only the inputs of fossil energies—directly as fuels and electricity, and indirectly in agricultural chemicals and machinery—could sustain both an expanding population and a higher per capita supply of food. Hybrid agricultures—with the first (indirect) fossil energy inputs—began to emerge first in the UK, then in Western Europe and the United States, with the adoption of iron and steel tools and machinery made from the metal smelted (first in 1709 in England) by using coke rather than charcoal. But even by 1850 Western farming was still overwhelmingly solar, and although metal devices and machines spread during the second half of the nineteenth century, fossil subsidies began to make a major difference only after 1910 with the diffusion of tractors, trucks, and synthetic nitrogenous fertilizers, advances I trace in the fifth chapter.

4   Preindustrial Prime Movers and Fuels

Most people in preindustrial societies had to spend their lives as peasants, laboring in ways that in some societies remained largely unchanged for millennia. But the inconsistent food surpluses that they produced with the aid of a few simple tools and the exertion of their muscles and the draft of their animals sufficed to support the unevenly advancing complexity of urban societies. Physically, these achievements were reflected above all in the construction of remarkable structures (ranging from ancient Egypt's pyramids to the Baroque churches of the early modern era), the rising capacities and increasing reach of transportation (ranging from slow wheeled transfers on land to faster ships capable of circumnavigating the planet), and improvements in a multitude of manufacturing techniques, spearheaded above all by advances in metallurgy.

The prime movers and fuels energizing these advances remained unchanged for millennia, but human ingenuity improved their performance in many remarkable ways. Eventually, some of these conversions became so powerful and so efficient that they were able to energize the initial stages of modern industrialization. Two principal roads led to higher outputs and better efficiencies. The first one was multiplication of small forces, primarily a matter of superior organization, especially with the application of animate energies. The second one was technical innovation, which introduced new energy conversions or increased the efficiencies of established processes. In practice, the two approaches often melded. For example, monumental structures built by virtually every old high culture demanded both massed labor and extensive applications of labor-easing devices, starting with simple levers and inclined planes and eventually including pulleys, cranes, windlasses, and treadwheels.

The differences between the first documented mechanical energy converters and their successors, used at the beginning of the industrial era, are often quite remarkable. Early designs of spoon tilt-hammers, the simplest machines powered by falling water, did not even involve continuous rotary motion; they were just repeatedly actuated simple levers (fig. 4.1). Later, vertical waterwheels turned trip hammers into reliable helpers in Asian and European forges, and some nineteenth-century water-powered forge hammers were impressive, complex, high-performance pieces of machinery (fig. 4.1).

Similar comparisons can be made for every class of water- and wind-powered prime movers. What a difference there is between rough-hewn medieval horizontal wooden waterwheels, whose power was just a few hundred watts (less than half a horsepower), the much better built seventeenth-century vertical machines, with power ratings easily ten times higher, and the Lady Isabella, England's largest iron overshot wheel, capable of delivering more than 400 kW, power equivalent to that of nearly 600 strong horses! Or between inefficient and heavy European medieval post mills that had to be laboriously turned into the wind, only to lose more than 80% of their potential power to poor sails and rough gearing, and their automatically regulated nineteenth-century American counterparts with spring sails and smooth transmissions, whose operation—they were often used to pump water—helped open up the Great Plains.

Contrasts are no less impressive for animate conversions and the combustion of phytomass fuels. A nineteenth-century heavy draft horse with iron horseshoes and a collar harness hitched to a light flat-top wagon on a hard-top road could easily pull a load 20 times as heavy as its much lighter, unshod, breast-harnessed ancestor linked to a heavy wooden cart on a muddy road could. And an eighteenth-century blast iron furnace consumed less than one-tenth the charcoal per unit of hot metal output than was needed by its early medieval predecessor (Smil 2016). The human capacity for heavy work, however, changed little between antiquity and the onset of industrialization. Even in those societies where average body weight increased over time, such a gain had only a marginal effect on maximum

---

Figure 4.1

Flowing water powered these three hammers, but their complexity and performance were vastly different. The primitive Chinese spoon tilt-hammer of the early fourteenth century was a simple lever actuated by falling water (top). European forge hammers of the late sixteenth century were energized by waterwheels whose rotary power was transferred by connecting rods (middle). Tilt-hammers at a nineteenth-century English foundry were high-performance, adjustable machines (bottom). Reproduced from drawings in Needham (1965) and Reynolds (1970).

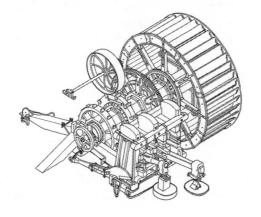

muscular exertion, and heavy exertions have always required the combined power of many individuals.

To move a 327 t Egyptian obelisk from the site where it was left by the Romans (Caligula had it placed at the central *spina* of his circus, the spot that is now just south of San Pietro) 269 m eastward, Domenico Fontana used huge (up to 15 m long) wooden levers and pulleys to lift it from its ancient base, and on September 10, 1586, when he raised it in the center of St. Peter's Square in Rome, he relied on 900 men and 75 horses to pull pulley-guided ropes and set it on a new foundation (Fontana 1590; Hemphill 1990). The entire project was accomplished in 13 months and the erection took one day. Later famous relocations of obelisks included the structures now standing at Place de la Concorde in Paris (completed in 1833), on the Thames Embankment (1878), and (since 1881) in New York's Central Park (Petroski 2011).

When the world's heaviest column—604 t of red Finnish granite erected to commemorate Russia's victory over Napoleon's invading army—was raised in Saint Petersburg on August 30, 1832, the French architect Auguste de Montferrand relied on 2,400 men (1,700 engaged in the actual pulling), who did the job in less than two hours (box 4.1). And the two essential devices that provided the necessary mechanical advantage for these two lifts and that have allowed men to execute many prodigious moves and lifts and emplacements, inclined planes and levers, have been with us not only since the time of ancient empires but, necessarily, long before that—or how else did Stonehenge's 40 t outer stones get raised?

In this chapter I first appraise the kinds, capacities, and limits of all traditional prime movers—human and animal muscles, wind, and water—as well as the combustion of phytomass fuels, mostly wood and charcoal made from it, but in deforested regions also many kinds of crop residues (particularly cereal straw) and on grasslands also dried dung. Afterward I look in some detail at the uses of prime movers and fuels in critical segments of traditional economies: in food preparation, in the provision of heat and light, in land and waterborne transportation, in construction, and in color and ferrous metallurgy.

Prime Movers

Animate labor and conversions of the kinetic energies of water and wind (by sails and mills) were the only prime movers in traditional societies before the diffusion of steam engines. Although the subsequent retreat of traditional prime movers was relatively rapid, waterwheels and windmills

---

Box 4.1

Raising Alexander's column

The large piece of red granite that became Alexander's Column was quarried at Virolahti in Finland, rolled onto a purpose-built barge able to carry 1,100 t (the column nearly fell into the water during loading), shipped about 190 km to the Neva's embankment in Saint Petersburg, offloaded onto a solid wood deck, moved up 10.5 m on an inclined plane, and positioned on a platform at a right angle to the pedestal in the center of Palace Square. A solid wooden scaffolding erected above the pedestal was 47 m tall, with pulley blocks hanging from five double oak beams. Montferrand built a 1/12 scale model of the scaffolding to guide the carpenters in its construction (Luknatskii 1936). Lifting was accomplished by 60 capstans mounted on the scaffolding in two staggered rows. Ratchets were iron drums mounted in a wooden frame, the upper blocks were hung from the double oak beams, and 522 ropes, each tested to lift 75 kg (triple the actual load), were attached to the shaft of the column. The total mass of the monolith with all devices was 757 t.

Lifting the column was done on August 30, 1832, and employed directly 1,700 soldiers and 75 officers, supervised by foremen who coordinated speed and a steady pace depending on the tension on the rope. Montferrand's assistants stood at the four corners of the scaffolding with 100 sailors who watched the blocks and ropes and kept them straight; 60 workers stood on the tower itself; and carpenters, stonecutters, and other craftsmen also stood in reserve. The total labor force for the lift was about 2,400 people, and the task was completed in just 105 minutes. Remarkably, the column has kept its upright position without being fastened to the pedestal: the 25.45 m tall, slightly conical (3.6 m diameter at the bottom, 3.15 m at the top) mass simply rests on its foundations.

---

retained (or even increased) their importance during the first half of the nineteenth century, sail ships became marginal means of ocean transportation only after 1880, and draft animals dominated even the most advanced Western agricultures until after World War I. The early phases of industrialization had actually increased the demand for human labor, ranging from some extremely heavy exertions in coal mining and the iron and steel industry to the myriad tiresome tasks involved in manufacturing, with child labor common in Western countries even at the beginning of the twentieth century: in 1900 about 26% of boys aged 10–15 years worked, and the percentage of children in agricultural employment was as high as 75% for girls (Whaples 2005).

And heavy exertions and child labor are still common in most rural areas of sub-Saharan African countries and in the poorest regions of Asia: in Africa, women carry heavy head loads of firewood; in India, women break stones with small hammers; in India, Pakistan, and Bangladesh, men dismantle massive ships on hot beaches (Rousmaniere and Raj 2007); in China, peasants dig coal in small rural mines. And millions of people are still subjected to different forms of forced and slave labor and human trafficking (International Labour Organization 2015). A continuing reliance on human labor (including its most offensive variants) is one of the most obvious marks of the great divide between the rich and poor worlds. But even in the West, heavy exertions (in underground coal mining, steelmaking, forestry, fishing) were not uncommon until the 1960s, and the use of animate prime movers is more than a matter of historical interest: it is the not so distant foundation of our present affluence.

This account of preindustrial prime movers would be incomplete without noting the medieval invention, diffusion, and historical importance of gunpowder. Awe of thunder and lightning can be seen in every old high culture. The aspiration to emulate their destructive power recurs in many narratives and fantasies (Lindsay 1975). But for millennia the only pale imitation was to attach incendiary materials to arrowheads, or to hurl them in containers from catapults. Sulfur, petroleum, asphalt, and quicklime were used in these incendiary mixtures. Only the invention of gunpowder combined the propulsive force with great explosive and inflammatory power.

Animate Power

Animate energies remained the most important prime mover for most of humanity until the middle of the twentieth century. Their limited power, circumscribed by the metabolic requirements and mechanical properties of animal and human bodies, restricted the reach of preindustrial civilizations. Societies that derived their kinetic energy almost solely (as in the case of ancient Mesopotamia or Egypt, with sail ships being the only exception) or largely from animate power—medieval Europe is an excellent example, with water and wind power limited only to certain tasks, and in rural China that was the case until two generations ago—could not provide a reliable food supply and material affluence to most of their inhabitants.

There were only two practical ways in which the delivery of useful animate power could be increased: either by concentrating individual inputs or by using mechanical devices to redirect and amplify muscular exertions.

The first approach soon runs into practical limitations, especially with the direct deployment of human muscles. Even unlimited labor force is of little use for directly grasping and moving a relatively small but very heavy object as only a limited number of people can fit around its perimeter. And while a group of people can carry a heavy object, lifting it first in order to insert slings or poles could be a challenging problem. Human capacities in lifting and moving loads are limited to weights substantially smaller than their body mass. Traditional sedan chairs (litters), used by most of the Old World societies, were carried by two men, each bearing at least 25 kg and up to 40 kg, with heavier loads supported by poles resting on their shoulders.

When unloading and loading ships and wagons, Roman *saccarii* lifted and carried (over short distances) sacks of 28 kg (Utley 1925). Heavier burdens were manageable only with the help of simple devices that conferred significant mechanical advantage, usually by deploying a lesser force over a longer distance. Five such devices were widely used during the Old World's antiquity: Philo (during the third century BCE) listed them as a wheel and axle, a lever, a system of pulleys, a wedge (an inclined plane), and an endless screw. Their common variations and combinations ranged from screws to treadwheels. By using these tools and simple machines, people could deploy smaller forces over longer distances, thereby enlarging the scope of human action (box 4.2). The three simplest aids providing mechanical advantage—levers, inclined planes, and pulleys—were used by virtually all old high cultures (Lacey 1935; Usher 1954; Needham 1965; Burstall 1968; Cotterell and Kamminga 1990; Wei 2012).

Levers are rigid, slender pieces of wood or metal. As they pivot around a fulcrum they convey an mechanical advantage that is easily calculated as a quotient of effort-arm and load-arm lengths (measured from the pivot point; the higher the number, the easier and faster the task). The ancient use of levers ranged from driving oared ships to moving heavy loads (fig. 4.2). Levers are classified according to the position of fulcrum (fig. 4.2). In the first class of lever the fulcrum is between the load and the applied force, which acts in the opposite direction to a displaced load. In the second class of lever the fulcrum is at one end and the force acts in the same direction as the load. Levers of the third class do not provide any mechanical advantage but increase the load's speed, as is clear from the operation of catapults, hoes, and scythes.

Common hand tools using levers of the first class are crowbars, scissors, and (a double lever) pliers. Wheelbarrows have been among the most often used levers of the second class (Needham 1965; Lewis 1994). Chinese

---

Box 4.2
Work, force, and distance

Work is done when a force—no matter whether provided by animate or inanimate prime movers—changes a body's state of motion. Its magnitude is equal to the product of exerted force and the displacement in the direction in which the force acts. In formal terms, a force of one newton and the displacement of one meter will require the energy of one joule ($J = Nm$). Just to get the feel for relevant orders of magnitude: lifting a 1 kg book from a desk (0.7 m above the floor) and placing it on a shelf (1.6 m above the floor) requires work of almost 9 J. Lifting an average stone (about 2.5 t) of Khufu's pyramid one course higher (about 75 cm) required about 18,000 J (18 kJ), or 2,000 times more energy than shelving the book.

Naturally, the same amount of work can be accomplished by applying a greater force over a shorter distance or a smaller force over a longer distance: any device that converts a small input force into a larger output force provides a mechanical advantage whose magnitude is measured simply as a dimensionless ratio of the two forces. This mechanical advantage has been exploited since the prehistoric era by using levers and inclined planes, and later also by deploying pulleys. There are countless examples of these actions in everyday life, from opening locks with a key (a row of wedges, that is, inclined planes, moving the pins inside a lock) to pulling a nail from a piece of wood by using a claw hammer (a lever action).

---

barrows, used since the Han dynasty, usually had a large (90 cm diameter) central wheel surrounded by a wooden framework. With the load right above the axle they could carry large loads (commonly 150 kg); they were used by peasants to take products to markets and sometimes also to transport people, who sat on the sides (Hommel 1937). Little sails could be erected to ease propulsion. European barrows are first convincingly documented during the high Middle Ages (late twelfth and early thirteenth centuries) and subsequently were used mostly in England and France, usually in construction and mining. Their fulcrum was at the end, which put more strain on people pushing them, but they still offered a considerable (typically threefold) mechanical advantage.

The wheel and axle form a circular lever, with the long arm being the distance between the axle and the wheel's outer rim and the short arm being the axle's radius, which produces a large mechanical advantage, even for heavy wheels on a rough surface. The first wheels (used in Mesopotamia before 3000 BCE) had solid wooden wheels; spoked wheels appeared about

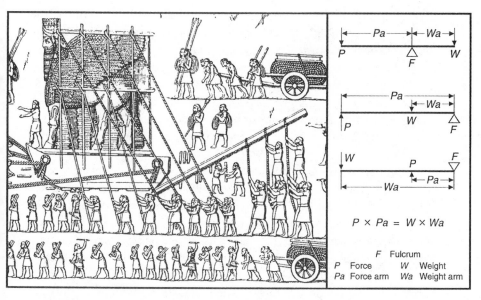

Figure 4.2
Three classes of levers are distinguished by the point at which the force is applied in relation to the object (whose weight, W, always acts in a downward direction) and the fulcrum (F). In levers of the first class, the force moves in a direction opposite to that of the object. In levers of the second class, the force moves in the same direction as the object, but both levers confer the same mechanical advantage: they gain power at the expense of distance. In levers of the third class, the force moves over a shorter distance than the object, resulting in a velocity gain. The first two classes of levers have had countless applications in lifting and moving objects and in machinery construction. A detail from a partially reconstructed Assyrian bas-relief at Kuyunjik (about 700 BCE) shows a large lever used to move a giant statue of a winged bull with a man's head. Reproduced from Layard (1853).

1,000 years later, first on chariots, and friction was reduced by iron rims. The wheel's enormous importance in the Old World can be seen in the rapid diffusion of the invention of wheeled vehicles and in their countless mechanical uses ever since. Curiously, the Americas had no native wheels, and the desert environments in many Muslim lands made pack camels more important than wheeled transport pulled by oxen (Bulliet 1975, 2016).

Neglecting friction, the mechanical advantage of an inclined plane is equal to the quotient of the length of slope and the height to which an object is raised. Friction can reduce this gain quite substantially, and that is why smooth surfaces and some form of lubrication (water being the easiest obtainable and the cheapest lubricant) were needed for the best practical performance. According to Herodotus, an inclined plane was the principal means of conveying heavy stones from the Nile shore to the building site of the great pyramids, and there has been much speculation on its further use during their actual construction (later in this chapter I explain why we should discount that choice). The most common modern use of inclined planes is as ramps, ranging from strong metal plates to load cargo on vehicles and ships to soft plastic surfaces to offload airplane passengers in an emergency.

Wedges are just double inclined planes exerting large sideways forces across small distances. They have commonly been used for splitting rocks, by means of pieces of wood inserted into stone cracks and wetted, and as the cutting edges of adzes and axes. Screws, first used in ancient Greek olive and grape presses, are nothing but circular inclined planes wrapped around a central cylinder. As already noted in the previous chapter, a screw design was also used for shallow water lifts. Their large mechanical advantage means that workers are able to exert high pressure with minimal effort. In many applications small screws (now mass-produced and usually tightened with a clockwise rotation) are used as irreplaceable fasteners.

A simple pulley, consisting of a grooved wheel guiding a rope or a cable, invented during the eighth century BCE, makes the handling of loads easier by redirecting the force, but it confers no mechanical advantage, and its use can result in accidental load falls. Ratchet and pawl take care of the last problem and multiple pulleys address the first deficiency, as the force required to lift an object is nearly inversely proportional to the number of deployed pulleys (fig. 4.3). *Mechanica*, attributed to but not written by Aristotle (Winter 2007), shows a clear understanding of the mechanical advantage afforded by such devices.

The ancient Chinese were such frequent users of pulleys that even palace entertainments could not do without them, and once a whole corps de ballet of 220 girls in boats was pulled up a slope from a lake (Needham 1965). But certainly the most famous ancient testimony to the efficacy of compound pulleys is Archimedes' demonstration to King Hiero, recorded in Plutarch's *Lives*. When Archimedes "declared that, if there were another world, and he could go to it, he could move this," Hiero asked him for a suitable demonstration of such powers.

> Archimedes therefore fixed upon a three-masted merchantman of the royal fleet, which had been dragged ashore by the great labours of many men, and after putting on board many passengers and the customary freight, he seated himself at a distance from her, and without any great effort, but quietly setting in motion with

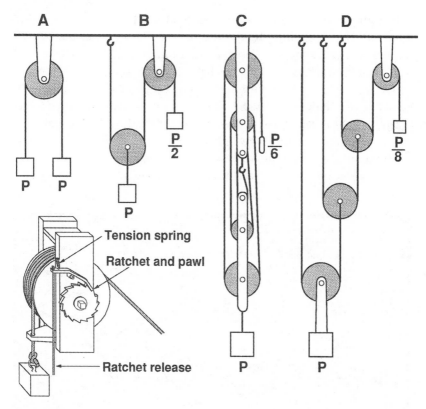

Figure 4.3
Equilibrium forces in pulleys are determined by the number of suspension cords. There is no mechanical advantage in A. In B the weight P is suspended by two parallel cords and hence the free end needs to be loaded only by $P/2$ to be in equilibrium, in C by $P/6$, and so on. A worker raising building materials with Archimedean potential pulley (D) could lift (ignoring friction) a 200 kg stone with a force of only 25 kg, but a lift of 10 m will require pulling 80 m of the counterweight cord. A ratchet and pawl can be used to interrupt this effort anytime.

his hand a system of compound pulleys, drew her towards him smoothly and evenly, as though she were gliding through the water. (Plutarch 1961, iv:78–79)

Three classes of mechanical devices—windlasses and capstans, tread-wheels, and gearwheels—became critical for applying continuous human power, needed in lifting, grinding, crushing, and pounding (Ramelli 1976 [1588]). Windlasses were commonly used not only in lifting water from wells and raising building materials with cranes but also in winding the most destructive stationary weapons of antiquity, the large catapults used in besieging towns and fortresses (Soedel and Foley 1979). Horizontal wind-lasses (winches), requiring the grip to be shifted four times during each revolution (fig. 4.4, left side), and vertical capstans (fig. 4.5) made it possible to transmit power by ropes or chains through simple rotary motion. Cranks, first used in China during the second century CE and introduced to Europe seven centuries later (fig. 4.4, right side), made this even easier, except that the speed of hand cranking (or foot treading) had to match the speed of a driven machine (often a lathe).

This limitation was eliminated by using a crank to power a large wooden or iron wheel (the great wheel) that was independently mounted on a

Figure 4.4
Miners using both a horizontal windlass (left) and a crank (right) to lift water from a shaft. A heavy wooden wheel, sometimes with pieces of lead fastened to its spokes, helped to conserve the momentum and make the lifting easier. Reproduced from Agricola, *De re metallica* (1912 [1556]).

Figure 4.5
Eight men rotating a large vertical capstan in a mid-eighteenth-century French workshop. The capstan winds a cord fastened to pincers, drawing a gold wire through a die. Reproduced from the *Encyclopédie* (Diderot and d'Alembert 1769–1772).

heavy shaft and whose rotation was transmitted to a lathe by a crossed belt. This allowed the use of many gear ratios, and the momentum of a large wheel helped maintain even revolutions even as muscular exertion rose and faded. This medieval innovation enabled accurate machining of wood and metal parts, used to construct a wide variety of precise mechanisms ranging from clocks to the first steam engines, but it could not eliminate the hard work needed when cutting hard metals (fig. 4.6). George Stephenson's workers, who used a great wheel to make parts for the first steam locomotive, had to rest every five minutes (Burstall 1968).

Figure 4.6
The great wheel powered by a crank, used to turn a metalworking lathe. The smaller wheel was used for working with larger diameters, and vice versa. In the background of this image a man works on a foot-powered lathe machining wood. Reproduced from the *Encyclopédie* (Diderot and d'Alembert 1769–1772).

Deployment of the body's largest back and leg muscles on treadwheels delivered much more useful power than hand-turned rotaries. The largest treadwheels (also called, confusingly, great wheels) were two wheels whose rims, connected by planking, formed a pavement trodden by men. A bas-relief in the Roman tomb of the Haterii (100 CE) is the first extant image of a large internal treadwheel (Greek *polyspaston*). Roman treadwheels could lift up to 6 t, and such large machines became a common sight during medieval and early modern Europe at major construction sites and docks and also at mines, where they were used to pump water (fig. 4.7).

The difference between the radius of the wheel and the radius of the axle drum gave these treadwheels a large mechanical advantage, and they could lift such heavy burdens as keystones, massive timbers, or bells to the tops of cathedrals and other tall buildings. In 1563 Pieter Bruegel the Elder painted such a crane lifting a large stone to the second level of his imaginary Tower of Babel (Parrott 1955; Klein 1978). His device, with

(a)

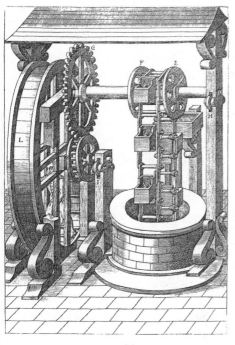

(b)

(c)

Figure 4.7
Details of treadwheels with different torque. a. Internal treadwheel. b. External treadwheel (maximum torque). c. Inclined treadwheel. Reproduced from Agricola, *De re metallica* (1912 [1556]).

treadwheels on both sides, was powered by six to eight men. Externally driven vertical wheels were less common, but they allowed the maximum torque when treading on a level with the axle (fig. 4.7). There were also inclined treadwheels, with laborers leaning against a bar (fig. 4.7), and in English prisons treadmills became common in the early nineteenth century (box 4.3, fig. 4.8).

All types of treadwheels could be also designed or adjusted for animal operation. All drumlike devices had the added advantage of relatively easy mobility: they could go from job to job by rolling along on a fairly flat surface. Until the introduction of steam-powered railway cranes they were the only practical way of tackling heavy lifting. Maximum power inputs on treadwheels were limited by their size and design. With a single worker the power output would be no more than 150–200 W during brief spells of hard effort, and no more than 50–80 W during episodes of sustained effort with tired muscles, while the largest treadwheels, powered by eight men, could operate briefly at about 1,500 W.

On the end of the exertion spectrum were the tasks powered by a single worker using cranks, treadles, pedals, or screws. This hand- or footwork-powered machines ranged from small wood-turning lathes and printing presses to sewing machines, whose first commercial models came during the 1830s but whose widespread use (both hand-turned and treadle-operated designs) began during the 1850s (Godfrey 1982). During the same period, large numbers of boys and men continued to flap (using pulleys) *punkha* (*pangkha* in Hindi), cloth or palm frond ceiling fans, the only means of making India's monsoon heat a bit more bearable for all those who could afford to pay a *punkhawallah,* who operated the fan.

The question of how much useful work a man could do in a single day remained unsettled for long time, and comparisons of the man-day effort with the work of horses had ranged widely, with extreme values differing as much as sevenfold (Ferguson 1971). Watt's definition of horsepower—equal to 33,000 pound-feet per minute, or 745.7 W (Dickinson 1939)—implied an equivalent of about seven workers. The first reliably measured rate was by Guillaume Amontons (1663–1705), who equated the work of glass polishers during a 10-hour shift with raising continuously a weight of 25 pounds at a speed of 3 ft/s (Amontons 1699). In modern scientific units that would amount to total useful work of 3.66 MJ at a rate of 102 W.

How powerful are people as prime movers, and how efficient? The first question was answered quite accurately long before the beginning of systematic energy studies of the nineteenth century. The early estimates

Box 4.3
Working on a treadmill

The largest treading devices operated during the nineteenth century in English prisons, where William Cubitt (1785–861) introduced them as a means of punishment, but soon they were deployed to grind grain and pump water, and sometimes were used simply for exercise (Mayhew and Binney 1862). These long inclined penal treadmills had wooden steps around a cylindrical iron frame and could accommodate as many as 40 prisoners standing side by side, holding on to a horizontal handrail for stability and being forced to step up at the same time. The use of penal treadmills was banned only in 1898.

But writing in 1823, a Devon prison governor, in answer to an inquiry, replied, "I consider the labour at the Tread Mill not as injurious, but conducive to the health of prisoners" (Hippisley 1823, 127). Millions of enthusiastic modern treadmill users might agree, and Landels (1980, 11–12), while noting that we cannot talk or even think about these machines unemotionally, stressed nevertheless that a well-designed treadwheel was not only a highly efficient mechanical device but also one most comfortable for the operator "in so far as any continuous, monotonous physical work can be comfortable."

Figure 4.8
Prisoners on a treadmill at the Brixton House of Correction (Corbis).

ranged from equating the labor of one horse with the exertion of just two to as many as 14 men (Ferguson 1971). Before 1800, rates converged on the correct maxima of 70–150 W for most adults steadily working for many hours. When working steadily at a rate of 75 W, ten men would be needed to equal the power of one standard horse.

In 1798, Charles-Augustin de Coulomb (1736–1806) took a more systematic look at the different ways in which men used their strength during their daily work (Coulomb 1799). These experiences ranged from climbing Tenerife (2,923 m) in the Canaries in a bit less than 8 hours to a day's work by wood carriers ascending 12 m 66 times a day with burdens of 68 kg. The former effort adds to a total work of 2 MJ and a power of 75 W, the latter to about 1.1 MJ and a power of about 120 W. All subsequent evaluations could only confirm the power range established by Coulomb's investigations: most adult men can sustain useful work at 75–120 W (Smil 2008a). During the early twentieth century, studies of human basal metabolic rate (BMR), led by Francis G. Benedict (1870–1957) of the Carnegie Institution in Boston, made it possible to formulate equations of expected energy expenditures and establish typical BMR multipliers for different levels of physical activity (Harris and Benedict 1919), and both are valid for a wide range of body types and ages (Frankenfield, Muth, and Rowe 1998).

As already noted, comparing the exertions of people and animals yields a wide range of men-to-horse ratios. Nicholson (1825, 55) concluded that "the worst way of applying the force of a horse, is to make him carry or draw up hill; for if the hill be steep, three men will do more than a horse. … On the other hand, … in a horizontal direction … a man … cannot exert above one-seventh part of the force of a horse employed to the same purpose." And employing animals was not always practical. As Coulomb (1799) noted, people require less space to work than animals and are easier to transport, and their efforts may be easier to combine.

The performance of small, often poorly fed beasts of antiquity and the early Middle Ages was much closer to human exertion than was the effort of powerful nineteenth-century draft horses. Animals were usually blindfolded (or blind) and harnessed directly to the beams, which were fastened to a central axle whose rotation would be used to mill (mostly grains, but also clays for tiles), extract (oil from seeds, juice from cane and fruits), or wind up a rope tied to a burden (when raising water, coal, ore, or men from mines). In some enterprises animals also rotated whims attached to geared assemblies to multiply the mechanical advantage.

Poor feeding and abuse of these animals forced to walk for hours in a small circle were common, as attested by Lucius Apuleius in his *Golden Ass* (second century CE, here in a classic William Adlington translation from 1566):

> But how should I speake of the horses my companions, how they being old and weake, thrust their heads into the manger: they had their neckes all wounded and worne away: they rated their nosethrilles with a continuall cough, their sides were bare with their harnesse and great travell, their ribs were broken with beating, their hooves were battered broad with incessant labour, and their skinne rugged by reason of their lancknesse. When I saw this dreadfull sight, I began to feare, least I should come to the like state.

This use of horses continued well into the nineteenth century: by the 1870s horses were powering thousands of whims (sweeps) in Appalachian states and throughout the U.S. South, both on farms (grinding grain, extracting oil, compacting cotton bales) and in water pumping and hoisting loads from mines (Hunter and Bryant 1991). They walked in circles often less than 6 m in diameter (see fig. 1.3; 8–10 m was more comfortable), and before the adoption of electric streetcars Western cities had many urban horses harnessed to omnibuses and carts (box 4.4; see also fig. 4.18).

The use of horses for transportation or in construction was constrained by the same factors that limited their employment as draft animals in farming. Neither good pastures nor a sufficient supply of feed grains were available in dry Mediterranean countries or in the densely populated lowlands of Asia, while poor harnessing converted their power quite inefficiently. In Eurasia's arid regions, much less demanding camels were used for many of the tasks done by oxen and horses in Atlantic Europe, but in Asia domesticated elephants (used in harvesting heavy timber, in construction, and in war) also put a considerable strain on feed resources (Schmidt 1996). A classic Indian source of elephant lore extols their effectiveness, but it also prescribes expensive feeding of newly caught elephants in training with boiled rice and plantains mixed with milk and sugar cane (Choudhury 1976). If the animals stayed healthy, such high energy costs were well compensated by their power and remarkable longevity.

Animals used for transport and stationary work ranged from small donkeys to massive elephants, and in some places dogs turned spits over kitchen fires or pulled small carts or wheelbarrows. But, not surprisingly, the modest nutritional demands of bovines—oxen and water buffaloes and yaks—made them the leading working animals both on farms and elsewhere. Yaks were invaluable as pack animals not because of any extraordinary power but because of their ability to walk in high mountains and in snow. The typical draft performance of bovines in transportation was at

---

Box 4.4
Draft horses in urban transportation

Draft horses were employed in cities to deliver food, fuel, and materials (pull-ing carts of different sizes) and for personal transport, pulling hackney car-riages and, since 1834, their modernized versions, cabs patented by Joseph Hansom (1803–1882) and widely known as hansoms. But as Western cities grew, the need for more efficient public transportation led to the introduction of horse-drawn omnibuses (horse-buses). Their use began in 1828 in Paris; a year later they appeared in London and in 1833 in New York, and then in most of America's large eastern cities (McShane and Tarr 2007). In New York their number peaked at 683 vehicles in 1853.

Horsecars (streetcars drawn by horses) on tracks made transport more effi-cient, and such lines were common before the introduction of electric street cars during the 1880s. Light omnibuses (with just a dozen passengers) were drawn by just two horses, but four horses were common, and carriages made for up to 28 passengers were often overcrowded. There were hourly departures, and many lines followed established suburban coach stages, reaching destina-tions 8–10 km from downtowns in about one hour. Hard-working horses had to be well fed, and data collected by McShane and Tarr (2007) show that typi-cal daily rations per animal were 5–8 kg of oats and a similar mass of hay. Supplying urban horses with this feed was an important service in all large nineteenth-century cities.

---

best moderate. For short spells on good roads they could pull loads as much as three or four times their body weight, but their steady work delivered no more than 300 W. Old and weak horses, which were often used for turning a whim, a beam attached to a central axle for work in small manufactures requiring steady rotary power, could not deliver much more, and before the introduction of steam engines many of them were replaced by much more powerful waterwheels and windmills.

Water Power

Antipater of Thessalonica, writing during the first century BCE, left the first literary reference to a simple water mill doing away with the hard work of manual grain milling (translated in Brunck 1776, 119):

> Set not your hands to the mill, O women that turn the millstone! Sleep sound though the cock's crow announces the dawn, for Ceres has charged the nymphs with the labours which employed your arms. These, dashing from the summit of a wheel, make its axle revolve, which by the help of moving radii, sets in action

the weight of four hollow mills. We taste anew the life of the first men, since we have learnt to enjoy, without fatigue, the produce of Ceres.

And with the notable exception of ancient sailing ships, the harnessing of wind started even later. Al-Masudi's report, dated 947, is one of the first reliable records of simple vertical shaft windmills (Forbes 1965; Harverson 1991). His description portrayed Seistan (in today's eastern Iran) as a land of winds and sand where the wind drove the mills and raised water from streams to irrigate gardens. The barely changed successors of these early mills—with plaited reed sails behind narrow openings in high mud walls creating faster wind flow—could be seen in the region well into the twentieth century. Both kinds of machines diffused fairly rapidly throughout the medieval world, but water mills were far more abundant.

Their ubiquity is attested by the Domesday Book count of 1086, when there were 5,624 mills in southern and eastern England, or one for every 350 people (Holt 1988). The earliest horizontal waterwheels are often referred to as either Greek or Norse wheels, but the origin of their design remains uncertain. They became common in many regions of Europe and everywhere east of Syria. The impact of flowing water, usually directed through a sloping wooden trough onto wooden paddles that were often fitted to a hub at an incline, rotated a sturdy shaft that could be directly attached to a millstone rotating above (fig. 4.9). This simple and relatively inefficient design was best suited for small-scale milling. Later designs, with water led through a wooden trough with a tapered bore (Wulff 1966), had efficiencies above 50% and a maximum power above 3.5 kW.

Vertical wheels supplanted the horizontal machines because of their superior efficiency. They turned the millstones by right-angle gears, and in the Western literature they became known as Vitruvian mills, after the Roman builder gave the first clear description of *hydraletae*, dated to 27 BCE. But Lewis (1997) thought the water mill originated during the first half of the third century BCE, most likely in the Ptolemaic Alexandria, and that by the first century CE water power was already in more common use. In any case, because of their eventual ubiquity and persistence we have a large body of literature dealing with their history, design, performance, and uses (Bresse 1876; Müller 1939; Moritz 1958; Forbes 1965; Hindle 1975; Meyer 1975; White 1978; Reynolds 1983; Wölfel 1987; Walton 2006; Denny 2007).

But one thing that is impossible to do is to estimate reliably the contribution of waterwheels to the overall primary energy supply of ancient and medieval societies. Wikander (1983) showed that waterwheels were more common during the Roman era than is usually assumed, and although only

Figure 4.9
The horizontal waterwheel, also called the Greek or Norse wheel.
The wheel was powered by the impact of running water and rotated directly the runner stone above. Reproduced from Ramelli
(1976 [1588]).

20 early medieval water mill sites have been identified, about 6,500 localities had mills in eleventh-century England (Holt 1988). But my estimates show that even with very liberal assumptions regarding the unit power and adoption of waterwheels throughout the Roman Empire, water power contributed only a fraction of 1% of the useful mechanical energy supplied by people and draft animals (Smil 2010c).

Vertical waterwheels are classified according to the point of impact. Undershot wheels were propelled by the kinetic energy of moving water (fig. 4.10). They worked well in a slow but steady flow, but location on swift-flowing streams was especially desirable because the maximum theoretical power of undershot wheels is proportional to the cube of the water speed: doubling the speed boosts the capacity eightfold (box 4.5). Where the stream flow was first impounded, undershots were used only with low heads of between 1.5 and 3 m. Radial boards were later fitted with backs to prevent water shooting over the floats.

The efficiency of undershot wheels could be further improved by forming the base below the waterwheel rim into a closely fitting breast over a 30° arc at the bottom center to increase the water retention. Their most efficient design, introduced around 1800 by Jean-Victor Poncelet (1788–1867), had curved blades and could convert about 20% of water's kinetic energy into useful power; later in the century the best performance rose to 35–45%. Wheel diameters were roughly three times as large as the head for paddle wheels and two to four times for Poncelet wheels.

Breast wheels were powered by a combination of water flow and gravity fall in streams with heads between 2 and 5 m. Close-fitting breastworks, preventing premature water spillage, were essential for good performance.

Box 4.5
Power of undershot wheels

The kinetic energy of flowing water (in joules) is $0.5\rho v^2$, one-half the product of its density ($\rho = 1,000$ kg/m^3) and squared velocity (v in m/s). The number of unit volumes of water impacting at waterwheel paddles per unit of time equals the flow speed, and hence the theoretical power of the stream is equal to its energy multiplied by velocity. Water flowing at a speed of 1.5 m/s and turning vanes with a cross section of about 0.15 m^2 (roughly 50 × 50 cm) could ideally develop just over 400 W of power—but an inefficient wooden medieval undershot could actually deliver no more than a fifth of this rate, or about 80 W, as useful rotary motion.

Figure 4.10
Engravings of a large undershot wheel running a French royal paper mill (top) and of an overshot wheel powering ore-washing machinery in a French forge (bottom). Reproduced from the *Encyclopédie* (Diderot and d'Alembert 1769–1772).

Low breast designs, with water entering below the elevation of the center shaft, had efficiencies no better than those of well-designed undershot wheels. High breast machines, with water impacting above the elevation of the center shaft, approached the outputs of overshot wheels. Traditional overshot wheels, powered largely by gravitational potential energy, operated with heads over 3 m, and their diameters were usually equal to about three quarters of the head (fig. 4.10). Water was fed through troughs or flumes into bucketlike compartments at rates of less than 100 L/s to more than 1,000 L/s and at speeds of 4–12 rpm. Because most of the rotary power was generated by the weight of the descending water, overshot wheels could be placed on slowly flowing streams (box 4.6).

This advantage was partially negated by the need for a well-directed and carefully regulated water supply, which required the frequent building of storage ponds and races. Overshot wheels operating with excess carrying capacity, that is, with reduced spillage from buckets, could be more efficient, though less powerful, than machines under full flow. Until the early decades of the eighteenth century, overshot wheels were considered to be less efficient than the undershots (Reynolds 1979). This error was disproved during the 1750s in the writings of Antoine de Parcieux, and Johann Albrecht Euler, and above all by careful experiments with scale models conducted by John Smeaton (1724–1792), who compared the waterwheel capacities with those of other prime movers (Smeaton 1759).

Smeaton's subsequent promotion of efficient overshot wheels helped slow down the diffusion of steam engines, and his experiments (when he correctly concluded that a wheel's power rises with the cube of the water's velocity) set an efficiency range for overshot wheels at 52–76% (average 66%), compared to 32% for the best undershot wheels (Smeaton 1759).

Box 4.6
Power of overshot waterwheels

The potential energy of water (in joules) is equal to mgh, the product of its mass (in kg), gravitational acceleration (9.8 m/s^2), and head (height in m). Consequently, an overshot wheel bucket containing 0.2 m^3 of water (200 kg) poised 3 m above the discharge channel has a potential energy of roughly 6 kJ. With a water flow rate of 400 kg/s, the wheel would have a theoretical power of nearly 12 kW. The useful mechanical power of such a machine would have ranged from less than 4 kW for a heavy wooden wheel to well over 9 kW for a carefully crafted and properly lubricated nineteenth-century metal machine.

Denny's (2004) modern theoretical analysis of waterwheel efficiency produced very similar results: 71% for overshot wheels, 30% for undershot
wheels (and about 50% for Poncelet wheels). In practice, properly designed
and well-maintained twentieth-century overshot wheels had potential
shaft efficiencies of nearly 90% and could convert up to 85% of water's
kinetic energy to useful work (Muller and Kauppert 2004), but a generally
achievable rate was 60–70%, while the best German all-metal undershot
wheels, designed and made during the 1930s, were up to 76% efficient
(Müller 1939).

Undershot wheels could be placed directly in a stream, but such a location naturally increased the chances of flood damages. Breast wheels and
overshot wheels needed a regulated water supply. The water bypass usually
consisted of a weir across a stream and a channel diverting the flow to the
wheel. In regions of low or irregular rainfall it was common to impound
water in ponds or behind low dams. No less attention had to be paid to
returning water to the stream. Backed-up water would have impeded wheel
rotation, and smooth tail races were also needed to prevent channel silt-
ing. Even in England, wheels, shafts, and gears were almost completely
wooden until the beginning of the eighteenth century. Afterward came a
growing use of cast iron for hubs and shafts. The first all-iron wheel was
built early in the nineteenth century (Crossley 1990). Besides fixed stream
waterwheels there were also the much less common floating wheels,
installed on barges, and tidal mills. Floating grain mills were successfully
used for the first time on the Tiber in the year 537 when Rome was
besieged by the Goths, who cut off the aqueduct water turning the milling
wheels.

They were a common sight in or near cities and towns in medieval
Europe, with many remaining until the eighteenth century. The use of
intermittent power from the sea was first documented in Basra during the
tenth century. During the Middle Ages small tidal mills were built in England, the Netherlands, Brittany, and on the Atlantic coast of the Iberian
Peninsula; later came the installations in North America and the Caribbean
(Minchinton and Meigs 1980). Perhaps the most important and long-lived
tide-powered machinery supplied drinking water for London. The first large
vertical tidal waterwheels, built after 1582, were destroyed by the 1666 fire,
but their replacements operated until 1822 (Jenkins 1936). Three wheels,
driven by water passing through the narrowed arches of the old London
Bridge, turned in either direction (other wheels usually worked only with
the ebbing tide) and powered 52 water pumps forcing 600,000 L of water to
a height of 36 m.

Grain milling remained the dominant use of water power: in medieval England it accounted for about 90% of all milling activity, with most of the rest used for cloth fulling (fluffing up and thickening woolens) and only 1% for other industrial activities (Lucas 2005). The late Middle Ages saw the widespread use of water power in ore crushing and smelting (blast furnace bellows) and in stone and wood sawing, wood turning, oil pressing, papermaking, tanning, wire pulling, stamping, cutting, metal grinding, blacksmithing, majolica glazing, and polishing. English waterwheels were also used for winding and water pumping in underground mines (Woodall 1982; Clavering 1995).

All of these tasks were done by waterwheels with a higher efficiency than people or animals could provide, and hence also with much increased labor productivity. Moreover, the unprecedented magnitude, continuity, and reliability of power provided by waterwheels opened up new productive possibilities. This was especially true in mining and metallurgy. Indeed, the energy foundations of Western industrialization rest to a significant degree on these specialized uses of waterwheels. Human and animal muscles could never convert energy at such high, concentrated, continuous, and reliable rates—but only such deliveries could increase the scale, speed, and quality of countless food-processing and industrial tasks. Yet it took a long time for typical waterwheels to reach capacities surpassing the power of large harnessed animal teams.

For centuries, the only way to achieve larger power outputs was to install a series of smaller units in a suitable location. The best-known example of this concentration is the famous Roman mill line at Barbegal, near Arles, which had 16 wheels, each with about 2 kW of capacity, for a total of just over 30 kW (Sellin 1983). Greene (2000, 39) called it "the greatest known concentration of mechanical power in the ancient world," and Hodge (1990, 106) described it as "something that, according to all the textbooks, never existed at all—an authentic, ancient Roman, power-driven, mass-production, assembly-line factory." A closer look shows a less impressive reality (box 4.7).

In any case, larger water mills remained rare for centuries to come. Even during the early decades of the eighteenth century European waterwheels averaged less than 4 kW. Only a few machines surpassed 7 kW, and crude finishing and poor (high-friction) gearing resulted in low conversion efficiencies. Even the most admired machines of that time—14 large waterwheels (12 m in diameter) built on the Seine at Marly between 1680 and 1688—fell short of the intended task of pumping water for 1,400 fountains and cascades in Versailles. The site's potential was nearly 750 kW,

Box 4.7
Barbegal waterwheels

Water for the Barbegal's 16 overshot wheels (most likely built during the early second century CE) was diverted from a nearby aqueduct into two parallel channels on 30° slope (Benoit 1940). Sagui (1948) used highly unrealistic assumptions (a water flow of 1,000 L/s, a speed of 2.5 m/s, an average productivity of 24 t of flour a day) to conclude that the establishment produced enough flour to make bread for about 80,000 people. But Sellin (1983) used more realistic numbers (a water flow of 300 L/s, a speed around 1 m/s), and estimated that each wheel had about 2 kW of useful power, hence a total of 32 kW and (with a 50% capacity factor) a daily output of 4.5 t of flour.

But Sellin adopted Sagui's assumption of 65% of the kinetic energy of water getting converted to kinetic energy of a rotating millstone—while Smeaton's (1759) careful calculations showed 63% to be the maximum efficiency of far better-designed eighteenth-century overshot wheels. The combination of a lower flow—Leveau (2006) argued for 240 or 260 L/s—and a lower efficiency (say, 55%) would translate to 1.5 kW/unit. That would be equal to the combined power of three (or four weak) Roman horses harnessed to a whim, and enough to produce daily about 3.4 t of flour to feed about 11,000 people: certainly a much higher performance than for typical mills of the second century CE but less than a prototype of mass production.

but inefficient transmission of rotary motion (through the use of long reciprocating rods) reduced the useful output to only about 52 kW, not enough to supply every fountain (Brandstetter 2005).

But even small waterwheels had a major economic impact. Even when assuming that flour supplied half of an average person's daily food energy intake, a small water mill, manned by fewer than 10 workers, would grind enough in a day (10 hours of milling) to feed some 3,500 people, a fair-sized medieval town, while hand milling would have required at least 250 laborers. And when combined with innovative machine design, the late eighteenth-century waterwheels made an enormous difference in productivity. A perfect example is the introduction of water-powered machinery for cutting and heading 200,000 nails a day, patented in the United States in 1795 (Rosenberg 1975). The widespread use of these machines brought nail prices down by nearly 90% during the next 50 years.

Waterwheels were the most efficient traditional energy converters. Their efficiencies were superior even when compared to the best steam engines, whose operation converted less than 2% of coal into useful power by 1780, and usually no more than 15% even by the end of the nineteenth century (Smil 2005). No other traditional prime mover could deliver so much continuous power. Waterwheels were indispensable during the early stages of both European and North American industrialization. Waterwheels reached their apogee—whether evaluated in terms of individual or total capacities or in terms of efficiency of design—during the nineteenth century, at the same time that steam engines were being adopted for new stationary and transportation uses, and the rise and eventual dominance of the new prime mover overshadowed the importance of water power.

But more water power capacity was added during the first six decades of the nineteenth century than ever before, and most of these machines continued to operate even as steam power, and later electricity, were conquering the prime mover markets. Daugherty (1927) estimated that in the United States in 1849, total installed capacity in waterwheels was nearly 500 MW (<7% of all prime movers, including working animals but excluding human labor) compared to about 920 MW installed capacity in steam engines. Comparing actual work performed is more revealing: Schurr and Netschert (1960) calculated that in 1850, U.S. waterwheels delivered about 2.4 PJ, or 2.25 times the total for coal-powered steam engines; that they were still ahead (by about 30%) in 1860; and that their useful work was surpassed by steam power only during the late 1860s. As recently as 1925, 33,500 waterwheels were in operation in Germany (Muller and Kauppert 2004), and some European wheels worked even after 1950.

The new nineteenth-century large textile plants were especially dependent on water power. For example, the Merrimack Manufacturing Company, America's first fully integrated clothmaker (mainly of calico fabrics), opened in 1823 in Lowell, Massachusetts, and relied on about 2 MW of water power from a large (10 m) drop in the Merrimack River (Malone 2009). By 1840 the largest British installation—the 1.5 MW Shaw's waterworks at Greenock, near Glasgow, on the Clyde—had 30 wheels built in two rows on a steep slope and fed from a large reservoir. The largest individual waterwheels had diameters around 20 m, a width of 4–6 m, and capacities well above 50 kW (Woodall 1982).

The world's largest wheel was the Lady Isabella, designed by Robert Casement and built in 1854 by the Great Laxey Mining Company on the Isle of Man to pump water from the Laxey mines. The wheel was a pitchback overshot machine (2.5 rpm) with a diameter of 21.9 m and a width of 1.85 m; its 48 spokes (9.75 m long) were wooden, but the axle and diagonal drawing rods were made of cast iron (Reynolds 1970). All the streams on the slope above the wheel were channeled into the collecting tanks, and the water was then piped into the base of the masonry tower and rose into a wooden flume. The power was transmitted to the pump rod, reaching 451 m to the bottom of the lead-zinc mine shaft, by the main-axle crank and by 180 m of timber connecting rods. The wheel's theoretical peak power was about 427 kW. In normal operation it generated about 200 kW of useful power. The wheel worked until 1926 and was restored after 1965 (Manx National Heritage 2015) (fig. 4.11).

But the era of giant waterwheels was short-lived. Just as these machines were being built during the first half of the nineteenth century, the development of water turbines brought the first radical improvement in water-driven prime movers since the introduction of vertical wheels centuries earlier. Benoît Fourneyron's first reaction turbine with radial outward flow was built in 1832 to power forge hammers in Fraisans. Even with a very low head of just 1.3 m and a rotor diameter of 2.4 m, it had a capacity of 38 kW. Five years later two improved machines working at the Saint Blaisien spinning mill rated at about 45 kW under heads of 108 and 114 m (Smith 1980).

Figure 4.11
The Great Laxey waterwheel after restoration (Corbis).

The performance of Fourneyron's machine was soon surpassed by an innovative design of an inward-flow turbine, a product that Layton (1979) called a prototypical industrial research product but now is generally known as the Francis turbine, named after James B. Francis (1815–1892), a British American engineer. Later came the jet-driven turbines of Lester A. Pelton (patented in 1889) and the axial flow turbines of the Viktor Kaplan type (in 1920). New turbine designs replaced waterwheels as the prime movers in many industries. For example, in Massachusetts they accounted for 80% of installed power by 1875. That was also the time of the greatest importance of water-driven machines in a rapidly industrializing society.

For example, each of the three leading textile mill centers on the lower Merrimack River in Massachusetts and southern New Hampshire, Lowell, Lawrence, and Manchester, had water machines totaling about 7.2 MW. The whole river basin had about 60 MW of installed capacity, averaging some 66 kW per manufacturing establishment (Hunter 1975). Even in the mid-1850s steam was still about three times more expensive as a prime mover in New England than water. The era of water turbines as direct prime movers rotating geared and belted shafts ended rather abruptly. By 1880 large-scale coal mining and more efficient engines had made steam cheaper than water power virtually anywhere in the United States. Before the end of the nineteenth century most water turbines had stopped delivering direct power and had started to turn electricity generators instead.

Wind Power

The history of harnessing wind for stationary power (as opposed to the much longer history of converting wind into motion through the clever use of sails) and the evolution of windmill designs toward complex and powerful machines of the early industrial era have been well covered by both general and specific national reviews. Notable contributions in the first category are those of Freese (1957), Needham (1965), Reynolds (1970), Minchinton (1980), and Denny (2007). Important national surveys are Skilton (1947) and Wailes (1975) on British mills, Boonenburg (1952), Stockhuyzen (1963), and Husslage (1965) on the much-talked-about Dutch designs, and Wolff (1900), Torrey (1976), Baker (2006), and Righter (2008) on the American machines, which played such a key but underappreciated role in opening the West. Windmills became the most powerful prime movers of the preindustrial era in flatlands where almost nonexistent water heads precluded the construction of small waterwheels (in the Netherlands,

Denmark, and parts of England) and in a number of arid Asian and European regions with seasonally strong winds.

The contribution of windmills to worldwide economic intensification was less decisive than that of waterwheels, mainly because their use eventually became common only in parts of Atlantic Europe. The first clear records of European windmills come from the last decades of the twelfth century. According to Lewis (1993), their use spread first from Persia to the Byzantine territory, where they were transformed into vertical machines, encountered by the Crusaders. Unlike the Eastern machines, whose sails rotated in a horizontal plane around a vertical axis, these mills were vertically mounted rotaries on a horizontal axis whose driving shafts could be turned into the wind. With the exception of Iberian octagonal sail mills working with with triangular cloth (an import from the eastern Mediterranean), early European machines were all post mills. Their wooden structure, housing gears, and millstones pivoted on a massive central post that was supported by four diagonal quarter-bars (fig. 4.12). Because they could not realign themselves once the wind direction changed, they had to be turned to face the wind. They were also unstable in high winds and vulnerable to storm damage, and their relatively low height limited their highest performance (box 4.8).

While post mills continued to work in parts of Eastern Europe until the twentieth century, in Western Europe they were gradually replaced by tower mills and smock mills. In both these designs only the top cap was turned into the wind, either from the ground or, with tall towers, from galleries. Smock mills had a wooden frame, usually octagonal, that was covered with clapboards or shingles. Tower mills were typically rounded, tapering stone structures. Only after 1745 did the English introduction of a fantail to power a winding gear start to turn the sails automatically into the wind. Curiously, the Dutch, with the largest number of windmills in Europe, adopted this innovation only in the early nineteenth century.

But Dutch millers were the first ones to introduce more efficient blade designs. They started to add canted leading-edge boards to previously flat blades around 1600. The resulting arching (camber) gave blades more lift while reducing drag. Later innovations included improvements in sail mounting, cast-metal gearings, and a centrifugal regulating governor. This device did away with the difficult and often dangerous task of adjusting the canvas to different wind speeds. By the end of the nineteenth century the English were starting to install true airfoils, aerodynamically contoured blades with thick leading edges. Grain milling and water pumping (also on ships, with small portable machines) were the most common applications.

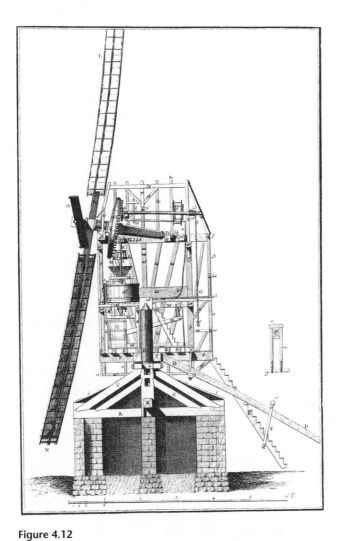

Figure 4.12
Post windmill. The main wooden, almost always oaken, post on
which the whole structure was balanced was held up by four
quarter-bars attached to massive cross-trees. Windmill rotations
were transferred to the millstone by a lantern-and-crown gear
and the only access was by ladder. Reproduced from the *Encyclopédie* (Diderot and d'Alembert 1769–1772).

Box 4.8

Wind energy and power

Average wind speed increases roughly with one-seventh the power of height. This means, for example, that it will be about 22% higher 20 m above the ground than at a height of 5 m. The kinetic energy of 1 m^3 of air (in joules) is equal to $0.5\rho v^2$, where ρ is air density (about 0.12 kg/m^3 near the ground) and v is its average wind speed (in m/s). Wind power (in watts) is the product of wind energy, an area perpendicular to the wind direction swept by the machine's blades (A, in m^2), and wind speed cubed: $0.5\rho A V^3$. As the wind power goes up with the cube of the average speed, doubling the speed increases the available power eightfold. Early (relatively heavy and poorly geared) windmills also needed winds of at least 25 km/h (7 m/s) in order to start milling or pumping; at lower speeds they just turned slowly, but sails had to be trimmed at wind speeds above 10 m/s (and furled at speeds above 12 m/s), providing only a narrow window (5–7 hours of daily rotation) for useful work (Denny 2007).

These realities obviously favored locations with sustained brisk winds. Later, more efficient, smoothly geared, and properly lubricated designs would work well with winds above 4 m/s, delivering 10–12 hours of useful operation a day. Preindustrial societies could capture only wind flowing near the ground, with the spans of most of the windmills being less than 10 m. Wind flows also have large temporal and spatial variation. Even in windy places the annual wind speed means fluctuate by up to 30%, and shifting a machine's location by just 30–50 m may cut or increase the average speed easily by half. The limited capacities of preindustrial land transport precluded location at the windiest sites, and the mills were often motionless. No wind machine can extract all of the available wind power: this would require a complete stopping of the airstream! The maximum extractable power is equal to 16/27, or nearly 60%, of the kinetic energy flux (Betz 1926). The actual performance was 20–30% for preindustrial windmills. An eighteenth-century tower mill with a blade diameter of 20 m thus had a theoretical power of about 189 kW with a velocity of 10 m/s—but it delivered less than 50 kW.

Windmills were also used both in Europe and in the Islamic world in grinding and crushing (of chalk, sugar cane, mustard, cocoa), papermaking, sawing, and metalworking (Hill 1984).

In the Netherlands they did all of these things, but their greatest contribution was in draining the country's low-lying land and reclaiming polders for crop fields. The first Dutch drainage mills date from the early fifteenth century, but they became common only in the sixteenth century. Hollow-post *wipmolen* turned big wooden wheels with scoops, and smaller mobile *tjasker* rotated Archimedean screws, but only efficient smock mills could deliver the power required for large-scale reclamation of polders. Zaanse Schans in North Holland had 600 windmills built after 1574, a few of them preserved (Zaanse Schans 2015). The tallest Dutch windmills (33 m) were in Schiedam (five of the original 30 are still standing), grinding grain for the production of jenever (Dutch gin).

Old American windmills, such as those in coastal Massachusetts, were often used for salt extraction, but their numbers remained low. New American windmills appeared right after the middle of the nineteenth century with the westward expansion across the Great Plains, where the scarcity of small streams and the erratic rainfall precluded use of small waterwheels but where the shortage of natural springs required pumping water from wells. Rather than extracting the power in the manner of heavy (and expensive) Dutch mills (that is, by using a few large and wide sails), American windmills were smaller, simple, affordable and yet efficient machines serving individual railway stations and farms.

They consisted usually of a large number of fairly narrow blades or slats that were fastened to solid or sectional wheels and were equipped either with the centrifugal or the side-vane governor and independent rudders. Placed on top of lattice towers 6–25 m tall, they were used to pump water for households, cattle, or steam locomotives (fig. 4.13). These windmills, barbed wire, and railroads were the iconic artifacts that helped open up the Great Plains (Wilson 1999). Daugherty's (1927) estimates show the U.S. nationwide capacity of windmills rising from about 320 MW in 1849 to nearly 500 MW in 1899 and peaking at 625 MW in 1919.

We have no information on the capacities of early windmills. The first reliable experimental measurements date from the late 1750s, when John Smeaton put the power of a common Dutch mill with 9 m sails equal to the power of ten men or two horses (Smeaton 1759). This calculation, based on measurements with a small model, was corroborated by actual performance in oilseed pressing. While the windmill-powered runners turned seven times a minute, two horses made scarcely 3.5 turns in the same time. A

Figure 4.13
Halladay windmill. During the last decade of the nineteenth
century, Halladay windmills were the most popular American
brand. They were a common sight at western railway stations,
where they pumped water for steam locomotives. Reproduced
from Wolff (1900).

typical large eighteenth-century Dutch mill with a 30 m span could develop about 7.5 kW (Forbes 1958). Modern measurements at a well-preserved 1648 Dutch drainage mill capable of lifting 35 m^3 with 8–9 m/s winds indicated a windshaft power of about 30 kW, but large transmission losses lowered the useful output to less than 12 kW.

All of these results confirm Rankine's comparison of traditional prime movers. He credited post windmills with 1.5–6 kW of useful power and tower mills with 4.5–10.5 kW (Rankine 1866). Measurements of American windmills put their useful power from a mere 30 W for 2.5 m mills to as much as 1,000 W for large 7.6 m machines (Wolff 1900). Typical ratings (in terms of useful power) were 0.1–1 kW for the nineteenth-century American designs, 1–2 kW for small and 2–5 kW for large post mills, 4–8 kW for common smock and tower mills, and 8–12 kW for the largest nineteenth-century machines. This means that typical medieval windmills were as powerful as contemporaneous waterwheels, but by the early nineteenth century many waterwheels were up to five times more powerful than the largest tower mills, and that difference only grew with the subsequent development of water turbines.

As in the case of waterwheels, the contribution of windmills as providers of stationary power peaked during the nineteenth century. In the UK their total reached 10,000 in 1800; during the late nineteenth century 12,000 worked in the Netherlands while18,000 worked in Germany; and by 1900 about 30,000 mills (with a total capacity of 100 MW) were installed in countries around the North Sea (De Zeeuw 1978). In the United States, several million units were erected between 1860 and 1900 during the country's westward expansion, and their numbers began to decline only in the early 1920s. By 1889 there were 77 manufacturers, with Halladay, Adams, and Buchanan being the leading brands (Baker 2006). Large numbers of American-type water-pumping windmills were used during the twentieth century in Australia, South Africa, and Argentina.

Biomass Fuels

Nearly all traditional societies could produce heat and light only by burning biomass fuels. Woody phytomass, the charcoal derived from it, crop residues, and dried dung provided all the energy needed for household heating, cooking, and lighting and for small-scale artisanal manufactures; later, in larger proto-industrial enterprises, those fuels were used in firing relatively large quantities of bricks and ceramics, making glass, and smelting and shaping metals. The only notable exceptions were found in ancient

China, where coal was used in the north in iron making and natural gas was burned in Sichuan to evaporate brines and produce salt (Adshead 1997), and in medieval England (Nef 1932).

The provision of biomass fuels could be as easy as making an everyday short trip to a nearby forest or bush or mountainside to collect fallen branches and break off dry twigs, or to gather dry grasses, or to gather some dry straw after the grain harvest and store it under house eaves. More often, however, it could entail long walks, mostly by women and children, to gather combustible biomass; laborious tree cutting; exhausting charcoal making; and the long-distance transport of the fuel in ox-drawn carts or in camel caravans to cities in the middle of long-deforested plains or in desert regions. The abundance or scarcity of fuel affected house design, as well as dressing and cooking practices. Provision of these energies was one of the principal reasons for traditional deforestation.

In West European countries this dependence diminished rapidly after 1850. The best reconstructions of the primary energy supply show that in France, coal began to provide more than half of all fuel energy by the mid-1870s, and in the United States, coal and oil (and a small volume of natural gas) surpassed the energy content of fuelwood by 1884 (Smil 2010a). But elsewhere the dependence on phytomass fuels continued well into the twentieth century: in the most populous nations of Asia it remained dominant until the 1960s or 1970s, and in sub-Saharan Africa it remains the single largest source of primary energy.

This ongoing use has allowed us to study the modalities and consequences of the inefficient combustion of traditional fuels and its widespread health impact, and observations and analyses done in recent decades (Earl 1973; Smil 1983; RWEDP 2000; Tomaselli 2007; Smith 2013) thus help us understand the long history of preindustrial biomass combustion. Many of these recent findings are perfectly applicable to preindustrial settings because the basic needs have not changed: for most people in traditional societies, energy needs have always amounted just to cooking two or three meals a day, in cold climates heating at least one room, and in some regions also preparing feed for animals and drying food.

Wood and Charcoal
Wood was used in any available form: as fallen, broken, or lopped-off branches, twigs, bark, and roots—but chopped stem wood became available only where good cutting tools—adzes, axes, later saws—were common. Wood variety made surprisingly little difference. There are thousands of woody plants, and though their physical differences are substantial—the

specific density of some oaks is almost twice as high as that of some poplars—their chemical composition is remarkably uniform (Smil 2013a). Wood is about two-fifths cellulose, roughly one-third hemicelluloses, and the rest is lignin; in elemental terms, carbon accounts for 45–56% and oxygen for 40–42% of the total mass. The energy content of wood rises with shares of lignin and resins (these contain, respectively, 26.5 MJ/kg and up to 35 MJ/kg, compared to 17.5 MJ/kg for cellulose), but the differences among common woody species are fairly small, mostly 17.5–20 MJ/kg for hardwoods and, because of their higher resin content, 19–21 MJ/kg for softwoods (box 4.9).

The energy density of wood should always refer to absolutely dry matter, but wood burned in traditional societies had a widely varying moisture content. Freshly cut mature hardwoods (leafy trees) are typically 30% water, while softwoods (conifers) are well over 40%. Such wood burns inefficiently as a significant part of the released heat goes into vaporizing released moisture rather than heating a cooking pot or a room. When wood has more than 67% of moisture it will not ignite. That is why dry fallen branches and twigs or hacked-off pieces of dead trees were always preferable to fresh wood, and why wood was usually air-dried before combustion. Cut wood was stacked, sheltered, and let dry for at least a few months, but even in dry climates it still retained about 15% moisture. In contrast, charcoal contains only a trace of moisture, and it was a biomass fuel always preferred by those who could afford its price.

This high-quality fuel is virtually smokeless, and its energy content, equal to that of good bituminous coal, is roughly 50% higher than that of

---

Box 4.9
Energy content of biomass fuels

Biomass fuel	Water content (%)	Energy content of dry matter (MJ/kg)
Hardwoods	15–50	16–19
Softwoods	15–50	21–23
Charcoal	<1	28–30
Crop residues	5–60	15–19
Dry straws	7–15	17–18
Dried dung	10–20	8–14

Source: Based on Smil (1983) and Jenkins (1993).

air-dried wood. Charcoal's other main advantage is its high purity. Because it is virtually pure carbon, it contains hardly any sulfur or phosphorus. This makes it the best possible fuel not only for indoor uses but also in kilns producing bricks, tiles, and lime and in the smelting of ores. A further advantage for smelting is charcoal's high porosity (with a specific density of merely 0.13–0.20 g/cm^3), facilitating the ascent of reducing gases in furnaces (Sexton 1897). But traditional production of this excellent fuel was very wasteful.

Partial combustion of the heaped wood inside primitive earth or pit kilns generates the heat necessary for carbonization. Consequently, there is no need for additional fuel, but both the quality and the quantity of the final products are difficult to control. Typical charcoal yields in such kilns were only between 15% and 25% those of air-dried wood. This means that about 60% of the original energy was lost in making charcoal, and in volumetric terms up to 24 m^3 of wood (and no less than 9–10 m^3) were required to make 1 t of charcoal (fig. 4.14). But the payoff was in the fuel's quality: its combustion could produce temperature of 900°C, and with a supplementary air supply, achieved most efficiently by using bellows, that could be raised to nearly 2000°C, more than enough to melt even iron ores (Smil 2013a).

The harvesting of wood for fuel (as well as for construction and shipbuilding) led to widespread deforestation, and the cumulative effect reached worrisome levels in previously heavily wooded regions. At the beginning of the eighteenth century about 85% of Massachusetts was covered by forests, but by 1870 only about 30% of the state was covered by trees (Foster and Aber 2004). Not surprisingly, on March 6, 1855, Henry David Thoreau (1817–1862) wrote in his *Journals* that

> our woods are now so reduced that the chopping this winter has been a cutting to the quick. At least we walkers feel it as such. There is hardly a wood-lot of any consequence left but the chopper's axe has been heard in it this season. They have even infringed fatally on White Pond, on the south of Fair Haven Pond, shaved off the top-knot of the Cliffs, the Colburn farm, Beck Stow's, etc., etc. (Thoreau 1906, 231)

Studies of traditional societies that remained dependent on biomass fuels into the second half of twentieth century indicate annual fuel requirements of less than 500 kg/capita in the poorest villages of tropical regions. Up to five times as much biomass was used in latitudes with pronounced winters and with a substantial wood-based production of bricks, glass, tiles, and metals and evaporation of brines. In Germany up to 2 t of wood (almost

Figure 4.14
Charcoal production started with leveling the ground and setting up the central pole; cut wood was stacked around it and covered up by clay before ignition. Reproduced from the *Encyclopédie* (Diderot and d'Alembert 1769–1772).

all of it burned to obtain potassium rather than to produce heat) were needed for to make 1 kg of glass, while evaporation of brines in large wood-heated iron pans consumed up to 40 kg of wood per 1 kg of salt (Sieferle 2001).

There are no records of typical biomass fuel consumption during antiquity, and only a few reliable quantities were recorded for some medieval societies. I estimated that average annual energy requirements in the Roman Empire around 200 CE added up to 650 kg/capita, that is, roughly 10 GJ, or about 1.8 kg/day (box 4.10). The best available reconstruction of firewood demand in medieval London (around 1300) resulted in an annual mean of about 1.75 t of wood, or roughly 30 GJ/capita (Galloway, Keene, and Murphy 1996). Estimates for Western Europe and North America just before their switch to coal show even higher average needs.

Those nineteenth-century northern European, New England, Midwestern, or Canadian communities heating and cooking only with

---

Box 4.10
Wood consumption in the Roman Empire

My conservative estimate accounts for all major wood consumption categories (Smil 2010c). Bread and stews were the Roman staple, and urban *pistrinae* and *tabernae* needed at least 1 kg of wood per day per capita. At least 500 kg of wood per year were needed for space heating, which was required for the roughly one-third of the empire's population that lived beyond the warm Mediterranean in temperate climates. To this must be added an average annual per capita consumption of 2 kg of metals, which required about 60 kg of wood per kilogram of metal. This adds up to 650 kg/capita (roughly 10 GJ, about 1.8 kg/day), but as the Roman combustion efficiencies were uniformly low (<15%), useful energy derived from burning that wood was only on the order of 1.5 GJ/year, an equivalent of nearly 50 L, or one tankful, of gasoline.

For comparison, when Allen (2007) constructed his two Roman household consumption baskets, he assumed an average consumption of nearly 1 kg of wood per capita per day for what he called a respectable alternative, and just 0.4 kg/capita for a bare-bones budget, but his rates excluded fuel used for metallurgy and artisanal manufacturing. And Malanima (2013a) put the average per capita wood consumption in the early Roman Empire at 4.6–9.2 GJ/year, half of the total energy use, with the other half split roughly 2:1 between food and fodder energy. His higher total was 16.8 GJ/capita, while my estimate for food, fodder, and wood was 18–19 GJ/capita (Smil 2010c).

wood consumed annually anywhere between 3 and 6 t of the fuel per capita. That was the range of German household consumption during the eighteenth century (Sieferle 2001). The Austrian mean in 1830 was close to 5 t/capita (Krausmann and Haberl 2002), and so was the American nationwide average in the middle of the nineteenth century (Schurr and Netschert 1960). Although that figure also included growing industrial (mainly metallurgical charcoal) and transportation uses, household combustion was still the leading consumer of American wood during the 1850s.

Crop Residues and Dung

Crop residues were indispensable fuels on deforested, densely settled agricultural plains and in arid, sparsely treed regions. Cereal straws and stalks were usually the most abundant, but many other residues were locally and regionally important. These included legume straws and tuber vines, cotton stalks and roots, jute sticks, sugar cane leaves, and branches and twigs pruned from fruit trees. Some crop residues needed drying before combustion. Ripe straws are only between 7% and 15% water, and their energy content is comparable to that of deciduous trees (hardwood).

But their density is obviously much lower, and so storing enough straw to last through the winter could never be as easy as stacking chopped wood. The low density of crop residues also meant that open fires and simple stoves had to be stoked almost constantly. Because of a number of competitive nonenergy uses, crop residues were often in short supply. Legume residues were an excellent high-protein feed and fertilizer. Cereal straw makes a good ruminant food and animal bedding; many societies (including England and Japan) used it for thatching house roofs, and it was also a raw material for manufacturing simple tools and domestic articles.

Consequently, every bit of combustible phytomass was often gathered for household use. Throughout the Middle East thorny shrubs were often burned, and date kernels were used to make charcoal. On the North China Plain women and children with rakes, sickles, baskets, and bags collected fallen twigs, leaves, and dry grasses (King 1927). And in the interior of Asia, as well as throughout the Indian subcontinent, parts of the Middle East, Africa, and both Americas, dried dung was the most important source of heat for cooking. Air-dried dung's heat value is comparable to that of crop residues or grasses (box 4.9).

A little-appreciated reality is dung's essential contribution to America's westward expansion (Welsch 1980). Wild buffalo and cattle dung made possible the early continental crossings and the subsequent colonization of the Great Plains during the nineteenth century. Travelers on the Oregon and Mormon Trails collected "buffalo wood," and the early settlers stacked winter supplies in igloo shapes or against house walls. Known as cow wood or Nebraska oak, the fuel burned evenly and with little smoke and odor, but its rapid combustion required almost continuous stoking. In South America llama dung was the principal fuel on the altiplano of the Andes, the core of the Inca Empire in southern Peru, eastern Bolivia, and northern Chile and Argentina (Winterhalder, Larsen, and Thomas 1974). Cattle and camel dung was used in the Sahelian region of Africa, as well as in Egyptian villages. Cattle dung was gathered in largest quantities in both arid and monsoonal Asia, and Tibetans always relied on yak dung. Only sheep dung has generally been avoided because its burning produces acrid smoke.

In India, where dung use is still common in many rural areas, both cow and water buffalo droppings are gathered regularly, mostly by *harijan* (untouchable) children and women, both for their own household use and for sale (Patwardhan 1973). Dung was (and is) collected either as dry chips or as a fresh biomass. Fresh dung is mixed with straw or chaff, hand-molded into patties and cakes, and sun-dried in rows, plastered on house walls, or stacked in piles (fig. 4.15). A recent survey of rural energy use in South Asia found that 75% of Indian, 50% of Nepali, and 47% of

Figure 4.15
Rows and piles of cow patties left to dry in Varanasi, Uttar Pradesh, India (Corbis).

Bangladeshi households are still using dung for cooking (Behera et al. 2015).

Household Needs

An ancient Chinese proverb had the right order of things that people cannot do without every day: firewood, rice, oil, salt, sauce, vinegar, and tea. In traditional agricultural societies where grains supplied most of the food energy, their cooking (by steaming, boiling, or baking) was necessary to make the hard seeds edible. But before the grains (stored in baskets, jars, or bins) could be cooked, they had to be processed, and grain milling has been a virtually universal, and also historically almost always the first, processing step; the extraction of oils by pressing a variety of seeds, fruits, and nuts came later. Tubers were processed to remove antinutritive factors or to enable their long-term storage, and sugar cane was crushed to express its sweet juice. In all of these tasks, human energy was only gradually augmented by animal labor.

As already noted, the first use of inanimate power in grain milling—horizontal waterwheels rotating small millstones—is about two millennia old.

Cooking required relatively little heat energy in East Asian stir-frying and steaming. In contrast, considerable fuel inputs were needed for baking bread, the staple throughout the rest of the Old World, and for the roasting done commonly in the Middle East, Europe, and Africa. In some societies fuel was also required to prepare feed for domestic animals, above all for pigs. Seasonal heating was necessary in the midlatitudes, but (save for sub-Arctic locations) preindustrial houses were usually heated only for short periods and to relatively low temperatures.

In some fuel-short regions there was no winter heating at all despite months of cold weather: there was no heating in the deforested lowlands of Ming and Qing China south of the Yangzi. But the northernmost parts of the Jiangnan (China south of the Yangzi) have mean January and February temperatures just between 2 and 4°C, with minima going below −10°C. And the chill of traditional English interiors, even after the introduction of coal stoves, is proverbial. The total household energy needs of East Asian or Middle Eastern societies were thus very low. The absolute fuel demand of some northern European and colonial North American societies was rather high, but low combustion efficiencies resulted in relatively low shares of useful heat. Consequently, even in nineteenth-century America, endowed with plenty of fuelwood, an average household claimed only a small fraction of the useful energy flows that became available to its twentieth-century counterpart.

Food Preparation

In light of the dominance of cereals in the nutrition of all high cultures, milling of grains was certainly the most important food-processing need in history. Whole grain is not very palatable; it is difficult to digest and, obviously, it could not be used for baking. Milling produces flours of various fineness that could be used for the preparation of highly digestible foods, above all breads and noodles. The evolutionary sequence of grain milling started with slightly hollowed rubbing stones and stone pestles and mortars. An oblong, oval saddlestone worked from a kneeling position was common in ancient Middle Eastern societies, as well as in preclassical Europe.

Push mills with hoppers and grooved bedstones were the first major innovation. The Greek hourglass mill had a cone-shaped hopper and a conical grinder. The productivity of muscle-driven processing was very low (Moritz 1958). Tedious labor with rubbing stones or mortars and pestles would yield no more than 2–3 kg of roughly ground flour per hour Two Roman slaves laboriously grinding flour with rotary *mola manualis* (used since the third century BCE) could produce less than 7 kg of coarse flour per hour. The more efficient *mola asinalis* (known as the Pompeian mill, confined to cities and towns) was made from rough volcanic rock, with the *meta* (the lower cylindrical part) covered by the hourglass *catillus*, which was rotated by a harnessed donkey walking in a tight circle, though in confined places slaves were commonly used, and slaves also powered dough-kneading machines in large bakeries: the empire's staple was paid for by terrible suffering (box 4.11).

Box 4.11

Lucius Apuleius (*Metamorphoses* IX, 12, 3.4) on Roman mill slaves

Ye gods, what a set of men I saw! Their skins were seamed all over with marks of the lash, their scarred backs were shaded rather than covered with tattered frocks. Some wore only aprons, all were so poorly clothed that their skin was visible through the rents in their rags! Their foreheads were branded with letters, their heads were half-shaved. They had irons on their legs. They were hideously sallow. Their eyes were bleared, sore, and raw, from the smoke of the ovens. They were covered with flour as athletes with dust! (J. A. Hanson translation)

A donkey-powered mill (energy input at the rate of 300 W) produced from less than 10 kg/h to 25 kg/h (Forbes 1965), while millstones driven by a small waterwheel (1.5 kW) would grind flour at rates between 80 and 100 kg/h. Flour would have been used to bake bread supplying at least half of all food energy in average dietary intake (but bread's share was often more than 70%). Consequently, a single mill would have produced enough flour in a 10-hour shift to feed 2,500–3,000 people, a fair-sized medieval town. Millstones could be rotated directly by horizontal waterwheels, but all vertical waterwheels and all windmills required the reasonably efficient transmission of rotary power by wooden gears. And no mill could produce good flour without accurately set and well-dressed millstones, the top runner and the stationary bedstone (Freese 1957). By the eighteenth century the stones were usually 1–1.5 m in diameter, up to 30 cm thick, weighed close to 1 t, and rotated 125–150 times a minute. Grain was fed from the hopper into the opening (eye) of the runner, and it was crushed and milled between lands, the stones' flat surfaces.

These massive stones had to be precisely balanced. If they rubbed against each other they could be badly damaged, and they could also spark a fire. If they were too far apart they produced rough meal rather than fine flour. Tolerance requirements were no more than the thickness of a heavy brown paper between the stones at the eye—and that of a tissue paper at the edge. Ground flour and milling by-products were channeled outward along incised grooves (furrows). Skilled craftsmen used sharp tools (mill bills) to deepen these furrows (dress the stone). They did this at regular intervals determined by the quality of the stone and the rate of milling, usually every two to three weeks. Solid granites or hard sandstones, or pieces of cellular quartz (buhrstones) cemented together and held by iron hoops, were the most common millstone choices, and none could do a perfect job in a single run. After the coarse bran was separated from the fine flour the intermediate particles were reground. The whole process could be repeated several times. Final sieving (bolting) separated flour from bran and the flours into different grades.

For centuries, milling with water or wind still required a great deal of heavy labor. Grain had to be unloaded and hoisted with pulleys to hoppers; the freshly milled flour had to be cooled by raking, sorted by sieving, and bagged. Sieves driven by water power were introduced during the sixteenth century. A fully automatic flour mill was first designed only in 1785 by an American engineer, Oliver Evans, who proposed using endless bucketed belts to lift grain and augers (Archimedean screws) to transport it horizontally and to spread the freshly ground flour for cooling. Evans's invention was not an immediate commercial success, but his self-published book on milling became a classic of the genre (Evans 1795).

The history of cooking shows remarkably few advances until the onset of the industrial era. Open hearths and fireplaces were used for roasting (in the fire or on spits, skewers, or gridirons), boiling, frying, and stewing. Braziers were used for boiling water and for grilling, and simple clay or stone ovens were used for baking. Flat breads were stuck to the sides of clay ovens (still the only way to bake proper Indian naan) and leavened breads were placed on flat surfaces. Fuel shortages contributed to the introduction of low-energy cooking methods. The Chinese used cooking pots on three hollow legs (*li*) already before 1500 BCE. Shallow sloping pans—Indian and Southeast Asian *kuali*, and the Chinese *kuo*, better known in the West as the Cantonese wok—sped up frying, stewing, and steaming (E. N. Anderson 1988).

The origin of kitchen stoves remains uncertain, but their wide acceptance obviously required the construction of chimneys. Even in the richest parts of Europe they were uncommon before the beginning of the fifteenth century as people continued to rely on smoky, inefficient fireplaces (Edgerton 1961). Many Chinese clay or brick stoves still did not have chimneys during the first decades of the twentieth century (Hommel 1937). Iron stoves fully enclosing the fire started to replace open fireplaces for cooking and heating only during the eighteenth century. Benjamin Franklin's famous stove, conceived in 1740, was not a self-standing device but rather a stove within a fireplace, able to cook and heat with a much higher efficiency (Cohen 1990). In 1798 Benjamin Thompson (Count Rumford, 1753–1814) designed a brick range with top openings for placing the pots and with a cylindrical oven; the range was first adopted by large kitchens (Brown 1999).

Heat and Light

The primitive nature and inefficiency of traditional heating and lighting are especially remarkable when contrasted with the often impressive mechanical inventions of ancient civilizations. The contrast is even greater in the context of the wide range of technical advances in post-Renaissance Europe. Open fires and simple fireplaces supplied generally inadequate heat during most of the early modern era (1500–1800). The glow of the fire and the flickering, weak flames of the (often smoking) oil lamps and of (mostly expensive) candles provided poor illumination for millennia of preindustrial evolution.

In heating, the much-needed transition from wasteful, unregulated open fires to more efficient arrangements was very slow. Merely moving an open fire into a three-sided fireplace brought only a marginal efficiency gain. Well-stoked fireplaces could keep an unattended fire overnight, but their heating efficiencies were poor. The best rates were close to 10%, but more typical performances were just around 5%. And often a working fireplace, warming its immediate vicinity with radiated heat but drawing the warm inside air outside, was actually causing an overall heat loss in the room. When this draft was impeded the combustion could produce dangerous, even lethal, levels of carbon monoxide.

The efficiencies of traditional brick or clay stoves varied not only with design (often mandated by cooking preferences) but also with the dominant fuel. Modern measurements of Asian rural stoves, whose design has not changed for centuries, make it possible to fix the highest practical efficiencies. Grated, massive brick stoves with long flues and tightly fitting tops, fueled with chopped wood, had efficiencies mostly around 20%. In less massive, drafty stoves with short flues, fueled with straw or grasses, the typical performance was closer to 15%, or even just 10%. But not all traditional heating arrangements were wasteful. At least three space-heating systems used wood and crop residues in ingeniously efficient ways while providing a great degree of comfort.

They were the Roman *hypocaust*, the Korean *ondol*, and the Chinese *kang*. The first two designs led hot combustion gases through raised room floors before exhausting them through a chimney. The hypocaust was a Greek invention, with the oldest remains found in Greece and Magna Graecia, the coastal areas of southern Italy settled by the Greeks, and dating to the third century BCE (Ginouvès 1962); the Romans used it first in the hot rooms (*caldaria*) of their public baths (*thermae*) and then to heat stone houses in colder provinces of the empire (fig. 4.16). Trials with a preserved hypocaust showed that just 1 kg of charcoal per hour could maintain a temperature of 22°C in a room 5 × 4 × 3 m when the outside temperature was 0°C (Forbes 1966). The third traditional heating setup is still found throughout North China. The kang, a large brick platform (at least 2 × 2 m and 75 cm tall) is warmed by the waste heat from the adjacent stove; it serves as a bed at night and as a resting place during the day (Hommel 1937).

Yates (2012) did a detailed engineering analysis of this traditional bed-stove (or heat exchanger) and offered suggestions for improving its efficiency. These arrangements conducted heat slowly over relatively large areas. In contrast, brazier heaters, common in most Old World societies,

Figure 4.16
Part of a Roman hypocaust (with the skeleton of a dog killed by fumes) displayed at Homburg-Schwarzenacker Roman Museum in Saarland. Photograph courtesy of Barbara F. McManus.

offered only limited point sources of warmth but could produce high concentrations of carbon monoxide. Japanese, great exploiters of Chinese and Korean inventions, could not introduce the ondol or kang into their flimsy wooden houses. They relied instead on charcoal braziers (*hibachi*) and on foot warmers (*kotatsu*). These small containers of charcoal, set into the floor and covered with wadded cloth, were used well into the twentieth century. They survive even today in the form of an electric kotatsu, a small heater built into a low table. And even the British House of Commons was heated by large charcoal pots until 1791.

Biomass fuels were also the principal sources of traditional lighting in all preindustrial societies. Fire glow, torches of resinous wood, and burning splinters were the simplest but also the least efficient and the least convenient solutions. The first fat-burning oil lamps appeared in Europe during the Upper Paleolithic, nearly 40,000 years ago (de Beaune and White 1993). Candles were used in the Middle East only after 800 BCE. Both oil lamps and candles offered inefficient, weak, and smoky illumination, but they were at least easily portable and safer to use. They burned a variety of animal and plant fats and waxes—olive, castor, rapeseed, and linseed oils and whale oil, beef tallow, and beeswax—with papyrus, rush pith, flaxen, or hempen wicks. Until the end of the eighteenth century artificial indoor light came only in units of one candle. Bright illumination was possible only through the massive multiplication of these tiny sources.

Candles convert only about 0.01% of their chemical energy into light. The bright spot in their flame has an average irradiance (rate of energy falling on a unit area) just 20% higher than clear sky. The invention of matches, dating to China of the late sixth century, made kindling fires and lighting lamps much easier than igniting tinder. The earliest matches were slender pinewood sticks impregnated with sulfur; they reached Europe only in the early sixteenth century. Modern safety matches, incorporating red phosphorus in the striking surface, were first introduced in 1844, and soon captured most of the market (Taylor 1972). In 1794, Aimé Argand introduced lamps that could be regulated for maximum luminosity using wick holders, with a central air supply and chimneys to draw in the air (McCloy 1952).

Soon afterward came the first lighting gas made from coal. Outside large cities, tens of millions of households around the world continued to depend for their light for more than half of the nineteenth century on an exotic biomass fuel, oil rendered from the blubber of sperm whales. The poorly paid, wearying, and dangerous hunt for these giant mammals—portrayed so unforgettably in Herman Melville's great book, *Moby-Dick* (1851)—reached its peak just before 1850 (Francis 1990). The American whaling fleet, by far the world's largest, had a record total of more than 700 vessels in 1846. During the first half of that decade about 160,000 barrels of sperm oil were brought each year to New England's ports (Starbuck 1878). The subsequent decline in sperm whale numbers and competition from coal gas and kerosene led to a rapid decline of the hunt.

Transportation and Construction

The preindustrial evolution of transportation and construction shows a highly uneven pattern of advances and stagnation, or even decline. Ordinary sail ships of the late eighteenth century were greatly superior to the best vessels of classical antiquity, both in their speed and in their ability to sail much closer into the wind. Similarly, well-upholstered coaches sitting on good springs and drawn by efficiently harnessed horses offered an incomparably more comfortable ride than travel on horseback or in unsprung carts. But at the same time, even in the richest European countries typical roads were hardly better, and often much worse, than during the last centuries of the Roman Empire. And the skills of the Athenian architects who designed the Parthenon, or of the Roman masons who finished the Pantheon, were hardly inferior to the abilities of their successors building late Baroque palaces and churches. Everything changed, and rather rapidly, only with the diffusion of a much more powerful prime mover and a superior construction material. The steam engine and cheap cast iron and steel revolutionized transportation as well as construction.

Moving on Land
Walking and running, the two natural modes of human locomotion, have accounted for most of the personal movements in all preindustrial societies. Energy costs, average speeds, and maximum daily distances have always depended primarily on individual fitness and on the prevailing terrain (Smil 2008a). The efficiency cost of walking increases both below and above the optimum speeds of 5–6 km/h, and uneven surfaces, mud, or deep snow will raise the costs of walking on the level by up to 25–35%. The cost of walking uphill is a function of both gradient and speed, and detailed studies show a nearly linear increase in energy needs across a broad range of speeds and inclines (Minetti et al. 2002).

Running requires power outputs mostly between 700 and 1,400 MW, equivalent to 10–20 times the basal metabolic rate. A slowly running 70 kg man will produce 800 W; the power of an accomplished marathoner running the race (32.195 km) in 2.5 hours will average about 1,300 W (Rapoport 2010); and when Usain Bolt set the world record for 100 m at 9.58 seconds, his maximum power (a few seconds into the run, and at time when his speed was only half of the maximum) was 2,619.5 W, that is, 3.5 horsepower (Gómez, Marquina, and Gómez 2013). The energy cost of running for humans is relatively high, but, as already noted (chapter 2), people have a unique capability of virtually uncoupling this cost from speed (Carrier 1984). Arellano and Kram (2014) showed that body weight support and forward propulsion account for about 80% of the total cost of running; leg swinging claims about 7%, and maintaining lateral balance about 2%—but arm swinging cuts the overall cost by about 3%.

Modern record performances in running improved steadily during the twentieth century (Ryder, Carr, and Herget 1976) and they are undoubtedly well above the best historical achievements. But there is no shortage of outstanding examples of long-distance running in many traditional societies. Pheidippides' fruitless run from Athens to Sparta just before the battle of Marathon in 490 BCE is, of course, the prototype of great running endurance. He covered the distance of 240 km in just two days (his average power output, assuming he weighed about 70 kg, would have been about 800 W, slightly more than 1 hp), only to find that the Spartans refused to help.

The domestication of horses not only introduced a new and more powerful and faster means of personal transport, it was also associated with the diffusion of Indo-European languages, bronze metallurgy, and new ways of warfare (Anthony 2007). Horses were ridden for a long time before they were first harnessed; the beginnings of horseback riding have been placed in the Asian steppes around the middle of the second millennium BCE. But Anthony, Telegin, and Brown (1991) concluded that it may have begun much earlier, around 4000 BCE, among the people of Sredni Stog culture in today's Ukraine.

They based the claim on still inconclusive evidence on the difference between the premolars of feral and domestic horses: animals that have been bitted show distinctive fractures and beveling on micrographs of their teeth. Similarly, Outram and co-workers (2009) used the signs of bitting damage (and other evidence) to conclude that the first horse domestication took place among the people of the Botai Culture and that some of those animals were bridled and perhaps ridden. When walking, bitted animals were no faster than humans, but trotting (in excess of 12 km/h) and cantering (up to 27 km/h) speeds easily covered the distance that would have require major human effort. Galloping horses have a great mechanical advantage: their muscular work is halved by storing and returning elastic strain energy in their spring-like muscles and tendons (Wilson et al. 2001).

Experienced riders on a fit animal had no difficulty riding 50–60 km/day, and by changing horses they could cover more than 100 km/day in emergencies. The longest distances ridden routinely in a day during the medieval era were those by the riders of the Mongolian *yam* (message delivery) service (Marshall 1993), and in the modern era William F. Cody (1846–1917) claimed that as a young rider with the Pony Express he covered, after his relief ride was killed, 515 km in 21 hours 40 minutes using 21 horses (Carter 2000). Minetti (2003) showed that the typical performances of long-distance services were carefully optimized. Relay postal systems preferred an average speed of 13–16 km/h and a daily distance of 18–25 km/animal to minimize the risk of damage to horses, and these optimum performance bands were followed by the ancient Persian service established by

Cyrus between Susa and Sardis after 550 BCE, by *yam* riders of the thirteenth century, and by the Overland Pony Express, which served California before the construction of the telegraph and railway links.

But riding a horse has always been a major physical challenge. Because a horse's fore-end contains about three-fifths of its body weight, the only way for the vertical planes intersecting rider's and animal's center of gravity to coincide is for the rider to sit forward. But an upright forward position leaves the rider's center of gravity much higher than that of the horse. This can produce a rapid lever action by the rider's back when the horse moves forward, jumps, or stops fast. Consequently, the most efficient position requires the rider to put his center of gravity not only forward but also low. The jockey's crouch ("monkey on a stick") is the best way to do it. Curiously, this optimum was irrefutably established only before the end of the nineteenth century by Federico Caprilli (Thomson 1987).

Pfau and co-workers (2009) found that major horse race times and records improved by up to 7% around 1900 when the crouched posture was first adopted. The posture isolates the rider from the movement of his mount: inevitably, the horse supports the rider's body weight but does not have to move the jockey through each cyclical stride path. Maintaining that posture requires substantial exertion, as reflected by the near-maximum heart rates of jockeys during racing. The forward-low position, used in the most exaggerated version in modern showjumping, differs radically from riding styles portrayed in historical sculptures and images. For a variety of reasons riders sat too far back and were too much extended to make the most efficient motion possible. Classical riders were even more disadvantaged because they did not have stirrups. Only the universal adoption of stirrups in early medieval Europe made armored riding, fighting, and jousting possible.

The simplest way of transporting loads is to carry them. Where roads were absent people could often do better than animals: their weaker performance was often more than compensated for by flexibility in loading, unloading, moving on narrow paths, and scrambling uphill. Similarly, donkeys and mules with panniers were often preferred to horses: steadier on narrow paths, with harder hooves and lower water needs they were more resilient. The most efficient method of carrying is to place the load's center of gravity above the carrier's own center of gravity—but balancing a load is not always practical. Poles slung over a shoulder and wooden yokes hung with loads or buckets are preferable to carrying with the hands or in the arms. Long-distance transfers in difficult terrain are best accomplished with backpacks fastened by good shoulder or head straps. Nepali Sherpas, carrying supplies of Himalayan expeditions, are generally acknowledged as the best porters. They can move between 30 and 35 kg (close to half their body weight) up to base camp, and less than 20 kg on steeper slopes in rarer air above it.

As already noted, the Roman saccarii who reloaded Egyptian grain at Ostia harbor from ships to barges carried sacks of 28 kg over short distances. In the light version of the traditional Chinese sedan chair two men carried a single customer, a load prorating to as much as 40 kg per carrier. These loads corresponded to as much as two-thirds of a carrier's body weight, and walking speeds usually did not exceed 5 km/h. In relative terms, people were better carriers than animals. Typical loads were only about 30% of an animal's weight (that is, mostly just 50–120 kg) on the level and 25% in the hills. Men aided by a wheel could move loads far surpassing their body weight. Recorded peaks are more than 150 kg in Chinese barrows where the load was centered right above the wheel's axle. European barrows, with their eccentric front wheel, were usually loaded with no more than 60–100 kg.

Massed applications of human labor, aided by simple mechanical devices, could accomplish some astonishingly demanding tasks. Undoubtedly the most taxing transport tasks in traditional societies were the deliveries of large-sized building stones or finished components to construction sites. Large stones were quarried, moved, and emplaced by every old high culture (Heizer 1966). A few ancient images offer firsthand illustrations of how this work was accomplished. Certainly the most impressive one is depicted in an already mentioned Egyptian painting from the tomb of Dje-hutyhotep at el-Bersheh, dated to 1880 BCE (Osirisnet 2015). The scene portrays 166 men dragging a colossus on a sledge whose path is lubricated by a worker pouring liquid from a vessel (fig. 4.17). With lubrication cutting the friction by about half, their massed labor, reaching a peak power of over 30 kW, could move a 50 t load. Yet even such efforts were greatly surpassed in a number of preindustrial societies.

Inca builders used enormous irregular stone polygons whose smoothed sides were fitted with amazing precision. Pulling a 140 t stone, the heaviest block at Ollantaytambo in southern Peru, up the ramp required the coordinated force of about 2,400 men (Protzen 1993. The brief peak power of this group would have been around 600 kW, but we know nothing of the logistics of such an enterprise. How were more than 2,000 men harnessed to pull in concert? How were they arranged to fit into the confines of narrow (6–8 m) Inca ramps? And how did the people in ancient Brittany handle the

Figure 4.17
Moving a massive (6.75 m tall, weighing more than 50 t) alabaster statue of Dje-
hutyhotep, Great Chief of the Hare Nome (Osirisnet 2015). The drawing reconstructs
a damaged wall painting in the tomb of Djehutyhotep at the site of el-Bersheh, Egypt
(Corbis).

Grand Menhir Brise (Niel 1961), at 340 t the largest stone erected by a European megalithic society?

The superiority of horses could be realized only with a combination of
horseshoes and an efficient harness. Performance in land transport also
depended on success in reducing friction and allowing higher speeds. The
state of roads and the design of vehicles were thus two decisive factors. The
differences in energy requirements between moving a load on a smooth,
hard, dry road and on a loose, gravelly surface are enormous. In the first
case a force of only about 30 kg is needed to wheel a 1 t load, the second
instance would call for five times as much draft, and on sandy or muddy
roads the multiple can be seven to ten times higher. Axle lubricants (tallow
and plant oils) were used at least since the second millennium BCE. Celtic
bronze bearings had inner grooves that contained cylindrical wooden rollers during the first century BC (Dowson 1973). Chinese rolling bearings
may be of even greater antiquity, but ball bearings are firmly documented
for the first time only in early seventeenth-century Europe.

The roads in ancient societies were mostly just soft tracks that seasonally turned into muddy ruts, or dusty trails. The Romans, starting with the
Via Appia (Rome to Capua) in 312 BCE, invested a great deal of labor and
organization in an extensive network of hard-top roads (Sitwell 1981).

Well-built Roman *viae* were topped with gravel concrete, cobblestones, or slabs set in mortar. By Diocletian's reign (285–305) the Roman system of trunk roads, the *cursus publicus*, had grown to some 85,000 km. The overall energy cost of this enterprise was equivalent to at least one billion labor-days. This large total prorates to easily manageable requirements over the centuries of ongoing construction (box 4.12) In Western Europe the Roman achievements in road building were surpassed only during the nineteenth century, in the eastern regions of the continent only during the twentieth.

The Muslim world had no roads network comparable to the Roman *cursus publicus*, although it had intense communication (Hill 1984). Its far-flung parts were connected by much-traveled caravan routes, which, technically, were mere tracks. This was the result of pack camels replacing wheeled transport in the arid region between Morocco and Afghanistan. This development, preceding the Muslim conquest, was driven largely by economic imperatives (Bulliet 1975). In comparison with oxen, pack camels are not only more powerful and faster, they also have greater endurance and longevity. They can move over a rougher ground, subsist on inferior forage, and tolerate longer spells of feed and water shortages. These economic advantages were strengthened with the introduction of the North Arabian saddle sometime between 500 and 100 BCE. The saddle provided an excellent riding and carrying arrangement, and allowed

---

Box 4.12
Energy cost of Roman roads

If we assume that the average Roman road was just 5 m wide and 1 m deep, construction of 85,000 km of trunk roads would have required the emplacement of about 425 Mm^3 of sand, gravel, concrete, and stone, after first removing at least some 800 Mm^3 of earth and rock for the roadbed, embankments, and ditches. Assuming that a worker handled only 1 m^3 of building materials a day, the tasks of quarrying, cutting, crushing, and moving stones, excavating sand for foundations, ditches, and roadbeds, preparing concrete and mortar, and laying the road would have added up to about 1.2 billion labor-days. Even if the maintenance and repair needs would have eventually tripled this requirement, prorating this grand total over 600 years of construction would result in an annual average of six million labor-days, an equivalent of some 20,000 full-time construction workers. This would represent (at 2 MJ/day) an annual energy investment of nearly 12 TJ of labor.

caravans to displace carts in the Old World's arid region before the Arab expansion.

The Incas, consolidating their empire between the thirteenth and fifteenth centuries, built an impressive road network by corvée labor. Its length totaled about 40,000 km, including 25,000 km of all-weather roads crossing culverts and bridges and equipped with distance markers. Of the two main royal roads, the one winding through the Andes was stone-surfaced. Its width ranged from up to 6 m on river terraces to just 1.5 m when cut through solid rock (Kendall 1973). The unsurfaced coastal link was about 5 m wide. Neither road had to support any wheeled vehicles, just caravans of people and pack llamas carrying 30–50 kg per animal and covering less than 20 km/day.

During the Qin and Han dynasties, the Chinese built an extensive road system totaling about 40,000 km (Needham et al. 1971). The contemporary Roman cursus was more extensive, both in its total length and in the road density per unit area, as well as more sturdily built. This is how Statius (Mozley 1928, 220) in his *Silvae* described the building of the Via Domitiana in 90 CE:

> The first labour was to prepare furrows and mark out the borders of the road, and to hollow out the ground with deep excavation; then to fill up the dug trench with other material, and to make ready a base for the road's arched ridge, lest the soil give way and a treacherous bed provide a doubtful resting-place for the o'erburdened stones; then to bind it with blocks set close on either side and frequent wedges. Oh! how many gangs are at work together ! Some cut down the forest and strip the mountain-sides, some plane down beams and boulders with iron; others bind the stones together, and interweave the work with baked sand and dirty tufa; others by dint of toil dry up the thirsty pools, and lead far away the lesser streams.

Chinese roads were constructed by tamping rubble and gravel with metal rammers. This provided a more elastic but less durable surface than the best Roman roads. An excellent messenger service survived the decline of the Han dynasty, but the land-borne transport of goods and people generally deteriorated. Only in some parts of the country was this decline more than made up for by the development of efficient canal transportation. Ox-drawn carts and wheelbarrows carried most of the goods. People were moved in two-wheeled carts and in sedan chairs well into the twentieth century. The first documented vehicles come from Uruk around 3200 BCE. They had heavy, solid-disk wheels up to 1 m in diameter made of dowelled and mortised planks. Their subsequent diffusion across different European cultures was remarkably rapid (Piggott 1983). Some early wheels rotated about a fixed axle; others turned together with it. Subsequent developments were in the direction of much lighter, free-turning spoked wheels (in the early second millennium BCE) and the use of a pivoted front axle in four-wheel vehicles, making sharp turns possible.

Inefficiently harnessed horses moving on poor roads were slow even when pulling relatively light loads. Maximum specifications restricted loads on Roman roads of the fourth century to 326 kg for horse-drawn, and up to 490 kg for slower ox-drawn, post carriages (Hyland 1990). The low speeds of this transport method limited its daily range to 50–70 km for passenger horse carts on good roads, 30–40 km for heavier horse-drawn wagons, and up to half those distances for oxen. Men with wheelbarrows would cover about 10–15 km/day. Of course, much longer distances were covered by messengers on fast horses. Recorded maxima on Roman roads are up to 380 km/day. Low speeds and low capacities of land transport translated into excessive costs, as illustrated by the figures in Diocletian's *edictum de pretiis*. In 301 CE it cost more to move grain just 120 km by road than to ship it from Egypt to Ostia, Rome's harbor. And after the Egyptian grain arrived at Ostia, just some 20 km away from Rome, it was reloaded onto barges and moved against the Tiber's stream rather than be hauled by ox-drawn wagons.

Similar limitations persisted in most societies well into the eighteenth century. At its beginning it was cheaper to import many goods into England by sea from Europe rather than to carry them by pack animals from the country's interior. Travelers described the state of English roads as barbarous, execrable, abominable, and infernal (Savage 1959). Rain and snow made poorly laid-out soft dirt or gravel roads impassable; in many cases their limited width allowed only pack traffic. Roads in continental Europe were in similarly bad shape, and coach horses harnessed in teams of four to six animals lasted on average less than three years. Fundamental improvements came only after 1750 (Ville 1990). Initially they included widening and better drainage of roads, and later their surfacing with durable finishes (gravel, asphalt, concrete). Heavy European horses could finally demonstrate their great performance in hauling. By the mid-nineteenth century the maximum allowable French load was increased to nearly 1.4 t, about four times the Roman limit.

In urban transportation horses reached the peak of their importance also only during the railway age, between the 1820s and the end of the nineteenth century (Dent 1974). While the railways were taking over long-distance shipments and travel, the horse-drawn transport of goods and people dominated in all rapidly growing cities of Europe and North America. Steam engine had actually expanded the deployment of horses (Greene

2008). Most railway shipments had to be collected and distributed by horse-drawn vans, wagons, and carts. These vehicles also delivered food and raw materials from the nearby countryside. Greater urban affluence brought many more private coaches and hansoms, cabs and omnibuses (first in London in 1829), and delivery wagons (fig. 4.18).

The stabling of the animals in mews and the provision and storage of hay and straw made an enormous demand on urban space (McShane and Tarr 2007). At the end of Queen Victoria's reign, London had some 300,000 horses. City planners in New York were thinking about setting aside a belt of suburban pastures to accommodate large herds of horses between the peak demands of rush-hour transport. The direct and indirect energy costs of urban horse-drawn transport—the growing of grain and hay, feeding and stabling the animals, grooming, shoeing, harnessing, driving, and removal of wastes to periurban market gardens—were among the largest items on the energy balance of the late nineteenth-century cities. This equine dominance ended rather abruptly. Electricity and internal combustion engines were becoming practicable just as the numbers of urban horses rose to record totals during the 1890s. In less than a generation,

Figure 4.18
Engraving from the *Illustrated London News* of November 16, 1872, captures perfectly the high density of horse-drawn traffic (hansom cabs, omnibuses, heavy wagons) in the rapidly industrializing cities of late nineteenth-century Europe.

horse-drawn city traffic was largely displaced by electric streetcars, automobiles, and buses.

Curiously, it was also only during that time that European and American mechanics came up with a practical version of the most efficient human-powered locomotive vehicle, the modern bicycle. For generations bicycles were clumsy, even dangerous, contrivances that had no chance of mass adoption as vehicles of convenient personal transport. Rapid improvements came only during the 1880s. John Kemp Starley and William Sutton introduced bicycles with equal-sized wheels, direct steering, and a diamond-shaped frame of tubular steel (Herlihy 2004; Wilson 2004; Hadland and Lessing 2014), and these designs have been closely followed by virtually all twentieth-century machines (fig. 4.19). The evolution of the modern bicycle was largely complete with the addition of pneumatic tires and the back-pedal brake in 1889.

Improved bicycles equipped with lights, various load carriers, and tandem seats became common for commuting, shopping, and recreation in a number of European nations, most prominently the Netherlands and Denmark. Later diffusion throughout the poor world multiplied European totals. The history of Communist China has been particularly closely connected with a massive use of the machine. Until the early 1980s there were no private cars in China, and until the late 1990s most commuters rode bicycles even in the country's large cities. The subsequent construction of subways in all major cities and a surge in car ownership reduced urban bicycle use (a shift that has been only partially offset by the rising popularity of e-bikes), but rural demand remains strong, and China is still the world's largest producer of bicycles, with more than 80 million units a year, of which more than 60% are exported (IBIS World 2015).

Oared Ships and Sail Ships

Human-powered waterborne movement achieved much higher power ratings than animate transport on land. Oared vessels were ingeniously designed to integrate the efforts of tens and even hundreds of oarsmen. Naturally, prolonged strenuous pulling of heavy oars required very hard labor, and when done in confined quarters below the deck it was extremely exhausting. Our admiration of the complex design and organizational mastery of large oared ships must be tempered by the realization of the human suffering exacted by their speedy motion. Ancient Greek oared ships have been particularly well studied (Anderson 1962; Morrison and Gardiner 1995; Morrison, Coates, and Rankov 2000). The vessels that took the Greek

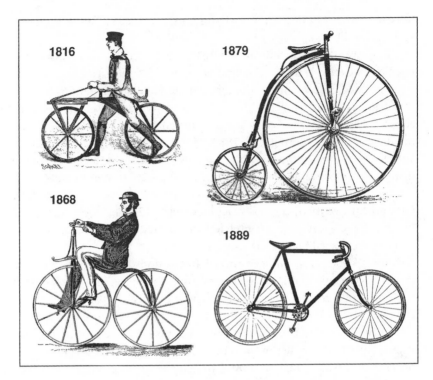

Figure 4.19
The development of the bicycle started surprisingly late and advanced rather slowly. Riders had to push themselves on Baron von Drais's 1816 clumsy draisine. Pedals were first applied to the axle of the drive wheel in 1855, an advance leading to the velocipedes of the 1860s. Subsequent design regression led to huge front wheels and plenty of accidents. Only the late 1880s brought the safety, efficiency, and simplicity of the modern bicycle. Adapted from Byrn (1900).

troops to Troy, *penteconteres* with 50 oarsmen, could receive briefly useful power inputs of up to 7 kW.

Triple-tiered *trieres* (Roman triremes), the best-performing warships of the classical era, were powered by 170 rowers (fig. 4.20). Strong oarsmen could propel them with more than 20 kW of power, enough to produce maximum speeds of close to 20 km/h. Even when moving at more common top speeds of 10–15 km/h, the highly maneuverable triremes were powerful fighting machines. Their bronze ram could hole the hulls of enemy ships with devastating effect. One of the decisive battles of Western history, the defeat of a larger Persian fleet by a smaller Greek force at Salamis (480 BCE), was won in this fashion by triremes. They were also the most important

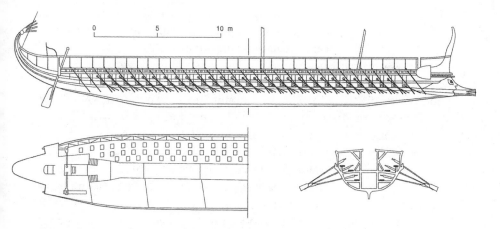

Figure 4.20
Side view, partial plan, and cross section of the reconstructed Greek trireme *Olympias*. Six files arranged in a V shape accommodate 170 rowers, and the topmost oars have their pivots on outriggers. Based on Coates (1989).

warships of republican Rome. A full-scale reconstruction was finally accomplished during the 1980s (Morrison and Coates 1986; Morrison, Coates, and Rankov 2000).

Larger ships—quadriremes, quinquiremes, and so on—followed in rapid succession after Alexander's death in 323 BCE. As there is no indication that any of these ships had more than three tiers, two or more men presumably powered a single oar. The end of this progression was reached with the construction of *tessarakonteres* during Ptolemaios Philopator's reign (222–204 BCE). The 126 m long ship was to carry more than 4,000 oarsmen and nearly 3,000 troops, and could theoretically be propelled with over 5 MW of power. But its weight, including the heavy catapults, made it virtually immovable, a costly shipbuilding miscalculation.

In the Mediterranean, large oared vessels retained their importance well into the seventeenth century: at that time the largest Venetian galleys had 56 oars, each crewed by five men (Bamford 1974; Capulli 2003). Large Maori dugout canoes were oared by almost as many warriors (up to 200). The general limits of aggregate human power in sustained rowing applications were thus between 12 and 20 kW. There were also ships powered by pedaling or stepping on treadmills. During the Sung dynasty the Chinese built increasingly larger paddle-wheel warships powered by up to 200 men treading pedals (Needham 1965). In Europe, smaller tugs powered by 40 men turning capstans or treadmills appeared in the middle of the sixteenth century. Animate power was also the principal prime mover for moving goods and people by canal boats and barges (box 4.13).

Canals were particularly important catalysts of economic development in the core area of the Chinese state (in the lower basin of the Huang He and on the North China Plain) beginning in the Han dynasty (Needham et al. 1971; Davids 2006). By far the longest and most famous of these transport arteries is *da yunhe*, the Grand Canal. Its first section was opened in the early seventh century, and its completion in 1327 made it possible to move barges from Hangzhou to Beijing. This is a latitudinal difference of 10° and an actual distance of nearly 1,800 km. Early canals used inconvenient double slipways on which oxen hauled boats to a higher level. The invention of the canal pound-lock in 983 made it possible to raise boats safely and without wasting water. A progression of locks raised the Grand Canal's highest point to just over 40 m above sea level. Chinese canal boats were pulled by gangs of laborers or by oxen or water buffaloes.

In Europe, canals reached their greatest importance during the eighteenth and nineteenth centuries. Horses or mules moving on adjoining tow paths pulled the barges with speeds of about 3 km/h when loaded, and up to 5 km/h when empty. The mechanical advantages of this form of transport are obvious. On a well-designed canal a single heavy horse could pull a load of 30–50 t, an order of magnitude more than a horse could manage on the best hard-top road. Steam engines gradually replaced barge-towing animals, but many horses still worked on smaller canals during the 1890s.

Box 4.13

Ancient canal transportation

The earliest description of their sluggish progress (snoring waterman, grazing mule) was left by Horace (Quintus Horatius Flaccus, 65–8 BCE) in his *Satires* (Buckley 1855, 160):

> While the waterman and a passenger, well-soaked with plenty of thick wine, vie with one another in singing the praises of their absent mistresses: at length the passenger being fatigued, begins to sleep; and the lazy waterman ties the halter of the mule, turned out a-grazing, to a stone, and snores, lying flat on his back. And now the day approached, when we saw the boat made no way; until a choleric fellow, one of the passengers, leaps out of the boat, and drubs the head and sides of both mule and waterman with a willow cudgel. At last we were scarcely set ashore at the fourth hour.

The European construction of transportation canals, an unmistakable import from China, started in North Italy during the sixteenth century. The 240 km long French Canal du Midi was completed by 1681. The longest continental and British links came only after 1750, and the German canal system actually postdated railroads (Ville 1990). Canal barges moved large quantities of raw materials and import commodities for expanding industries and growing cities, and they also took out their wastes. They handled a large share of European traffic just before the introduction of railways and for a few decades afterward (Hadfield 1969).

In contrast to canal shipping and to warships, the long-distance seaborne transport of goods and people was dominated by sail ships from the very beginning of high civilizations. The history of sail ships may be understood primarily as a quest for the better conversion of the kinetic energy of the wind into the efficient motion of vessels. Sails alone could not do this, but they were obviously the key to nautical success. They are basically fabric aerofoils (as they are inflated by wind they form a foil shape) designed to maximize lift force and minimize drag (box 4.14). But this force from the sail's foil shape must be combined with the balancing force of the keel; otherwise the vessel will drift downwind (Anderson 2003).

Square sails set at right angles across the ship's long axis were efficient energy converters only with the wind astern. Roman ships pushed by the northwesterlies could make the Messina-Alexandria run in just 6–8 days, but the return could take 40–70 days. Irregular sailings, substantial seasonal differences, and the cessation of all travel during winter (shipping between Spain and Italy was closed between November and April) make it almost impossible to say what speeds were typical (Duncan-Jones 1990). Longer voyages against the wind were primarily the result of lengthy course changes. All ancient ships were rigged with square sails, and there was a long interval before the introduction and widespread diffusion of radically different designs (fig. 4.21).

Ships with fore-and-aft rigging had sails aligned with the vessel's long axis, and their masts were pivots for the sails to swing around and to catch the wind. They could change direction much more easily by simply turning into the wind and proceeding on a zig-zag course. The earliest fore-and-aft rigging most likely came from Southeast Asia in the form of a rectangular canted sail. Modifications of this ancient design were eventually adopted both in China and, through India, in Europe. The characteristic batten-strengthened Chinese lug sails were in use since the second century BCE. The canted square sail became common in the Indian Ocean during the

Box 4.14
Sails and sailing near the wind

When wind strikes a sail, the difference in pressure generates two forces: lift, whose direction is perpendicular to the sail, and drag, which acts along the sail. With wind astern, the lift force will obviously be much stronger than the drag force, and a ship will make good progress. With wind on the beam, or slightly ahead it, the force pushing the vessel sideways is obviously stronger than the force propelling it forward. If the ship were to try to steer even closer to the wind, the drag would surpass the lift and the vessel would be pushed backward. The maximum capabilities for sailing near the wind have advanced by more than 100° since the beginning of sailing. Early Egyptian square-sailed ships could manage only a 150° angle, while medieval square rigs could proceed slowly with the wind on their beam (90°), and their post-Renaissance successors could move at an angle of just about 80° into the wind. Only the use of asymmetrical sails mounted more in line with the ship's long axis and capable of swiveling around their masts made sailing closer to the wind possible.

Ships combining square sails with triangular mizzens could manage 60°, and fore-and-aft rigs (including triangular, lug, sprit and gaff sails) could come as close as 45° to the wind. Modern yachts come very close to 30°, the aerodynamic maximum. The only way to circumvent the earlier limits was to proceed under the best manageable angle and keep changing the course. Square-rigged ships had to resort to wearing, or making a complete downwind turn. Ships with fore-and-aft sails tried tacking, turning their bows into the wind, and eventually catching the wind on the opposite side of the sail.

third century BCE, a clear precursor of the triangular (lateen) sails that were so typical of the Arab world after the seventh century.

Viking expansion (which eventually reached as far west as Greenland and Newfoundland) was made possible by the deployment of a large number of massive rectangular or square woolen sails. Production of those large sails was very labor-intensive (a single 90 m² one-ply sail took one craftsman, using vertical warp and horizontal weft, up to five years to produce), and the need to convert land for pasture and to maintain extensive sheep herds in order to produce enough wool for large Norse fleets were likely based on slave labor (Lawler 2016). After the Viking voyages ended, large woolen sails were used in the Northeastern Atlantic (between Iceland and Scandinavia, including the Hebrides and the Shetlands) until the nineteenth century (Vikingeskibs Museet 2016).

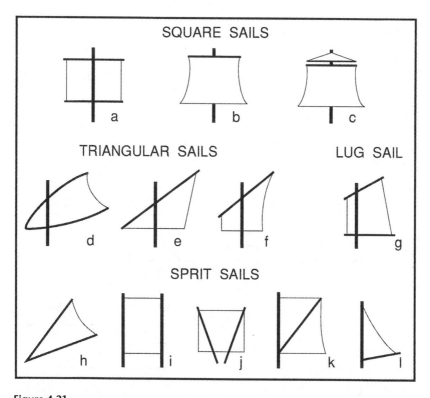

Figure 4.21
Principal types of sails. Square sails, straight (a) or flared (b), are the oldest types.
Triangular sails include the Pacific boom (d), and lateens without or with a luff edge
(e, f). Sprit sails (h) were common in Polynesia, Melanesia (i), the Indian Ocean (j),
and Europe (k, l). Masts and all supporting structures (booms, sprits, gaffs) are drawn
with thicker lines, and the sails are not shown to scale. Based on Needham and co-
workers (1971) and White (1984).

In Europe, only the late medieval combination of square rigging and
triangular sails made sailing close to the wind possible. Gradually, these
ships were rigged with a larger number of loftier and better adjustable sails
(fig. 4.22). Better and deeper hull designs, a stern-post rudder (in use in
China since the first century CE, in Europe only a millennium later), and a
magnetic compass (in China after 850, in Europe around 1200) turned
them into uniquely efficient wind energy converters. This combination was
made almost irresistibly powerful by the addition of accurate heavy guns.
The gunned ship, developed in Western Europe during the fourteenth and
fifteenth centuries, launched the era of unprecedented long-distance expansion. In Cipolla's (1965, 137) apt characterization, the ship

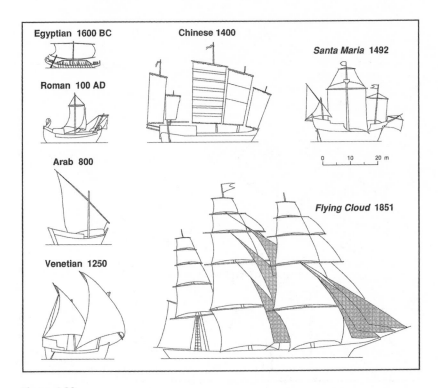

Figure 4.22
Evolution of sail ships. Ancient Mediterranean societies used square-rigged sails. Before they were adopted by Europeans, triangular sails were dominant in the Indian Ocean. A large seagoing junk from Jiangsu typifies efficient Chinese designs. Columbus's *Santa Maria* had square sails, a foretopsail, a lateen on mizzen, and the spritsail under the bowsprit. *Flying Cloud*, a famous mid-nineteenth-century record-breaking U.S. clipper, was rigged with triangular jibs fore, a spanker aft, and lofty main royal and skysails. Simplified outlines are based on images in Armstrong (1969), Daumas (1969), and Needham and co-workers (1971) and are drawn to scale.

was essentially a compact device that allowed a relatively small crew to master unparalleled masses of inanimate energy for movement and destruction. The secret of the sudden and rapid European ascendancy was all there.

These ships reached their largest sizes and became equipped with an increasing number of guns during the late eighteenth and the early nineteenth centuries. The French-British naval rivalry eventually ended in a clear British maritime supremacy, but it was an original French design of a large two-decked battleship (about 54 m long at the gun deck, with 74 guns and crew of 750 men) that became the dominant class of sailing vessels before being displaced by steam-powered ships. The British Royal Navy eventually commissioned nearly 150 of those large ships (Watts 1905; Curtis 1919), and they ensured the country's naval dominance before and after the Napoleonic age. Starting in the early fifteenth century the simplest ships of this innovative design carried the audacious Portuguese sailors on longer voyages (box 4.15).

In 1492 the Atlantic was crossed to America by three Spanish ships captained by Christopher Columbus (1451–1506). In 1519 Ferdinand Magellan (1480–1521) traversed the Pacific, and after his death in the Philippines his *Victoria* was captained by Juan Sebastián Elcano (1476–1526), who completed the first circumnavigation of the world. Rich historical records enable us to chart the progress in tonnages and speeds of both typical and best sail ships used during the colonial expansion and for the rising volume of maritime trade (Chatterton 1914; Anderson 1926; Cipolla 1965; Morton 1975; Casson 1994; Gardiner 2000). Although the Romans built ships with capacities of more than 1,000 t, their standard cargo vessels carried less than 100 t.

More than a millennium later, Europeans embarked on their explorations with ships nearly as small. In 1492 Columbus's *Santa Maria* had a capacity of 165 t, and the *Trinidad,* Magellan's ship, had a mere 85 t. A century later vessels of the Spanish Great Armada (sailing in 1599) averaged

---

Box 4.15
Portuguese voyages of discovery

Portuguese sailors advanced first southward, along the western coast of Africa: the mouth of the Senegal Rover was reached in 1444, the equator was crossed in 1472, today's Angola was sighted in 1486, and in 1497 Vasco da Gama (1460–1524) rounded the Cape of Good Hope and crossed the Indian Ocean to India (Boxer 1969; Newitt 2005). Luís de Camões (1525–1580), in his great epic poem *Os Lusíadas,* published in 1572 and cited here in Richard Burton's translation (Burton 1880, 11), captured their progress:

> They walked the water's vasty breadth of blue,
> parting the restless billows on their way;
> fair favouring breezes breathed soft and true,
> the bellying canvas bulging in their play:
> The seas were sprent with foam of creamy hue,
> flashing where'er the Prows wide open lay
> the sacred spaces of that ocean-plain
> where Proteus' cattle cleave his own domain.

515 t. By 1800 British ships in the Indian fleet had capacities of about 1,200 t. And while Roman cargo ships could not go faster than 2–2.5 m/s, the best mid-nineteenth-century clippers could surpass 9 m/s. In 1853 the Boston-built and British-crewed *Lightning* logged the longest daily run under sail: its 803 km prorate to an average speed of 9.3 m/s (Wood 1922). And in 1890 the *Cutty Sark*, perhaps the most famous tea clipper, ran 6,000 km in 13 consecutive days, averaging 5.3 m/s (Armstrong 1969).

Too many questionable assumptions would have to be made to calculate the total energy needed to move either individual ships on long voyages or harnessed annually by a nation's merchant or military fleets. According to Unger (1984), the contribution of sail ships to the nationwide energy use during the Dutch Golden Age was about equal to the output of all Dutch windmills—but that was equivalent to less than 5% of the country's huge peat consumption (box 4.16). While it may be elusive to quantify aggregate energies in sailing, there is no doubt that an expansion of shipping

---

Box 4.16
Contribution of sail ships to Dutch energy use

Information on tonnages and speeds that would allow us to calculate the energies needed to move individual ships on long voyages, or to come up with the aggregate annual contributions of wind power harnessed by merchant or military fleets, is inadequate. Critical variables—hull designs, sail areas and cuts, cargo weights, and utilization rates—are far too heterogeneous to allow for estimation of meaningful averages. Still, Unger (1984) made a set of assumptions to calculate the contribution of sail ships to the Dutch nation's energy use during the Dutch Golden Age and ended up with an annual total of roughly 6.2 MW during the seventeenth century. For comparison, this is almost exactly equal to the total power of all Dutch windmills as estimated by De Zeeuw (1978)—but it was only a small fraction (<5%) of the country's huge peat consumption.

But such quantitative comparisons are misleading: no amount of peat would have made the trips to the East Indies possible; useful energy gained from the peat was most likely less than a quarter of its gross heat value; and, of course, there is the fundamental contrast in comparing limited and nonrenewable (or not renewable on a historical time scale) deposits of a young fossil fuel and an abundant and renewable resource constantly recharged by differences in atmospheric pressure. Comparisons of aggregate power thus make no greater sense than those of specific conversion efficiencies (in this case, contrasting the efficiency of a sail with the performance of a peat stove).

(preceding that of the economy as a whole) and its rising productivity were critical contributions to Europe's economic growth between 1350 and 1850 (Lucassen and Unger 2011).

Buildings and Structures

The enormous variety of building styles and ornaments can be reduced to only four fundamental structural members: walls, columns, beams, and arches. Only human labor aided by a few simple tools was needed to create them from the three basic building materials of the preindustrial world, timber, stone, and bricks, either sun-dried or kiln-burned. Trees could be cut and roughly shaped with axes and adzes. Stone could be quarried with only hammers and wedges and shaped with chisels. Sun-dried bricks could be made with readily available alluvial clays. Shortages of large trees limited the use of timber in many regions, and the expensive transport of stone restricted its choice largely to local varieties. Subsequent, and often very elaborate, fine shaping and surface detailing of timber and stone could greatly increase the energy cost of using these building materials.

Sun-dried mud bricks, common throughout the Middle East and Mediterranean Europe, were the least energy-intensive building blocks. Their output reached prodigious quantities even in the earliest settled societies. This how the Sumerian capital Uruk is described in the Sumerian epic *Gilgamesh*, one of the first preserved literary documents, from before 2500 BCE (Gardner 2011): "One part is city, one part orchard, and one part claypits. Three parts including the claypits make up Uruk." They were made from loams or clays, water, and chaff or chopped straw, sometimes with the addition of dung and sand; the mixture was compacted, rapidly shaped in wooden molds (up to 250 pieces per hour), and left to dry in the sun. Dimensions ranged from chunky square Babylonian pieces (40 × 40 × 10 cm) to slimmer, oblong (45 × 30 × 3.75 cm) Roman bricks. Mud bricks are poor heat conductors, helping to keep buildings cool in hot arid climate. They also had an important mechanical advantage: building a mud-brick vault requires no wooden beams for support (Van Beek 1987). With suitable clays and requisite labor they could be produced in prodigious quantities.

Burned bricks were used in ancient Mesopotamia, and later became common in both the Roman Empire and Han China. For centuries, most of the firings were in unenclosed piles or pits, resulting in a great waste of fuel and uneven baking. Later, firing in regular mounds or stacks could reach temperatures up to 800°C, yielding a more uniform product with a much higher efficiency. Completely enclosed horizontal kilns ensured better consistency and a higher combustion efficiency. They had properly spaced flues, and the rising hot gases were reflected downward from domed roofs—but they needed wood or charcoal for their operation. In Europe, these needs increased during the sixteenth century when bricks started to replace wattle-and-daub or timber studding, and when they began to be more commonly used for foundations as well as for walls.

Regardless of their principal materials, preindustrial structures demonstrate a skillful integration of large numbers of men (including some experienced builders), or men and animals, accomplishing tasks that appear extraordinarily demanding even by the standards of today's mechanized world. All quarrying was done by hand. Animal teams transported quarried stone to a site, and animals were sometimes used to power hoisting machines used to lift heavy pieces to a higher elevation, but otherwise traditional construction relied solely on human labor. Craftsmen used saws, axes, hammers, chisels, planes, augers, and trowels and worked compound pulleys or cranes and treadwheels for lifting timber, stones, and glass (Wilson 1990).

Cranes powered by men turning capstans or windlasses or treading drums could do that task readily, if slowly, and some machines—including Filippo Brunelleschi's (1377–1446) ox-powered hoist, used to raise masonry materials for building the spectacular cupola of the Cathedral of Santa Maria del Fiore in Florence, and a rotary crane to set the lantern top (Prager and Scaglia 1970)—were designed for specific demanding tasks (box 4.17). Some projects were completed speedily: the Parthenon in just 15 years (447–432 BCE), the Pantheon in about eight (118–125 CE), and Constantinople's Hagia Sophia, a high-vaulted Byzantine church later converted into a mosque, in less than five years (527–532).

Several types of large construction projects stand out. By far the best known are various ceremonial structures, above all funerary monuments and places of worship. The most remarkable structures in the first group, pyramids and tombs, are distinguished by their massiveness, while temples and cathedrals combine monumentality with complexity and beauty. Among the preindustrial utilitarian structures I would single out aqueducts because of their length and the combination of canals, tunnels, bridges, and inverted siphons. No accurate energy accounts can be prepared for the construction of any ancient structures, and even the energy cost of building medieval projects is not easy to estimate. But approximate calculations reveal substantial differences in total energy requirements, and even greater differences in average power flows.

Box 4.17

Brunelleschi's ingenious machines

Filippo Brunelleschi's work on the Cathedral of Santa Maria del Fiore is a perfect demonstration of the roles played by ingenious inventions in deploying the needed amount of energy in a suitable manner. Draft animals and laborers were readily available to supply the requisite power, but the record size of the cathedral's cupola (an inner span of 41.5 m in diameter) and, even more, the unprecedented manner of its construction (without any ground-based scaffolding) could not have been accomplished without Brunelleschi's new ingenious machines (Prager and Scaglia 1970; King 2000; Ricci 2014). Those machines were dismantled once the construction ended, but fortunately, their drawings have been preserved in Buonaccorso Ghiberti's *Zibaldone*.

They included ground-supported and elevated cranes and a reversible hoist, a rotary crane used in the construction of the lantern, elaborate winches, and, perhaps the most ingenious machine of all, a load positioner (not necessarily Brunelleschi's original invention but certainly an excellent execution of the idea). Materials for the cupola were lifted by a central (ox-powered) hoist. Bricks were easily moved to masons building the ascending curved structure, but the heavy stone blocks used for the tie rings (needed to arrest any spreading of the structure) could not be moved from the central elevated location to their precisely predetermined sites by pulling or pushing: the task was done by the load positioner with two horizontal screw-actuated slideways mounted on a vertical rod and using a counterweight.

Impressive funerary or religious structures requiring huge and sustained energy flows—long-range planning, outstanding organization, and large-scale labor mobilization—were built by every preindustrial high culture (Ching, Jarzombek, and Prakash 2011). These tombs and temples express the universal human striving for permanence, perfection, and transcendence (fig. 4.23). I would very much like to say something definite about the construction process and energy requirements of building the Egyptian pyramids, the grandest structures of the ancient world. We know that their construction required a meshing of long-range planning, efficient grand-scale logistics, effective supervision and servicing, and admirable, though nearly completely obscured, technical skills.

The largest pyramid, the pharaonic tomb of Khufu of the Fourth Dynasty, best embodies all these qualities. Built of nearly 2.5 million stones weighing on average about 2.5 t, this mass of over 6 Mt within a volume of 2.5 Mm^3 was assembled with remarkable precision, and with admirable speed. From

GIZA

TEOTIHUACAN

ANURADHAPURA

ELAM

Figure 4.23
Khufu's pyramid at Giza, Pyramid of the Sun at Teotihuacan, the Jetavana stupa at
Anuradhapura, and the Choga Zanbil ziggurat at Elam. Detailed information about
these structures is available in Bandaranayke (1974), Tompkins (1976), and Ching,
Jarzombek, and Prakash (2011).

the orientation of the Great Pyramid (using the alignment of two circumpolar stars, Mizar and Kochab) we can narrow the beginning of its construction to between 2485 and 2475 BCE (Spence 2000), and the structure was
completed in 15–20 years. Egyptologists concluded that core stones were
quarried at the Giza site, that the facing stones had to be brought from Tura
quarries across the Nile, and that the most massive granite blocks, those
forming the corbel roof inside the pyramid (the heaviest one weighing
nearly 80 t), had to be shipped from southern Egypt (Lepre 1990; Lehner
1997).

All of that seems quite intelligible. Ancient Egyptians mastered the craft of stone quarrying, in terms of both the mass output of similarly seized blocks and extracting massive monoliths. They could also move heavy objects on land and on boats. A well-known painting shows how a 50 t colossus from a cave at el-Bersheh (1880 BCE) was moved by 127 men (developing peak useful power of over 30 kW) on a sledge whose friction was reduced by a worker pouring water from a vessel. And that very large stones were transported on boats is attested by a unique image from Deir el-Bahari: two 30.7 m long Karnak obelisks were carried on a 63 m long barge pulled by about 900 oarsmen in 30 boats (Naville 1908).

But beyond quarrying and moving stones to the building site, all is conjecture; we still do not know how the largest pyramids were actually built (Tompkins 1971; Mendelssohn 1974; Hodges 1989; Grimal 1992; Wier 1996; Lehner 1997; Edwards 2003). The Egyptian hieroglyphic and pictorial record, so rich in many other aspects, provides no contemporaneous depictions or descriptions. The most common modern assumptions specify the use of clay, brick, and stone ramps, with no consensus about their form (a single inclined plane, multiple planes, an encircling ramp?) or slope (with suggested ratios as high as 1:3 and as low as 1:10). But such disagreements do not matter as it is highly unlikely that any construction ramps were used (Hodges 1989).

A single inclined plane would have to be completely rebuilt after every layer of stonework was finished, and with a manageable slope of 10:1 its volume would have far surpassed that of the pyramid itself. Ramps encircling the pyramid would have been narrow; quite difficult to build, buttress, and maintain under heavy use; and perilous if not impossible to negotiate. Pivoting ropes at right angles around corner posts was suggested as a solution, but we have no proof that Egyptians could do that or that it would actually work. In any case, there are no remnants of vast volumes of the ramp-building rubble anywhere on the Giza Plateau.

The earliest description of pyramid building was written by Herodotus (484–425 BCE) two millennia after their completion. During his Egyptian travels he was told that

> for the making of the pyramid itself there passed a period of twenty years; and the pyramid is square, each side measuring eight hundred feet, and the height of it is the same. ... This pyramid was made after the manner of steps, which some call "rows" and others "bases": and when they had first made it thus, they raised the remaining stones with machines made of short pieces of timber, raising them first from the ground to the first stage of the steps, and when the stone got up to this it was placed upon another machine standing on the first stage, and so from

this it was drawn to the second upon another machine; for as many as were the courses of the steps, so many machines there were also, or perhaps they transferred one and the same machine, made so as easily to be carried, to each stage successively, in order that they might take up the stones; for let it be told in both ways, according as it is reported. However that may be, the highest parts of it were finished first, and afterwards they proceeded to finish that which came next to them, and lastly they finished the parts of it near the ground and the lowest ranges.

Might this be the description of the actual construction method? Proponents of lifting think so, and they have offered many solutions as to how the work could have been done with the help of levers or simple but ingenious machines. Hodges (1989) argued for the simplest method of using wooden levers to lift stone blocks and then rollers to emplace them. Objections to this process rest above all on the large number of vertical transfers required for every block placed in higher rows and on the need for constant vigilance and accuracy to prevent accidental falls during the manipulation of stones weighing 2–2.5 t.

Construction specifics aside, first principles allow us to quantify the total energy required to build the Great Pyramid and hence to estimate the required labor force: my calculations (erring on a generous side rather than assuming theoretical minima) show that it could have been as low as 10,000 people (box 4.18). One of the few certainties regarding the pyramid construction is that order-of-magnitude-higher claims of required labor force are indefensible exaggerations. Feeding very large numbers of workers, most of them concentrated on the Giza Plateau, might have been a factor as limiting as or even more limiting than delivering and lifting the stones.

Other ancient structures that required a long-term labor commitment included the Mesopotamian stepped temple towers (ziggurats) built after 2200 BCE, and stupas (or *dagobas*), monuments built to honor Buddha and often housing relics (Ranaweera 2004). Falkenstein (1939) calculated that the construction of the Anu ziggurat near Warqa in Iraq required at least 1,500 men working 10 hours a day for five years, adding up to embodied energy of nearly 1 TJ. And Leach (1959) estimated that Jetavanaramaya, the largest Anuradhapura stupa (122 m tall, built with about 93 million roughly laid baked bricks), needed about 600 laborers for100 days a year for 50 years, or just over 1 TJ of useful energy (see fig. 4.23).

Mesoamerican pyramids, especially those at Teotihuacan (built during the second century CE) and Cholula, are also quite imposing. Teotihuacan's flat-topped Pyramid of the Sun was the tallest, probably just over 70 m, including the temple (see fig. 4.23). Its construction was much easier than the building of the three stone structures at Giza. The pyramid's core is

Box 4.18

Energy cost of the Great Pyramid

The Great Pyramid's potential energy (required to lift the mass of 2.5 Mm^3 of stones) is about 2.5 TJ. Wier (1996) got that total right, but his assumption of 240 kJ/day of average useful work was too low. These are my conservative assumptions. To cut 2.5 Mm^3 of stone in 20 years (the length of Khufu's reign) would require 1,500 quarrymen working 300 days per year and producing 0.25 m^3 of stone per capita by using copper chisels and dolerite mallets. Even assuming that three times as many stonemasons were needed to square and dress the stones (although many interior blocks were only rough-hewn) and to move them to the construction site, the total labor force supplying the building material would be on the order of 5,000 men.

With net daily inputs of useful energy at 400 kJ/capita, lifting the stones would have required about 6.25 million workdays, and prorated over 20 years and 300 workdays per year it could have been accomplished by about 1,000 workers. If the same number were needed to emplace the stones in the rising structure, and even if that number were doubled, reflecting the additional labor needed as organizers and overseers and for transport, repair of the tools, delivery of food, cooking of meals, and washing of clothes, the grand total would still be fewer than 10,000 men. During the peak labor periods pyramid workers at the Giza site were investing collectively at least 4 GJ of useful mechanical energy every hour, that is, an overall power of 1.1 MW, and to maintain this effort they consumed every day an additional 20 GJ of food energy, the equivalent of nearly 1,500 t of wheat.

Wier (1996) calculated the maximum of 13,000 pyramid builders for the 20-year period. Hodges (1989) calculated that 125 teams could have jacked up all the stones into position during 17 years of work, and that the numbers would add up to only about 1,000 permanent workers for stone lifting; he also allowed three years for the dressing of casing stones in place, proceeding from the top. In contrast, Herodotus was told of 100,000 men working for three months a year for 20 years, while Mendelssohn (1974) estimated the total at 70,000 seasonal laborers and perhaps as many as 10,000 permanent masons. Both of these are indefensible exaggerations.

made up of earth, rubble, and adobe bricks, and only the exterior was faced with cut stone, which was anchored by projecting catches and plastered with lime mortar (Baldwin 1977). Still, its completion could have required the work of up to 10,000 laborers for more than 20 years.

In contrast to our conjectures about the construction of the largest pyramids, there is little mystery about the way such classical structures as the Parthenon or the Pantheon were built (Coulton 1977; Adam

1994; Marder and Jones 2015). The Pantheon's remarkable design is often cited as an ingenious use of concrete, but the often repeated claim that the Romans were the first builders to use this material is inaccurate. Concrete is a mixture of cement, aggregates (sand, pebbles), and water, and cement is produced by high-temperature processing of a carefully formulated and finely ground mixture of lime, clay, and metallic oxides in an inclined rotating kiln—and there was no cement in the Roman *opus caementicium* used to build the Pantheon or in any other building until the 1820s (box 4.19).

We know that massive architraves (such as those at the Parthenon, weighing almost 10 t) had to be lifted by a crane (and could be rolled to the site encased in circular frames), and very similar crane designs were used nearly two millennia later in building cathedrals, the most elaborate

---

Box 4.19
Rome's Pantheon

The Roman *opus caementicium* was a mixture of aggregates (sand, gravel, stones, often also broken bricks or tiles) and water, but its bonding agent was not cement (as it is in concrete) but lime mortar (Adam 1994). The mixture was prepared on a building site, and the unique combination of slaked lime and volcanic sand—excavated near Puteoli (modern-day Pozzuoli, just a few kilometers west of Mount Vesuvius) and known as *pulvere puteolano* (later as *pozzolana*)—made a sturdy material that could harden even under water. Although inferior in comparison with modern concrete, the pozzolanic aggregate and a high-quality lime produced material that was strong enough not only for massive and durable walls but also for large vaults and domes (Lancaster 2005).

The Roman use of *opus caementicium* reached its design apogee in the Pantheon, whose construction was completed in 126 CE during Hadrian's rule. The large dome, 43.3 m in diameter (the structure's interior could fit into a sphere of the same diameter), was never topped by any preindustrial builders, although Saint Peter's dome, designed by Michelangelo and completed in 1590, came close, at 41.75 m (Lucchini 1966; Marder and Jones 2015). Besides its obvious visual appeal, the dome's most remarkable property is its vertically decreasing specific mass: the five rows of square coffered ceiling not only diminish in size as they converge on the central oculus but are built of progressively thinner layers of masonry using lighter aggregates, from travertine at the bottom to pumice at the top (MacDonald 1976). The entire dome weighs about 4,500 t.

structures of the European Middle Ages. Their builders included many experienced craftsmen and required the use of many special tools (Wilson 1990; Erlande-Brandenburg 1994; Recht 2008; Scott 2011). Much of the labor was seasonal, but a typical need would be an equivalent of hundreds of full-time workers—lumbermen, quarrymen, wagon drivers, carpenters, stonemasons, glass workers—engaged for one to two decades. The total energy investment was thus two orders of magnitude smaller than in pyramid building, with peak labor flows only a few hundred kilowatts.

Although some cathedrals were completed quickly (Chartres took only 27 years, the original Notre-Dame de Paris took 37), construction was commonly interrupted by epidemics, labor disputes, regime changes, money shortages, and intra- and international conflicts. As a result, the building of a cathedral usually lasted several generations, and in some cases even centuries were needed for the completion: Prague's Saint Vitus cathedral, begun by Charles IV in 1344, was abandoned in the early fifteenth century, and the unfinished (and provisionally walled-off) structure was completed (with the erection of two Gothic spires) only in 1929 (Kuthan and Royt 2011).

Extensive waterworks, including dams, canals, and bridges, are well documented from Jerusalem, Mesopotamia, and Greece. But Roman achievements are certainly the best-known examples of bold engineering solutions to an urban water supply. Virtually every sizable Roman town had a well-planned water supply. This accomplishment was surpassed only by industrializing Europe. Roman aqueducts were especially impressive (fig. 4.24). Pliny in his *Historia Naturalis* called them "the most remarkable achievement anywhere in the world."

Starting with the Aqua Appia in 312 BCE, the water supply system eventually comprised 11 lines, totaling almost 500 km (Ashby 1935; Hodge 2001). By the end of the first century CE the total daily water supply was just above 1 Mm^3 (1 GL)/day, averaging more than 1,500 L/capita, while by the end of the twentieth century Rome (with a population of about 3.5 million) averaged (including all industrial use) about 500 L/capita (Bono and Boni 1996). Equally impressive was the scale of Rome's underground sewage canal system, with *cloaca maxima* arches about 5 m in diameter.

Throughout the Roman Empire, aqueducts consisted of a number of common structural elements (fig. 4.24). Starting from springs, lakes, or artificial impoundments, water channels had a rectangular cross section and were built of stone slabs or concrete lined with fine cement. Channels with the usual gradient of no less than 1:200 followed slopes in order to avoid tunneling whenever possible. Where an underground course was

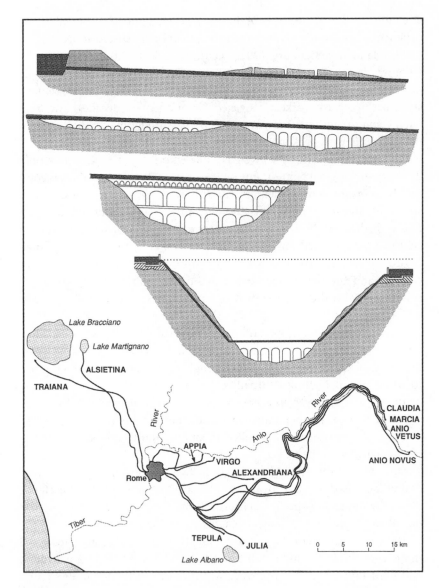

Figure 4.24
Roman aqueducts carried water from rivers, springs, lakes, or reservoirs by combining at least two or three of the following structures (from the top): shallow rectangular channels running on a foundation, tunnels accessible by shafts, embankments pierced by arches, single- or double-tiered arched bridges, and inverted lead-pipe siphons to take water across deep valleys. Rome's aqueducts, supplying about 1 Mm^3/day of water, formed an impressive system built over a period of more than 500 years. Based on Ashby (1935) and Smith (1978). Aqueduct slope is exaggerated.

unavoidable, the channel could be accessed from above by shafts. Only in valleys too long to skirt or too deep for simple embankments did the Romans resort to bridges. No more than about 65 km of Rome's aqueducts were carried on (sometimes shared) arches. The Augustan bridges at Gard (more than 50 m high), Merida, and Taragona are the finest examples of this art. The cleaning and repair of channels, tunnels, and bridges, which were often threatened by erosion, was a continuous task.

If crossing a valley would have required a bridge taller than 50–60 m, Roman engineers opted for an inverted siphon. Its pipes connected a header tank on one side of the valley with a slightly lower-lying receiving tank on the opposite side (Hodge 1985; Schram 2014). Crossing the stream at the valley bottom still required building a bridge. The high energy cost of these structures primarily reflected the large amounts of lead needed for high-pressure pipes—they could withstand up to 1.82 MPa (18 atmospheres)—and the cost of transporting the metal over often considerable distances from its smelting centers. For example, the total amount of lead for nine siphons in the Lyon water supply was about 15,000 t.

Metallurgy

The beginnings of all old high cultures are marked by the use of color (nonferrous) metals. Besides copper, early metallurgists also recognized tin (which was combined with copper to make bronze), iron, lead, mercury, and the two precious elements, silver and gold. Mercury is a liquid at ambient temperatures, while gold's relative scarcity and softness precluded its use beyond the minting of coins and ornamental items. Although much more abundant, silver was also too rare for mass-produced items. The softness of lead and tin limited their unalloyed uses largely to pipes and food containers. Only copper and iron were relatively abundant and possessed, especially when alloyed, great tensile strength and hardness. The combination of their abundance and their properties made them the only two practical choices for the mass production of durable items. Copper and bronze dominated the first two millennia of recorded history, while iron and its alloys (an enormous variety of steels) are now dominant more than ever.

Charcoal fueled the smelting of both nonferrous and iron ores, as well as the subsequent refining and finishing of crude metals and metallic objects. Hard human labor did all ore mining and crushing, tree cutting and charcoal making, furnace building and charging, casting, refining, and repeated forging. In many societies, ranging from sub-Saharan Africa to Japan, metallurgy remained solely manual until the introduction of modern industrial methods. In Europe and later in North America, animals and above all water power took over such repetitive, exhausting tasks as ore crushing, water pumping from mines, and metal forging. The availability of wood, and later also the accessibility and reliability of water power resources needed to energize larger bellows and hammers, were thus the key determinants of metallurgical progress.

Nonferrous Metals

Copper tools and weapons bridged the stone and the iron eras of human evolution. The first copper uses, datable to the sixth millennium BCE, did not involve any smelting. Pieces of naturally pure metal were merely shaped by simple tools or worked by annealing, alternating heating and hammering (Craddock 1995). The earliest evidence of exploiting native metal (in the form of beads of malachite and copper in southeastern Turkey) goes as far back as 7250 BCE (Scott 2002). Smelting and casting of the metal became common after the middle of the fourth millennium BCE in a number of regions with rich and relatively accessible oxide and carbonate ores (Forbes 1972). Numerous copper objects—rings, chisels, axes, knives, and spears—were left behind by the early Mesopotamian societies (before 4000 BCE), predynastic Egypt (before 3200 BCE), the Mohenjodaro culture in the Indus valley (2500 BCE), and the ancient Chinese (after 1500 BCE).

The copper-mining centers of antiquity included most notably Egypt's Sinai peninsula, North Africa, Cyprus, today's Syria, Iran, and Afghanistan, the Caucasus region, and Central Asia. Italy, Portugal, and Spain later became production regions. Because of the metal's relatively high melting point (1083°C), the production of pure copper was fairly energy-intensive. Ore reduction was done with wood or charcoal, first just in clay-lined pits and later in simple, low shaft clay furnaces with a natural draft. The first clear evidence of bellows use comes from Egypt of the sixteenth century BCE, but their use is almost certainly older than that. Impure metal was refined by heating in small crucibles, after which it was cast into stone, clay, or sand molds. Castings were fashioned into utilitarian or ornamental products by hammering, grinding, piercing, and polishing.

Much higher technical skills were necessary for producing the metal from abundant sulfide ores (Forbes 1972). They had to be first crushed and roasted in heaps or furnaces to remove sulfur and other impurities (antimony, arsenic, iron, lead, tin, and zinc), which would change the metal's properties. For millennia, the crushing of ores was done by hand hammering, a practice common in Asia and Africa until the twentieth century. In Europe, waterwheels and horses harnessed to whims gradually took over this work. The roasting of crushed ores needed relatively little fuel. The smelting of roasted ores in shaft furnaces was followed by smelting of the coarse metal (only 65–75% of copper) and its resmelting to produce nearly pure (95–97%) blister copper. This product could be further refined by oxidation, slagging, and volatilization. The sequence added up to high fuel requirements.

Calculating the annual and cumulative fuel demand of ancient smelting operations is an inherently uncertain exercise, strongly influenced by the estimates of total slag mass and by assumptions about the length of the extraction and the energy intensity of smelting. All these uncertainties are perfectly illustrated by the ancient world's largest smelting concentration, Rio Tinto, in southwestern Spain, less than 100 km west of Seville (box 4.20). In any case, the extent of Roman smelting operations remained unsurpassed for another 1,500 years. Summaries of the late medieval metallurgical expertise (Agricola 1912 [1556]; Biringuccio 1959 [1540]) describe copper smelting in ways that hardly differ from the Rio Tinto practices.

Box 4.20
Fuelwood needs for Roman copper and silver smelting at Rio Tinto

The first mapping of Rio Tinto's enormous slag heaps resulted in estimates of 15.3 Mt of slag from lead and silver mining and 1 Mt of slag from copper mining; these estimates led Salkield (1970) to conclude that the Romans had to cut down 600,000 mature trees a year to fuel the smelting, an impossible total for southern Spain. New mapping (based on extensive drilling) ended with about 6 Mt of slag, and although copper was the main product during the Roman era, there was an extensive pre-Roman smelting of silver (Rothenberg and Palomero 1986). With a 1:1 slag:charcoal ratio and a 5:1 wood:charcoal ratio, the production of 6 Mt of slag would have needed 30 Mt of wood, or 75,000 t/ year during 400 years of large-scale operations.

Supplying this fuel by cutting down natural forest (storing no more than 100 t/ha) would have required clearing annually about 750 ha of forest, the equivalent of a circle with a radius of about 1.5 km: that would have been a major but manageable undertaking, and one resulting in extensive deforestation. Similarly, centuries of copper smelting on Cyprus (starting around 2600 BCE) left behind more than 4 Mt of slag. Clearly, ancient smelting was a major cause of deforestation in the Mediterranean region, as well as in Transcaucasia and Afghanistan, and local wood shortages eventually limited the extent of smelting.

From the very beginning of copper smelting some of the metal was incorporated into bronze, the first practical alloy, chosen by Christian Thomsen for his now classical division of human evolution into Stone, Bronze, and Iron Ages (Thomsen 1836). This is a highly generalized division. Some societies, most notably Egypt before 2000 BCE, went through a pure copper era, while others, above all in sub-Saharan Africa, moved directly from the Stone Age to the Iron Age. The first bronzes came out of the inadvertent smelting of copper ores containing tin. Later they were produced by cosmelting of the two ores, and only after 1500 BCE were they made by smelting two metals together. With its low melting point of a mere 231.97°C, tin was produced with relatively little charcoal from its crushed oxide ores. The total energy cost of bronze was thus lower than that of pure copper, but it was an alloy with superior properties.

As the shares of tin varied anywhere between 5% and 30% (and, consequently, the melting points ranged between 750°C and 900°C), it is impossible to speak of a typical bronze. An alloy preferred for casting guns, composed of 90% copper and 10% tin, had both the tensile strength and hardness about 2.7 times higher than the best cold-drawn copper (Oberg et al. 2012; box 4.21). The availability of bronze thus brought the first good metallic axes, chisels, knives, and bearings, as well as the first reliable swords, of both cutting and thrusting type. Bronze bells usually were 25% tin.

Brass has been the other historically important copper alloy combining the element (<50% up to about 85% of the total) with zinc. As with bronze,

---

Box 4.21
Tensile strength and hardness of common metals and alloys

Metal or alloy	Tensile strength (MPa)	Hardness (Brinell scale)
Copper		
Annealed	220	40
Cold drawn	300	90
Bronze (90% Cu, 10% Sn)	840	240
Brass (70% Cu, 30% Zn)	520	150
Cast iron	130–310	190–270
Steel	650–>2,000	280– >500

Source: Based on Oberg and co-workers (2012).

its production needs less energy than the smelting of pure copper (zinc's melting point is only 419°C). A higher zinc content improves the alloy's tensile strength and hardness. For a typical brass they are about 1.7 times higher than for cold-drawn copper, but there is no reduction of the alloy's malleability and corrosion resistance. The first uses of brass date to the first century BCE. The alloy became widely used in Europe only during the eleventh century, and became common only after 1500.

Iron and Steel

The replacement of copper and bronze by iron proceeded slowly. Small iron objects were produced in Mesopotamia during the first half of the third millennium BCE, but ornaments and ceremonial weapons became more common only after 1900 BCE. The extensive use of iron dates only to after 1400 BCE, and the metal became truly abundant after 1000 BCE. Egypt's iron era dates from the seventh century BCE, China's from the sixth. African iron making is also ancient, but the metal was never smelted by any New World society. Iron smelting was necessarily bound up with the large-scale production of charcoal. Iron melts at 1535°C; an unaided charcoal fire can reach 900°C, but a forced air supply can raise its temperature close to 2000°C. Charcoal thus fueled all iron ore smelting in every traditional society except China (where coal was also used since the Han dynasty), but the efficiency of its production and of its metallurgical use had gradually improved.

The development of iron making began with fires enclosed in shallow, and often clay- or stone-lined, pits where crushed iron ore was smelted with charcoal. These primitive hearths were commonly located on hilltops to maximize the natural draft. Later a few narrow clay tubes (tuyères) were used to deliver a blast into the hearth, first from small hand-operated leather bellows, then from larger bellows driven by a treadle or a rocking bar, and throughout Europe eventually powered by waterwheels. Simple clay walls were erected to contain the smelting: they were just a few decimeters to more than a meter tall, but in some parts of the Old World (including Central Africa) they eventually reached more than 2 m (van Noten and Raymaekers 1988).

Archaeologists have unearthed thousands of these temporary structures throughout the Old World, from the Iberian peninsula to Korea and from northern Europe to Central Africa (Haaland and Shinnie 1985; Olsson 2007; Juleff 2009; Park and Rehren 2011; Sasada and Chunag 2014). The temperature inside these small charcoal-fueled furnaces reached no more than 1100–1200°C, high enough to reduce iron oxide but far below iron's melting point (pure Fe liquefies at 1535°C): their final product was a bloom (whose typical medieval mass was just 5–15 kg, later 30–50 kg or even more than 100 kg), a spongy mass of iron, and an iron-rich slag full of nonmetallic impurities (Bayley, Dungworth, and Paynter 2001).

This bloomery iron contained 0.3–0.6% carbon and had to be repeatedly reheated and hammered to produce a lump of tough and malleable wrought iron containing less than 0.1% carbon. Wrought iron was used in making objects and tools ranging from nails to axes. The European demand for bloomery iron began to rise in the eleventh century thanks to the adoption of iron mail and the increased production of hand weapons and helmets, as well as of common tools and implements ranging from sickles and hoes to hoops and horseshoes. Metal bands were also used in the building of cathedrals, and the new papal palace, the Palais des Papes, in Avignon, whose construction began in 1252, required 12 t of the construction metal (Caron 2013).

Han dynasty (207 BCE–220 CE) Chinese were the first craftsmen to produce liquid iron. Their furnaces, built with refractory clays and often strengthened by vine cables or heavy timbers, eventually reached just over 5 m. They could take a charge of nearly 1 t of iron ore, and produced cast iron in two tappings a day. The high phosphorus content, which lowered the iron's melting point, and the invention of double-acting bellows, which delivered a strong air blast, were critical ingredients of this early success (Needham 1964). Later came the use of coal packed around the batteries of tubelike crucibles containing the ore and the powering of larger bellows by waterwheels. Casting into interchangeable molds was used commonly to mass produce iron tools, thin-walled cooking pots and pans, and statues before the end of the Han dynasty (Hua 1983). There were few subsequent substantial improvements, and China's small blast furnaces did not start the lineage of today's huge structures.

These originated in the slow evolution of European shaft furnaces, from the simple Catalan forge to the rock-lined osmund furnaces of Scandinavia to the *Stuckofen* of Styria. Higher stacks and better shapes lowered fuel consumption. Higher temperatures and longer contact between the ore and the fuel produced liquid iron. European blast furnaces originated most likely in the lower Rhine valley region just before 1400. Blast furnaces produce cast, or pig, iron, an alloy with 1.5–5% of carbon that cannot be directly forged or rolled. Its tensile strength is no higher than copper's (and it can be up to 55% weaker) but it is two to three times harder (Oberg et al. 2012; box 4.21).

The number of blast furnaces grew steadily during the sixteenth and seventeenth centuries. Perhaps the most notable improvement of that time was the introduction of larger leather bellows. Their tops and bottoms were made of wood, their sides of bull hides. After 1620 came double bellows operated alternatively by the cams on the waterwheel axle, as well as a gradual elongation of the stack. Both these trends soon ran into limits imposed by the maximum power of waterwheels and by charcoal's physical properties. By 1750 the largest waterwheels were delivering up to 7 kW of useful power. But during the summer smelting campaigns there often was not enough water to generate the maximum output. Charcoal's main disadvantage is its high friability: it crushes under heavier loads, and its use limited the mass of charged ore and limestone and hence the height of blast-furnace stacks to less than 8 m (Smil 2016; fig. 4.25). Before 1800 both of these limits were removed, the first one by Watt's steam engine, the other one by the use of coke.

Medieval bloomery hearths needed 3.6–8.8 times more fuel than the mass of the charged ore (Johannsen 1953). Even with ores containing about 60% Fe, bloomeries would have needed at least 8 and as much as 20 kg of charcoal per kilogram of hot metal. By the end of the eighteenth century typical charcoal/metal ratios were around 8:1; they fell to just around

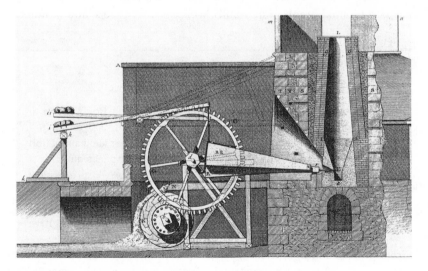

Figure 4.25
Charcoal-fueled blast furnace of the mid-eighteenth century, with bellows powered by an overshot waterwheel. Reproduced from the *Encyclopédie* (Diderot and d'Alembert 1769–1772).

1.2 by 1900, and to 0.77 in Swedish charcoal furnaces (Campbell 1907; Greenwood 1907). A good late nineteenth-century charcoal-fueled furnace thus needed only about one-tenth the energy of its medieval counterpart. The high energy requirements of pre-1800 charcoal-fueled smelting inevitably caused extensive deforestation around furnace sites. A typical early eighteenth-century English furnace needed about 1,600 ha of trees for a sustainable supply (box 4.22).

The total national wood requirements of charcoal-based iron making can be fairly well estimated for England of the early 1700s, before the industry began to adopt coke-based smelting: a sustainable supply would have required harvesting coppiced or natural wood growth from about 1,100 km^2 of wood groves or forests (box 4.22). A century later the United States had no problems energizing its iron ore smelting with charcoal made from wood from its rich natural forests, but by the beginning of the twentieth century it would have been impossible, and only the use of coke allowed the country to become the world's larger producer of pig iron (box 4.23).

Not surprisingly, during the wooden era communities surrounded by traditional iron mills and forges found themselves in a desperate situation. Already in 1548 anguished inhabitants of Sussex wondered how many towns were likely to decay if furnaces continued working: they would have no wood to build houses, water mills, wheels, barrels, piers, and hundreds

---

Box 4.22
Fuel needs of an eighteenth-century English blast furnace

Early eighteenth-century English blast furnaces worked only from October to May, and during that time their average output was just 300 t of pig iron (Hyde 1977). Assumptions of as little as 8 kg of charcoal per kilogram of iron and 5 kg of wood per kilogram of charcoal translate into annual requirements of some 12,000 t of wood for a single furnace. After 1700, nearly all accessible natural forest growth was gone, and the wood was cut in 10- to 20-year rotations from coppicing hardwoods, whose annual harvestable increment would be between 5 and 10 t/ha. A medium productivity of 7.5 t/ha would have required about 1,600 ha of coppicing hardwoods for perpetual operation. For comparison, a much less efficient, large seventeenth-century English furnace in the Forest of Dean needed about 5,300 ha of coppice growth, while the smaller Wealden ironworks needed around 2,000 ha for each furnace-forge combination (Crossley 1990).

Box 4.23

Energy needs in British and American iron production

In 1720, 60 British furnaces produced about 17,000 t of pig iron, requiring, with 40 kg of wood per kilogram of metal, about 680,000 t of wood. Forging the metal to produce 12,000 t of bars added, with 2.5 kg of charcoal per kilogram of bars, another 150,000 t, for a total annual consumption of some 830,000 t of charcoaling wood. With an average productivity of 7.5 t/ha, this would have required about 1,100 km^2 of forests and coppiced growth for sustainable harvests.

For the United States, the earliest available pig iron total is for the year 1810, when about 49,000 t of the metal needed (assuming 5 kg of charcoal, or at least 20 kg of wood per kilogram of hot metal) about 1 Mt of wood. At that time, all that wood could come from clear-cutting natural old-growth hardwood forests, rich ecosystems that stored around 250 t/ha (Brown, Schroeder, and Birdsey 1997), and if all aboveground phytomass were used in charcoaling, an area of about 4,000 ha (a square with a side of 6.3 km) would have to be cleared every year to sustain that level of production. The rich U.S. forests could support an even higher rate, and by 1840 all U.S. iron was still smelted with charcoal, but after a subsequent rapid switch to coke energized nearly 90% of iron production by 1880 and future increases in iron production could not be based on charcoal: in 1910—with an iron output at 25 Mt, and even with much reduced charges of 1.2 kg of charcoal and 5 kg of wood per kilogram of hot metal—the country would have required 125 Mt of wood a year.

Even when assuming a high average increment of 7 t/ha in secondary-growth forests, a sustainable supply of that wood would have required annual harvests from nearly 180,000 km^2 of forest, an area equal to Missouri (or a third of France), represented by a square whose side would reach from Philadelphia to Boston or from Paris to Frankfurt. Obviously, even a forest-rich America could not afford to energize its iron ore smelting with charcoal.

of other necessities—and they asked the king to close down many of the mills (Straker 1969; see also Smil 2016). The limiting role of energy in traditional iron smelting was thus unmistakable. When a single furnace could strip each year a circle of forest with a radius of about 4 km, it is easy to appreciate the cumulative impact of scores of furnaces over a period of many decades.

This effect was necessarily concentrated in wooded mountainous regions. There the radius of animal-drawn charcoal transport could be kept to a minimum (a restriction further aggravated by the fuel's fragility), and the need to power furnace and forge bellows was readily satisfied by install-ing waterwheels. Proximity to the ore was also important, but because the ore charge was only a fraction of charcoal's weight, it was easier to trans-port. Deforestation was the inevitable environmental price paid for making nails, axes, and horseshoes, as well as mail shirts, lances, guns, and cannon-balls. The early expansion of iron making and the limited supply of domes-tic wood led to a clear energy crisis in Britain during the seventeenth century. That situation was further aggravated by the high timber demand in the country's burgeoning shipbuilding industry.

While iron was relatively abundant in many preindustrial societies, steel was available only for special uses. Like cast iron, steel is also an alloy, but one containing only between 0.15% and 1.5% of carbon and often also small amount of other metals (mainly nickel, manganese and chromium). The metal is superior to cast iron or to any copper alloys: the best tool steels have a tensile strength an order of magnitude higher than either copper or iron (Oberg et al. 2012; box 4.21). Some simple ancient smelting tech-niques could produce directly relatively high-quality steel, but only in small amounts. Traditional East African steelmakers used low (< 2 m), circular, cone-shaped charcoal-fueled slag and mud furnaces built over a pit of charred grass. Eight men operating goatskin bellows connected to ceramic tuyères were able to achieve temperatures above 1800°C (Schmidt and Avery 1978). This method, apparently known since the early centuries of the Common Era, made it possible to produce directly small amounts of good-quality medium carbon steel.

But preindustrial societies usually followed one of the two effective ancient routes toward steel: either by carburizing wrought iron or by decarburizing cast iron. The first, older technique entailed prolonged heat-ing of the metal in charcoal, resulting in gradual inward diffusion of car-bon. Without further forging this method produced a hard steel layer over a core of softer iron. This was a perfect material for plowshares—or for body armor. Repeated forging distributed the absorbed carbon fairly evenly and produced excellent sword blades. Decarburization, the removal of carbon from cast iron by oxygenation, was done in China already during the Han dynasty and produced the metal for such exacting applications as chains for suspension bridges.

The spreading availability of iron and steel gradually led to a number of profound social changes. Iron saws, axes, hammers, and nails sped up house construction and improved its quality. Iron kitchenware and a vari-ety of other utensils and objects, ranging from rings to rakes, from grates to graters, made it easier to cook and run a household. Iron horseshoes and plowshares were instrumental in advancing the intensification of cropping. On the destructive side, warfare was revolutionized first by flexible chain-mail suits, helmets, and heavy swords, later by guns, iron cannonballs, and more reliable firearms. These trends were greatly accelerated with the introduction of coke-based iron smelting and the emergence of the steam engine.

Warfare

Armed conflicts have always had a formative role in history: they require the mobilization of energy resources (often on an extraordinary scale, be it by marshalling masses of foot soldiers equipped with simple weapons or by producing highly destructive explosives and machines and laying down supplies for prolonged wars), and they have repeatedly resulted in the most concentrated and the most devastating releases of destructive power. Moreover, the basic energy supplies, whether of food or fuel, of populations exposed to armed conflicts are affected not only during the conflict's duration (through the acquisition of food for roaming armies, the destruction of crops, or the interruption of normal economic activities with the mobilization of young males and the damage inflicted on settlements and infrastructures) but often for years after its end.

All historical conflicts have been fought with weapons, but weapons are not the prime movers of war: two exceptions aside, until the invention of gunpowder the only prime movers of wars were human and animal muscles. The first exception was the use of incendiary materials; the second, of course, was the use of wind-powered sails to speed up and facilitate naval maneuvers. Traditional mechanical weapons—handheld (daggers, swords, lances) and projectile (spears, arrows, heavy weights discharged by catapults and trebuchets)—were designed to maximize physical damage through the sudden release of kinetic energy. Only the invention of gunpowder introduced a new, and much more powerful, prime mover. The explosive reaction of chemicals could propel projectiles faster and farther and increase their destructive impact. For centuries this impact was limited by the clumsy designs of personal weapons (front- and breach-loading rifles), but gunpowder gained ever greater importance as the propellant of cannonballs.

Animate Energies

All prehistoric land warfare and all conflicts of ancient and early medieval eras were powered solely by human and animal muscles. Warriors wielded daggers, axes, and swords in close combat, on foot or on horseback. They used spears and lances, and they drew bows and much more powerful crossbows (both the Chinese and the Greeks used them since the fourth century BCE) to shoot arrows whose impact would injure and kill unprotected enemies as far as 100–200 m away. The antiquity of archery warfare is attested by the fact that the Egyptian hieroglyph for soldier is a man kneeling on his left knee, bow in his outstretched right arm and a quiver on his left shoulder (Budge 1920). Animate energies also wound winches of massive catapults and took advantage of leveraged gravity to hurl massive weights by trebuchets in order to breach city walls and destroy castle fortifications.

Handheld weapons could cause grievous injuries, and well-aimed cuts or thrusts could kill instantly, but they required a commingling of fighting forces, and their power was obviously limited by the capabilities of a warrior's muscles. Bows and arrows allowed the separation of fighting forces, and master archers achieved admirable accuracy over relatively long distances, but archery battles wasted too many arrows through inaccurate targeting and the relatively low kinetic energy of light arrows (box 4.24), and the time needed to reset the arrows between successive discharges limited the magnitude and frequency of injuries that could be inflicted by these weapons. The limits of human performance also determined the daily range of advancing armies, and even if well-rested and well-fed men were able to

---

Box 4.24
Kinetic energy of swords and arrows

Even heavy medieval swords weighed no more than 2 kg, usually less than 1.5 kg. Kinetic energy increases with the square value of speed: it is only 9 J for a 2 kg sword clumsily wielded at just 3 m/s, 75 J for a 1.5 kg Japanese katana the (traditional Japanese curved, slender, 60–70 cm long single-edged sword) swung by an expert swordsman at 10 m/s. This appears to be rather low, but the impact of a slashing cut was highly concentrated, aimed at a narrow part of a body (neck, shoulder, arm), while the piercing thrust penetrated deeply into soft body tissues. A typical lightweight arrow weighing just 20 g and launched by a good archer from a compound bow would fly at up to 40 m/s (Pope 1923) and its kinetic energy would be 16 J. Again, this may seem low, but the projectile's impact is basically punctiform and hence deeply penetrating: flint- or metal-tipped arrows could easily penetrate a coat of mail when shot from distances of up to 40–50 m and, when well aimed, could kill unprotected men from more than 200 m away.

move swiftly, the army's progress was often limited by the speed of its supply train, made up of slow-moving animals.

The two most powerful military machines of antiquity and the early Middle Ages relied on the mechanical advantage of levers. Catapults were large-scale mechanized bows powered by a sudden release of the elastic deformation of twisted ropes or sinews (fig. 4.26). They were in use from the fourth century BCE (Soedel and Foley 1979; Cuomo 2004). They could shoot arrows or throw objects; mangonel catapults, used in city sieges, were third-class levers: their base was a fulcrum, force was provided by tension bands, and the load was thrown with a speed unattainable by the direct deployment of human muscles. Typical medieval catapults throwing stones of 15–30 kg could do only limited damage to city walls, however.

In contrast, trebuchets, invented in China before the third century BCE, were first-class levers with beams that pivoted around an axle and a projectile loaded at the end of the throw arm that was four to six times the length of the short arm (Hansen 1992; Chevedden et al. 1995). The earliest, smaller trebuchets were operated by men pulling ropes attached to the short arm; later large machines had massive counterweights and were able to throw objects weighing hundreds of kilograms (with record weights around or even surpassing 1 t) farther than the range of early medieval artillery. They were also used in defense against sieges, with trebuchets positioned on the high ramparts of castles or city walls, ready to hurl massive stones at any siege construction within their reach.

Animals in preindustrial warfare had two distinct roles: as the enablers of rapid and long-distance thrusts and as an indispensable means of transportation, making it possible to field larger armies whose supplies were carried by pack or draft beasts. In the earliest recorded depictions horses were harnessed to light chariots with spoked wheels (first used around 2000 BCE). No other traditional military innovation preceding the use of gunpowder was as consequential, owing to the combination of speed and the possibility of rapid tactical adjustments, as archers on horseback. Riding small horses and shooting arrows with powerful compound bows, mounted archers (first Assyrians and Parthians, later Macedonians and Greeks) were a formidable and highly mobile fighting force centuries before the introduction of stirrups (Drews 2004).

These simple pieces of metal providing footholds for a rider were first used in China in the early third century CE and then diffused westward; they gave riders unprecedented support and stability in the saddle (Dien 2000). Without them a fighter clad in armor could not even mount a larger (and sometimes also partially armored) horse, and would have been unable to fight effectively with a lance or a heavy sword while on its back. This

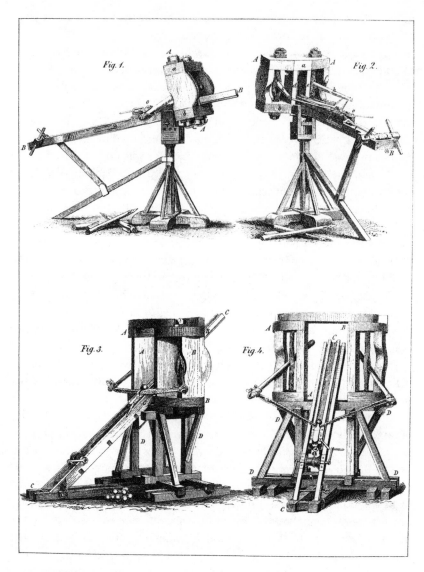

Figure 4.26
Roman catapults (Corbis).

does not mean that riders equipped with stirrups enjoyed easy supremacy in battle. Asian riders, unarmored and with small but extraordinarily hardy horses, created especially effective fighting units: they could move with high speed and enjoyed superior maneuverability.

This combination brought the Mongolian horsemen from eastern Asia to the center of Europe between 1223 and 1241 (Sinor 1999; Atwood 2004; May 2013) and enabled several empires of the steppes to survive in Central Asia until the early modern era (Grousset 1938; Hildinger 1997; Amitai and Biran 2005; Perdue 2005). The most spectacular series of long-distance forays by armored riders brought the Crusaders from many European countries to the eastern Mediterranean, where they established (between 1096 and 1291) temporary rule over fluctuating coastal and interior areas of what is now Israel and parts of Jordan, Syria, and Turkey (Grousset 1970; Holt 2014).

The importance of horses, both in cavalry units and harnessed to heavy wagons and field artillery, persisted in all major Western conflicts of the early modern era (1500–1800), as well in the epoch-defining Napoleonic Wars. Large armies projected far from their home base had to rely on animals to move their supplies: pack animals (donkeys, mules, horses, camels, llamas) were used in difficult terrain; draft animals (mostly oxen, in Asia also elephants) pulled heavy supply wagons and increasingly heavier field guns. The mass requirements of large military campaigns relying on draft power are well illustrated by the list of provisions and animals that occupied Prussia agreed to supply to Napoleon's armies for their invasion of Russia in 1812 (box 4.25). Without oxen—44,000 of them—pulling the supply wagons, the army could not have advanced.

Western armed conflicts fought after 1840 would use the first modern inanimate prime mover, the steam engine, to mobilize troops and animals and move them to the front lines by trains (or, in the case of troops sent to colonial wars on other continents, to embarkation ports to be loaded on steamships), but movement on the battlefield was still solely powered by human and animal muscles. And although World War I saw the first deployment of new inanimate mechanical prime movers (internal combustion engines powering trucks, tanks, ambulances and airplanes) in or near combat zones, horses remained indispensable.

By late 1917 the British armies on the Western front relied on 368,000 horses (with two-thirds of them engaged in transporting supplies, the rest in cavalry units), and although the Wehrmacht's advances in France (in spring 1940) and Russia (in summer 1941) are often cited as textbook examples of rapid mechanized tank-led warfare, Germany mobilized 625,000

Box 4.25

Prussian supplies and animals for the Russian invasion

Opening the road to Russia to Napoleon: that is how Philip Paul, comte de Ségur (1780–1873), one of Napoleon's young generals and perhaps the most famous chronicler of the disastrous Russian invasion, described the Prussian contribution:

> By this treaty, Prussia agreed to furnish two hundred thousand quintals of rye, twenty-four thousand of rice, two million bottles of beer, four hundred thousand quintals of wheat, six hundred and fifty thousand of straw, three hundred and fifty thousand of hay, six million bushels of oats, forty-four thousand oxen, fifteen thousand horses, three thousand six hundred waggons, with harness and drivers, each carrying a load of fifteen hundred weight; and finally, hospitals provided with everything necessary for twenty thousand sick. (Ségur 1825, 17)

---

horses for its invasion of Russia, and by the war's end the Wehrmacht had about 1.25 million animals (Edgerton 2007). Similarly, the Soviet armies deployed hundreds of thousands of horses in their advance from Moscow and Stalingrad to Berlin (fig. 4.27). Hay and oats thus remained in the category of strategic matériel until the end of World War II.

Explosives and Guns

The only inanimate energies used in the pregunpowder warfare were incendiary materials prepared by combining sulfur, asphalt, petroleum, and quicklime and either fastened to arrowheads or hurled at targets across moats and walls from catapults and trebuchets. Gunpowder's origins stem undoubtedly from the long experience of Chinese alchemists and metallurgists (Needham et al. 1986; Buchanan 2006). They worked with the three ingredients—potassium nitrate (KNO_3, saltpeter), sulfur, and charcoal—long before they started to combine them. The first incipient gunpowder formula comes from the mid-ninth century; clear directions for preparing gunpowders were published in 1040. The early mixtures consisted of only about 50% saltpeter and were not truly explosive. Eventually the mixtures capable of detonation settled at 75% saltpeter, 15% charcoal, and 10% sulfur.

Unlike in ordinary combustion, where oxygen must be drawn from the surrounding air, ignited KNO_3 readily provides its own oxygen, and the gunpowder rapidly produces a roughly 3,000-fold expansion of its volume in gas. When appropriately confined and directed in rifled gun barrels, a small amount of gunpowder can impart to bullets kinetic energy an order

Figure 4.27
Soviet cavalry in Red Square, Moscow, on November 7, 1941, a week before the start of the German offensive to reach Moscow (Corbis).

of magnitude greater than that of a heavy crossbow arrow used to propel heavy projectiles, and larger charges can propel heavy projectiles from field artillery. Not surprisingly, the diffusion and perfection of firearms and cannons followed rather rapidly after their initial introduction.

Artillery developments started with Chinese fire-lances of the tenth century. These bamboo, and later metal, tubes ejecting bits of materials evolved first into simple bronze cannons inaccurately hurling loosely fitting stones. The first true guns were cast in China before the end of the thirteenth century, with Europe only a few decades behind (Wang 1991; Norris 2003). The pressures of frequent armed conflicts led to rapid rates of innovation, resulting in more powerful and more accurate guns. Already by 1400 the longest guns measured 3.6 m with a 35 cm caliber; the Mons Meg cannon, built in France in 1499 and donated to Scotland, was nearly 4.06 m long, could fire a 175 kg shot, and weighed 6.6 t (Gaier 1967). The destructive power increased with the general replacement of stone balls by iron projectiles.

The strategic implications of gunpowder-powered warfare were immense, both on the land and on the seas. There was no need to mount prolonged and often desperate sieges of seemingly impregnable castles.

The combination of accurate artillery and iron cannonballs, whose higher density made them much more destructive than their stone predecessors, made them indefensible. Attackers able to destroy sturdy stone structures from far beyond the range of archers put an end to the defensive value of traditionally built castles and walled cities, and the medieval practice of building relatively compact fortresses with thick stone walls was superseded by new designs of low spreading star-shaped polygons with massive earthen embankments and huge water ditches.

These projects consumed enormous amounts of materials and energy. The fortifications of Longwy in northeastern France, the largest project of the famous French military engineer Sebastien Vauban (1633–1707), required moving 640,000 m^3 of rock and earth (a volume equal to about a quarter of Khufu's pyramid) and emplacement of 120,000 m^3 of masonry (M. S. Anderson 1988). But they, too, went out of fashion with the more mobile warfare of the eighteenth century when sieges became much less common. During the Napoleonic Wars light Gribeauval guns (including a 12-pounder model firing 5.4 kg projectiles, and weighing, including carriage, just below 2 t, compared to nearly 3 t for British guns) made faster maneuvers easier (Chartrand 2003).

On the seas, gunned ships (once equipped with two other Chinese innovations, compasses and good rudders) became the principal carriers of European technical supremacy, the tools of aggressive expansion to distant locations during all but the closing decades of the expansive colonial era: their dominance ended only with the introduction of naval steam engines, a process that began only in the 1820s. In the continent's waters, long-range guns gave the English captains a decisive advantage over the Spanish Armada in 1588 (Fernández-Armesto 1988; Hanson 2011). A century later large men-of-war were fitted with up to 100 guns, and the British and Dutch ships engaged during the battle of La Hogue in 1692 carried a total of 6,756 guns (M. S. Anderson 1988). The concentrated discharges of destructive energy reached levels that were not surpassed until the middle of the nineteenth century with the introduction of nitrocellulose-based powders (during the 1860s) and dynamite (patented by Alfred Nobel in 1867).

5 Fossil Fuels, Primary Electricity, and Renewables

Fundamentally, no terrestrial civilization can be anything else but a solar society dependent on the Sun's radiation, which energizes a habitable biosphere and produces all of our food, animal feed and wood. Preindustrial societies used this solar energy flux both directly, as incoming radiation (insolation)—every house has always been a solar house, passively heated—and indirectly. Indirect uses included not only the cultivation of field crops and trees (be it for fruits, nuts, oil, wood, or fuel) and the harvesting of natural arboreal, grassy, and aquatic phytomass but also conversions of wind and water flows to useful mechanical energy.

Wind and water flows are almost immediate transformations of solar radiation: atmospheric pressure gradients arise rapidly from the differential heating of Earth's surfaces, and evaporation and evapotranspiration constantly drive the global water cycle. Solar radiation is converted to food and feed and to some biomass fuels with delays ranging between a few days (for animal dung) and a few months (for crop residues typically 90–180 days). And it takes only a few years for domestic animals to reach their working ages, while children in traditional societies started to help with adult work as soon as they were five or six years old. Only when mature trees were cut down and their wood burned or turned into charcoal was the use of solar radiation postponed for several decades (later, when large saws made it possible to cut down giant trees in old-growth rain forests, for several centuries).

The origins of fossil fuels are also in the transformation of solar radiation: peat and coals arose from the slow alteration of dead plants (phytomass), hydrocarbons from more complex transformations of marine and lacustrine single-celled phytoplankton (mostly cyanobacteria and diatoms), zooplankton (mostly foraminifera), and some algae, invertebrates, and fish (Smil 2008a). Pressure and heat were the dominant transforming processes; they lasted at least a few thousand years for the youngest peat to as long as hundreds of millions of years for hard coals. This origin dictates their high carbon content, which, combined with the low content of water and incombustible impurities, translates into high energy densities (box 5.1)

But only a small fraction of initially sedimented biomass carbon was transformed into fossil fuels (Dukes 2003). During the formation of coal up to 15% of plant carbon ended up as peat, up to 90% of that could be preserved in coal, and in open-cast mines up to 95% of coal in place can be extracted from thick seams. As a result, up to 13% of the original ancient plant carbon can be extracted as coal; inversely, this means that roughly eight units of ancient carbon ended up in marketed coal (the range is mostly 5–20 units). In contrast, the overall carbon recovery factor is much lower for crude oil and natural gas. These fuels arose from organisms buried in marine and lacustrine sediments, and the production of fossil hydrocarbon recovers at best close to 1% but commonly just 0.01% of the carbon that was initially present in the ancient biomass whose transformation yielded oil and gas. The recovery rate of 0.01% means that 10,000 units of ancient carbon were required to produce one unit of carbon sold as crude oil or natural gas.

But a society using fossil fuels simply as substitutes for traditional uses of phytomass—that is, burning them inefficiently just to produce heat and light—would look like a richer version of eighteenth-century Europe or China. The transition to fossil fuels has also entailed two classes of fundamental qualitative improvements, and only their accumulation and combination have produced the energetic foundations of the modern world. The first category of these advances was the invention, development, and eventually mass-scale diffusion of new ways to convert fossil fuels: by introducing new prime movers—starting with steam engines and progressing to internal combustion engines, steam turbines, and gas turbines—and by coming up with new processes to transform raw fuels, including the production of metallurgical coke from coal, the refining of crude oils to produce a wide range of liquids and nonfuel materials, and the use of coals and hydrocarbons as feedstock in new chemical syntheses.

The second class of inventions used fossil fuels to produce electricity, an entirely new kind of commercial energy. Any solid, liquid, or gaseous fuel could be burned, its released heat used to convert water into steam, and the steam used to rotate turbogenerators and produce electricity. But since the very beginning of electricity generation we have also used the kinetic energy of water, rather than that of expanding steam, to produce electricity. Hydroelectricity is thus classified as a form of primary electricity (as opposed to electricity derived from fuel combustion); later additions to this category

Box 5.1
Fossil fuels

Plant mass is 45–55% C, anthracite is nearly 100% C, good bituminous coals have in excess of 85% C, crude oils are mostly between 82% and 84% C, and methane (CH_4), the dominant constituent of natural gas, is 75% C. Bituminous (black) coals make up the bulk of global solid fuel extraction. Because they nearly always contain some ash and sulfur, their combustion generates fly ash and SO_2. Until after World War II these were two common sources of industrial and urban air pollution, causing visible deposition of particulate matter and dry acid deposition and acidification of precipitation (Smil 2008a). Crude oils are mixtures of complex hydrocarbons whose refining produces gasolines, jet and diesel fuel for transportation, fuel oils for heat and steam generation, lubricants, and paving materials. Natural gases, the cleanest fossil fuels, are the lightest hydrocarbons. Hydrocarbons can be also produced from coal. "Town gas" was widely used for lighting during the nineteenth century, and modern coal gasification produces synthetic gas akin to that of natural gases. Synthetic liquid fuels were produced for the first time on a large scale by Germany during World War II.

Energy densities vary widely in coals but are relatively uniform for hydrocarbons. Crude oils are always superior sources of energy: they contain nearly twice as much energy per unit mass as do common bituminous coals. International energy statistics use one of three common denominators: standard coal equivalent (fuel containing 29.3 MJ/kg), oil equivalent (worth 42 MJ/kg), or values in standard energy units (joules) or in two traditional units, calories (cal) and British thermal units (Btu).

Fuel	Energy density	
	MJ/kg	MJ/m^3
Coals		
Anthracites	31–33	
Bituminous coals	20–29	
Lignites	8–20	
Peats	6–8	
Crude oils	42–44	
Natural gases		29–39

have included electricity generated in geothermal plants, by nuclear fission, and, most recently, by large wind turbines and photovoltaic cells or concentrated solar radiation.

The long-term trend is clear: we have been converting a steadily rising share of fossil fuels to thermal electricity, and we have also been expanding the capacities for primary electricity generation because electricity is the most convenient, most versatile, and, at the point of its use, the cleanest form of modern energy. In the first part of this chapter I describe many key developments of the great transition from phytomass fuels and animate energies to fossil fuels and inanimate prime movers; in the second part I trace important technical innovations that have combined to create the efficiency, reliability, and affordability characteristic of modern high-energy societies.

The Great Transition

In some countries fossil fuels were used, though in relatively small quantities, for centuries before the beginning of a rapid displacement of biomass fuels and animate labor. Coal and natural gas in China and coal in England are the best-known examples. Chinese used coal on a small industrial scale during the Han dynasty (206 BCE–220 CE), and England, Wales, and Scotland had many locations where coal was outcropping and was dug up easily, some of it already during the Roman rule, more of it during the Middle Ages. But as Nef (1932, 12) noted,

> Until the sixteenth century coal was hardly ever burned, either in the family hearth or the kitchen, at distances of more than a mile or two from the outcrops, and, even within the area thus circumscribed it was used only by the poor who could not afford to buy wood.

Coal was generally the dominant fossil fuel in the European transition. The most notable exception energized one of the continent's most influential economies of the early modern era: during the seventeenth and the eighteenth centuries the Dutch Golden Age was fueled largely by domestic peat. The extent of its recovery is illustrated by De Zeeuw's (1978) estimate that of some 175,000 ha of high peatlands in the Netherlands, only about 5,000 ha remained in a more or less undisturbed state. U.S. and Canadian transitions also started with coal, but unlike in Europe, those two economies switched sooner and faster to oil and natural gas (Smil 2010a). Similarly, Russia was one of the pioneers of large-scale commercial oil production, and later also took advantage of its enormous natural gas resources.

And while most of Europe reduced its dependence on biomass fuels to very low levels during the nineteenth century, the shift away from phytomass fuels is still under way in many low-income countries. If patterns differ, so do the fuels. We should use plurals to emphasize their variability. Coals, crude oils, and natural gases have a wide range of properties (see box 5.1). Heat released by their combustion can be used directly for cooking, heating, or smelting metals and indirectly for energizing various prime movers. Steam engine became the leading inanimate prime mover of the nineteenth century. Internal combustion engines and steam turbines started to make commercial inroads during the 1890s. Before 1950 gasoline and diesel engines became the dominant prime movers in transportation and steam turbines in the large-scale generation of electricity (Smil 2005); the widespread use of gas turbines (stationary for electricity generation or powering jetliners and ships) came only after 1960 (Smil 2010a).

Recent studies of energy transitions demonstrate many commonalities governing these gradual shifts and identify major factors that have promoted or impeded the process (Malanima 2006; Fouquet 2010; Smil 2010a; Pearson and Foxon 2012; Wrigley 2010, 2013). These have ranged from technical imperatives, with prolonged periods of experimentation followed by a peak growth phase and upscaling (Wilson 2012), to some considerably earlier and faster transitions in small energy consumers (Rubio and Folchi 2012). Moreover, some smaller countries skipped the coal stage, even those with abundant coal deposits, and became rapidly dependent on domestic or, more commonly, imported crude oil. But in all cases the eventual outcome was a substantial increase in per capita consumption of primary energy as societies previously limited by harvests of phytomass fuels and the deployment of animate energies entered a new era of diversifying fossil fuel supply and the mass-scale deployment of mechanical prime movers.

The Beginnings and Diffusion of Coal Extraction

The beginnings of coal utilization go back to antiquity, when the most important use of the fuel was by the Han dynasty Chinese in iron production (Needham 1964). European records show the first extraction in Belgium in 1113, the first shipments to London in 1228, the first exports from the Tynemouth region to France by 1325, and England being the first country to accomplish the shift from plant fuels to coal during the sixteenth and seventeenth centuries (Nef 1932). After 1500 the country's serious regional wood shortages led to increases in the cost of fuelwood, charcoal, and lumber. These shortages worsened during the seventeenth century owing to the growing demand for iron and the huge timber requirements for shipbuilding. They were only temporarily alleviated by higher imports of bar iron and timber (Thomas 1986). Rising domestic coal extraction was the obvious solution: almost all of the country's coalfields were opened up between 1540 and 1640.

By 1650 annual English coal output had passed 2 Mt; 3 Mt/year were extracted in the early eighteenth century and more than 10 Mt/year by its end. The rising use of coal required solving many technical and organizational problems connected with its mining, transport, and combustion. The depletion of outcropping seams led to the development of deeper pits. While the pits were rarely more than 50 m deep during the late seventeenth century, the deepest shafts surpassed 100 m shortly after 1700, 200 m by 1765, and 300 m after 1830. By that time daily production was between 20 and 40 t per mine, compared to just a few tonnes a century earlier. Deeper pits required more water pumping, and more energy was also needed for mine ventilation, for hoisting the coal from deeper shafts, and for its distribution. Waterwheels, windmills, and horses powered these needs. Coal mining itself was energized by heavy human labor.

Hewers, wielding picks, wedges, and mallets, extracted coal from seams in positions ranging from standing upright to crawling through narrow tunnels. Putters filled woven baskets with coal and dragged them on wooden sledges to the pit bottom, where onsetters hung them on ropes. Windsmen hauled them up and banksmen dumped them on piles. Adult men did most of the extraction, but boys as young as six to eight years were employed for lighter tasks. In many pits some of the heaviest work was done by women or teenage girls. They had to carry coal to the surface by ascending steep ladders with heavy baskets on their back fastened with straps to the forehead (fig. 5.1). In 1812 Robert Bald, a Scottish civil engineer and mineral surveyor, published an inquiry into the lives of those women, and an extensive quotation is worthwhile not only because of its painful description of hardships endured but also for its accurate appraisal of actual physical exertion (box 5.2).

Bald's vivid description is also a perfect illustration of a fundamental fact of energetics, an impressive example of how every transition to a new form of energy supply has to be powered by the intensive deployment of existing energies and prime movers: the transition from wood to coal had to be energized by human muscles, coal combustion powered the development of oil, and, as I stress in the last chapter, today's solar photovoltaic cells and wind turbines are embodiments of fossil energies required to smelt the requisite metals, synthesize the needed plastics, and process other materials requiring high energy inputs.

Figure 5.1
Coal carriers in a Scottish mine in the early nineteenth century (Corbis).

In deeper pits, horses were employed for turning the whims hoisting coal or pumping water. After 1650 they, and donkeys, were also used underground. Horse-drawn wagons, sometimes on rails, were used to distribute coal over shorter distances and move it to rivers or harbors for loading onto canal boats or ships. By the beginning of the seventeenth century coal was commonly used by households and in heating forges, firing bricks, tiles, and earthenware, making starch and soap, and extracting salt. But the transfer of impurities to the final product made its direct use impossible in glassmaking, malt drying, and, most important, in iron smelting. The glassmaking problem was solved first, around 1610, with the introduction of reverberating (heat-reflecting) furnaces, where the raw materials were heated in closed vessels. Only the availability of coke satisfied the other needs (see the next section).

Another important indirect use of coal came with production of coal, or town, gas by the carbonization of bituminous coal, that is, by high-temperature heating of the fuel in ovens with a limited oxygen supply (Elton 1958). The first practical installations were done independently in

Box 5.2
An Inquiry Into the Condition of the Women Who Carry Coals Under Ground in Scotland, Known by the Name of BEARERS

This was the title of the addition to *A General View of the Coal Trade in Scotland,* published in 1812; here are the key findings (Bald 1812, 131–132, 134):

> The mother ... descends the pit with her older daughters, when each, having a basket of suitable form, lays it down, and into it the large coals are rolled; and such is the weight carried, that it frequently takes two men to lift the burden upon their backs ... The mother sets out first, carrying a lighted candle in her teeth; the girls follow ... with weary steps and slow, ascend the stairs, halting occasionally to draw breath. ... It is no uncommon thing to see them, when ascending the pit, weeping most bitterly, from the excessive severity of labor. ... The execution of work performed ... in this way is beyond conception. ... The weight of coals thus brought to the pit top by a woman in a day, amounts to 4080 pounds ... and there have been frequent instances of two tons being carried.

Assuming a body weight of 60 kg, a daily lifting of 1.5 t of coal from a depth of about 35 m would alone need about 1 MJ, and including the cost of carrying the coal horizontally, or on a slight incline—belowground to the pit's bottom and aboveground to the distribution point—and the cost of return trips would bring the daily total to about 1.8 MJ. Assuming a labor efficiency of 15%, an adult female bearer would expend about 12 MJ of energy, averaging about 330 W during a ten-hour day. Modern measurements of energy expenditures in heavy labor confirmed that work at the rate of 350 W is sustainable during an eight-hour shift, but only rarely can it be exceeded (Smil 2008a). Clearly, the bearers were operating, day after day, many for years—they entered this work when seven years old, and frequently continued till they were upward of 50—near the maximum of human capacity.

1805–1806 in English cotton mills. A company to provide centralized gas supply for London was chartered in 1812. Better retorts, the removal of sulfur from the gas, a new technique to make small-diameter wrought iron pipes, and more efficient burners ensured the rapid diffusion of gas lighting. Its use did not end with the introduction of light bulbs. An incandescent gas mantle, patented in 1885 by Carl Auer von Welsbach, enabled the gas industry to compete for a few more decades with electrical lights.

Outside England, the diffusion of coal mining was rather slow during the eighteenth century. Major output came first from northern France, the Liege and Ruhr regions, and from parts of Bohemia and Silesia. North American coal extraction started to matter nationwide only during the early nineteenth century. Historical coal production statistics and the best available (and much less reliable) estimates of nationwide fuel wood consumption make it possible to narrow down, in a few cases even to pinpoint, the dates when coal surpassed wood and began to contribute more than half of a nation's primary energy supply (Smil 2010a). In England and Wales this took place exceptionally early and the timing of this earliest energy transitions can only be approximated.

Warde (2007) concluded that choosing a precise date for the tipping point between wood and coal would be arbitrary, but his reconstructions show that the most likely date when coal surpassed biomass as the source of heat was around 1620, or maybe even a little earlier. By 1650 coal's share was up to 65%, by 1700 about 75%, by 1800 about 90%, and by the 1850s more than 98% (the last two shares are for the UK). British coal supremacy lasted for another century: in 1950 coal supplied 91% of the country's primary energy and still 77% by 1960. As a result, coal dominated (more than 75%) the country's energy use for 250 years, much longer than in any other country.

Early Napoleonic France derived more than 90% of its primary energy from wood, and the share was still about 75% in 1850, before declining to below 50% by 1875 (Barjot 1991). Coal remained France's dominant fuel until the late 1950s when imported oil became the leader. Coal mining in colonial America began in 1758 in Virginia, and by the early nineteenth century Pennsylvania, Ohio, Illinois, and Indiana had become coal-producing states (Eavenson 1942). The fuel supplied just 5% of the total primary energy by 1843, but the subsequent rapid rise of extraction brought the share to 20% by the early 1860s, and in 1884 extracted coal contained more energy than the country's large fuel wood consumption (Schurr and Netschert 1960). In 1880, the starting date of Japan's historical statistics, wood (and charcoal derived from it) supplied 85% of the country's primary energy, but by 1901 intensive modernization had lifted coal's share above 50%, and to a peak of 77% by 1917 (Smil 2010a).

The Russian Empire, with its large boreal forest in its northern European part and in Siberia, was a quintessential wooden society. According to Soviet historical statistics, fuelwood supplied 20% of all primary energy in 1913 production (TsSU 1977)—but that obviously referred only to commercially produced fuel, which was only a fraction of the energy needed to heat Russian interiors: even a small house would have required no less than 100 GJ/year. My best estimate is that wood provided 75% of all energy in by 1913, and that oil and coal began to supply more than half of all primary energy only during the early 1930s (Smil 2010a).

The latest leading economy to accomplish the transition from phyto-mass to coal was China, where the process was delayed by the endless twentieth-century crises: they started with the collapse of the imperial rule in 1911, continued with the protracted civil war between the Communists and the Kuomintang (1927–1936, 1945–1950) and the war with Japan (1933–1945), and afterward came decades of Maoist economic mismanagement, including the world's largest, Mao-made, Great Famine (1958–1961) and the insanely named Cultural Revolution (1966–1976). As a result, it was only in 1965 that biomass fuels began to supply less than half of China's primary energy; by 1983 their share had fallen below 25% and by 2006 below 10% (Smil 2010a).

From Charcoal to Coke

The replacement of charcoal by metallurgical coke in pig (or cast) iron smelting belongs undoubtedly to the greatest technical innovations of the modern era as it accomplished two fundamental changes, severing the industry's dependence on wood (and hence requiring furnace locations in or near forested regions) and allowing much larger furnace capacities and hence a rapid increase in annual production. Moreover, it was also a replacement by a superior metallurgical fuel. Pyrolysis (destructive distillation) of coal—the heating of bituminous coals (with low ash and low sulfur content) in the absence of oxygen—produces a nearly pure carbon matrix with a low apparent density ($0.8-1$ g/cm^3) but a high energy density ($31-32$ MJ/kg) that is also much stronger in compression than charcoal and hence can support the heavier charges of iron ore and limestone in taller blast furnaces (Smil 2016).

Coke was used in England already during the early 1640s for drying malt (coal would not do as its combustion emitted soot and sulfur oxides), but its metallurgical use began only in 1709, when Abraham Darby (1678–1717) pioneered the practice at Coalbrookdale. Coking offered a virtually unlimited supply of a superior metallurgical fuel, but the process was initially wasteful and costly, and its widespread adoption came only after 1750 (Harris 1988; King 2011). English ironmasters of the first half of the eighteenth century did not immediately follow Darby's example largely because of depressed prices of bar iron, with domestic production competing with Swedish imports. Once the market improved in the mid-1750s the English ironmasters began building new coke-fueled furnaces, and by 1770 coke was used to produce 46% of the British iron (King 2005). This epochal change ended the unsustainable pressure on wood resources, felt as much in the UK (see box 4.22) as it was on the continent: for example, in 1820

52% of Belgium's forested area was used to produce metallurgical charcoal (Madureira 2012).

The American situation was not as dire during the early nineteenth century (see box 4.23), and by 1840 all U.S. pig iron was still smelted with charcoal, but the subsequent expansion of the industry brought a rapid switch, first to anthracite, then to coke, which became dominant by 1875. For generations, coke was produced wastefully in enclosed beehive ovens (Sexton 1897; Washlaski 2008). Radical improvement came only with the adoption of by-product coking ovens: they recover CO-rich gases for fuel, chemicals (tar, benzol, toluol) to be used as raw materials, and ammonium sulfate as fertilizer. Their use began in Europe in 1881, in the United States in 1895, and their improved designs remained the mainstay of modern coking (Hoffmann 1953; Mussatti 1998).

The first coke-fueled blast furnaces were only as tall (about 8 m) and as voluminous (<17 m^3) as their large contemporary charcoal-fired structures, but by 1810 coke-fueled furnaces were commonly around 14 m tall (>70 m^3). After 1840 Lowthian Bell (1816–1904), a leading British metallurgist, introduced a major redesign, and by the end of the nineteenth century large blast furnaces were nearly 25 m and had internal volumes of around 300 m^3 (Bell 1884; Smil 2016). Larger furnaces capable of much higher productivities (<10 t/day for the best charcoal-fueled furnaces vs. >250 t/day for coke-fueled furnaces by 1900) made it possible to boost the global pig iron output from only about 800,000 t in 1750 to about 30 Mt in 1900, laying the foundation for the post-1860 development of the modern steel industry and providing the key metal of industrialization (Smil 2016).

Steam Engines

Steam engine was the first new prime mover successfully introduced since the adoption of windmills, which preceded it by more than 800 years. The machine was the first practical, economic, and reliable converter of coal's chemical energy into mechanical energy, the first inanimate prime mover energized by fossil fuel rather by an almost instant transformation of solar radiation. The first engines of the early eighteenth century delivered only reciprocating motion suitable for pumping, but before 1800 came designs delivering more practical rotary motion (Dickinson 1939; Jones 1973). Undoubtedly, the engine's adoption was of a profound importance for global industrialization, urbanization, and transportation, and much has been written about these impacts (von Tunzelmann 1978; Hunter 1979; Rosen 2012).

At the same time, commercialization and the widespread adoption of steam engines advanced slowly, taking more than a century, and even during the years of their rapid diffusion, after 1820, they had to compete (as already noted in chapter 4) with waterwheels and turbines. While their use eliminated many kinds of animate labor (water pumping from mines, numerous manufacturing tasks), absolute reliance on human labor and animal draft kept on increasing during the entire nineteenth century. These realities have led to a reexamination of a widely held understanding that almost equates the adoption of steam engines with the process that is generally but misleadingly known as the Industrial Revolution.

The dominant perception of the era as a time of epoch-making economic and social change (Ashton 1948; Landes 1969; Mokyr 2009) has been questioned by those who see it as a much more restricted, even localized, phenomenon, with technical changes affecting only some industries (cotton, iron making, transportation) and leaving other economic sectors in premodern stagnation until the middle of the nineteenth century (Crafts and Harley 1992). Some critics have gone further, arguing that the change was so small relative to the entire economy that the very name of the Industrial Revolution is a misnomer (Cameron 1982)—indeed, that the entire notion of a British industrial revolution is a myth (Fores 1981).

More specifically, British data show that to relate the nineteenth-century economic growth primarily to steam is a misconceived conclusion (Crafts and Mills 2004). Steam engines notwithstanding, "the British economy was largely traditional 90 years after 1760" (Sullivan 1990, 360), and "the typical British worker in the mid-nineteenth century was not a machine-operator in a factory but still a traditional craftsman or labourer or domestic servant" (Musson 1978, 141). But the judgment is clearer when the process is seen in terms of overall energy consumption: its enormous increase—Wrigley's (2010) annual totals for England and Wales are about 117 PJ in 1650–1659, 231 PJ a century later, and 1.83 EJ in 1850–1859, or roughly a 15-fold increase in 200 years—made exponential economic growth possible, and undoubtedly, the steam engine was the key mechanical driver of industrialization and urbanization.

But its full impact came only after 1840, with the rapid construction of railroads and steamships and with more installations as a centralized producer of kinetic energy (transmitted by belts to individual machines) in manufacturing enterprises. The engine's practical evolution began with Denis Papin's (1647–1712) experiments with a tiny model built in 1690. Soon after Papin's toylike machine came Thomas Savery's (1650–1715) small (about 750 W, or a single horsepower) steam-driven pump operating without a piston. By 1712 Newcomen (1664–1729) had built a 3.75 kW engine to power mine pumps (Rolt 1963). Because this engine, working at atmospheric pressure, condensed steam on the underside of the piston, it had a very low efficiency, at best 0.7% (fig. 5.2). By 1770 John Smeaton, whose work on the comparative power of prime movers was noted in chapter 4, had improved the design and doubled the low efficiency.

Newcomen's engines began to spread to English mines after 1750, but their poor performance could be tolerated only with on-site access to fuel, and that made them impractical where the fuel would have to be transported. James Watt (1736–1819) captured the intent of his famous redesign in the very title of his 1769 patent: *A New Invented Method of Lessening the Consumption of Steam and Fuel in Fire Engines* (Watt 1855 [1769]). The patent was granted on April 25, 1769, and Watt's systematic listing of improvements makes it clear how the new machine differed from its predecessors (box 5.3).

(a)   (b)

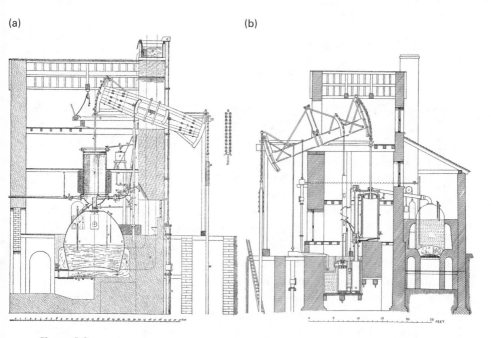

Figure 5.2
Newcomen's and Watt's steam engines. a. In Newcomen's engine, built by John Smeaton in 1772, the boiler was placed underneath the cylinder and the steam was condensed inside the cylinder by injecting water from the pipe leading to its lower right side. b. In Watt's engine, built in 1788, the boiler was placed in a separate enclosure, the cylinder was enveloped by an insulating steam jacket, and a separate condenser was connected to an air pump maintaining vacuum. Reproduced from Farey (1827).

Box 5.3
Watt's 1769 patent

This is how Watt explained his improved design:

My method of lessening the consumption of steam, and consequently fuel, in fire engines consists of the following principles: First, that vessell in which the powers of steam are to be employed to work the engine, which is called the cylinder in common fire engines, and which I call the steam vessell, must during the whole time the engine is at work kept as hot as the steam that enters it. ...

Secondly, in engines that are to be worked wholly or partially by condensation of steam, the steam is to be condensed in vessells distinct from the steam vessells or cylinders, although occasionally communicating with them. These vessells I call condensers, and whilst the engines are working, these condensers ought at least to be kept as cold as the air in the neighbourhood of the engines by application of water or other cold bodies.

Thirdly, whatever air or other elastic vapour is not condensed by the cold of the condenser, and may impede the working of the engine, is to be drawn out of the steam vessells or condensers, by means of pumps wrought by the engines themselves, or otherwise. (Watt 1855 [1769], 2)

A separate condenser was clearly the most important innovation (fig. 5.2). Later Watt also introduced a double-acting engine (in which steam moved the piston on both the up-stroke and the down-stroke) and a centrifugal governor that would maintain constant speeds with varying loads. In a thoroughly modern arrangement, Watt and his financial partner, Matthew Boulton (1728–1809), were paid not for a delivered engine but for its improved performance based on a comparison with the common Newcomen machine. Coal mining and steam engines reinforced each other's development. The need to pump more water from deeper mines was a key reason for developing steam engines. The availability of cheaper fuel led to their proliferation, and thus to a further expansion of mining. Soon the engines also powered winding and ventilating machinery.

Watt's improved steam engine was an almost instant commercial success, and it was easy to see its eventual impact beyond coal mining, in manufacturing and transportation (Thurston 1878; Dalby 1920; von Tunzelmann 1978). At the same time, it was a success measured against the industrial backdrop of the late eighteenth century: the total deployment of improved machines was miniscule when judged on the scale of modern mass production. By 1800, when the 25-year extension (by the Steam Engine Act of 1775) of the original patent expired, the company owned by Watt and Boulton had completed about 500 engines, 40% of them for water pumping. The engines' average capacity of about 20 kW was more than five times higher than the mean for typical contemporary watermills and nearly three times larger than that for windmills.

Watt's largest units (just over 100 kW) matched the most powerful existing waterwheels. But the location of waterwheels was inflexible, while steam engines could be sited with incomparably greater freedom, particularly near any port or along canals, where cheap transport by ships or boats could bring the requisite fuel. Though Watt's inventions ushered in the engine's industrial success, the extension of his patent actually impeded further innovation. Was a safety-minded Watt reluctant to use high-pressure steam, or was Watt trying to prevent any similar patents while his extension lasted? Watt and Boulton not only did not make any attempt to develop steam-driven transportation, they discouraged William Murdoch (1754–1839), the principal erector of their engines, from developing such a machine, and when Murdoch persisted, Boulton persuaded him not to file a patent (box 5.4)

But perhaps that made no difference for the future development of steam-powered transportation because even the best conceivable road carriage built in 1800 would have been unacceptably heavy, a burden made

---

Box 5.4

Watt and Boulton delay the development of a steam carriage

In 1777, when he was 23 years old, William Murdoch walked nearly 500 km to Birmingham to take a job with James Watt's steam engine company. Both Watt and his partner, Matthew Boulton, soon found him a great asset. Boulton's skilled installations of new machines ensured their efficient and profitable operation. By 1784 Murdoch had made a small model of a steam carriage, a three-wheeler with a boiler amid two back wheels. Another model followed, and Murdoch eventually decided to have his steam carriage patented (Griffiths 1992).

But on the way to London to do so he was intercepted in Exeter by Boulton, who convinced him to return home without filing the patent: he thought of Murdoch's persistence as a disorder. As Boulton wrote to Watt:

> He said He was going to London to get Men but I soon found he was going there with his Steam Carg to shew it & to take out a patent. He having been told by Mr W. Wilkn what Sadler had said & he had likewise read in the news paper Simmingtons puff which had rekindled all Wms fire & impations to make Steam Carriages. However, I prevailed upon him readily to return to Cornwall by the next days diligence & he accordingly arivd here this day at noon ... I think it is fortunate that I met him, as I am persuaded I can either cure him of the disorder or turn evil to good. At least I shall prevent a mischief that would have been the consequence of his journey to London. (Griffiths 1992, 161)

even less acceptable by a near absence of well-paved roads able to withstand heavy traffic. The only practical way to make it go was to put it on rails— and even there it took a few decades from conceiving the idea to making it commercial, once the expiry of Watt's patent in 1800 led to an intense period of innovation. The first essential advance was the introduction of high-pressure boilers by Richard Trevithick (1771–1833) in England in 1804 and by Oliver Evans (1755–1819) in 1805 in the United States. Other milestones included a uniflow design, introduced by Jacob Perkins (1766–1849) in 1827, the invention of a regulating valve gear in 1849 by George Henry Corliss (1817–1888), and French improvements of compound locomotive engines after the mid-1870s. A score of basic engine types gave rise to a large variety of specialized designs (Watkins 1967).

The engine's original uses for pumping and winding in mines (fig. 5.3) were soon extended to a large variety of stationary and mobile applications. By far the most notable uses were for belt drives in countless factories, and in revolutionizing nineteenth-century transportation on land and on water. The development of steamships and steam locomotives proceeded concurrently. The first steamboats were built during the 1780s in France, the United States, and Scotland, but the first commercially successful ships came only in 1802 in England (Patrick Miller's *Charlotte Dundas*) and in 1807, with Robert Fulton's *Clermont*, in the United States.

Figure 5.3
The C Pit of the Hebburn Colliery was a typical English coal mine of the early steam engine era. The mine's steam engine was housed in the building with a stack and powered the winding and ventilation machinery. Reproduced from Hair (1844).

All early rivergoing ships were propelled by paddle wheels (astern or amidship), as were still fully rigged ships for sea travel. The first Atlantic crossing was the Quebec-London trip by the *Royal William* in 1833 (Fry 1896). The first westward run was the race between the paddle wheelers *Sirius* and *Great Western* in 1838, the year in which John Ericsson deployed the first successful screw propeller. Gradually, larger and faster steamships displaced sail ships from the busiest passenger and cargo runs across the North Atlantic, and later from long-distance routes to Asia and Australia. They transported most of the 60 million emigrants who left the continent between 1815 and 1930 for overseas destinations, above all for North America (Baines 1991). At the same time, coal-fired oceangoing ships became important instruments of U.S. foreign policy (Shulman 2015).

The adoption of overland steam transport showed a similarly slow start, followed by the speedy diffusion of railways. Richard Trevithick's 1804 experiment with a machine on cast-iron rails was followed by a number of small private railroads. The first public railway, from Liverpool to Manchester, opened only in 1830, its train pulled by George Stephenson's (1781–1848) *Rocket*. A profusion of new designs brought more efficient and faster machines. By 1900 the best locomotive engines operated at pressures up to five times higher than in the 1830s, and with efficiencies of more than 12% (Dalby 1920). Speeds over 100 km/h became common, and during the 1930s streamlined locomotives approached, and even surpassed, 200 km/h (fig. 5.4).

Starting with the first 56 km intercity link, between Liverpool and Manchester, in 1830, British railroads reached about 30,000 km by 1900, when the European total was nearly 250,000 km. Elsewhere most of the railroad expansion took place during the three closing decades of the century: by 1900 the Russian network had reached 53,000 km (but the trans-Siberian railroad to the Pacific was completed only in 1917), the U.S. system was more than 190,000 km (including three transcontinental routes), and the worldwide total (with much of the remainder in British India) was 775,000 km (Williams 2006). As a result, railway expansion was the main reason for an unprecedented demand for steel during the second half of the nineteenth century.

Of course, metal was needed in ever larger quantities by many new industrial markets: by the steel industry itself (to provide metal for new iron- and steel-making capacities), by the new electrical industry (for boilers and steam turbogenerators, transformers, and electrical wires), in oil and gas extraction and transportation (for drilling pipes, drill bits, well casings, pipelines, and storage tanks), in shipping (for new steel-hulled

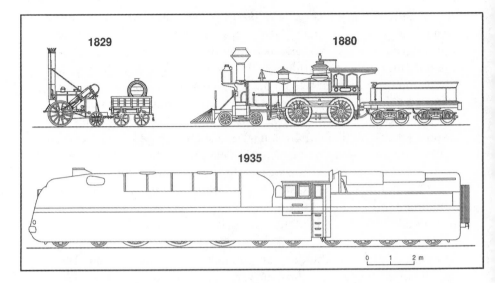

Figure 5.4
Notable machines of the steam locomotive age. Stephenson's 1829 Rocket, the first commercial machine, introduced two features of every subsequent design: separate cylinders on each side that moved the wheels by short connecting rods, and an efficient multitube boiler. Standard American designs have dominated U.S. railways since the mid-1850s. The streamlined German Borsig design reached 191.7 km/h in 1935. Based on Byrn (1900) and Ellis (1983).

vessels), in manufacturing (for machines, tools, and components), and in traditional textile and food-processing industries. But rails (previously made form wrought iron) were the most important finished product made from affordable Bessemer steel (for more, see chapter 6) introduced in the late 1860s, and remained so for the rest of the century (Smil 2016).

The engine's apogee came more than a century after Watt's improved patent: by the early 1880s its widespread adoption had laid the energetic foundation of modern industrialization, and the affordable availability of such concentrated power transformed both manufacturing productivity and long-distance land and marine transportation. In turn, these changes led to extensive urbanization, the rise of incipient affluence, growth of international trade, and shifts in national leadership. The cumulative technical progress was remarkable: the largest machines designed during the 1890s were about 30 times more powerful than those in 1800 (3 MW vs. about 100 kW), and the efficiency of the best units was ten times better, 25% versus 2.5% (fig. 5.5). This huge performance gain, translating into

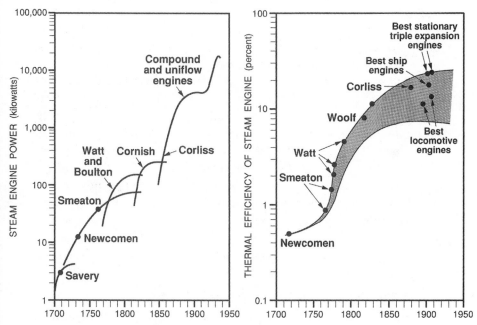

Figure 5.5
Rising power and improving efficiency of the best steam engines, 1700–1930. Plotted from data in Dickinson (1939) and von Tunzelmann (1978).

large fuel savings and less air pollution, came mainly from a hundredfold rise in operating pressures, from 14 kPa to 1.4 MPa.

These advances—combined with the engine's suitability for many manufacturing, construction, and transportation uses, owing to its ruggedness and durability—turned the machine into the inanimate prime mover of nineteenth-century industrialization. Its stationary uses ranged from tasks previously performed by animate prime movers, water wheels, or windmills (such as water pumping, wood sawing, or grain grinding) to new duties in expanding factories (powering the belt drives to run drilling, turning, or polishing machinery, compressing air)—and some of the largest steam engines ever built were used to turn the dynamos of the first electricity-generating stations during the 1880s and 1890s (Smil 2005).

The engine's mobile uses had (correct appraisal, not a commonly hyperbolic statement) revolutionized both land and waterborne transport with the rapid extension of railways and the launching of new steamships. Other mobile applications that made heavy tasks easier included steam cranes, pile drivers, and excavators (the first steam shovel was patented already in

1839). The Panama Canal could not have been completed so fast (1904–1914) without the deployment of about 100 large Bucyrus and Marion steam shovels (Mills 1913; Brodhead 2012), and steam engines found their way even into the American fields with cable plows.

But steam engines became victims of their own success: as their typical efficiencies rose and their highest capacities reached unprecedented levels (orders of magnitude above any traditional prime mover), they began to encounter their inherent limitations as more was demanded of them than they could deliver (Smil 2005). Even after more than a century of improvements the most commonly used steam engines remained highly inefficient: by 1900 the typical steam locomotive wasted 92% of coal fed into its boiler. And they remained heavy, limiting mobile uses beyond water and rails where it was easier to support their mass (box 5.5)

Box 5.5
Mass/power ratio of steam engines and a *Megatherium*

A medium-heavy horse weighing 750 kg and delivering one horsepower (745 W) will have a mass/power ratio of almost exactly 1,000 g/W—and so will an 80 kg man working steadily at the rate of 80 W. The first steam engines of the eighteenth century were exceedingly massive, with ratios (600–700 g/W) nearly as high as those of people and draft animals. By 1800 the ratio had declined to about 500 g/W, and by 1900 the best locomotive steam engines weighed just 60 g/W. But that was still too heavy for two very different but fundamental applications: to power land vehicles and to rotate large dynamos in new electricity-generating stations.

In 1894 a new Daimler-Maybach gasoline engine installed in a car that won the Paris-Bordeaux race rated less than 30 g/W (Beaumont 1902), leaving no place for steam engines in road transportation. And even the first commercial design of Charles Parsons's small steam turbines—a 100 kW unit built in 1891—rated only 40 g/W, and before the beginning of World War I the ratio fell below 10 g/W and their efficiencies surpassed 25%, far above the 11–17% for the best steam engines (Smil 2005). Consequently, 16 large Westinghouse-Corliss steam engines installed in New York's Edison plant in 1902 were already outdated, and yet three years later London's Council Tramway Greenwich station installed "a megatherium of the engine world" (Dickinson 1939, 152), the first of the 3.5-MW compound steam engines, in a cathedral-like space. The massive Greenwich machines were nearly as high (14.5 m) as they were wide—while a Parsons's generator of the same power would be only 3.35 m wide and 4.45 m high.

When steam engines reached their highest efficiency, greatest power, and the lowest mass/power ratio, these were not achievements paving the way for further dominance: despite its impressive improvements and its recently achieved ubiquity in industry, on railroads, and in waterborne transportation, the dominant prime mover of the nineteenth century could not continue to be the leading prime mover of the twentieth century. That role was split, as steam turbines had rapidly filled the demand for the most powerful prime movers in electricity generation, and internal combustion engines (first the gasoline-fueled machines, starting in the 1880s, then diesel engines) finally provided a prime mover that was light, powerful, and affordable to energize road transport. The ascendance of internal combustion engines was made possible by the availability of inexpensive liquid fuels refined from crude oil: they had higher energy density than coal, were cleaner to burn, and were easier to move and store, a combination that still makes them the best fuel choice for transportation.

Oil and Internal Combustion Engines
The beginnings of large-scale crude oil extraction and utilization were concentrated in just a few decades of the late nineteenth century. Of course, hydrocarbons (crude oils and natural gases) were well known for millennia from seepages, bitumen pools, and "burning pillars," especially common in the Middle East (particularly in northern Iraq)—but also found elsewhere. The Schedule of Property attached to George Washington's will describes a burning spring in the Kanawha River valley of West Virginia:

> The tract, of which the 125 acres is a moiety, was taken up by **General Andrew Lewis** and myself, for and on account of a bituminous spring which it contains, of so inflammable a nature as to burn as freely as spirits, and is nearly as difficult to extinguish. (Upham 1851, 385)

But the use of hydrocarbons in antiquity was limited almost solely to building materials or protective coatings. Their use for combustion, including the heating of Constantinople's *thermae* during the late Roman Empire, was rare (Forbes 1964). A remarkable exception was the Chinese burning of natural gas to evaporate brines in the landlocked Sichuan province (Adshead 1992). Used at least since the beginning of the Han dynasty (200 BCE), this process was made possible by the Chinese invention of percussion drilling (Needham 1964). Heavy iron bits attached to long bamboo cables from bamboo derricks were raised rhythmically by two to six men jumping on a lever. The deepest recorded boreholes started with only 10 m during the Han dynasty, reached 150 m by the tenth century, and culminated in the 1 km deep Xinhai well in 1835 (Vogel 1993). Natural gas, distributed by bamboo pipelines, was used to evaporate brines in huge cast-iron pans.

This Chinese practice remained isolated, and the beginning of the worldwide hydrocarbon age had to wait for two millennia. In North America oil was collected from natural seeps in western Pennsylvania during the late eighteenth century and sold as a medicinal "Seneca oil," and in France oil sands were exploited since 1745 in Alsace, near Merkwiller-Pechelbronn, where the first small refinery was built in 1857 (Walther 2007). But there was only one place in the preindustrial world with a long history of collecting crude oil, Baku's Absheron peninsula on the Caspian Sea in today's Azerbaijan.

The area's oil pools and wells were described in medieval sources, and a 1593 inscription marks a 35 m deep well that was dug manually in Balakhani (Mir-Babaev 2004). By 1806, when czarist Russia took control, the Absheron peninsula had many shallow wells from which lighter oil was collected and distilled to produce kerosene for local lighting and for export by camels (in skin bags) and in wooden barrels. The world's first commercial oil-distilling factory was built by the Russians in 1837 in Balakhani, and in 1846 they sank the world's first (21 m deep) exploratory oil well in Bibi-Heybat and thus began the exploitation of one of the world's giant oil fields, still producing today.

Western histories of oil industry either omit these Baku developments or mention them only after they describe the beginnings of the U.S. oil industry in Pennsylvania. The American search for oil was motivated by finding a replacement of expensive whale oil rendered on board ships from the blubber of sperm whales and burned in lamps (Brantly 1971). Americans had the world's largest whaling fleet—the total peaked in 1846 at more than 700 vessels—and during the early 1840s they brought about 160,000 barrels of sperm oil to New England's ports every year (Starbuck 1878; Francis 1990).

But the first North American well was dug manually in Canada in 1858, by Charles Tripp and James Miller Williams near Black Creek in Lambton County in southwestern Ontario, bringing the world's first oil boom and the renaming of the hamlet as Oil Springs (Bott 2004). The famous first drilled rather than dug well, described in all histories of oil production, was completed under the supervision of Edwin Drake (1819–1880), a former train conductor hired by George Henry Bissell (1821–1884), who set up the Pennsylvania Rock Oil Company (Dickey 1959). Its site was an oil seep at Oil Creek near Titusville, Pennsylvania, and the drillers struck oil at a depth of 21 m on August 27, 1859, the date commonly seen as the beginning of the modern oil era. The task was accomplished with the percussion drill powered by a small steam engine.

During the 1860s three countries—the United States, Canada, and Russia—had new, growing oil industries. Canadian output rose temporarily with the world's first gusher at Oil Springs in 1862 and with new discoveries in nearby Petrolea in 1865, but well before the century's end it became negligible, and Canada rejoined the ranks of leading oil producers only after World War II with discoveries of giant oil fields in Alberta. In contrast, American production kept on rising, first from numerous small fields in the Appalachian basin (from New York through Pennsylvania to West Virginia), then, starting in 1865, in California. Extraction in the Los Angeles basin began in 1880, in the San Joaquin basin in 1891 (its giant Midway-Sunset and Kern River fields are still producing more than a century later), and in Santa Barbara County after 1890 (including the world's first offshore wells drilled from wooden piers).

Kansas joined the oil-producing states in 1892, Texas (Corsicana field) in 1894, Oklahoma in 1897, and in 1901 Anthony Francis Lukas made a spectacular discovery of the Spindletop field near Beaumont in southern Texas with a gusher that produced 100,000 barrels/day on January 10, 1901 (Linsley, Rienstra, and Stiles 2002; fig. 5.6). The nascent Russian oil industry received plenty of foreign investment, notably from Ludwig and Robert Nobel through their Nobel Brothers Petroleum Company, set up in 1875, and from the Rothschild brothers' Caspian and Black Sea Oil Industry and Trade Society, established in 1883. By 1890 Russia was producing more energy as oil than as coal, and in 1899, before the discovery of oil in southern Texas, it became briefly the world's largest producer of crude oil, just over 9 Mt/year (Samedov 1988). Most of this fuel was exported by foreign investors. Baku production began to decline after 1900, and by 1913 Russia's coal consumption was more than twice as large as its oil use. Other substantial pre-1900 oil discoveries were made in Romania, Indonesia (Sumatra in 1883), and Burma (production began in 1887). Mexico joined the ranks of oil producers in 1901; in 1908 came the first major Middle Eastern discovery at Masjid-e-Soleiman in Iran; Trinidad oil was first produced in 1913; and Venezuela's giant Mene Grande field on Lake Maracaibo's coast began deliveries in 1914.

Most of these discoveries were hydrocarbon fields containing both crude oil and associated natural gas, but gas was rarely used during the early decades of the hydrocarbon industry because without compressors and steel pipes it could not be moved over long distances, and it was simply vented. In contrast, the high energy density of liquid fuels refined from

Figure 5.6
Oil gushing from the Spindletop well near Beaumont, Texas, in January 1901 (Corbis).

crude oil (gasoline, kerosene, and fuel oils) and their easy portability made them the superior energy source for transportation, and the invention and rapid adoption of the internal combustion engine opened up a huge new market for their use.

Crude oil extraction was to produce a more affordable source of energy for lighting, but less than 25 years after its U.S. beginnings, commercial electricity generation and light bulbs (see the next section) began to offer a superior alternative. When the oil extraction began to expand during the 1860s there were no commercial internal combustion engines capable of powering vehicles, but again within about 25 years, two German engineers had built their first practical automotive engines and created a new fuel demand that has yet to see its global peak more than 130 years later.

The development of the internal combustion engine, a new prime mover burning fuel within the cylinder, proceeded very rapidly. The designs perfected during the first generation of its commercial use, between 1886 and 1905, remained fundamentally unchanged (though they were much improved) for most of the twentieth century (Smil 2005). After several decades of failed experiments and abandoned designs, the first commercially successful internal combustion engine was patented in 1860 by Jean Joseph Étienne Lenoir (1822–1900). But his engine was completely unsuitable for any mobile use: it was a horizontal double-acting machine that burned an uncompressed mixture of illuminating gas and air ignited with an electric spark, and its efficiency was just 4% (Smil 2005).

In 1862, Alphonse Eugène Beau (Beau de Rochas, 1815–1893) conceptualized the workings of a four-stroke engine, but it took another 15 years before Nicolaus August Otto (1832–1891) patented such a machine, in 1877, and subsequently sold nearly 50,000 units (averaging 6 kW, with a compression ratio of just 2.6) to small workshops that could not afford a steam engine (Clerk 1909). This slow and coal gas–fueled engine could not serve as a prime mover in transportation. Such an engine was designed by Gottlieb Daimler (1834–1900), a former employee of Otto's company, and Wilhelm Maybach (1846–1929), to burn gasoline in their workshop in Stuttgart (Walz and Niemann 1997). Gasoline has an energy density of 33 MJ/L (about 1,600 times that of the town gas used by Otto) and a very low flashpoint (−40°C), which makes engines easy to start.

Daimler and Maybach built the first prototype in 1883. In November 1885 they used a subsequent air-cooled version to power the world's first motorcycle, and in March 1886 their larger (0.462 L, 820 W), 600 rpm, water-cooled design was mounted on a wooden-wheeled coach (Walz and Niemann 1997). At the same time, working in Mannheim, about 120 km north of Stuttgart, Karl Friedrich Benz (1844–1929) designed his first two-stroke gasoline engine in 1883 and (after the expiry of Otto's patent) a four-stroke engine, which he patented in January 1886. He mounted the 500 W, 250 rpm engine on a three-wheel chassis and showed the vehicle publicly on July 3, 1886. The combination of Daimler's high-revolution engine, Benz's electrical ignition, and Wilhelm Maybach's float-feed carburetor supplied the key components of modern road vehicles, and as the new century began, the leading German manufacturer designed the first essentially modern car (box 5.6, fig. 5.7).

DMG may have embodied the highest car quality, but at the beginning of the twentieth century it was typical in its focus on affluent market. Two decades after its mid-1880s German debut, the passenger car remained an expensive machine, made in small series by artisanal production methods. And there was nothing special about American cars: a leading British car

---

Box 5.6
The first modern car

The car was a German invention, but a French engineer, Emile Levassor (1844–1897), designed the first vehicle that was not merely a horseless carriage—though it had the best German engine. Levassor was shown the German V engine made by Daimler Motoren Gesselschaft (DMG) in 1891 and designed a new chassis to accommodate it. During the 1890s cars with DMG engines kept on winning European car races, but the company's history-making car had a thoroughly commercial origin (Robson 1983; Adler 2006). When Emil Jellinek (1853–1918), a businessman and the Austro-Hungarian consul general in Monaco, established a dealership for Daimler cars on April 2, 1900, he ordered 36 vehicles, and soon afterward doubled the order. In return for this lucrative order he asked for exclusive selling rights in the Austro-Hungarian Empire, France, Belgium, and the United States and for the cars to carry the Mercedes trademark, the given name of his daughter.

For this unique order Maybach designed a car that his successor company, Mercedes-Benz, described as the first true automobile and that was called the "first modern car in all essentials" (Flink 1988, 33). The Mercedes 35 was conceived as a race car with a lengthened profile; it had a very low center of gravity and a total weight of 1,200 kg. The car had an exceptionally (for its time) powerful four-cylinder engine (5.9 L, 26 kW or 35 hp, 950 rpm) with two carburetors and with mechanically operated inlet valves. Maybach cut the engine's weight to just 230 kg by using an aluminum block and reducing its mass/power ratio to below 9 g/W, 70% lower than in the best DMG engine made in 1895. The new car soon set the world speed record (64.4 km/h), and an even more powerful Mercedes 60, with fancier bodywork, followed in 1903, the beginning of a marquee that has lost little of its appeal 125 years later.

---

expert wrote in 1906 that "progress in the design and manufacture of motor vehicles in America has not been distinguished by any noteworthy advance upon the practice obtaining in either this country or on the Continent" (Beaumont 1906, 268). Two years later all that changed when Henry Ford (1863–1947) introduced his mass-produced, affordable Model T, built to meet the rigors of American driving: his achievement and legacy are explained in the next chapter.

And two unlikely pioneers—Wilbur (1867–1912) and Orville (1871–1948) Wright, bicycle makers from Dayton, Ohio—were the first innovators to power the first successful flight by a light internal combustion engine when their airplane lifted briefly above the dunes at North Carolina's Kitty

Figure 5.7
The Mercedes 35, designed by Wilhelm Maybach and Paul Daimler in 1901. Photograph from the Daimler website.

Hawk on December 17, 1903 (McCullough 2015). They were not the first ones to try. Just nine days before their successful flight, Charles M. Manly made a second attempt to launch *Aerodrome A* by a catapult from a barge on the Potomac. The plane was built with a U.S. government grant received by Samuel Pierpoint Langley (1834–1906), the secretary of the Smithsonian Institution, and was equipped with a powerful (39 kW, 950 rpm) five-cylinder radial engine. But, just as during Manly's first attempt on October 7, 1903, the plane immediately plunged into the water.

Why did the Wrights succeed, and why did they do it less than five years after they, without any previous knowledge, wrote to the Smithsonian to request information on flight? After refusals from engine manufacturers to build a machine to their specifications they designed it themselves, and their mechanic, Charles Taylor, built it in only six weeks. The engine had aluminum body, no carburetor, and no spark plugs, but its four steel cylinders displaced 3.29 L and were to deliver 6 kW (Gunston 1986). The finished engine, weighing 91 kg, actually developed as much as 12 kW in flight, for a mass/power ratio of 7.6 g/W. But this light, powerful engine was only one key component of their success. The brothers studied aerodynamics and came to understand the importance of balance, stability, and control in flight, and solved this challenge in their 1902 glider (Jakab 1990). They combined their engineering experience with rigorous and systematic tests of airfoils and wing shapes and with experimental glider flights. Their first flights, on December 17, 1903, are well documented (box 5.7, fig. 5.8).

The patent (U.S. 821,393) was granted only in May 1906, and it was widely infringed as designers in many countries began to build their airplanes. Progress in flight control and duration was rapid. On September 20, 1904, the Wrights flew the first complete circle, and on November 9, 1904, they covered three miles (McCullough 2015). Less than five years later, after a period of intensifying international competition, Louis Charles Joseph Blériot (1872–1936), who had previously built the world's first monoplane, made the first English Channel crossing on July 25, 1909 (Blériot 2015), and by 1914 the major powers had nascent air forces, which were deployed and enlarged in World War I.

When gasoline spark engines were on the road to commercial success, an entirely different mode of fuel ignition was introduced by Rudolf Diesel's (1858–1913) invention, patented in 1892 (Diesel 1913). In diesel engines fuel injected into the cylinder is ignited spontaneously by high

---

Box 5.7
The first flights

Nine days after Manly's second dunking, the Wright brothers were ready to test the *Flyer* at Kitty Hawk. The airplane was a fragile canard biplane (with tailplane ahead of the wings) with a wooden (spruce) frame and a fine-woven cotton cloth covering; its wing span was 12 m, and it weighed just 283 kg. A sprocket chain drive powered two propellers rotating in opposite directions. During the first flight, at about 10:35 a.m., Orville was the pilot, lying on his front on the lower wing and steering by moving a cradle that pulled wires attached to the wings and to the rudder. The first flight was more like a jump of 37 m, with the pilot airborne for just 12 seconds.

The second flight, made after a skid damaged during the first landing was repaired, went 53 m, the third one 61 m. During the fourth attempt the plane began to pitch up and down before Wilbur regained control, then it suddenly crash-landed and broke its front rudder frame—but not before staying up 57 seconds in the air and traveling 260 m. Before they began their return trip to Dayton the brothers sent a telegram to their father, Reverend Milton Wright: "Success four flights Thursday morning all against twenty one mile wind started from Level with engine power alone average speed through air thirty one miles longest 57 second inform Press home Christmas" (World Digital Library 2014).

Figure 5.8
The first flight of a self-propelled machine heavier than air at Kitty Hawk, North Carolina, at 10:35 a.m. on December 17, 1903, with Orville Wright at the controls. Library of Congress photograph.

temperatures generated by compression ratios of 14–24, compared to just 7–10 in the Otto cycle gasoline engines. This requires a higher engine mass and lower speed, but diesels are inherently more efficient. Even during the engine's first certification tests in February 1897 the prototype had an efficiency above 25% (compared to 14–17% for the day's best gasoline engines). By 1911 their rate had reached 41%, and now the best large diesel engines rate just above 50%, double the rate for gasoline engines (Smil 2010b). Moreover, the engines use heavier and cheaper fuel. Diesel oil is nearly 14% heavier than gasoline (820–850 g/L vs. 720–750 g/L) and their energy density per mass is similar, which means that the diesel engine's energy density per volume is, at almost 36 MJ/L, about 12% higher.

Diesel had resolved to design a more efficient internal combustion engine already during his university studies, and in December 1892 he was finally (after two rejections) granted a patent

> for an internal combustion engine characterized by the fact that in a cylinder of pure air ... is so highly compressed by the piston that the resulting temperature is far above the ignition temperature of the fuel ... and addition of fuel ... occurs so

gradually that combustion takes place without essential pressure or temperature increase because of the outward moving piston and the expansion of compressed air ... (Diesel 1893a, 1)

As filed, the patent could not be converted into a working engine; the second patent was granted in 1895, and Diesel then secured practical help from Heinrich von Buz (1833–1918), general director of the Maschinenfabrik Augsburg, the country's leading mechanical engineering enterprise, and the leading steel producer Friedrich Alfred Krupp (1854–1902), whose companies spent considerably to develop a working machine. The official certification test with a 13.5 kW engine on February 17, 1897, indicated a net efficiency of 26.2% and a maximum pressure of 34 atmospheres, one-tenth of Diesel's original specifications (Diesel 1913). By the fall of 1897 the performance was up to 30.2%. And so Diesel got a better machine, and his ambition was largely realized, but his initial hopes for the machine's social impact were completely misplaced—yet another instance of the unintended consequences of a technical advance (box 5.8).

Commercialization of the new engine was slower than initially anticipated, with fewer than 300 units sold by the end of 1901 (Smil 2010b). In 1903 the first diesel-powered vessel, the small oil tanker *Vandal*, began to operate on the Caspian Sea and on the Volga; in 1904 the first diesel-powered electricity-generating station opened in Kiev, and the French *Aigrette* became the first submarine propelled by a diesel. But the great advance came in February 1912 when the Danish *Selandia* (6,800 dwt freight and passenger ship) became the first oceangoing ship powered by diesel engines. A year before his death, in mid-1912, Diesel noted, "There is now a new verb in naval circles: 'to diesel'. We do nothing but diesel ... they now say everywhere" (Diesel 1937, 421).

And yet the rapid success of internal combustion engines—powering road vehicles, airplanes, and ships and beginning to take over agricultural tasks as tractors, combines and irrigation pumps started to displace draft animals in Western farming—did not end the steam era. Yet another prime mover became commercially available before the end of the nineteenth century, and its subsequent development determined much of the twentieth-century industrial advances. That momentous invention was the steam turbine, soon deployed as a superior prime mover to rotate generators producing electricity in increasingly larger central stations.

Electricity

A systematic understanding of the fundamental properties and laws of electricity was gained through the labors of many European and American

Box 5.8
Diesel's engine: The intent and the outcome

Diesel's ambition was to produce a light, small (about the size of a contemporary sewing machine), cheap engine that would be bought by independent entrepreneurs (machinists, watchmakers, restaurant owners) and would enable extensive decentralization of industry, one of his great social dreams:

> It is undoubtedly better to decentralize small industry as much as possible and to try to get it established in the surroundings of the city, even in the countryside, instead of centralizing it in large cities where it is crowded together without air, light, or space. This goal can only be achieved by an independent machine, the one proposed here, which is easy to service. Undoubtedly, the new engine can give a sounder development to small industry than the recent trends which are false on economic, political, humanitarian, and hygienic grounds. (Diesel 1893b, 89)

A decade later, in *Solidarismus: Natürliche wirtschaftliche Erlösung des Menschen* (Diesel 1903), he promoted worker-run factories and dreamed about the age of honesty, justice, brotherhood peacefulness, compassion, and love, and saw workers' cooperatives as beehives and the workers themselves as bees with ID cards and contracts. But only 300 of 10,000 copies of the book were sold, and modern societies have not organized themselves around worker cooperatives. Diesel told his son that his "chief accomplishment is that I have solved the social question" (Diesel 1937, 395)—but his engines did not find their most important uses in small workshops but in heavy machinery, trucks, and locomotives, and, after World War II, in large tankers, bulk carriers, and container ships, helping to create the very opposite of Diesel's vision, an unprecedented concentration of mass-scale manufacturing and the inexpensive distribution of its products in a new global economy (Smil 2010b).

scientists and engineers during the latter part of the eighteenth century and the first six decades of the nineteenth century, and in many cases their pioneering contributions were acknowledged by using their surnames for basic physical units. Famous eighteenth-century milestones included Luigi Galvani's (1737–1798) experiments with frog legs during the 1790s (and hence his mistaken notion of "animal electricity"), Charles Augustin Coulomb's (1736–1806) studies of electric force (the coulomb is now the standard unit of electric charge), and Alessandro Volta's (1745–1827) construction of the first electric battery (the volt is the unit of electric potential).

In 1819 Hans Christian Ørsted (1777–1851) uncovered the magnetic effect of electrical currents (the orsted is now a unit of magnetic field), and during the 1820s André-Marie Ampère (1775–1836) formulated the concept of a complete circuit and quantified the magnetic effects of electric current (the ampere is the unit of electric current). None of these early nineteenth-century discoveries was more important than Michael Faraday's (1791–1867) demonstration of electromagnetic induction (fig. 5.9). Faraday set out to answer a simple question—if, as Ørsted demonstrated, electricity induces magnetism, can magnetism induce electricity?—and we have an exact date and his detailed description of the answer he discovered (box 5.9).

Figure 5.9
Michael Faraday. Wellcome Library, London, photograph.

Box 5.9
Faraday's discovery of electromagnetic induction

Faraday, a self-educated assistant at the Royal Institution, working mostly with Humphry Davy (1778–1829), the first scientist to describe the electric arc created by a slight separation of two carbon electrodes, published his first major work on electricity (on electromagnetic rotation) in 1821 when he outlined the principle of the electric motor. He began a new series of experiments in 1831, which eventually led to his discovery of electromagnetic induction on October 17, 1831. Concerned that his results might be an artifact of experimental design, he ran the final experiment using a different technique, producing continuous current. Faraday presented the results in a Royal Society lecture on November 24, 1831. This is how he described them in *Experimental Researches in Electricity* (Faraday 1832, 128):

> In the preceding experiments the wires were placed near to each other, and the contact of the inducing one with the battery made when the inductive effect was required; but as some particular action might be supposed to be exerted at the moments of making and breaking contact, the induction was produced in another way. Several feet of copper wire were stretched in wide zigzag forms, representing the letter W, on one surface of a broad board; a second wire was stretched in precisely similar forms on a second board, so that when brought near the first, the wires should everywhere touch, except that a sheet of thick paper was interposed. One of these wires was connected with the galvanometer, and the other with a voltaic battery. The first wire was then moved towards the second, and as it approached, the needle was deflected. Being then removed, the needle was deflected in the opposite direction. By first making the wires approach and then recede, simultaneously with the vibrations of the needle, the latter soon became very extensive; but when the wires ceased to move from or towards each other, the galvanometer needle soon came to its usual position.
>
> As the wires approximated, the induced current was in the contrary direction to the inducing current. As the wires receded, the induced current was in the same direction as the inducing current. When the wires remained stationary, there was no induced current.

Faraday's demonstration that mechanical energy can be converted into electricity (to generate alternating current) and vice versa opened the way for the practical production and conversion of electricity that would not be dependent on (and limited by) heavy, low-energy-density batteries. But decades still had to pass before the combined efforts turned this possibility into commercial reality. When Jules Verne (1828–1905) published his *Twenty Thousand Leagues under the Sea*, he had Captain Nemo explain to Aronnax that "there is a powerful agent, obedient, rapid, easy, which conforms to every use, and reigns supreme on board my vessel. Everything is done by means of it. It lights it, warms it, and is the soul of my mechanical apparatus. This agent is electricity," but in 1870 that was still science fiction, as electricity could not be generated on a large scale and the capacities of electric motors were restricted by power delivered from small batteries.

This delay is not at all surprising because the generation of electricity, its transmission, and its conversion to heat, light, motion, and chemical potential represented an unparalleled achievement among energy innovations. Previously, new energy sources and new prime movers had been designed to do specific tasks faster, cheaper, or with more power, and they could easily be used within existing productive arrangements (for example, millstones were turned by a waterwheel instead of by animals). In contrast, the introduction of electricity required the invention, development, and installation of a whole new system required to generate it reliably and affordably, to transmit it safely over long-distance transmission and distribute it locally to individual consumers, and to convert it efficiently in order to deliver the final forms of energy desired by users.

The commercialization of electricity began with the quest for better lights. As already noted, Davy demonstrated the arc effect in 1808, but the first electric arc lights were lit briefly in Place de la Concorde in December 1844 and then in the portico of London's National Gallery in November 1848. In 1871 Z. T. Gramme (1826–1901) presented the first ring-wound armature dynamo—he called it the new *machine magnéto-electrique produisant de courant continu*—to the Académie des Sciences in Paris (Chauvois 1967). That design eventually opened the way to arc lights powered by electricity generated by dynamos: since 1877 arcs have illuminated some famous public places in Paris and London, and by the mid-1880s they had spread to many European and American cities (Figuier 1888; Bowers 1998). But they required controls to maintain a steady arc as the current consumes the positive electrode, they were unsuited for indoor uses, and the resupply of spent electrodes was a major logistical challenge: for 500 W arc spaced 50 m apart, every kilometer of urban road would have required annually 3.6 km of 15 and 9 mm thick carbon electrodes (Garcke 1911).

The quest for indoor lighting produced by glowing filaments spanned four decades—from Warren de La Rue's 1830s experiments with a platinum coil to 1879, when Edison unveiled his first durable carbon filament lamp (Edison 1880)—and it involved about two dozen prominent (but now forgotten) inventors from the UK, France, Germany, Russia, Canada, and the United States (Pope 1894; Garcke 1911; Howell and Schroeder 1927; Friedel and Israel 1986; Bowers 1998). I must note at least Hermann Sprengel, who invented the mercury vapor pump to produce a high vacuum in 1865; Joseph Wilson Swan (1828–1914), who started his work in 1850 and eventually obtained the UK patent for a carbon filament lamp in 1880; and the Canadians Henry Woodward and Matthew Evans, whose 1875 patent served as the basis of Edison's work. So why did Edison's achievements far surpass those of his many predecessors and competitors?

Edison succeeded because he realized that the race is not just to have the first reliable light bulb but to put an entire practical commercial system of electric lighting in place—and that included reliable electricity generation, transmission, and metering (Friedel and Israel 1986; Smil 2005). As a result, the creation of the electrical industry was driven, more than in any other case of nineteenth-century innovation, by one man's vision. This required accurate identification of technical challenges, solving them through tenacious interdisciplinary research and development, and rapid introduction of the resulting innovations into commercial use (Jehl 1937; Josephson 1959). There were other contemporary inventors of light bulbs or large generators, but only Edison had both the vision of a complete system and the determination and organizational talent to make the whole work (box 5.10, fig. 5.10).

This remains undeniable: Edison was an exceptionally inventive and hard-driving man (his mental commitment was surpassed only by his legendary physical endurance) whose contradictory qualities of a rational, dedicated innovator and a self-promoter offering dubious claims could inspire as well as alienate those working with him. And he could not have achieved so much without generous financing by some of the era's richest businessmen—but he made good use of this investment as his Menlo Park laboratory explored many new concept and options, deserving to be seen as a precursor of the corporate R&D institutions whose innovations have done much to create the twentieth century.

Edison's filament of a carbonized cotton sewing thread in a high vacuum gave off steady light in Edison's first durable light bulb on October 21, 1879, and he demonstrated 100 of his new light bulbs in Menlo Park, New Jersey, on December 31, 1879, by illuminating his laboratory, nearby streets, and the railway station. Although the first light bulbs were very inefficient, their performance was superior to that of any contemporary light source. They were about ten times brighter than gas mantles and a hundred times brighter than a candle. These huge advances in lighting were no less important for industrial modernization and for a higher quality of life than the introduction of better prime movers.

A durable light bulb was a mere beginning: in the three years after its unveiling Edison filed nearly 90 new patents for filaments and lamps, 60 dealing with magneto- or dynamo-electric machines, 14 for the system of

Box 5.10
Edison's electrical system

The first durable electric lamp, demonstrated by Joseph Swan in Newcastle-on-Tyne on December 18, 1878, had the same key components as Edison's first longer-lasting lamp, patented ten months later: platinum lead wires and a carbon filament (Electricity Council 1973; Bowers 1998). But Swan's filaments had a very low resistance (<1–5 Ω), and their large-scale use would have required very low voltages and hence very high currents and massive transmission wires. Moreover, pre-Edisonian lamps were connected in series and energized by a constant current from a dynamo, making it impossible to switch on the lights individually and shutting down the entire system with a single interruption. Edison realized that a commercially viable lighting system would have to minimize electricity consumption by using high-resistance filaments connected in parallel in a constant-voltage system.

This understanding completely contradicted the technical consensus of the day (Jehl 1937), but a simple comparison illustrates the practical consequences of the two approaches. Common pre-Edisonian settings—a lamp of 100 W and 2 Ω—required 7 A. Edison's choice of 140 Ω required just 0.85 A, and greatly reduced the cost of copper conduits (Martin 1922). As Edison put it in his patent application submitted on April 12, 1879: "By the use of such high-resistant lamps I am enabled to place a great number in multiple arc without bringing the total resilience of all the lamps to such a low point as to require a large main conductor; but, on the contrary, I am enabled to use a main conductor of very moderate dimensions" (Edison 1880, 1). And Ohm's law requires that Edison's specifications demanded a 118 V supply, and that voltage (110–120 V) is still the North American (and Japanese) standard, with Europe running at 240 V.

But the verdict is hardly unanimous. I agree with Hughes (1983, 18) that "Edison was a holistic conceptualizer and determined solver of the problems associated with the growth of systems. … Edison's concepts grew out of his need to find organizing principles that were powerful enough to integrate and give purposeful direction to diverse factors and components." But Friedel and Israel (1986, 227) concluded that "the completeness of that system was more the product of opportunities afforded by technical accomplishments and financial resources than the outcome of a purposeful systems approach."

Figure 5.10
Thomas A. Edison in 1882, the year his first coal-fired electricity-generating station began operating in lower Manhattan. Library of Congress photograph.

lighting, 12 for electricity distribution, and 10 for electric and electric meters and motors (Thomas Edison Papers 2015). And concurrently, he and his co-workers translated these ideas into practical realities in an astonishingly short period. The first electricity-generating plant, built by Edison's London company at Holborn Viaduct, started to transmit power on January 12, 1882. New York's Pearl Street Station, commissioned on September 4 of the same year, was the first American thermal power plant. A month after its opening it energized some 1,300 light bulbs in the city's financial district, and a year later more than 11,000 lights were wired.

I find two realities particularly remarkable. The first is the combination of insights and the quality of finished work that made Edison's system so successful and so complete that its basic parameters are still with us. Despite critics and questions (see box 5.9), those who understood the intricacies and complexities of designing such a system de novo had always appreciated the achievement. Perhaps the greatest tribute came from Emil Rathenau, founder of Allgemeine Elektrizitäts Gesselschaft, Germany's largest maker of electrical equipment and a leading pioneer of European electrical industry. In 1908 he recalled his impressions after seeing the display at the Paris Electrical Exhibition of 1881:

> The Edison system of lighting was as beautifully conceived down to the very details, and as thoroughly worked out as if it had been tested for decades in various towns. Neither sockets, switches, fuses, lamp-holders, nor any of the other accessories necessary to complete the installation were wanting; and the generating of the current, the regulation, the wiring with distribution boxes, house connections, meters, etc., all showed signs of astonishing skill and incomparable genius. (In Dyer and Martin 1929, 318–319)

And the second reality is perhaps even more remarkable. As wide-ranging and as fundamental as Edison's work had been, it alone would not have sufficed to create a complete, durable, and efficient modern electricity system: all of the required innovations had to fall into place not only in a very short period of time (nearly all during the miraculous 1880s) but also in a near optimal manner. After more than 120 years the dominant constituents of our pervasive electrical systems—steam turbogenerators, transformers, and high-voltage alternating current (AC) transmission—have grown in efficiencies, capacities, and reliabilities, but their fundamental design and properties remain the same, and their originators would easily recognize the latest variations on the themes they created.

And although incandescent lights have been surpassed by fluorescents (commercialized during the 1930s), and more recently by even more efficient light sources (sodium vapor, sulfur lamps, light-emitting diodes), electric motors, another key component of the system from the 1880s, are ever more common parts of the global electric system. That is why I must take a closer look at the four critical, non-Edisonian, inventions or innovations that helped translated electricity's immense theoretical potential into universal economic and social reality: steam turbines, transformers, electric motors, and transmission using AC.

I have already noted the high mass/power ratios of steam engines and their limited power ratings. These prime movers, which were also bulky and fairly inefficient, were abandoned soon after Charles Parsons (1854–1931) patented the more efficient, smaller, and lighter steam turbine in 1884 (Parsons 1936). Parson's company installed a 75 kW turbine in a Newcastle station in 1888 and progressed to 1 MW unit at Germany's Elberfeld station by 1900; Parson's largest pre–World War I machine, installed in Chicago in 1912, rated 25 MW (Smil 2005). While steam engines rarely rotated faster than a few hundred rpm, modern turbines reach 3,600 rpm and can work under pressure of up to 34 MPa and with steam superheated to 600°C, resulting in efficiencies of up to 43% (Termuehlen 2001; Sarkar 2015). They can be also built in capacities ranging from a few kilowatts to more than 1 GW, able to fill niches ranging from the small-scale conversion of waste heat to electricity to massive turbogenerators in nuclear power stations.

Transformers would probably win a contest for a device that is as common and as indispensable for the modern world as it is absent from public consciousness (Coltman 1988). Usually hidden (underground, inside structures, behind tall fencing), silent and stationary, it made inexpensive, centralized electricity generation possible. The earliest direct current (DC) transmission of electricity from plants to customers had a limited reach. Extending the transmission beyond the quarter of a square mile limit would have necessitated installing massive connectors, which, as Siemens (1882, 70) concluded, "could no longer be accommodated in narrow channels placed below the kerb stones, but would necessitate the construction of costly subways—veritable *cava electrica*." The only other option would have been the construction of numerous stations serving limited local areas—and either option would have been very expensive. Transformers of AC provided an inexpensive and reliable solution (box 5.11)

As already noted, transformers work by electromagnetic induction, a process discovered by Faraday, and their development was not the result of a breakthrough invention but the outcome of gradual improvements based on Faraday's fundamental insight. An early influential design by Lucien H. Gaulard (1850–1888) and John D. Gibbs was introduced in 1883, and subsequently three Hungarian engineers improved it by using closed iron cores, but it was William Stanley (1858–1916), a young engineer employed by Westinghouse, who in 1885 developed a prototype of the device we still use today, and which makes it possible to transmit high-voltage AC from power plants with relatively low losses and distribute it at low voltages to households and industries (Coltman 1988).

Box 5.11

Transforming electricity and transmission losses

Electricity is most efficiently generated and most conveniently converted to final uses at low voltages, but because the power loss in transmission rises as the square of the current transmitted, it is best to use high voltages to limit transmission losses. Transformers convert one electric current into another, by either reducing or increasing the voltage of the input flow, and do so with hardly any loss of energy and across a large range of voltages (Harlow 2012). Simple calculations illustrate the advantage. The power of transmitted electricity is the product of current and voltage (watts = amperes × volts); voltage is the product of current and resistance (Ohm's law, $V = A\Omega$), and hence power is the product of $A2\Omega$.

Power loss (resistance) thus declines as the inverse square of voltage: boost voltage 10-fold, and the line resistance will become only 1/100 when transmitting electricity at the same rate. This would always favor the highest conceivable voltage, but in practice, its increases are limited by other considerations (corona discharge, increased insulation requirements, transmission tower sizes), though high-voltage (HV) and extra-high-voltage (EHV) transmission are now done routinely at 240,000–750,000 V (240–750 kV), with losses limited usually to less than 7% of the transmitted electricity.

As with other components of new electrical systems, the capacities and voltages of transformers grew rapidly during the remainder of the nineteenth century and before World War I. I cannot offer a better appraisal of this simple but ingenious device than Stanley's remarks, presented in a 1912 address to the American Institute of Electrical Engineers:

It is such a complete and simple solution for a difficult problem. It so puts to shame all mechanical attempts at regulation. It handles with such ease, certainty, and economy vast loads of energy that are instantly given to or taken from it. It is so reliable, strong, and certain. In this mingled steel and copper, extraordinary forces are so nicely balanced as to be almost unsuspected. (Stanley 1912, 573)

Transformers were essential for the emergence of AC as the standard choice of the new electric networks. DC was a logical choice for the earliest isolated minigrids serving sections of cities, and there were some undeniable concerns about the safety of high-voltage AC (HVAC). But neither justified Edison's aggressive anti-AC campaign, which began in 1887 and included the electrocutions of stray dogs and cats on a sheet of metal charged with 1 kV from an AC generator (to demonstrate the mortal perils of AC) and personal attacks on George Westinghouse (1846–1914), America's leading industrialist, Stanley's employer, and an early promoter of AC.

Even in 1889 Edison wrote that "My personal desire would be to prohibit entirely the use of alternating currents. They are as unnecessary as they are dangerous ... and I can therefore see no justification for the introduction of a system which has no element of permanency and every element of danger to life and property" (Edison 1889, 632). In his opposition to AC, Edison found a surprising ally in the UK in Lord Kelvin, a world's leading physicist. But a year later Edison spoke as a DC advocate, and David (1991) offered the best explanation of these developments, arguing that Edison's seemingly irrational opposition was actually a rational choice made to support the value of Edison's enterprises, which were committed to producing components of DC-based systems, and hence to improve the terms of selling his remaining shares. Once he had divested, the conflict ceased abruptly.

But this famous "battle of the systems" had a foregone conclusion: fundamental physics favored AC, and after 1890 new systems were AC-based (the switch was further helped by the introduction of an accurate and cheap AC meter in 1889), while the existing DC systems, which by 1891 were supplying more than half of U.S. urban lighting, could be switched to AC thanks to the invention of the rotary converter, patented by Charles S. Bradley, Edison's former employee, in 1888: the converter made it possible to use existing DC-generating equipment while transmitting polyphase HVAC over larger areas. The spread of HVAC was accelerated by a few large-scale projects of the 1890s, including London's large Deptford Station, which served more than 200,000 lights, and the development of the world's largest AC link, from a Niagara Falls hydroelectric plant to Buffalo (Hunter and Bryant 1991). In 1900 came the first public supply using the three-phase current, and the highest transmission voltages rose to 60 kV by 1900 and to 150 kV by 1913. Thus all the component of modern electricity generation and transmission were in place before World War I.

Three years after Stanley's transformer, Nikola Tesla patented the first practical polyphase induction motor running on AC (Cheney 1981; fig. 5.11). Much as with the development of incandescent lights, however, this invention came after decades of experiments, trials, and even the commercial deployment of DC motor designs powered by batteries, starting in the late 1830s, and eventually, by the late 1870s, also by dynamos (Hunter and Bryant 1991). The high operating cost and limited battery capacity made small DC motors inferior prime movers compared to steam engines.

Figure 5.11
Nikola Tesla in 1890. Photograph by Napoleon Sarony.

The first commercially successful small DC electric motor (thousands were sold) was also powered by a bulky battery and patented by Edison in 1876; it was to be mounted on top of a stylus to drive a stencil-making pen used for the mechanical duplication of monuments (Pessaroff 2002). Once large dynamos became available there were also attempts to use small DC motors to power streetcars (first in Germany) and for many industrial tasks (above all in the United States). The prospects changed fundamentally only with Nikola Tesla's (1857–1943) invention, conceptualized in Europe and turned into a working machine after the young Serbian engineer emigrated to the United States.

Tesla claimed that his original idea came in 1882, but after his emigration to the United States, Edison, his first American employer, had little interest in AC; Tesla, however, had no difficulty securing financing, opening his own company in 1887 and filing all the requisite patents—40 of them between 1887 and 1891. In designing his polyphase motor Tesla aimed at

> a greater economy of conversion that has heretofore existed, to construct cheaper and more reliable and simple apparatus, and, lastly, the apparatus must be capable of easy management, and such that all danger from the use of currents of high tension, which are necessary to an economical transmission, may be avoided. (Tesla 1888, 1)

Westinghouse bought all of Tesla's AC patents in July 1888, and in 1889 the company had its first electrical device powered by a Tesla motor: a small fan (125 W), powered by a 125 W AC motor; by 1900 it had sold nearly 100,000 units (Hunter and Bryant 1991). Tesla's first patent was for a two-phase machine, and the first three-phase design was built by Mikhail Osipovich Dolivo-Dobrowolsky (1862–1919), a Russian engineer working for AEG. Three-phase motors (with each phase offset by 120°) ensure that one of the three phases is always near or at its peak, resulting in more even power output than a two-phase design and being almost as good as a four-phase machine, which would require another wire. The conquest of the market by three-phase motors was rapidly creating, as I explain in the next section, a major transformation of manufacturing.

Technical Innovations

The great transition from phytomass fuels to fossil fuels and from animate to mechanical prime movers brought unprecedented changes, both in terms of their new, truly epoch-making qualities and in terms of the pace of their adoption. In 1800 the inhabitants of Paris, New York, or Tokyo lived in a world whose energetic foundations were no different not only from those of 1700 but also from those of 1300: wood, charcoal, hard labor, and draft animals powered all of those societies. But by 1900 many people in major Western cities lived in societies whose technical parameters were almost entirely different from those that dominated the world in 1800 and that were, in their fundamental features, much closer to our lives in the year 2000. As the historian Lewis Mumford (1967, 294) summed up, "Power, speed, motion, standardization, mass production, quantification, regimentation, precision, uniformity, astronomical regularity, control, above all control—these became the passwords of modern society in the new Western style."

Examples of these changes abound, and from them I have selected just a few global accomplishments to illustrate the magnitude of those rapid gains. At the most fundamental level, in 1800 the world consumed about 20 EJ of energy (an equivalent of less than 500 Mt of crude oil), of which 98% was phytomass, mostly wood and charcoal; by 1900 the total primary energy supply had more than doubled (to about 43 EJ, equal to just over 1 Gt of crude oil), and half of it came from fossil fuels, mostly from coal. In 1800 the most powerful inanimate prime mover, Watt's improved steam engine, had a capacity just above 100 kW; in 1900 the largest steam engines rated 3 MW, or 30 times as much. In 1800 steel was a rarity; by 1850, even in the UK, it "was known in commerce in comparatively very limited quantities" (Bell 1884, 435), and only a few hundred thousand tonnes of it were produced worldwide—but by 1900 the global output was 28 Mt (Smil 2016).

But note my hedging and qualifying terms—"almost" and "in their fundamental features"—when describing the world of 1900. The accomplished shift, both in qualitative and quantitative terms, was profound and its pace was frequently astonishing; at the same time, the world of fossil fuels and inanimate prime movers was still new, far from mature, often highly inefficient, and associated with many negative environmental impacts. By 1900 the United States and France were already overwhelmingly fossil-fueled societies, but the world as a whole still derived half of its primary energy from wood, charcoal, and crop residues—and even in the United States the year of the peak draft horse numbers was still 17 years in the future. And though incandescent lights, electric motors, and telephones were making rapid inroads, most of the electricity used by urban families in the United States or Germany powered just a few light bulbs.

The foundations of a new energy world were firmly in place, but during the twentieth century all components of this new system were greatly transformed by a combination of further rapid growth and qualitative improvements, that is, mainly by gains in efficiency, productivity, reliability, safety, and environmental impact. This progression was interrupted by World War I and then by the economic crisis of the 1930s. World War II sped up the development of nuclear energy and the introduction of gas turbines (jet engines) and rocket propulsion. Renewed growth after 1945 in all energy industries reached new peaks by the early 1970s, and subsequently many energy techniques reached unmistakable size, and often also performance, plateaus. Notable examples include the capacities of steam turbines, the tonnages of typical large oil tankers, and the ratings of dominant HVAC transmission lines.

This leveling off has not been largely a matter of technical limits but rather a result of the prohibitive costs and unacceptable environmental impacts. Another important factor that contributed to moderating the pace of energy advances was the two rounds (1973–1974, 1979–1980) of rapid OPEC-led oil price rises and their dampening effect on energy consumption. As a result, greater efficiency, reliability, and environmental compatibility became new engineering goals. But energy prices eventually stabilized, and the U.S. economy, still the world's largest, experienced another decade of strong expansion during the 1990s as it became ever more engaged with China.

After decades of Maoist misery the world's most populous country embraced reform policies that quadrupled its per capita use of primary energy between 1980 and 2010: in 2009 China became the world's largest consumer of energy (by 2015 it was about 30% ahead of the United States). Its 2015 average per capita energy use of about 95 GJ was similar to that of France in the early 1970s, but industrial use is still dominant, and China's residential use remains lower than in the West at a comparable stage of development. By 2015 the growth rates of China's economy and energy demand had, inevitably, moderated, but there are billions of people in India, Southeast Asia, and Africa hoping to replicate China's success, and more than two billion people will be added to the 2015 total by 2050.

That the demand for energy will continue to rise is a truism, but none of us can foresee how it will be met in a world full of economic inequalities and concerns about the global environment. Forecasts and scenarios abound, but the history of energy advances has shown that failure to foresee is their most common trait (Smil 2003). In this section I review and summarize major trends that have determined the expansion, maturation, and transformation of fossil fuel extraction, processing, and delivery, of advances in both thermal and renewable electricity generation, and the changing composition and performance of mechanical prime movers, but before I do so in some specific detail I should point out several commonalities that have characterized the production of fossil fuels, the generation of electricity, and the diffusion of prime movers.

The post-1900 extraction of fossil fuels has been marked by three notable trends. First, the global expansion of coal mining and hydrocarbon production raised the annual extraction of fossil carbon roughly 20-fold between 1900 and 2015: from 500 Mt in 1900 to 6.7 Gt a century later and to about 9.7 Gt in 2015 (Olivier 2014; Boden and Andres 2015; to express those totals in terms of CO_2, multiply them by 3.67). Because of the unequal distribution of fossil fuels, this expanded extraction has led inevitably to the emergence of a truly global trade in easily transportable crude oil and to rising exports of coals and natural gas (both by pipelines and in liquefied natural gas tankers). But a closer look reveals some important qualifications and exceptions as the global rise has subsumed many complex national trajectories, including those with notable production declines in or the complete cessation of fuel extraction.

Second, numerous technical advances have been the most important enablers of this expansion, resulting in cheaper and more productive extraction, transportation, and processing methods as well as in reduced specific pollution rates (and, in one remarkable case, even in an absolute global emission decline). Third, there has been a clear secular shift toward higher-quality fuels, that is, from coals to crude oil and natural gas, a process that has resulted in relative decarbonization (a rising H:C ratio) of global fossil fuel extraction, while absolute levels of CO_2 emitted to the atmosphere have been rising except for a few temporary slight annual declines. The H:C ratio of wood combustion varies but is no higher than 0.5, while the ratios are 1.0 for coal, 1.8 for gasoline and kerosene, and 4.0 for methane, the dominant constituent of natural gas.

When compared on the basis of energy content, high-carbon fuels (wood and coal) supplied 94% of the world's energy in 1900 and 73% in 1950, but only about 38% by the year 2000 (Smil 2010a). As a result, the average carbon intensity of the world's fossil fuel supply kept on declining: when expressed in terms of carbon per unit of the global total primary energy supply, it fell from nearly 28 kg C/GJ in 1900 to just below 25 in 1950 and to just over 19 in 2010, roughly a 30% decrease; subsequently, as a result of China's rapidly rising coal output, it rose a bit during the first decade of the twenty-first century (fig. 5.12). At the same time, global emissions of carbon from the combustion of fossil fuels rose from just 534 Mt C in 1900 to 1.63 Gt in 1950, 6.77 Gt in 2000, and 9.14 Gt C in 2010 (Boden, Andres, and Marland 2016).

The generation of electricity has combined technical improvements with large-scale spatial expansion, with the latter process surprisingly delayed even in parts of the United States and still far from completed in many low-income populous nations. This process started with small, isolated networks and has advanced to massive grids: in Europe they span the entire continent, Russia has an extensive network, since 1990 China has constructed many new long-distance interconnections, and among the high-income economies only the United States and Canada do not have any integrated nationwide grids. The latest transformation affecting the industry is the installation of wind turbines, photovoltaic cells, and central solar power stations: these new renewables (as opposed to hydroelectricity,

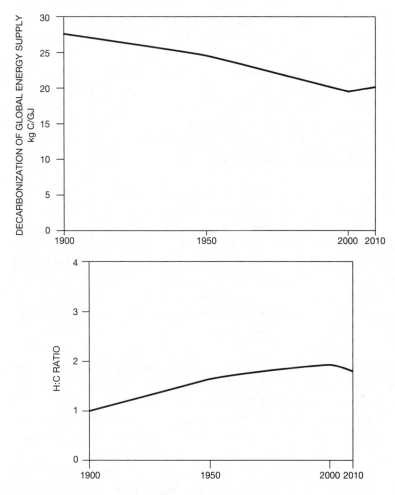

Figure 5.12
Decarbonization of the world's primary energy supply, 1900–2010. Plotted from data in Smil (2014b).

the old form of renewable generation) are often heavily promoted and subsidized, and they have seen some rapid capacity additions, but their inherent intermittency and their low-capacity factors pose nontrivial problems in integrating them into existing grids.

Coals

The two universal trends in coal production have been the growing mechanization of underground extraction and the rising share of surface mining. American productivities, highest in the world, rose from less than 1 t of coal per miner per shift in 1900 to a nationwide hourly average of about 5 t per worker, with specific rates ranging from 2 to 3 t/h in Appalachian underground mines to about 27 t/h in surface mines of the Powder River Basin in Montana and Wyoming (USEIA 2016a). High productivities also mark the surface extraction of thick lignite (brown coal) seams in Australia and Germany. Coal from such large mines has been increasingly burned in large adjacent (mine-mouth) power plants. Its transport to distant markets is done by special unit trains made up of up more than 100 large, lightweight, permanently coupled hopper cars pulled by powerful locomotives (Khaira 2009).

Coal consumption has seen two principal trends as losses of its traditional industrial, household, and transportation markets were more than made up by gains resulting from coal-based generation electricity (and, to a much lesser extent, by the rising production of metallurgical coke and the use of coal as a feedstock for chemical syntheses). The coal burned by households for heating and cooking has been displaced by cleaner and more efficient alternatives, now dominated by natural gas and electricity. Coal remained the leading transportation fuel during the first half of the twentieth century, but the conversion of locomotives and ships to diesel engines (begun, respectively, in the century's first and third decade) accelerated after World War II, and all new rapid trains (first Japan's *shinkansen* in 1964, then the French TGV in 1978 and other European and Asian versions) have been powered by electric motors.

Coal combustion launched the thermal electricity generation during the 1880s and in every traditional coal-mining country, and that dependence only grew as large central power plants were constructed after World War II, when the increasing share of surface extraction made coal even more affordable. During the 1950s coal combustion provided the largest share of electricity generation in the United States, the UK, Germany, Russia, and Japan. Fuel oil gained importance during the 1960s, but most countries stopped using it in electricity generation after OPEC raised oil prices during the 1970s, and a dependence on coal remains high in China, India, and the United States. The specific use of metallurgical coke (kg of coke/kg of hot metal) has been declining for decades, but the worldwide expansion of pig iron smelting, from about 30 Mt in 1900 to about 1.2 Gt by the year 2015, pushed the coking coal output to roughly 1.2 Gt (Smil 2016).

National coal histories have shown many predictable and some surprising developments, including the mining's end in the country that pioneered the fuel's extraction (fig. 5.13). British output reached a peak of 292 Mt in 1913, and coal energized not only Britain's industries but also the

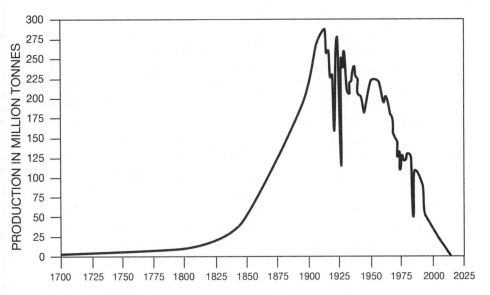

Figure 5.13
British coal production, 1700–2015. Plotted from data in Nef (1932) and Department
of Energy & Climate Change (2015).

nineteenth-century expansion of its colonial empire and, through its dominance of naval forces and commercial shipping, also the functioning of its
trade empire. In 1947, at the time the Labour government nationalized the
industry and created the National Coal Board, it was still almost 200 Mt
(Smil 2010a). The postwar peak came in 1952 (and again in1957) at 228 Mt,
but then the rising imports of crude oil and, after 1970, the availability
of the North Sea oil and natural gas halved the country's coal dependence
by 1980.

During a long coal miners' strike of 1984 the total output fell to 51 Mt,
and then recovered only brief before resuming its fall which continued
after re-privatization in 1994 (Smil 2010a). In the year 2000 the output was
only about 31 Mt, and in July 2015 UK Coal Holdings announced the
immediate shutdown of its Thoresby colliery and end of operations at the
last British mine, Kellingley, in December of 2015 (Jamasmie 2015). After
400 years of energizing the country the industry that made Britain economically and strategically great (and whose peak employment during the
early 1920s was 1.2 million workers or about 7% of total labor force) now
amounts to a few museums and underground guided tours (National Coal
Mining Museum 2015).

American coal extraction reached 508 Mt in 1950 and peaked at 1.02 Gt in 2001. During that period it lost all of its transportation and nearly all of its household markets, and the production of coking coal also declined but, exports have expanded. More than 90% of all shipments are now burned in thermal power plants: in 1950 the United States generated 46% of its electricity from coal, and the share rose to 52% by 1990 and remained that high for more than a decade; by 2010 it was still at 45%, but by 2015 (with closures of old coal-fired plants and plenty of inexpensive natural gas) it had slipped to 33% (USEIA 2015a). The U.S. coal output was surpassed by the Chinese production in 1985, and coal has been by far the most important energizer of China's extraordinary economic growth (USEIA 2015b; box 5.12).

Until 1983 the USSR was a larger coal producer than the United States, but after the state collapsed, Russian coal extraction declined as natural gas and crude oil filled the need. India is now the third world's largest producer (in 2014 only one-sixth of the Chinese output), but its coal is of much lower quality than the Chinese and U.S. deposits and productivity of its extraction is still dismal. Indonesia and Australia (both major exporters) complete the top five, followed by Russia, South Africa, Germany, Poland, and Kazakhstan, while some former major coal producers, including Germany and the UK, have been substantial coal importers.

Because the fuel generates more CO_2 per unit of released energy than any other fossil fuel—the rates are typically more than 30 kg C/GJ for coal, about 20 kg C/GJ for liquid hydrocarbons, and less than 15 kg C/GJ for natural gas—its future in a world concerned about rapid global warming is uncertain. A high dependence on coal for electricity generation in China, India, and at least a dozen other nations precludes any rapid abandonment of the fuel, but in longer run, coal may be the first major energy resource whose extraction, despite of its still very abundant resources, will be limited because of environmental concerns.

Hydrocarbons
At the beginning of the twentieth century oil was produced in quantity in just a handful of countries, and the fuel provided just 3% of all energy from fossil fuels; by 1950 that share was about 21%, energy content of crude oil surpassed that of coal by 1964 and it peaked in 1972 at about 46% of all fossil fuels. The two common impressions—that the twentieth century was dominated by oil, much as the nineteenth century was dominated by coal—are both wrong: wood was the most important fuel before 1900 and, taken as a whole, the twentieth century was still dominated by coal (Smil

Box 5.12

Chinese coal production

Once the Chinese Communist Party established a new regime on October 1, 1949, it energized its Stalinist-type industrialization with China's abundant but very unevenly distributed coal deposits. During the subsequent decades the country's relative dependence on coal has declined but the totals have grown to record levels (Smil 1976; Thomson 2003; China Energy Group 2014; World Coal Association 2015). Coal output rose from only about 32 Mt in 1949 to 130 Mt in 1957 and was claimed to be nearly 400 Mt in 1960, during the infamous (famine-inducing) Great Leap Forward, launched by Mao Zedong to surpass Britain in 15 years or less in the output of iron, steel, and other major industrial products (Huang 1958). After the Leap collapsed, a more orderly progress raised the output to more than 600 Mt by 1978, when Deng Xiaoping began his far-reaching economic reforms, which would eventually transform China into the world's largest exporter of manufactured goods and raised the living standards of its nearly 1.4 billion people.

Two things that have not changed is the party's firm control of the state and the economy's dependence on coal. Its relative dependence has declined from more than 90% in 1955 to 67% in 2010, and the share of China's electricity generated from coal has also declined, though it remains above 60%. But China's total coal output more than quadrupled between 1980 (907 Mt) and 2013 (3.97 Gt), when it accounted for almost as much as the rest of the world production put together. The year 2014 was the first year when the extraction showed a dip of 2.5%, and in 2015 there was another decline of 3.2%, but real totals remain uncertain: in September 2015 China's National Bureau of Statistics raised, without any explanation, its previous data on annual coal extraction between 2000 and 2013. The enormous coal production has been a leading source of China's occupational fatalities and the largest source of extremely high levels of air pollution, with levels of small particulate matter (<2.5 μm) repeatedly reaching an order of magnitude above the desirable maxima (Smil 2013b).

2010a). My best calculations show coal about 15% ahead of crude oil (approximately 5.2 YJ vs. 4 YJ), and even when nonenergy applications of processed oil (in lubricants, paving materials) are included, coal would still be just ahead of liquid hydrocarbons, or, because of the inherent uncertainties when converting extracted masses to common energy equivalents, the twentieth century's cumulative production of the two fuels would be roughly equal.

But liquid fuels separated by refining from crude oil are superior to coals, and while the twentieth-century coal market (as just shown) gradually contracted to just two major sectors, electricity generation and coke, the market for liquid hydrocarbons was steadily expanding, both through substitutions and through the rise of new major consuming sectors. Major substitutions resulted in coal getting replaced by fuel oil and diesel oil in shipping (starting before World War I, accelerating during the 1920s) and then in railroads (starting during the 1920s), by fuel oil (and then by natural gas) in industrial, institutional, and household heating, and by liquid and gaseous hydrocarbons as feedstocks for the petrochemical industry (after World War II).

The first new large market was created by the introduction of affordable automobiles, starting before World War I with Ford's Model T, and by the rapid post–World War II rise in car ownership; the second one began with the introduction of jet engines in commercial aviation during the 1950s, an innovation that changed flying from a very expensive and a rare experience to a mass global industry (Smil 2010b). The oil industry could meet this expanding demand because of a multitude of technical advances that have affected every aspect of its operation. Even a list restricted to just the key twentieth-century improvements has more than a dozen items (Smil 2008a).

The list must start with advances in geophysical prospecting: they include the idea of electrical conductivity measurements (1912), electrical resistivity well log (1927) to identify hydrocarbon-bearing subsurface structures, spontaneous potential (1931) and induction log (1949), introduced by Conrad Schlumberger (1878–1936) and his relatives and subsequently perfected by the eponymous company and other oil-and-gas explorers (Smil 2006). Advances in extraction must include first the universal adoption of rotary drilling (used for the first time at the Spindletop gusher in Beaumont, Texas, in 1901; see fig. 5.6), then the introduction of the rolling cutter rock bit by Howard Hughes (1905–1976) in 1909, the invention of the tricone bit in 1933, and improvements in monitoring and regulating oil flow and preventing well blow-ups. An increasing reliance on secondary and tertiary recovery methods (using water and other liquids or gases to force more oil to the surface) has prolonged the life span of wells and increased their traditionally very low productivity (as little as 30% of oil-in-place used to be recovered).

An increasing share of oil production has been coming from offshore wells. Near-shore drilling from piers was common in California by 1900, but the first well completed out of sight of land was drilled in 1947 off Louisiana. Offshore rigs (mostly semisubmersible designs) work in waters more than 2,000 m deep. Production platforms installed at major offshore fields are among the most massive structures ever built. And the most recent production advance is the rising extraction from nonconventional sources of crude oil, including heavy oils (in many places around the world), oil embedded in tar sands (Alberta, Venezuela), and extraction by hydraulic fracturing to produce oil from shales. This technique, pioneered in the United States, has proved so successful that America became once again the world's largest producer of crude oil and other petroleum liquids—but if only crude oil is considered, Saudi Arabia was still a tiny bit ahead in 2015, producing 568.5 Mt compared to 567.2 Mt for the United States.

The transportation of crude oil has been transformed by seamless steel pipes fashioned into large-diameter trunk pipelines eventually capable of spanning continents. These pipelines are the most compact, most reliable, cleanest, and safest mode of bulk transportation on land. American lines carrying crude oil from the Gulf to the East Coast, built during World War II, were surpassed during the 1970s by the world's longest system, designed to move the Western Siberian crude oil to Europe. The Ust-Balik-Kurgan-Almetievsk line (diameter 120 cm, length 2,120 km) carries annually up to 90 Mt of crude oil from the supergiant Samotlor oil field to European Russia, and then about 2,500 km of branching large-diameter lines move that oil to European markets as far west as Germany and Italy. The post–World War II demand for oil imports to Europe and Japan led to a rapid growth in oil tanker sizes (Ratcliffe 1985). This turned oil into an affordable global commodity as the distance between its origin and the final user has become only a minor economic consideration and as annual intercontinental crude oil sales surpass 2 Gt (box 5.13).

The single most important advance in refining was catalytic cracking of crude oil. Thermal cracking was the norm until 1936, when Eugène Houdry (1892–1962) began to produce high-octane gasoline, the principal automotive fuel, at Sun Oil's Pennsylvania refinery in the first catalytic cracking unit. Catalytic cracking has made it possible to produce higher shares of more valuable (lighter) products (gasoline, kerosene) from intermediate and heavy compounds. Shortly afterward a new moving-bed catalyst could be regenerated without shutting down production, and even better yields of high-octane gasoline became possible with an airborne powdered catalyst (Smil 2006). During the 1950s fluid catalytic cracking was supplemented by hydrocracking at relatively high pressures, and the two techniques are still the mainstays of modern refining. Refining has also benefited from the desulfurization of liquid fuels, which has made even such proverbially polluting fuels as diesel oil acceptable for low-emission passenger cars (CDFA 2015).

Box 5.13
Giant oil tankers

The first tanker, the British-built German *Glückauf,* launched in 1886, was rated at just 2,300 gross tons (Tyne Built Ships 2015). Subsequent growth brought the maximum size to about 20,000 deadweight tons (dwt) by the early 1920s. During the war the most commonly deployed U.S. tankers (T-2) had a capacity of 16,500 dwt, and the rapid capacity rise began only with the expansion of the global oil trade (shipments to Europe and Japan) during the late 1950s. The *Universe Apollo* was the first 100,000 dwt ship, in 1959; in 1966 the *Idemitsu Maru* reached 210,000 dwt, and when OPEC quintupled its oil prices in 1973, the largest vessel could carry more than 300,000 t (Kumar 2004).

Building ships capable of carrying a million tonnes of crude was technically possible but impractical for many reasons: their draft restricts their routes and ports of call (they cannot go through the Suez Canal or the Panama Canal), they need long distances to stop, they are very expensive to insure, and they have caused such catastrophic oil spills as those of the *Amoco Cadiz* (France 1978), *Castillo de Belver* (South Africa 1983), and the *Exxon Valdez* (Alaska 1989). The world's largest tanker, the *Seawise Giant*, was built in 1979, enlarged to 564,763 dwt, hit in 1988 during the Iran-Iraq War, relaunched as the (nearly 459 m long) *Jahre Viking* (1991–2004), renamed *Knock Nevis* and used as a floating storage and offloading unit off Qatar (2004–2009), then sold to Indian ship-breakers and renamed *Mont* for its final journey to Alang in Gujarat (Konrad 2010).

All of this has brought four notable outcomes. First, global oil production grew roughly 200-fold during the twentieth century; by 2015 it was (at over 4.3 Gt) about 20% higher than in the year 2000, and since 1964, when its energy content surpassed that of coal extraction, it has been the world's mostly commonly used fuel. Second, oil is now produced on every continent and from offshore wells in every ocean except the high Arctic seas and Antarctica, and from fields as deep as 7 km belowground on land, while Brazil's Tupi deposits are 2.1 km below the surface of the Atlantic and then nearly 5 km below the ocean's bottom. Third, oil is the single most valuable traded commodity: in 2014 (averaging about $93 a barrel for West Texas Intermediate), its annual output was worth about $3 trillion, in 2015 (as prices declined to about $49 a barrel) it was about $1.6 trillion (BP 2016).

Finally, although oil extraction is widely distributed, the world's largest oil fields were discovered on land, in the Persian Gulf region between 1927

(Kirkuk in Iraq) and 1958 (Ahwaz in Iran). Al-Ghawar, the world's largest oil field in the eastern province of Saudi Arabia, has been producing since 1951, and the second largest field, Kuwaiti al-Burqan, has been in operation since 1946 (Smil 2015b; fig. 5.14). Nothing can change this fundamental reality: in 2015 almost half of known reserves of conventional (liquid) oil were in the region, which is, unfortunately, also the world's most prominent source of complex conflicts and chronic political instability (BP 2016).

For decades, natural gas remained a minor contributor to global energy supply: in 1900 it supplied merely 1% of all fossil energies and by 1950 its share was still only about 10%, but afterward three major demand trends lifted its global share to nearly 25% of all fossil energies by the year 2000, and the twentieth century saw a 375-fold increase in total energy derived annually from this cleanest of all fossil fuels (Smil 2010a). Relatively the smallest but very important new market was the use of natural gas as both feedstock and fuel for the synthesis of ammonia—the most important nitrogenous fertilizer, now mostly used as a feedstock to produce solid urea (Smil 2001; IFIA 2015)—and for the production of plastics.

Figure 5.14
Wells of al-Burqan oilfield (on the right, eastern, side of the image) were set on fire by the retreating Iraqi army in 1991. Image produced on April 7,1991 by earthobservatory.nasa.gov.

The largest new global market has developed in response to high levels of urban air pollution experienced in most Western cities during the period of accelerated post–World War II industrialization: replacement of coal and fuel oil by natural gas for industrial, institutional, and household heating (and cooking) eliminated emissions of particulate matter and almost eliminated the generation of SO_2 (it is not difficult to remove sulfurous compounds from gas before combustion). Cities in the rapidly modernizing countries of Latin America and Asia have followed the trend, although many of them, including Tokyo and other Japanese conurbations, Seoul, Guangzhou, Shanghai, and Mumbai, had to do it with expensive imported liquefied natural gas (LNG). The latest trend boosting natural gas use has been the fuel's efficient use for generating electricity by gas turbines and, even more efficiently, by combined-cycle gas turbines (see the next section). Post-2005 hydraulic fracturing has not only stopped the further decline of American natural gas extraction but also it made the country, again, the world's largest producer.

Pipeline transport of natural gas is inherently more expensive than moving liquids, and long-distance pipelines became economical only with the introduction of large-diameter steel pipes (up to 2.4 m in diameter) and efficient gas turbine compressors (Smil 2015a). The United States and Canada have had integrated gas pipeline systems since the 1960s, but the most extensive international network has evolved in Europe since the late 1960s. The longest lines—a 4,451 km line from Urengoy to Uzhgorod station on the Ukrainian-Slovak border, and a 4,190 km line from Yamal to Germany—now bring Siberian gas to Central and Western Europe, where the lines connect to supplies from the Netherlands, the North Sea, and North Africa.

The first shipments of LNG during the 1960s were very expensive, and for the next three decades the limited trade was supplying mostly East Asian nations (Japan, Taiwan, South Korea) that had no domestic gas resources. New gas discoveries and the introduction of larger LNG tankers brought a relatively sudden expansion of this trade, and by 2015 almost exactly one-third of all exported gas was shipped by tankers (BP 2016). Japan is still the largest importer, but before too long China will become the world's largest buyer, and, in a major reversal of roles, the United States, traditionally a large importer of Canadian pipeline gas, is developing many new LNG facilities, hoping to become a leading exporter of LNG, perhaps even as a future rival of Qatar, a small, rich country selling LNG from the world's largest gas field in the Persian Gulf (Smil 2015a).

Electricity

Advancing electrification has required exponential increases in the ratings of all system components. The earliest, relatively small boilers were stoked with lump coal that burned on moving grates. Starting in the 1920s, they were replaced by multistory units burning pulverized fuel injected into the combustion chamber and heating water circulating in steel pipes lining boiler walls. Fuel oil and natural gas also became common choices of fuel for large central power plants, but use of the former (except in Russia and Saudi Arabia) was discontinued following OPEC's second round of oil price increases in 1979–1980, while natural gas for electricity generation is now burned mostly in gas turbines, not only in gas-rich nations but also in countries that have to import expensive LNG. In the United States, gas-generated electricity went from 12% of the total in 1990 to 33% by 2014, while in Japan LNG's share was 28% in 2010, rising to 44% in 2012 after the closure of nuclear power plants following the Fukushima disaster (The Shift Project 2015).

Large boilers supply steam to turbogenerators whose top ratings are three orders of magnitude higher than they were in 1900 (the largest unit, at France's Flamanville Nuclear Power Plant, rates 1.75 GW), and their higher operating pressures and temperatures raised the best efficiencies from less than 10% in 1900 to just over 40% (fig. 5.15). Even higher efficiencies, on the order of 60%, are possible by using a combination of gas turbines (the largest machines now rate above 400 MW) and steam turbines (using hot gas leaving a gas turbine to produce steam). Not surprisingly, using combined-cycle gas turbines has become a favored way of generating electricity, especially to cover the need generated during peak periods of demand (Smil 2015b). Large diesel engines have been the most economical choice for electricity generation in remote locations as well as for providing standby capacities to deliver uninterrupted power during emergencies.

The expansion of utilities from urban to national systems began slowly after World War I and accelerated after World War II; it has included the following universal components (Hughes 1983): the pursuit of economies of scale; building larger stations in or near large cities; the development of high-voltage links to transmit electricity from remote hydro stations; the promotion of mass consumption; and the interconnection of smaller systems to improve supply security and lower installed and reserve capacities. After 1950 concerns about air pollution led to new, large generating stations being located close to sources of fuel. This shift to mine-mouth power plants further increased the need for high-voltage transmission.

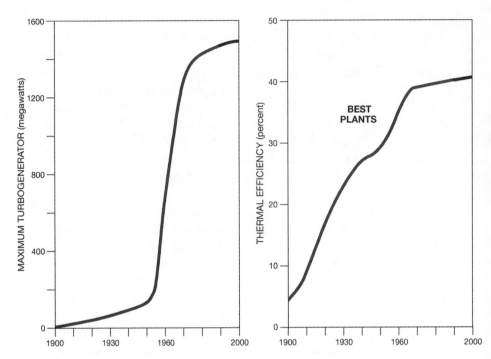

Figure 5.15
Maximum capacities of steam turbogenerators and the efficiencies of the best thermal power plants, 1900–2000. Plotted from data in Dalby (1920), Termuehlen (2001), and Smil (2008a).

Consequently, the power of the largest transformers has grown 500 times, and the highest transmission voltages have risen more than 100-fold, since the 1890s. Transmission started with wooden poles and solid copper wires. Eventually it advanced to steel towers carrying steel-reinforced aluminum cables charged with up to 765 kV, and the highest DC voltage is now ± 800 kV carrying 6.4 GW between the Xiangjiaba hydropower station and Shanghai. Household service grew from a handful of sockets to systems including commonly more than 50 switches and outlets in a dwelling. Higher capacities and rising generation have been accompanied by increased reliability of the service, a particularly important consideration in a world full of electronic devices and controls (box 5.14)

Nuclear fission's arrival as the other major way to raise steam for thermal electricity generation was accelerated by World War II. The first demonstration of the phenomenon, by Lise Meitner and Otto Frisch in December

---

Box 5.14

Reliability of electricity supply

The reliability of the electricity supply is often expressed in terms of nines, the percentage of time in standard 365-day year that a particular grid is working properly and is able to supply the needed demand. A system with four nines, with electricity available 99.99% of all time, may seem highly reliable—but the total annual outage would be almost 53 minutes. Five nines would cut the total outage to just over 5 minutes, and the industry's goal has been to achieve 99.9999% (six nines) reliability, leaving a system without electricity for just 32 s/year. The current U.S. performance is about 99.98%, with outages caused not only by weather (tornadoes, hurricanes, snowstorms, extreme cold) but also by vandalism or interruptions of the fuel supply (Wirfs-Brock 2014; North American Electric Reliability Corporation 2015).

Electronic communication, controls, and storage of information underlie every sector of modern economies, ranging from dispatching and monitoring truck shipments of food to automated production of microchips, and from stock trading to air traffic control and the only way to ensure uninterrupted service is to install emergency systems (batteries and generators capable of rapid response. Even brief losses of electricity supply can be very expensive, with costs reaching more than $10 million per hour for some service and industrial operations, and between 2003 and 2011 the nationwide U.S. loss ranged between $18 and $75 billion (in 2008, with Hurricane Ike) (Executive Office of the President 2013). And electrical grids are the prime candidates for cyber-attacks by terrorist groups or adversary governments.

1938, was followed by the first sustained chain reaction, at the University of Chicago on December 2, 1942. The first nuclear bomb was tested in July 1945, and two bombs were dropped three days apart in August 1945 (Kesaris 1977; Atkins 2000). The continuous development of more powerful nuclear weapons aside (see the section on weapons and war in the next chapter), the first major postwar U.S. nuclear program was to develop nuclear reactors for the propulsion of submarines: the *Nautilus* was launched in January 1955, and almost immediately Hyman Rickover (1900–1986), leader of the nuclear submarine program, was put in charge of reconfiguring the reactor for commercial electricity generation (Polmar and Allen 1982). The first U.S. nuclear power station, Shippingport, in Pennsylvania, began operating in December 1957, more than a year after the British Calder Hall was started, in October 1956.

In retrospect, this was not the best possible choice of reactor design, but it became the dominant type worldwide. Although not a superior design, its early adoption made it entrenched by the time other reactors were ready to compete (Cowan 1990). In mid-2015 277 of the world's 437 operating nuclear reactors were pressurized water reactors with most of them in the United States and France. Looking back after nearly half a century of commercial nuclear generation, I called nuclear electricity a successful failure (Smil 2003), and that verdict has only been reinforced by subsequent developments. It was successful because in 2015 it supplied 10.7% of the world's electricity, and before the recent Chinese surge of coal-fired power plant construction that share was about 17%. Many national shares are higher, including nearly 20% in the United States, 30% in South Korea (and also in pre-2011 Japan), and 77% in France. It is a failure because its enormous early promise (during the 1970s it was widely expected that by the century's end, nuclear power would be the dominant mode of global electricity generation) remains largely unfulfilled.

The technical weaknesses of dominant designs, the high construction costs of nuclear plants and chronic delays in their completion, the unresolved problem of long-term disposal of radioactive wastes, and widespread concerns about operation safety (including, even after 60 years of commercial experiences, some grossly exaggerated claims of possible health impacts) have prevented further rapid growth of the nuclear industry. Safety concerns and public perception of intolerable risks were strengthened by the 1979 Three Mile Island accident and even more so by the 1986 Chornobyl disaster and by the 2011 explosion of three Fukushima Dai-ichi reactors in the wake of a major earthquake and tsunami (Elliott 2013).

As a result, some countries have refused to allow any construction of nuclear stations (Austria, Italy), others have plans for their complete closure in the near future (Germany, Sweden), and most nations with operating plants either stopped adding new capacities decades ago (Canada, UK) or have been building only few new stations, far below the number needed just to replace their aging plants. The United States and Japan are the two most prominent countries in this last category: by mid-2015 there were 437 reactors operating worldwide, and of the 67 reactors under construction 25 were in China, nine in Russia, and six in India (WNA 2015b). The West has essentially given up on this clean, carbon-free way of electricity generation.

Renewable Energies

A rising dependence on fossil fuels has made biofuels relatively much less important, but because of the rapid growth of populations in rural areas of low-income countries (where there has been no, or a very limited, access to modern energies), the world now consumes more fuelwood and charcoal than ever. According to my best estimates, gross energy in traditional biofuels reached about 45 EJ in the year 2000, almost exactly twice as much as in 1900 (Smil 2010a), and during the first 15 years of the twenty-first century the total decreased only marginally. This means that in the year 2000 biofuels supplied roughly 12% of the world's primary energy and by 2015 that share declined to about 8% (whereas in 1900 they provided 50%).

Unfortunately, even that harvest, equivalent to about 1 Gt of oil, has not been enough: with hundreds of millions of people in rural areas of low-income countries of Africa, Asia, and Latin America still burning biomass fuels, the demand for fuelwood and charcoal has been a leading cause of deforestation, most acutely in Sahelian Africa, Nepal, India, interior China, and much of Central America. The most effective way to lessen this degradation is to introduce new, efficient (25–30% compared to traditional 10–15%) stoves: this substitution has been most successful in China, where efficient stoves reached about 75% of rural households before the century's end (Smil 2013).

At the same time, it is wrong to think about wood only in relation to forests because in many low-income countries large shares of woody matter are gathered by families (in most cases by women and children) from small tree groves and bushes, from tree plantations (rubber, coconut), and from roadside and backyard trees. Surveys in Bangladesh, Pakistan, and Sri Lanka showed that such nonforest wood accounted for more than 80% of all combustion (RWEDP 1997). At least a fifth of all crop residues produced in low-income countries is still burned, and dried dung remains important in parts of Asia, but charcoal has become a preferred biofuel. As expected, China and India are the world's largest consumers of traditional biofuels, followed by Brazil and Indonesia, but in relative terms there is no match for sub-Saharan Africa, where at the twentieth century's end some countries derived more than 80% of rural energy from wood and crop residue, compared to 25% in Brazil and less than 10% in China (Smil 2013a). In per capita terms these uses range from 5 GJ to as much as 25 GJ/year.

The closing decades of the twentieth century saw the emergence of relatively large-scale ethanol production. Experiments with ethanol for passenger cars predate World War II (Henry Ford was among the proponents), but modern large-scale ethanol production began in 1975 with Brazil's ProÁlcool program fermenting the fuel from sugar cane (Macedo, Leal, and da Silva 2004; Basso, Basso, and Rocha 2011), and the U.S. corn-based ethanol production started in 1980 (Solomon, Barnes, and Halvorsen 2007).

Brazilian output has been stagnating since 2008, and the U.S. output, whose rising volume was mandated in 2007 by the U.S. Congress, is unlikely to increase. There is also a much smaller biodiesel industry, making liquid fuel from such oil-rich phytomass as soybeans, rapeseed, and oil palm fruit (USDOE 2011). The global production of liquid biofuels reached about 75 Mt of oil equivalent in 2015, accounting for about 1.8% of energy extracted annually from crude oil (BP 2016). Scaling this industry to supply a significant share of the world's liquid biofuels is, bluntly put, delusionary (Giampietro and Mayumi 2009; Smil 2010a).

Using the potential and kinetic energies of water to generate electricity is the world's second most important source of renewable energy, following traditional and modern biofuels. Hydroelectricity generation began in 1882, concurrently with thermal generation, when a small waterwheel on the Fox River in Appleton, Wisconsin, powered two dynamos to produce 25 kW for 280 weak lights (Dyer and Martin 1929). Before the century's end, increasingly taller dams were being built in alpine countries, in Scandinavia, and in the United States. But the first large AC station, built at Niagara in 1895, was small (37 MW) compared to the projects built during the 1930s with state support in the United States (the New Deal's Tennessee Valley Authority, U.S. Bureau of Reclamation) and in the USSR as a part of Stalinist industrialization during the 1930s (Allen 2003). The largest U.S. projects of the era were the Hoover Dam on the Colorado River (1936; 2.08 GW) and the Grand Coulee Dam on the Columbia River, whose first stage was completed in 1941 (eventually 6.8 GW).

Three post-1945 decades made hydropower the source of nearly 20% of the world's electricity, with large projects completed in Brazil, Canada, the USSR, Congo, Egypt, India, and China. In most countries the construction of new projects has slowed down or stopped since the 1980s, but not in China, where the world's largest dam—Sanxia, the Three Gorges Dam (installed capacity of 22.5 GW in 34 units)—was completed in 2012 (Chincold 2015). In 2015 water turbines supplied about 16% of the world's electricity, with shares as high as 60% in Canada and nearly 80% in Brazil, and even higher shares in a number of smaller African countries.

Two renewable energy conversions that have received a great deal of attention have been wind and solar electricity. The interest owes to their rapid expansion—between 2010 and 2015 global wind-driven generation had increased about 2.5 times, while solar generation had nearly octupled—and to exaggerated expectations of their future rate of adoption. Rapid expansion is a common attribute of early stages of development, and the contribution of these two sources of electricity remains negligible on the global scale (in 2015 wind generated about 3.5% and direct solar radiation produced 1% of the world's electricity). The integration of larger flows of these intermittent energies (many wind turbines work only 20%–25% of the time, some offshore farms manage 40%) into current grids presents many challenges (J.P. Morgan 2015).

Modern wind energy development was launched by the U.S. tax credits of the early 1980s and ended abruptly with their expiration in 1985 (Braun and Smith 1992). Europe became a new leader during the 1990s as several governments—Denmark, the UK, Spain, and above all Germany as a part of its *Energiewende*—adopted policies designed to accelerate the transition to renewable electricity. Costs have declined, and larger machines (now up to 8 MW, commonly 1–3 MW) and larger wind farms (including offshore installations) have boosted the recent growth from less than 2 GW of installed capacity in 1990 to 17.3 GW by the year 2000 and 432 GW by the end of 2015 (Global Wind Energy Council 2015).

The photovoltaic (PV) effect (electricity generation using metal electrodes exposed to light) was discovered by Edmund Becquerel (1852–1908) in 1839, but it was only in 1954 that Bell Laboratories produced expensive, low-efficiency (initially just 4.5%, later 6%) silicon solar cells, which were first used in 1958 to power (mere 0.1 W) the *Vanguard 1* satellite; four years later, in 1962, *Telstar 1*, the first commercial telecommunications satellite, had PV cells rated at 14 W, and in 1964 the Nimbus satellites carried cells capable of 470 W (Smil 2006). Space applications, where cost is not a primary consideration, have been thriving for decades, but land-based uses for electricity generation were limited by high costs, and the industry began to grow only during the late 1990s. In terms of peak power (which is available, even when sunny, for only a few hours a day), just 50 MW of PV cells were shipped in 1990, 17 GW in 2010, and about 50 GW in 2015, when the cumulative capacity reached 227 GW (James 2015; REN21 2016).

But PV generation has even lower capacity factors than wind (without tracking in cloudier climates just 11–15%, even in Arizona about 25%), and in 2015 the global electricity generation was only about 30% of the total delivered by wind turbines (fig. 5.16). Again, the industry's growth has not been a gradual, organic process but a promotion driven by government subsidies: nothing illustrates this better than the fact that in 2015, cloudy Germany produced nearly three times as much PV electricity as sunny Spain (BP 2016). Water heating, using small home rooftop heaters as well as large industrial arrays, predates the expansion of PV generation. By the end of 2012 the installed capacity of heaters was about 270 GW, mostly in China and Europe (Mauthner and Weiss 2014). Concentrated solar power

288   Chapter 5

(CSP), in which mirrors are used to concentrate solar radiation in order to heat water (or salt) for electricity generation, is a useful alternative to PV electricity, but only a few installations (total capacity of less than 5 GW) were operating by 2015.

Compared to the big four, biofuels, hydro, wind, and PV electricity, other renewable conversions remain negligible on the global scale, although some of them are important on a national or regional level, none more so than geothermal energy. Hot springs and wells have been used since prehistoric times, and deeper wells now supply hot water for space heating and industrial processes in many countries. But places where this energy can recovered as natural hot steam and used to generate electricity are much less common. The world's first geothermal plant began operating in Italy's Larderello field in 1902; New Zealand's Wairakei came on line in 1958, and California's Geysers in 1960. By 2014 the global installed capacity was 12 GW. The United States has the highest installed capacity, and Iceland is most dependent on this renewable energy (Geothermal Energy Association 2014).

None of the long-standing plans for large tidal power plants have been realized; only a few small installations work, in France and China. Relying on new plantations of fast-growing trees (willows, poplars, eucalypti,

Figure 5.16
Lucaneina de las Torres photovoltaic power plant in Spanish Andalusia (Corbis).

leucaenas, or pines) to be harvested for woodchips destined for electricity generation is a choice beset by many environmental problems, and crop residues and other organic wastes are now also used for large-scale biogas production (above in all in Germany and China), but its contribution makes a difference only a local scale. Despite many renewable options, some rapid advances, and many contradictory claims, the basic verdict is clear: as with all other energy transitions, moving away from fossil fuels will be a protracted process, and we will have to wait to see how different conversions will evolve to claim key roles in a new energy world.

Prime Movers in Transportation

In light of the importance of mobility of both people and goods in modern civilization, the final section of my survey of technical advances that determine the current energy foundations of modern societies will deal with prime movers in transportation, spanning their entire range from humble small engines to powerful rockets. The development of Otto cycle engines (now overwhelmingly fueled by gasoline, with ethanol and natural gas making some inroads) has been rather conservative since the first decade of the twentieth century when they entered mass production. The most important changes have included an approximate doubling of compression ratios and their lower weight and a rising power, resulting in a falling mass/power ratio: it declined from nearly 40 g/W in 1900 to just around 1g/W a century later. America's first mass-produced car, Ransom Olds's Curved Dash, had a single-cylinder, 5.2 kW (7 hp) engine. The engine of Ford's Model T, whose production ended in 1927 after 19 years and 16 million units, was three times as powerful.

An increase in the average power of American cars was interrupted by OPEC's oil price rises of the 1970s but resumed during the 1980s: the average car power rose from about 90 kW in 1990 to about 175 kW in 2015 (USEPA 2015). But "car" is actually a wrong term because in the United States, about 50% of all light-duty vehicles used for personal transportation are vans, pickup trucks, and SUVs (one of the greatest misnomers ever: where is the sport and what is the utility of driving these heavy minitrucks to a shopping center?). Diesel engines, too, got both relatively lighter and much more powerful, and these improvements made them dominant in several key transportation markets (Smil 2010b). The first diesel engine trucks appeared in Germany in 1924, the first heavy passenger cars to use that engine (also in Germany) in 1936. Just before World War II most new European trucks and buses had diesel engines, and after the war, this became the global norm. Diesel bus engines, rated as high as 350 kW, have mass/power ratios of 3–9 g/W and can operate up to 600,000 km without any major overhaul.

The mass/power ratios of automotive diesel engines eventually declined to as low as 2 g/W, which means that the engines in passenger cars are only slightly heavier than their gasoline-fueled counterparts (Smil 2010b). Lower fuel costs made diesel cars common in the EU, where they now account for more than 50% of new registrations (ICCT 2104). But the cars remain rare in the United States: in 2014 they accounted for fewer than 3% of all vehicles. And the brand's image has suffered heavily since the fall of 2015, when Volkswagen was forced to admit that many diesel models sold since 2008 contained illegal software that produced false readings in engines tested for emissions in order to pass U.S. environmental regulations for nitrogen oxides.

Diesel-powered locomotives (rated up to 3.5 MW) pull (and push) freight trains on all nonelectrified railways around the world. As already noted, diesel engines began their conquest of marine shipping even before World War I, and they have become indispensable prime movers of globalization because all waterborne trade in energy resources, raw materials, recyclable waste, food and feed, and manufactured products is powered by these massive, efficient machines (Smil 2010b). The most powerful marine diesels in supertankers and large bulk carriers, designed in Europe by MAN and Wärtsilä and built in South Korea and Japan, have ratings up to almost 100 MW.

Reciprocating aircraft engines improved very rapidly. Those powering Boeing's 1936 Clipper (a large hydroplane flying scheduled service between the U.S. West Coast and East Asia) were about 130 times more powerful than Wright's 1903 machine, whose weight/power ratio was more than ten times as high (fig. 5.17). Gas turbines—entirely new prime movers that have revolutionized flying as well as the performance of many industries—were conceptualized in some detail at the turn of the twentieth century, but their first practical designs emerged only during the late 1930s. Frank Whittle in England and Hans Pabst von Ohain in Germany independently built their experimental gas turbines for military planes, but the first jet-powered fighter planes entered service too late to affect the course of World War II (Constant 1981; Smil 2010b).

Rapid development of the new prime mover followed after 1945. The speed of sound was surpassed for the first time on October 14, 1947, with the Bell X-1 plane, and scores of supersonic fighter and bomber plane designs have been introduced since the late 1940s, with the fastest fighter plane, the MiG-35, reaching a maximum speed of 3.2 Mach. The

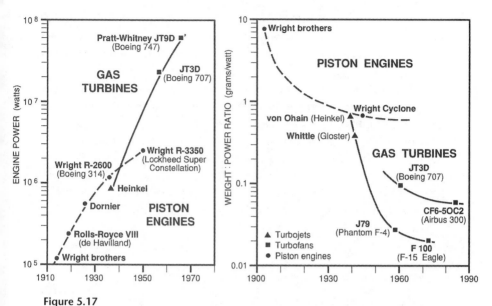

Figure 5.17

Increasingly more powerful yet lighter aeroengines have made the continuing progress in flying possible. Just before piston engines reached their limit of performance, jet engines began their spectacular advances. Those powering large Boeing and Airbus planes now weigh less than 0.1 g/W, a 100-fold improvement compared to the Wrights' pioneering piston design. Engines in military jets are lighter still. Plotted from data in Constant (1981, Gunston (1986), Taylor (1989), and Smil (2010b).

deployment of gas turbines made affordable intercontinental flight possible: their low mass/power ratio (with a thrust of 500 kN, it is just 0.06–0.07 g/W), high thrust/weight ratio (>6 for commercial engines, 8.5 for the best military engines), and high bypass ratio (at 12:1, currently the highest value, 92% of air compressed by an engine bypasses its combustion chamber; this lowers specific fuel consumption and reduces engine noise) have distinguished the evolving design of these increasingly more powerful and more efficient prime movers (fig. 5.17). And gas turbines in flight also became so reliable that two-engine aircraft now fly not only across the Atlantic but also on many transpacific routes (Smil 2010b).

As is often the case with mature industries, the global market for jet engines has become dominated by only four makers. Rolls Royce was the first maker to market the commercial engines in 1953, followed by two American companies, General Electric and Pratt & Whitney, and CFM International, a joint company established by GE and French Snecma Moteurs in 1974 that has specialized in making engines for short- and medium-range aircraft (CFM International 2015). On the other hand, flights of the supersonic Concorde (commercially first in 1976) have been too expensive to capture that market, and the transatlantic service ended in 2003 (Darling 2004).

In 1952 the British Comet became the first passenger jet, but structural defects rather than engine problems led to three fatal accidents and to its withdrawal from service. The redesigned aircraft flew again in 1958, but it was not a commercial success (Simons 2014). The first successful commercial jet plane was the Boeing 707, introduced in 1958 (fig. 5.18). The first wide-bodied Boeing 747 began flying in 1969: the iconic wide-body jet was powered by large turbofan engines that developed more than 200 kN of thrust and could deliver a peak combined thrust of about 280 MW during

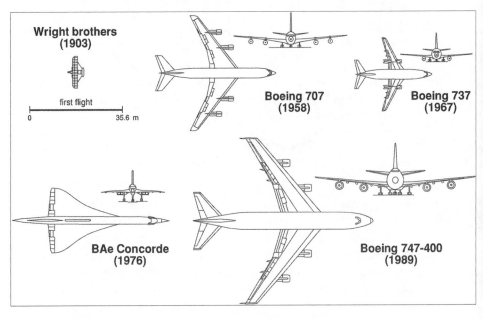

Figure 5.18
Plans and front views of notable jet planes. The Boeing 707 (1957) was based on an in-flight refueling tanker. The Boeing 737 (1967) is the all-time best-selling jet aircraft (nearly 9,000 planes had been delivered by the end of 2015 and 13,000 more had been ordered). The supersonic French-British Concorde, which flew limited routes between 1976 and 2003, was an expensive oddity. The Boeing 747 (in service since 1969) was the first wide-body long-haul aircraft. For comparison with these scaled drawings the Wright brothers' plane and its total flight path on December 7, 1903, are shown. Based on Boeing and Aerospatiale/BAe publications and on Jakab (1990).

take-off (Smil 2000c). By 2015 the most powerful jet engine, the GE 90–115B, was rated at 513 kN of thrust.

The only prime movers that can deliver more power per unit of weight than gas turbines are the rocket engines launching missiles and space vehicles. The founders of modern rocket science—Konstantin Tsiolkovsky (1857–1935) in Russia, Hermann Oberth (1894–1989) in Germany, and Robert H. Goddard (1882–1945) in the United States—correctly envisaged the eventual success of the ancient idea of rocket propulsion, which was translated by modern engineering into the world's most powerful prime movers (Hunley 1995; Angelo 2003; Taylor 2009). Rapid advances began during World War II: in 1942 ethanol-powered German V-2 missile designed by Wernher von Braun (1912–1977) reached the sea-level thrust of 249 kN (equivalent to about 6.2 MW, with a mass/power ratio of 0.15 g/W), a maximum speed of 1.7 km/s. Its range, 340 km, was long enough to attack the UK (von Braun and Ordway 1975).

The superpowers' space race started with the launch of Earth's first artificial satellite, the Soviet Sputnik, in 1957, and produced increasingly more powerful, and also more accurate, intercontinental ballistic missiles. On July 16, 1969, eleven kerosene- and hydrogen-burning engines of America's Saturn C-5 rocket (whose principal designer was also Wernher von Braun) started the Apollo spacecraft on its journey to the Moon. They were fired for just 150 seconds, and their combined thrust reached nearly 36 MN, an equivalent of about 2.6 GW and a mass/power ratio (including the weight of the fuel and three booster rockets) of just 0.001 g/W (Tate 2009).

6 Fossil-Fueled Civilization

The contrast is clear. Preindustrial societies tapped virtually instantaneous solar energy flows, converting only a negligible fraction of practically inexhaustible radiation income. Modern civilization depends on extracting prodigious energy stores, depleting finite fossil fuel deposits that cannot be replenished even on time scales orders of magnitude longer than the existence of our species. Reliance on nuclear fission and the harnessing of renewable energies (adding wind- and photovoltaic-generated electricity to more than 130-year-old hydrogeneration, and turning to new ways of converting phytomass to fuels) have been increasing, but by 2015 fossil fuels still accounted for 86% of the world's primary energy, just 4% less than a generation ago, in 1990 (BP 2016).

By turning to these rich stores we have created societies that transform unprecedented amounts of energy. This transformation brought enormous advances in agricultural productivity and crop yields; it has resulted first in rapid industrialization and urbanization, in the expansion and acceleration of transportation, and in an even more impressive growth of our information and communication capabilities; and all of these developments have combined to produce long periods of high rates of economic growth that have created a great deal of real affluence, raised the average quality of life for most of the world's population, and eventually produced new, high-energy service economies.

But the use of this unprecedented power has had many worrisome consequences and has resulted in changes whose continuation might imperil the very foundations of modern civilization. Urbanization has been a leading source of inventiveness, technical advances, gains in the standard of living, expanded information, and instantaneous communication, but it has also been a key factor behind deteriorating environmental quality and worrisome income inequality. The political implications of an uneven distribution of energy resources have both intra- and international consequences ranging from regional disparities to the perpetuation of corrupt, and often intolerant or outright violent, regimes.

Modern high-energy weapons have raised the destructive powers of nations by many orders of magnitude when compared to preindustrial capacities, and hence modern armed conflicts have seen commensurate increases not only in military but also in civilian casualties. Above all, the development of nuclear weapons has created, for the first time in history, the possibility of, if not destroying, then greatly crippling the entire civilization. At the same time, some of the most intractable means of modern aggression and warfare do not require superior command of concentrated energies as they rely on time-tested ways of individual terrorism. But even if modern civilization were guaranteed to avoid a large-scale thermonuclear conflict, it would still face profound uncertainties. Certainly the most worrisome challenge is the widespread environmental degradation. This rapid change arises from the extraction and conversion of both fossil fuels and nonfossil energies, as well as from industrial production, rapid urbanization, economic globalization, deforestation, and improper practices in crop cultivation and animal husbandry.

The cumulative effects of these changes have already gone well beyond local and regional problems to the destabilizing effects of global biospheric change, above all the many unwelcome consequences of relatively rapid global warming. Modern civilization has engineered a veritable explosion of energy use and has extended human control over inanimate energies to previously unthinkable levels. These gains have made it fabulously liberating and admirably constructive—but also uncomfortably constraining, horribly destructive, and, in many ways, self-defeating. All these changes have brought generations of strong economic growth and expectations that this process, fed by incessant innovation, need not end anytime soon—but its continuation is by no means certain.

Unprecedented Power and Its Uses

Even though interrupted by two world wars and the world's worst economic crisis during the 1930s, the growth of global energy proceeded at unprecedented rates during the first seven decades of the twentieth century. Afterward there came a slowdown precipitated by OPEC's quintupling of oil prices between October 1973 and March 1974; the growth would have moderated even without that jolt because the absolute levels had grown too large to support rates of growth that are possible at lower aggregate levels. But (at a slower pace) enormous quantitative changes have continued, and they have been accompanied by some new and remarkable qualitative gains. The best compilations of global statistics show the sustained exponential growth of fossil fuel production since the large-scale extraction of such fuels began during the nineteenth century (Smil 2000a, 2003, 2010a; BP 2015; fig. 6.1).

Coal mining grew 100-fold, from 10 Mt to 1 Gt, between 1810 and 1910; it reached 1.53 Gt in 1950, 4.7 Gt in the year 2000, and 8.25 Gt in 2015 before it declined a bit to about 7.9 Gt in 2015 (Smil 2010c; BP 2016). Crude oil extraction rose about 300-fold, from less than 10 Mt in the late 1880s to just over 3 Gt in 1988; it was 3.6 Gt in the year 2000 and almost 4.4 Gt in 2015 (BP 2016). Natural gas production rose 1,000-fold, from less than 2 Gm^3 in the late 1880s to 2 Tm^3 by 1991; it was 2.4 Tm^3 in 2000 and 3.5 Tm^3 in 2015. During the twentieth century the global extraction of fossil energies rose 14-fold in aggregate energy terms.

But a much better way to trace this expansion is to express the growth in terms of useful energy, the actually delivered heat, light, and motion. As we have already seen, early conversions of fossil fuels were rather inefficient (<2% for incandescent light, <5% for steam locomotives, <10% for thermal generation of electricity, <20% for small coal stoves), but improvements in coal-fired boilers and stoves soon doubled those efficiencies, and still left a great potential for further gains. Liquid hydrocarbons burned in household furnaces and in industrial and power plant boilers are converted with higher efficiencies, and only gasoline-fueled internal combustion engines in passenger cars are relatively inefficient. The combustion of natural gas, be it in furnaces, boilers, or turbines, is highly efficient, commonly in excess of 90%, as are the conversions of primary electricity.

Consequently, in 1900 the average weighted efficiency of global energy use was no higher than 20%; by 1950 it was more than 35%, and by the year 2015 the global mean of converting fossil fuels and primary electricity had reached 50% of total commercial inputs: the International Energy Agency (IEA 2015a) accounts for 2013 show worldwide primary supply of 18.8 Gt of oil equivalent and a final consumption of 9.3 Gt of oil equivalnet, with the highest losses, predictably, in thermal electricity generation and transportation. Even more remarkably, in a key consumption sector, household heating, entire populations have experienced a complete efficiency transition just in a matter of a few decades (box 6.1).

While the total supply of all fossil energies was up 14-fold during the twentieth century, the steady progress of efficiencies supplied more than 30 times as much useful energy as was available in 1900. As a result, affluent nations, where fossil fuel already dominated the overall supply by 1900,

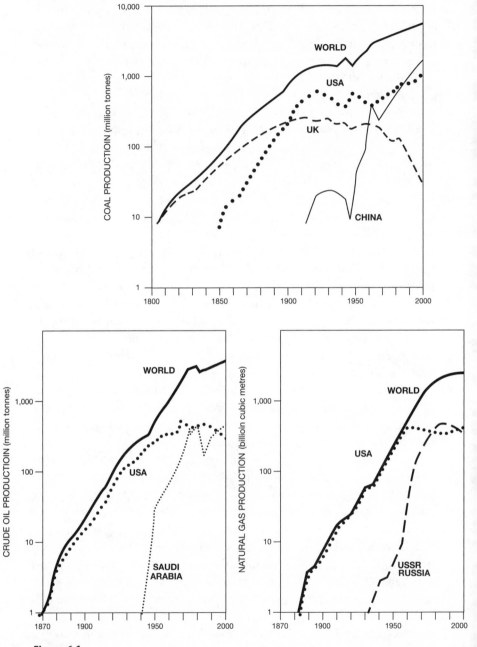

Figure 6.1
Production of the three principal fossil fuels: global totals and annual outputs for the largest producers. Plotted from data in United Nations Organization (1956), Smil (2010a), and BP (2015).

Box 6.1

Efficiency of household heating

In less than 50 years I have lived in homes heated by four different fuels and have seen the conversion efficiency of this key energy service tripled (Smil 2003). During the late 1950s, living in a village surrounded by forests close to the Czech-Bavarian border, we heated our house, as did most of our neighbors, with wood. My father ordered pre-cut logs of spruce or fir, and it was my summer duty to chop them into ready-to-stoke pieces of wood (and also into some finer pieces for kindling) and then to stack them in a sheltered place to air-dry. The efficiency of our woodstoves was not more than 30–35%. When I studied in Prague, all energy services—space heating, cooking, electricity generation—depended on lignite, and the coal stove I had in my room, in a thick-walled former monastery, had an efficiency of about 45%. After moving to the United States we rented the upper floor of a suburban house that was heated by fuel oil delivered by truck and burned in a furnace with no more than 60% efficiency. Our first Canadian house had a natural gas furnace rated at 65%, and when I designed a new super-efficient home, I installed a natural gas furnace rated at 94%—and have since replaced it with one rated at 97%.

now derive more than twice or even three times as much useful energy per unit of primary supply than they did a century ago, and because traditional biomass energies were converted with very low efficiencies (<1% for light, <10% for heat), those low-income nations where modern energies became dominant only during the latter half of the twentieth century now derive commonly five to ten times as much useful energy per unit of primary supply than they did a century ago. In per capita terms—with the global population at 1.65 billion in 1900 and 6.12 billion in the year 2000—the global increase in useful energy supply was more than eightfold, but this mean hides large national differences (more on this topic later in the chapter in the discussion of economic growth and the standard of living).

Another way to appreciate the aggregate size of modern energy flows is to compare them with traditional uses, in both absolute and relative terms. Best estimates show the worldwide total of biomass fuel consumption rising from around 700 Mt in 1700 to about 2.5 Gt in the year 2000. This would be from about 280 Mt to 1 Gt in terms of oil equivalent, less than quadrupling in three centuries (Smil 2010a). During the same time, the extraction of fossil fuels rose from less than 10 Mt to about 8.1 Gt of oil equivalent, about an 800-fold expansion (fig. 6.2). In gross energy terms, the global supply of biofuels and fossil fuels was about the same in 1900

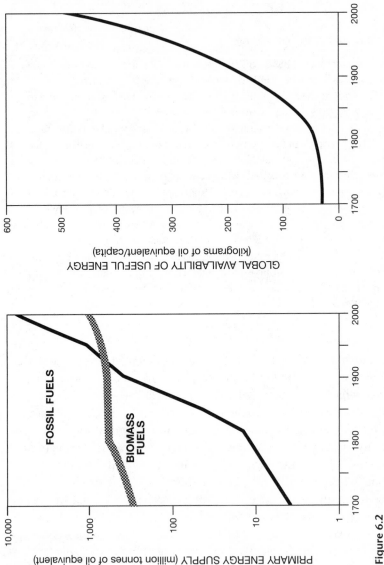

Figure 6.2

Global output of fossil fuels had surpassed the total supply of traditional biomass energies just before the end of the nineteenth century (left). Increase of useful energy was more than twice as high as the increase of the total primary supply (right). Plotted from data in United Nations Organization (1956) and Smil (1983, 2010a).

(both at roughly 22 EJ); by 1950 fossil fuels supplied nearly three times as much energy as wood, crop residues, and dung; and by the year 2000 the difference was nearly eightfold. But when adjusted for actually delivered, useful energy, the difference in the year 2000 was nearly 20-fold.

Surges in energy use raised average per capita consumption levels to unprecedented heights (fig. 6.3). The energy needs of foraging societies were dominated by the provision of food, and their annual consumption averages did not go above 5–7 GJ/capita. Ancient high cultures added slowly rising energy use for better shelters and clothing, for transportation (energized by food, feed, and wind), and for a variety of manufactures (with charcoal prominent). New Kingdom Egypt averaged no more than 10–12 GJ/capita, and my best estimate for the early Roman Empire is about 18 GJ/capita (Smil 2010c). Early industrial societies easily doubled

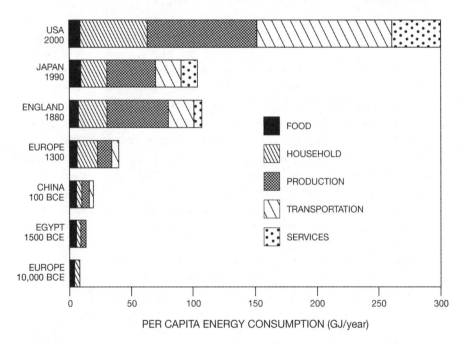

Figure 6.3
Comparisons of typical annual per capita energy consumption during different stages of human evolution. Large increases in absolute consumption have been accompanied by growing shares of energy used by households, industries, and transportation. Pre-nineteenth-century values are only approximations based on Smil (1994, 2010c) and Malanima (2013a); later figures are taken from specific national statistical sources.

the traditional per capita energy use. Most of that increase went into coal-fueled manufactures and transportation. Malanima (2013b) put the European averages at about 22 GJ/t in 1500, followed by a stagnation at 16.6–18.1 GJ/t until 1800.

Afterward came a pronounced differentiation among the industrializing nations and countries and those whose economies remained largely agrarian. Kander (2013) puts the mean for England and Wales rising from 60 GJ/capita in 1820 to 153 GJ/capita by 1910, while during the same period the German rate more than quintupled (from 18 to 86 GJ/capita) and the French rate tripled (18 to 54 GJ/capita), but the Italian rates rose by only 20% (from 10 to just 22 GJ/capita). For comparison, the average U.S. rate rose from less than 70 GJ to about 150 GJ between 1820 and 1910 (Schurr and Netschert 1960). A century later all richer European countries were above 150 GJ/capita, the United States was above 300 GJ/capita, and as the average consumption rates rose, their composition changed (fig. 6.3).

In foraging societies food was the only energizer; my estimate puts food and fodder at about 45% of all energy in the early Roman Empire (Smil 2010c). In preindustrial Europe food and fodder ranged from 20% to 60%, but by 1820 the mean was no more than about 30%; by 1900 it was less than 10% in the UK and Germany. By the 1960s fodder energy had declined to a negligible level and food had become no more than 3% and even less than 2% of all energy supply in the most affluent societies, whose consumption became dominated by industrial, transportation, and household uses of fuels and electricity (fig. 6.3). Per capita deliveries of electricity have risen by two orders of magnitude in high-income economies, by 2010 becoming around 7 MWh/year in Western Europe and about 13 MWh/year in the United States. Contrasts between energy flows controlled directly by individuals are no less impressive.

When in 1900 a Great Plains farmer held the reins of six large horses while plowing his wheat field, he controlled—with considerable physical exertion, perched on a steel seat, and often enveloped in dust—no more than 5 kW of animate power. A century later his great-grandson, sitting high above the ground in the air-conditioned comfort of his tractor cabin, controlled effortlessly more than 250 kW of diesel engine power. In 1900 an engineer operating a coal-fired locomotive pulling a transcontinental train at close to 100 km/h commanded about 1 MW of steam power, the maximum performance permitted by manual stoking of coal (Bruce 1952; fig. 6.4). By the year 2000, pilots of a Boeing 747 retracing the transcontinental route 11 km aloft could choose an auto-mode for a large part of the journey as four gas turbines developed up to about 120 MW and the plane flew at 900 km/h (Smil 2000a).

Figure 6.4
Stoking a late-nineteenth-century steam locomotive (top) and piloting a Boeing jetliner (bottom). The two pilots control two orders of magnitude more power than did the stoker and the engineer in their locomotive. Locomotive from VS archive; Boeing cockpit from http://wallpapersdesk.net/wp-content/uploads/2015/08/2931_boeing_747.jpg.

This concentration of power also demands much greater safety precautions because of the inevitable consequences of errors in control. Drivers sitting atop coaches used on inter-city journeys until the nineteenth century usually controlled steady power of no more than 3 kW (four harnessed horses) deployed to transport 4–8 people; pilots of intercity jetliners control 30 MW developed by jet engines to fly 150–200 passengers. Temporary inattention or an error of judgment will obviously have vastly different consequences when an operator is in control of 3 kW or 30 MW, a four-orders-of-magnitude difference. An obvious way to cut such risks is to deploy electronic controls.

The world's safest public transportation system ever—Japan's *shinkansen* between Tokyo and Osaka, which celebrated 50 years of accident-free operation on October 1, 2014 (Smil 2014b)—centralized electronic controls from its very beginning: an automatic train control keeps a proper distance between trains and instantly engages the brakes if the speed exceeds the indicated maximum; a centralized traffic control carries out route control; and the earthquake detection system senses the very first seismic waves reaching Earth's surface and can halt or slow trains before the main shock arrives (Noguchi and Fujii 2000). Modern jetliners have been highly automated for decades, and advance controls are now becoming common in passenger cars. Electronic controls and continuous monitoring—whose penetration now ranges from room thermostats to the operation of large blast furnaces, and from anti-lock braking in cars to ubiquitous CCTV in cities—have emerged, together with the mass adoption of computers and portable electronics, as a major new category of electricity demand.

The twentieth-century growth of global electricity output was even faster than the expansion of fossil fuel extraction, whose annual average was about 3% (fig. 6.5). Less than 2% of all fuel was converted to electricity in 1900; the share was still less than 10% in 1945, but by the century's end it had risen to about 25%. In addition, new hydroelectric stations (on a large scale after World War I) and new nuclear capacities (since 1956) further expanded electricity generation. As a result, the global electricity supply went up by about 11% a year between 1900 and 1935, and by more than 9% annually thereafter until the early 1970s. For the remainder of the century the growth of electricity generation declined to about 3.5% annually, largely because of lower demand in high-income economies and higher conversion efficiencies. New ways of electricity generation from renewable sources such as solar energy and wind have shown notable advances only since the late 1980s.

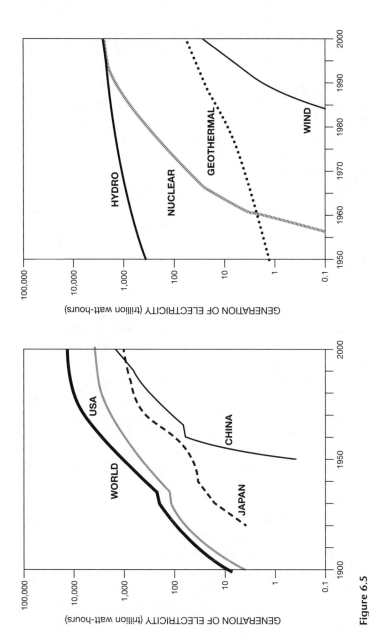

Figure 6.5
Global electricity generation has been growing considerably faster than the supply of fossil fuels. The largest economies have been always the leading producers, and thermal generation (now largely coal- and natural gas–based) continues to dominate the global output (left). Hydroelectricity and nuclear generation remain, respectively, in the second and third places, while wind and solar electricity have seen rapid post-2000 gains (right). Plotted from data in United Nations Organization (1956), Palgrave Macmillan (2013), and BP (2015).

No gain enabled by this new power has been more fundamental than the substantial rise in global food production, which has made it possible to provide adequate nutrition to nearly 90% of the world's population (FAO 2015b). No change has molded modern societies more than the process of industrialization, and no new developments have contributed more to the emergence of an interdependent global civilization than the evolution of mass transportation and the enormous expansion of our capacities for amassing information and engaging in communication with a frequency and an intensity that have no historical precedents. But these impressive gains have not been shared equally, and I will examine the extent to which the benefits of global economic growth have gone disproportionately to a minority of the world's population, and will also note considerable intranational inequities. Even so, there have been many universal improvements.

Energy in Agriculture

Fossil fuels and electricity have become indispensable inputs in modern farming. They are used directly to power machines and indirectly to build them, to extract mineral fertilizers, to synthesize nitrogenous compounds and a still expanding variety of protective agrochemicals (pesticides, fungicides, herbicides), to develop new crop varieties, and most recently to energize the electronics used in many functions that now support precision farming. Fossil fuels and electricity have brought higher and more reliable yields. They have displaced virtually all draft animals in all rich countries and greatly reduced their importance in the poor ones, and the replacement of muscles by internal combustion engines and electric motors sustained the reduction of labor started by preindustrial farming advances.

Indirect fossil fuel subsidies in agriculture began already (on a very small scale) in the eighteenth century when the smelting of ion ores was converted from charcoal to coke, expanded with widespread adoption of steel machinery during the latter half of the nineteenth century, and reached new highs with the introduction of larger and more powerful field machines, irrigation pumps, and crop-processing and animal husbandry equipment during the twentieth century. But machinery's embedded energy cost is a fraction of the energy used directly to run tractors, combines, and other harvesters, to pump water, to dry grain, and to process crops. Because of their inherently high efficiency, diesel engines have come to dominate most of these uses, but gasoline and electricity are also major energy inputs.

The use of internal combustion engines in field machinery started first in the United States, in the same decade that passenger cars finally became a mass-produced commodity (Dieffenbach and Gray 1960). The first American tractor factory was set up in 1905; power takeoff for attached implements was introduced in 1919; and power lifts, diesel engines, and rubber tires were introduced in the early 1930s. Until the 1950s mechanization proceeded much more slowly in Europe. In the populous countries of Asia and Latin America it really started only during the 1960s, and the shift is still under way in many poor countries. The mechanization of field work has been the main reason behind the rising labor productivity rise and the reduction of agricultural populations: a strong early twentieth-century Western horse worked at a rate equal to the labor of at least six men, but even early tractors had power equivalent to 15–20 heavy horses, and today's most powerful machines working on Canadian prairies rate up to 575 horsepower (Versatile 2015).

In chapter 3 I showed how rising productivity reduced average labor inputs to American wheat farming from about 30 h/t of grain in 1800 to less than 7 h/t in 1900; by the year 2000 the rate was down to about 90 minutes. Inevitably, this released labor found its way to cities, resulting in a worldwide decline of rural populations and a still continuing rise of urbanization (reviewed later in this chapter). American statistics illustrate the resulting displacements. The country's rural labor fell from more than 60% of the total workforce in 1850 to less than 40% in 1900; the share was 15% in 1950, and in 2015 it was just 1.5% (USDOL 2015). For comparison, agricultural labor in the EU is now about 5% of the total, but in China it is still around 30%.

American draft horses reached their highest number in 1915, at 21.4 million animals, but mule numbers peaked only in 1925 and 1926, at 5.9 million (USBC 1975). During the second decade of the twentieth century total draft power was about ten times as large as that of all the newly introduced tractors; by 1927 the two kinds of prime movers had equal power capacity, and the peak animal total was halved by 1940. But mechanization alone could not have released so much rural labor. Higher crop yields, brought about by new crop varieties responding to higher fertilization, applications of herbicides and pesticides, and more widespread irrigation, were also necessary.

The importance of a well-balanced supply of plant nutrients was formulated by Justus von Liebig (1803–1873) in 1843 and became widely known as Liebig's law of the minimum: the nutrient in the shortest supply will determine the yield. Of the three macronutrients (elements required in relatively large quantities)—nitrogen, phosphorus, and potassium—the latter two were not difficult to secure. In 1842 John Bennett Lawes (1814–1900) introduced the treatment of phosphate rocks by diluted sulfuric acid to produce ordinary superphosphate, and this led to discoveries of large phosphate deposits in Florida (1888) and Morocco (1913), while potash (KCl) could be mined at many sites in Europe and North America (Smil 2001).

Supplying nitrogen, the macronutrient always needed in the largest quantities per unit of cropped land, was the greatest challenge. Until the 1890s the only inorganic option was to import Chilean nitrates (discovered in 1809). Then relatively small amounts of ammonium sulfate began to be recovered from new by-product coking ovens; the expensive cyanamide process (coke reacting with lime produced calcium carbide, whose combination with pure nitrogen produced calcium cyanamide) became commercial in Germany in 1898; and at the very beginning of the twentieth century an electric arc (the Birkeland-Eyde process, 1903) was used to produce nitrogen oxide, to be converted to nitric acid and nitrates. None of these methods could supply fixed nitrogen on a mass scale, however, and the outlook for feeding the world changed fundamentally only in 1909 when Fritz Haber (1868–1934) invented a catalytic, high-pressure process to synthesize ammonia from its elements (Smil 2001; Stoltzenberg 2004).

Its rapid commercialization (by 1913) took place in the BASF plant in Ludwigshafen under the leadership of Carl Bosch (1874–1940). But the first practical use of the process was not to produce fertilizers but ammonium nitrate to make explosives during World War I. The first synthetic nitrogen fertilizers were sold in the early 1920s. The pre–World War II output remained limited, and even by 1960 more than a third of American farmers did not use any synthetic fertilizers (Schlebecker 1975). The synthesis of ammonia and subsequent conversions to liquid and solid fertilizers are energy-intensive processes, but technical advances lowered the overall energy cost and enabled the worldwide applications of nitrogenous compounds to reach an equivalent of about 100 Mt N by the year 2000, accounting for about 80% of the compound's total synthesis (box 6.2, fig. 6.6).

No other energy use offers such a payback as higher crops yields resulting from the use of synthetic nitrogen: by spending roughly 1% of global energy, it is now possible to supply about half of the nutrient used annually by the world's crops. Because about three quarters of all nitrogen in food proteins come from arable land, almost 40% of the current global food supply depends on the Haber-Bosch ammonia synthesis process.

Box 6.2
Energy costs of nitrogenous fertilizers

The energy requirements of the Haber-Bosch synthesis include fuels and electricity used in the process and energy embodied in the feedstocks. The coke-based Haber-Bosch synthesis process in BASF's first commercial plant required more than 100 GJ/t NH_3 in 1913; before World War II the rate was down to around 85 GJ/t NH_3. After 1950 natural gas–based processes lowered the overall energy cost to 50–55 GJ/t NH_3; centrifugal compressors and high-pressure reforming of steam and better catalysts lowered the requirements first to less than 40 GJ/t by the 1970s, and then to around 30 GJ/t by the year 2000, when the best plants needed only about 27 GJ/t NH_3, close to the stoichiometric energy requirement (20.8 GJ/t) for ammonia synthesis (Kongshaug 1998; Smil 2001). Typical new natural gas–based plants use 30 GJ/t NH_3, about 20% more when using heavy fuel oil, and up to about 48 GJ/t NH_3 for coal-based synthesis (Rafiqul et al. 2005; Noelker and Ruether 2011).

The average performance was about 35 GJ/t in 2015; the last rate corresponds to about 43 GJ/t N. But most farmers do not apply ammonia (a gas under normal pressure) and prefer liquids or solids, especially urea, which has the highest share of nitrogen (45%) among solid compounds that are easily applied even to small fields. Converting ammonia to urea, packaging, and transportation bring the overall energy cost to 55 GJ/t N. Using this rate as the global average means that in 2015, with about 115 Mt N used in agriculture, the synthesis of nitrogenous fertilizers claimed about 6.3 EJ of energy, or just over 1% of the global energy supply (Smil 2014b).

Stated in reverse, without Haber-Bosch synthesis the global population enjoying today's diets would have to be almost 40% smaller. Western nations, using most of their grain as feed, could easily reduce their dependence on synthetic nitrogen by lowering their high meat consumption. Populous low-income countries have more restricted options. Most notably, synthetic nitrogen provides about 70% of all nitrogen inputs in China. With over 70% of the country's protein supplied by crops, roughly half of all nitrogen in China's food comes from synthetic fertilizers. In its absence, average diets would sink to a semistarvation level—or the currently prevalent per capita food supply could be extended to only half of today's population.

The mining of potash (10 GJ/t K) and phosphates and the formulation of phosphatic fertilizers (altogether 20 GJ/t P) would add another 10% to that total. The total energy cost of other agricultural chemicals is much lower.

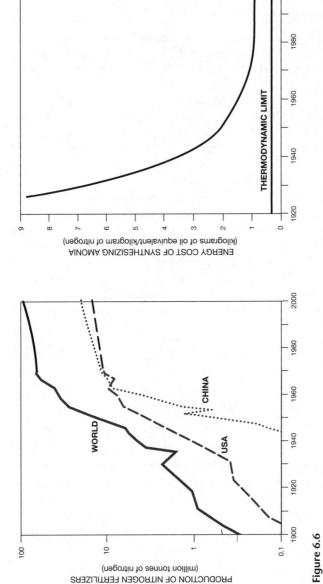

Figure 6.6
An exponential increase in the global production of nitrogenous fertilizers (left) has been accompanied by an impressive decline in the energy costs of ammonia synthesis (right). Plotted from data in Smil (2001, 2015b) and FAO (2015a).

The post–World War II growth of fertilizer applications was accompanied by the introduction and expanding use of herbicides and pesticides, chemicals that reduce weed, insect, and fungal infestations of crops. The first commercial herbicide, marketed in 1945, was 2,4-D, which kills many broadleaved plants without serious injury to crops. The first insecticide was DDT, released in 1944 (Friedman 1992). The global inventory of herbicides and pesticides now contains thousands of compounds, mostly derived from petrochemical feedstocks: their specific syntheses are much more energy-intensive than the production of ammonia (commonly >100 GJ/t, and some well above 200 GJ/t), but the quantities used per hectare are orders of magnitude lower.

The global extent of irrigated farmland roughly quintupled during the twentieth century, from less than 50 Mha to more than 250 Mha, reaching about 275 Mha by 2015 (FAO 2015a). In relative terms this means that about 18% of the world's harvested cropland is now irrigated, about half of it with water pumped mostly from wells, with about 70% of the irrigated land being in Asia. Where irrigation draws water from aquifers, the energy cost of pumping (using mostly diesel engines or electric pumps) invariably accounts for the largest share of the overall (direct and indirect) energy cost of crop cultivation. Irrigation still supplies most of the withdrawn water to furrows, but much more efficient, and more expensive, sprinklers (above all the circular center pivots) are used in many countries (Phocaides 2007).

Only approximate calculations can be made to trace the rise of the direct and indirect use of fossil fuels and electricity in modern farming. During the twentieth century, as the world population grew 3.7 times and the harvested area expanded by about 40%, anthropogenic energy subsidies soared from only about 0.1 EJ to almost 13 EJ. As a result, in 2000 an average hectare of cropland received roughly 90 times more energy subsidies than it did in 1900 (fig. 6.7). Or, skipping the numbers, we could simply say, with Howard Odum (1971, 115–116):

> A whole generation of citizens thought that the carrying capacity of the earth was proportional to the amount of land under cultivation and that higher efficiencies in using the energy of the sun had arrived. This is a sad hoax, for industrial man no longer eats potatoes made from solar energy, now he eats potatoes partly made of oil.

But this transformation has changed the global availability of food in several profound ways. In 1900 the gross global crop output (before storage and distribution losses) provided only a small margin above the average

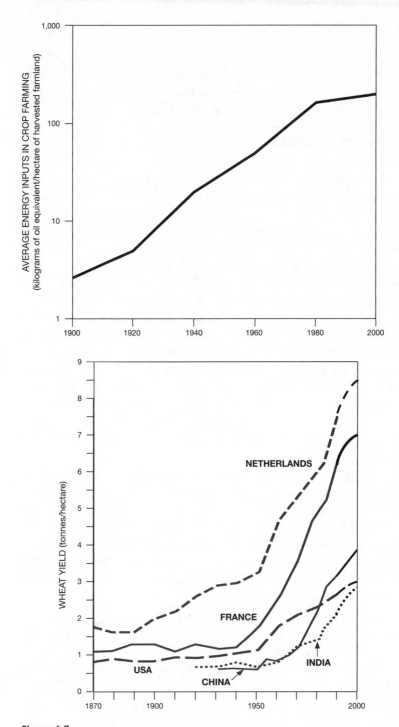

Figure 6.7
Total (direct and indirect) energy subsidies in modern farming (top), total harvests and rising wheat yields (bottom). Plotted from data in Smil (2008b), Palgrave Macmillan (2013), and FAO (2015a).

human food needs, which meant that a large share of humanity had only barely adequate or inadequate nutrition, and that the share of the harvest that could be used to feed animals was minimal. Greatly increased energy subsidies allowed the new staple cultivars (hybrid corn, introduced during the 1930s, and short-stalked wheat and rice cultivars, first adopted during the 1960s) to reach their full potential, resulting in rising yields of all staples and in an overall sixfold increase in harvested food energy (Smil 2000b, 2008).

At the beginning of the twenty-first century, the global harvest provided a daily supply averaging (for a population nearly four times as large as in 1900) about 2,800 kcal/capita, more than adequate if it were equitably accessible (Smil 2008a). The roughly 12% of the world's population that is still undernourished does not have enough to eat because of limited access to food, not because of its unavailability, and the food supply in affluent countries is now about 75% higher than the actual need, resulting in enormous food waste (30–40% of all food at the retail level) and high rates of overweight and obesity (Smil 2013a). Moreover, plenty of grain (50–60% in affluent countries) is fed to domestic animals. Chickens are the most efficient converters of feed (about three units of concentrate feed for a unit of meat); the feed:meat ratio for pork is about 9, and the production of grain-fed beef is most demanding, requiring up to 25 units of feed for a unit of meat.

This inferior ratio is also the function of the meat:live weight ratio: for chicken it is as high as 0.65 and for pork it is 0.53, but for large beef animals it is as low as 0.38 (Smil 2013d). But the energy loss in conversion to meat (and milk) has its nutritional rewards: the rising consumption of animal foods has brought high-protein diets to all rich nations (evident in taller statures) and has assured, on average, adequate nutrition even in most of the world's largest poor populous countries. Most notably, the energy content of China's average per capita diet is now, at about 3,000 kcal/day, about 10% ahead of the Japanese mean (FAO 2015a).

Industrialization

Critical ingredients of the industrialization process include a large number of interconnected changes (Blumer 1990), and this has been true at every scale of the unfolding process. By far the most important change on the factory floor was the introduction of electric motors powering individual machines, allowing precise and independent control by displacing generations of central drives transmitting the power of steam engines via leather belt and line shafts, but even this fundamental transformation would have had a limited impact if high-speed tools and better-quality steels had not been available to produce superior machines and finished components. As already noted, the intensifying international trade could not have happened without new, powerful prime movers, but their development in turn depended not only on advances in the technical design of machines but also on huge amounts of new liquid fuels being supplied, fuels produced by crude oil extraction and complex refining.

Similarly, the rising share of machine production concentrated in factories governed by hierarchical control required the location of workers near these establishments (hence new forms of urbanization) and the development of new skills and occupations (hence an unprecedented expansion of apprentice training and technical education). The utilization of a money economy and the mobility of labor and capital established new contractual relations that fostered the growth of migration and banking. Quests for mass output and low unit cost created new large markets whose functioning was predicated on reliable and inexpensive transportation and distribution.

And, contrary to common belief, the rising availability of coal-derived heat and mechanical power produced by steam engines was not necessary to initiate these complex industrialization changes. Cottage and workshop manufacturing, based on cheap countryside labor and serving national and even international markets, had been going on for generations before the beginning of coal-energized industrialization (Mendels 1972; Clarkson 1985; Hudson 1990). This proto-industrialization had a considerable presence not only in parts of Europe (Ulster, the Cotswolds, Picardy, Westphalia, Saxony, Silesia, and many more). Voluminous artisanal production for domestic and export markets had also been present in Ming and Qing China, in Tokugawa, Japan, and in parts of India.

A notable example is the carburization of wrought iron to produce Indian *wootz* steel, whose best-known transformation took the form of Damascene swords (Mushet 1804; Egerton 1896; Feuerbach 2006). Its production in some Indian regions (Lahore, Amritsar, Agra, Jaipur, Mysore, Malabar, Golconda) was on an almost industrial scale for exports to Persia and the Turkish Empire. The partially mechanized and relatively large-scale manufacturing of textiles based on water power was frequently the next step in the European transition from cottage production to centralized manufacturing. In a number of locations, industrial waterwheels and turbines competed successfully with steam engines for decades after the introduction of the new inanimate prime mover.

Nor was mass consumption a real novelty. We tend to think of materialism as a consequence of industrialization, but in parts of Western Europe, especially in the Netherlands and France, it was a major social force already during the fifteenth and sixteenth centuries (Mukerji 1981; Roche 2000). Similarly, in Tokugawa Japan (1603–1868), the richer inhabitants of cities, and especially of Edo, the country's capital, began to enjoy diversions ranging from buying illustrated books *(ehon)* to eating out (this is when sushi became popular), attending theater performances, and collecting colorful prints *(ukiyoe)* of landscapes and actors (Sheldon 1958; Nishiyama and Groemer 1997). The tastes and aspirations of increasing numbers of wealthier people provided an important cultural impetus to industrialization. They sought access to goods ranging from assortments of mundane cooking pots to exotic spices and fine textiles, from fascinating engraved maps to delicate tea services.

The term "industrial revolution" is as appealing and deeply entrenched as it is misleading. The process of industrialization was a matter of gradual, often uneven, advances. This was the case even in the regions that moved rather rapidly from domestic manufactures to concentrated large-scale production for distant markets. A spuriously accurate timing of these changes (Rostow 1965) ignores the complexity and the truly evolutionary nature of the whole process. Its English beginnings should go back at least to the late sixteenth century, but full development in Britain came only before 1850 (Clapham 1926; Ashton 1948). Even at that time traditional craftsmen greatly outnumbered machine-operating factory workers: the 1851 census showed the UK still had more shoemakers than coal miners, more blacksmiths than ironworkers (Cameron 1985).

To view the worldwide industrialization process largely as waves imitative of English developments (Landes 1969) is no less misleading. Even Belgium, whose advances resembled most closely the British progress, followed a distinct path. There was a much greater stress on metallurgy and a much lower importance of textiles. Critical national peculiarities resulted in far from uniform industrialization patterns. They included a French emphasis on water power, America's and Russia's longlasting reliance on wood, and Japan's tradition of meticulous craftsmanship. Coal and steam were initially not revolutionary inputs. Gradually they came to provide heat and mechanical power at an unprecedented level and with great reliability. Industrialization could then be broadened and accelerated at the same time, eventually becoming synonymous with an ever higher consumption of fossil energies.

Coal mining was not necessary for industrial expansion—but it was certainly critical for speeding it up. A comparison of Belgium and the Netherlands illustrates this effect. The highly urbanized Dutch society, equipped with excellent shipping and with relatively advanced commercial and financial capabilities, fell behind coal-rich, although otherwise poorer, Belgium, which became the most industrialized continental country in mid-nineteenth century Europe (Mokyr 1976). Other European regions whose coal-based economies took off early included the Rhine-Ruhr region, Bohemia and Moravia in the Habsburg Empire, and both Prussian and Austrian Silesia.

The pattern was repeated outside Western and Central Europe. In the United States, Pennsylvania, with its high-quality anthracites, and Ohio, with its excellent bituminous coal, emerged as the early leaders (Eavenson 1942). In pre–World War I Russia it was the discovery of the rich Donets coal deposits in Ukraine and the development of the Baku oil fields during the 1870s that ushered in the subsequent rapid industrial expansion (Falkus 1972). Japan's quest for modernity during the Meiji era was energized by coal from northern Kyushu. The country's first integrated modern iron and steel plant began its production only 48 years after the country's opening to the world, in 1901, with the blowing-in of the Higashida No. 1 blast furnace at Yawata Steel Works (the predecessor of Nippon Steel) in northern Kyushu (Yonekura 1994). India's largest commercial empire grew from J. Tata's blast furnace using Bihari coke in Jamshedpur beginning in 1911 (Tata Steel 2011).

Once energized by coal and steam power, traditional manufacturers could turn out larger volumes of good-quality products at lower prices. This achievement was a necessary precondition for mass consumption. The availability of an inexpensive and reliable supply of mechanical energy also allowed increasingly sophisticated machining. In turn, this led to more complex designs and greater specialization in the manufacture of parts, tools, and machines. New industries energized by coal, coke, and steam were set up to supply national and international markets with unprecedented speed. The making of high-pressure boilers and pipes started after 1810. The production of rails and railway locomotives and wagons rose rapidly after 1830, as did the making of water turbines and screw propellers after 1840. Iron hulls and submarine telegraph cables found large new markets after 1850, and commercial ways of making inexpensive steel—first in Bessemer converters after 1856, then in open-hearth (Siemens-Martin) furnaces during the 1860s (Bessemer 1905; Smil 2016)— found new large markets for finished products ranging from cutlery to rails and from plows to construction beams.

Rising fuel inputs and the replacement of tools by machines reduced human muscles to a marginal source of energy. Labor turned increasingly to supporting, controlling, and managing the productive process. The trend is well illustrated by an analysis of one and a half centuries of the England and Wales census and the Labour Force Survey (Stewart, De, and Cole 2015). In 1871 about 24% of all workers were in "muscle power" jobs (in agriculture, construction, and industry) and only about 1% were in "caring" professions (in health and teaching, child and home care, and welfare), but by 2011 caring jobs claimed 12% and muscle jobs only 8% of the labor force, and many of today's muscle jobs, such as cleaning and domestic service and routine factory line jobs, involve mostly mechanized tasks.

But even as the importance of human labor was declining, new systematic studies of individual tasks and complete factory processes demonstrated that labor productivity could be greatly increased by optimizing, rearranging, and standardizing muscular activities. Frederick Winslow Taylor (1856–1915) was the pioneer of such studies. Starting in 1880 he spent 26 years quantifying all key variables involved in steel cutting, reduced his findings to a simple set of slide-rule calculations, and drew general conclusions for efficiency management in *The Principles of Scientific Management* (Taylor 1911); a century later its lessons continue to guide some of the world's most successful makers of consumer products (box 6.3).

A radically new period of industrialization came when steam engines were eclipsed by electrification. Electricity is a superior form of energy, and not only in comparison with steam power. Only electricity combines instant and effortless access with the ability to serve very reliably every consuming sector except flying. The flip of a switch converts it into light, heat, motion, or chemical potential. Its easily adjustable flow allows a previously unsustainable precision, speed, and process control. Moreover, it is clean and silent at the point of consumption. And once proper wiring is in place, electricity can accommodate an almost infinite number of growing or changing uses—yet it requires no inventory.

These attributes made electrification of industries a truly revolutionary switch. After all, steam engines replacing waterwheels did not change the way of transmitting mechanical energy powering various industrial tasks. Consequently, this substitution did little to affect general factory layout. Space under factory ceilings remained crowded with mainline shafts linked to parallel countershafts transferring the motion by belts to individual machines (fig. 6.8). A prime mover's outage (whether caused by low water or by an engine failure) or a transmission failure (be it a line shaft crack or a slipped belt) disabled the whole setup. Such arrangements also generated large frictional losses and allowed only limited control of power at individual workplaces.

Box 6.3

From experiments with steel cutting to Japan's car exports

Frederick Winslow Taylor's main concern was with wasted labor, that is, with the inefficient use of energy—those "awkward, inefficient, or ill-directed movements of men" that "leave nothing visible or tangible behind them"— and argued for optimized physical exertion. Taylor's critics saw this as nothing but a stressful way of exploitation (Copley 1923; Kanigel 1997), but Taylor's effort was based on understanding the actual energetics of labor. He opposed excessive quotas (if the "man is overtired by his work, then the task has been wrongly set and this is as far as possible from the object of scientific management") and stressed that the combined knowledge of managers falls "far short of the combined knowledge and dexterity of workmen under them," and hence called for "the intimate cooperation of the management with the workmen" (Taylor 1911, 115).

Taylor's recommendations were initially rejected (Bethlehem Steel fired him in 1901), but his *Principles of Scientific Management* eventually became a key guide for global manufacturing. In particular, the global success of Japanese companies has been founded on a continuous effort to eliminate unproductive labor and excessive workloads, to eliminate an uneven pace of work, to encourage workers to participate in the production process by making suggestions for its improvement, and to minimize labor-management confrontation. Toyota's famous production system—an alliterative trio of *muda mura muri* (reducing non-value-adding activities, an uneven pace of production, and an excessive workload)—is nothing but pure Taylorism (Ohno 1988; Smil 2006).

The first electric motors powered shorter shafts for smaller groups of machines. After 1900 unit drives rapidly became the norm. Between 1899 and 1929 the total installed mechanical power in American manufacturing roughly quadrupled, while the capacities of industrial electrical motors grew nearly 60-fold and reached over 82% of the total available power, compared to less than 5% at the end of the nineteenth century (USBC 1954; Schurr et al. 1990). Afterward the share of electric power changed little: the substitution of steam- and direct water-powered drive by motors was practically complete just three decades after it began during the late 1890s. This efficient and reliable unit power supply did much more than remove the overhead clutter, with its inevitable noise and risk of accidents. The demise of the shaft drive freed up ceilings for the installation of superior illumination and ventilation and made possible a flexible plant design and easy

Figure 6.8
Interior of the main lathe workshop of the Stott Park Bobbin Mill in Finthswaite, Lakeside, Cumbria, showing the typical arrangement of overhead belts transmitting power from a large steam engine to individual machines. The mill produced wooden bobbins used by Lancashire's spinning and weaving industries (Corbis).

capacity expansion. The high efficiency of electric motors, combined with precise, flexible, and individual power control in a better working environment, brought much higher labor productivities.

Electrification also launched vast specialized industries. First came the manufacturing of light bulbs, dynamos, and transmission wires (after 1880) and steam and water turbines (after 1890). High-pressure boilers burning pulverized fuel were introduced after 1920; the construction of giant dams using large quantities of reinforced concrete started a decade later. The widespread installation of air pollution controls came after 1950, and the first nuclear power plants were commissioned before 1960. The rising demand for electricity also stimulated geophysical exploration, fuel extraction, and transportation. A great deal of fundamental research in material properties, control engineering, and automation was also necessary to produce better steels, other metals, and their alloys and to increase the reliability and extend the lifetime of expensive installations for extracting, transporting, and converting energies.

The availability of reliable and cheap electricity has transformed virtually every industrial activity. By far the most important effect on manufacturing was the widespread adoption of assembly lines (Nye 2013). Their classic, and now outdated, rigid Fordian variety was based on a moving conveyor introduced in 1913. The modern, flexible Japanese kind relies on just-in-time delivery of parts and on workers capable of doing a number of different tasks. The system, introduced in Toyota factories, combined elements of American practices with indigenous approaches and original ideas (Fujimoto 1999). The Toyota production system (*kaizen*) rested on continuous product improvement and dedication to the best achievable continuous quality control. Again, the fundamental commonality of all of these actions is minimizing energy waste.

The availability of inexpensive electricity has also created new metal-producing and electrochemical industries. Electricity allowed the large-scale smelting of aluminum by the electrolytic reduction of alumina (Al_2O_3) dissolved in an electrolyte, mainly cryolite (Na_3AlF_6). Starting in the 1930s electricity has been indispensable for the synthesis and shaping of an increasing variety of plastics and, most recently, for the introduction of a new class of composite materials, above all carbon fibers. The energy cost of these materials is about three times as high as that of aluminum, and their largest commercial use has been in replacing aluminum alloys in commercial aircraft construction: the latest Boeing 787 is about 80% composite by volume.

While new lightweight materials have been widely substituted for steel, steelmaking itself is increasingly done using electric arc furnaces, and new lighter but stronger steels have found many uses, particularly in the auto industry (Smil 2016). And before terminating this list, which could run for pages, I must stress that without electricity there could be no large-scale micromachining producing parts with exacting tolerances for such now ubiquitous applications as jet engines or medical diagnostic devices, and, of course, there would be neither accurate electronic controls nor the omnipresent computers and billions of telecommunication devices now in global use.

Although manufacturing's shares (as a percentage of the labor force or GDP) have been steadily declining in virtually all rich countries—in early 2015, those shares were just over 10% of workers and about 12% of the U.S. GDP (USDOL 2015)—industrialization continues, but its configuration has changed. Mass flows of energy and materials will remain at its foundation;

metals remain quintessential industrial materials; and iron, now used mostly in many kinds of steel, retains its dominance among metals. In 2014 steel production was nearly 20 times larger than the combined total output of the four leading nonferrous metals, aluminum, copper, zinc, and lead (USGS 2015). The smelting of iron ore in blast furnaces, followed by steelmaking in basic oxygen furnaces, and the use of recycled steel in electric arc furnaces dominate steel production. The massive growth of steel production would have been impossible without the much larger and more efficient blast furnaces (box 6.4, fig. 6.9).

Similarly, steelmaking techniques have become more efficient, not only because of reduced energy use but also because of the rising product yield (Takamatsu et al. 2014). The early Bessemer converters turned first

---

Box 6.4
Growth and mass and energy balances of blast furnaces

Few production structures with a medieval pedigree remain as important for the functioning of modern civilization do blast furnaces. As noted in chapter 5, Bell's 1840 redesign quintupled their internal volume, bringing it to 250 m^3. By 1880 the largest furnace surpassed 500 m^3; it reached 1,500 m^3 by 1950, and by 2015 the record inner volumes were between 5,500 and 6,000 m^3 (Smil 2016). The resulting increases in productivity brought the output of hot metal from 50 t/day in 1840 to more than 400 t/day by 1900. The 1,000 t/day mark was approached before World War II, and today's largest furnaces produce around 15,000 t/day, with the record rate at POSCO's Pohang 4 furnace, in South Korea, about 17,000 t/day.

The mass and energy flows needed to operate large blast furnaces and associated oxygen furnaces are prodigious (Geerdes, Toxopeus, and Van der Vliet 2009; Smil 2016). A blast furnace producing daily 10,000 t/day of iron and supplying an adjacent basic oxygen furnace will need 5.11 Mt of ore, 2.92 Mt of coal, 1.09 Mt of flux materials, and nearly 0.5 Mt of steel scrap. A large integrated steel mill thus receives every year nearly 10 Mt of materials. Modern furnaces now produce hot metal continuously for 15–20 years before their refractory brick interior and their carbon hearth are relined. These productivity gains have been accompanied by declines in specific coke consumption. In 1900 typical coke requirements were 1–1.5 t/t of hot metal, while by 2010 nationwide rates were about 370 kg/t in Japan and less than 340 kg/t in Germany (Lüngen 2013). The energy cost of coke-fueled iron smelting thus fell from about 275 GJ/t in 1750 to about 55 GJ/t in 1900, close to 30 GJ/t in 1950, and between 12 and 15 GJ/t by 2010.

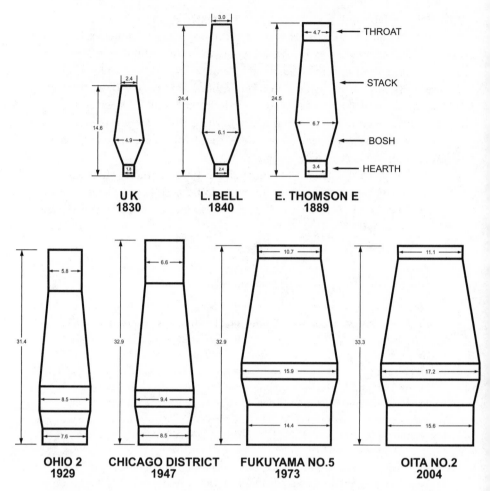

Figure 6.9
Changing designs of blast furnaces, 1830–2004. The principal trends have included taller and wider stacks, larger hearths, and lower and steeper boshes. The largest furnaces now produce more than 15,000 t of hot metal a day. Reproduced from Smil (2016).

less than 60% and later just above 70% of iron into steel. Open-hearth furnace eventually converted about 80%, and today's best basic oxygen furnaces, first introduced during the 1950s, yield as much as 95%, with electric arc furnaces converting up to 97%. And electric arc furnaces now consume less than 350 kWh/t kWh/t of steel, compared to more than 700 kWh/t in 1950; moreover, these gains have been accompanied by reduced emission rates: between 1960 and 2010 specific U.S. rates (per tonne of hot metal) fell by nearly 50% for CO_2 emissions and by 98% for dust emissions (Smil 2016). The energy cost of steel has been further lowered by continuous casting of the hot metal. This innovation supplanted the traditional production of ingots, which required reheating before further processing.

The resulting production increases have been large enough to translate into order-of-magnitude gains even in per capita terms: in 1850, before the beginning of modern steel production, fewer than 100,000 t of the metal were produced annually in artisanal ways, a mere 75 g/year/capita. In 1900, with 30 Mt, the global mean was 18 kg/capita; in the year 2000, with 850 Mt, the mean rose to140 kg/capita; and by 2015, with 1.65 Gt, it reached about 225 kg/capita, roughly 12 times the rate in 1900. My calculations show that in 2013 the worldwide production of iron and steel required at least 35 EJ of fuels and electricity, or less than 7% of the total of the world's primary energy supply, making it the world's largest energy-consuming industrial sector (Smil 2016). This compares to 23% for all other industries, 27% for transportation, and 36% for residential use and services. But if the sector's energy intensity had remained the same as in the 1960s, then the industry would have consumed at least 16% of the world's primary energy supply in 2015, an impressive illustration of continuing efficiency improvements.

By far the most important innovation in nonferrous metallurgy was the development of aluminum smelting. The element was isolated in 1824, but an economical process for its large-scale production was devised only in 1866. The independent inventions of Charles M. Hall in the United States and P. L. T. Héroult in France were based on electrolysis of aluminum oxide. The minimum energy needed to separate the metal is more than six times higher than that needed to smelt iron. Consequently, aluminum smelting advanced only slowly even after the beginning of large-scale electricity generation. During the 1880s specific electricity requirements were more than 50,000 kWh/t of aluminum, and subsequent steady improvements of the Hall-Héroult process lowered this rate by more than two-thirds by 1990 (Smil 2014b).

Aluminum's uses expanded first with advancing aviation. Metal bodies displaced wood and cloth during the late 1920s, and the demand rose sharply during World War II for the construction of fighters and bombers. Since 1945 aluminum and its alloys have become a substitute for steel wherever the design has required a combination of lightness and strength. These uses have ranged from automobiles to railway hopper cars to space vehicles, but this market is now also served by new lightweight steel alloys. And since the 1950s titanium has been replacing aluminum in high-temperature applications, above all in supersonic aircraft. Its production is at least three times as energy-intensive as aluminum's production (Smil 2014b).

Though the fundamental importance of mass-produced metals is often overlooked in a society preoccupied with the latest electronic advances, there is no doubt that modern manufacturing has been transformed by its continuing fusion with modern electronics, a union that has greatly enhanced available design options, introduced unprecedented precision controls and flexibility, and changed marketing, distribution, and performance monitoring. An international comparison showed that in the United States in 2005 services purchased by manufacturers from outside firms were 30% of the value added to finished goods, with similar shares (23–29%) in major EU economies, while in 2008 service-related occupations added up to a slight majority (53%) of all jobs in the U.S. manufacturing sector, to 44–50% in Germany, France, and the UK, and to 32% in Japan (Levinson 2012). And though many products do not look that different from their predecessors, they are actually very different hybrids (box 6.5).

Cars are just one prominent example of an industry that now finds research, design, marketing, and servicing no less important than the actual production of goods. Even if a specific embedded energy use (per vehicle, computer, or a production assembly) has increased (owing to the use of more energy-intensive materials, a larger mass, or better performance), remained the same, or declined, concerns other than a preoccupation with produced quantity have become very important, chief among them appearance, brand distinction, and quality considerations. This trend has major implications both for future energy use and for the structure of labor force, but not necessarily in any simple, unidirectional way (for more on this topic, see chapter 7).

Transportation
Several attributes apply to all forms of fossil-fueled, or electrified, transport. In contrast to traditional ways of moving people and goods they are much faster, often almost incredibly so: every year tens of millions of people now cross the Atlantic in 6–8 hours, though a century ago a crossing took nearly six days (Hugill 1993) and half a millennium ago the first crossing took five weeks. Transport conveyances are also incomparably more reliable: even the best coaches drawn by the strongest horse teams found it challenging to cross Alpine passes, succumbing to broken axles, crippled animals, and blinding storms; now hundreds of flights daily overfly the range and trains speed through deep tunnels. As for the expense, just before World War I the cost of a transatlantic crossing averaged $75 (Dupont, Keeling, and Weiss

---

Box 6.5

Cars as mechatronic machines

There is no better example of the fusion of mechanical and electronic components than a modern passenger car. In 1977 GM's Oldsmobile Toronado was the first production car with an electronic control unit (ECU) to govern spark timing. Four years later GM had about 50,000 lines of engine control software code in its domestic car line (Madden 2015). Now even inexpensive cars have up to 50 ECUs, and some premium brands (including the Mercedes-Benz S class) have up to 100 networked ECUs supported by software containing close to 100 million lines—compared to 5.7 million lines of software needed to operate the F-35, the U.S. Air Force's Joint Strike Fighter, or 6.5 million lines for the Boeing 787, the latest model of the company's commercial jetliners (Charette 2009).

Car electronics are getting more complex, but comparing lines of code is a misleading choice. The main reason for bloated software in cars is to cover the excessive number of options and configurations offered with luxury models, including those for infotainment and navigation that have nothing to do with actual motoring; there is a great deal of re-used, auto-generated, and redundant code. Even so, electronics and software now represent up to 40% of the cost of premium vehicles: cars have been transformed from mechanical assemblies into mechatronic hybrids, and every addition of a useful control function—such as a lane-departure warning, automatic braking to a avoid rear-end collision, or advanced diagnostics—expands the software requirements and adds to the cost of a vehicle. The trend has been clear, but completely autonomous, self-driving vehicles are not coming as soon as many uncritical observers believe.

2012), or about $ 1,900 in 2015 monies. The return trip of nearly $4,000 in current dollars compares to about $1,000 for an average (undiscounted) London–New York flight.

While the early nineteenth century saw some important gains, in terms of both unit capacities and efficiencies and in the stationary harnessing of natural kinetic energies by water wheels and windmills, land transport, powered solely by animate muscles, had changed very little since the antiquity. For millennia, no mode of traveling on land was faster than riding a good horse. For centuries, no conveyance was less tiring than a well-sprung coach. By 1800 some roads had better hard tops, and many coaches were well sprung, but all of these were differences of degree, not of kind. Railways removed these constants in a matter of years. They not only shrank distances and redefined space, they did so with unprecedented comfort. Speed of a mile a minute (96 km/h) was first reached briefly on a scheduled English run in 1847; that was also the year of the greatest railway-building activity in the country, which laid a dense network of reliable links within just two generations (O'Brien 1983).

The large-scale construction of railways with trains pulled by increasingly powerful coal-fueled steam engines was accomplished in Europe and North America in less than 80 years: the 1820s were the decade of experimentation; by the 1890s the fastest trains traveled along some sections at more than 100 km/h. Very soon after their introduction passenger cars ceased to be merely carriages on rails and acquired heating and washrooms. For a higher price passengers also enjoyed good upholstery, fine meal services, and sleeping arrangements. Faster and more comfortable trains carried not only visitors and migrants to cities but also urbanites to the countryside. Thomas Cook offered railway holiday packages starting in in 1841. Commuter rail lines made the first great wave of suburbanization possible. Increasingly capacious freight trains brought bulky resources to distant industries and speedily distributed their products.

The total length of British railways was soon surpassed by American construction, which began in 1834 in Philadelphia. By 1860 the United States had 48,000 km of track, three times the UK total. By 1900 the difference was nearly tenfold. The first transcontinental link came in 1869, and by the end of the century there were four more such lines (Hubbard 1981). The Russian development also progressed fairly rapidly. Fewer than 2,000 km of track were laid by 1860, but the total rose to over 30,000 by 1890 and to nearly 70,000 km by 1913 (Falkus 1972). The transcontinental link across Siberia to Vladivostok, begun in 1891, was not fully completed until 1917. When the British withdrew from India in 1947 they left behind about 54,000 km of track (and 69,000 km in the whole subcontinent). No other mainland Asian country built a major railway network before World War II.

Since the end of World War II, competition from cars, buses, and planes reduced the relative importance of railways in most industrialized countries, but during the latter half of the twentieth century the Soviet Union, Brazil, Iraq, and Algeria were among the vigorous builders of new lines, and China was the Asian leader, with over 30,000 km of track added between 1950 and 1990. But the most successful innovation of the post–World War II period has been fast long-distance electrical train. The Japanese *shinkansen*, first run in 1964 between Tokyo and Osaka, reached a maximum of 250 km/h, and its latest trains (*nozomi*) go 300 km/h (Smil 2014a; fig. 6.10).

Figure 6.10
Shinkansen N700 Series at Kyoto Station in 2014, the 50th year of accident-free
operation of Japan's rapid trains on the Tokaido line. Photograph by V. Smil.

French *trains a grand vitesse* (TGV) have been operating since 1983; the
fastest scheduled trip is at nearly 280 km/h. Similarly rapid links now also
exist in Spain (AVE), Italy (*Frecciarossa*) and Germany (Intercity), but China
has become the new record-holder in the overall length of high-speed rail:
in 2014 it had 16,000 km of dedicated track (Xinhua 2015). In contrast,
America's solitary *Acela* (Boston-Washington, averaging just over 100 km/h)
does not even qualify as a modern high-speed train.

If the count starts from the introduction of the first practical gasoline
engines in the late 1880s, then the second transportation revolution on
land, the progress of road vehicles powered by internal combustion engines,
did not take less time. In the higher-income countries of Europe and North
America it was twice interrupted by world wars. And while the United States
had a high rate of car ownership already during the late 1920s, a comparable stage in Europe and Japan came only during the 1960s, and in China
the age of mass car ownership began only in the year 2000, but, owing to
the country's large population and rapid investment in new factories, China's car sales surpassed the U.S. total in 2010. By that time the world had

about 870 million passenger cars and a total of more than one billion road vehicles (fig. 6.11).

The economic, social, and environmental changes brought by cars rank among the most profound transformations of the modern era (Ling 1990; Womack, Jones, and Roos 1991; Eckermann 2001; Maxton and Wormald 2004). In country after country (first in the United States during the mid-1920s), car making emerged as the leading industry in terms of product value. Cars have also become major commodities of international trade. Their exports from Germany (after 1960) and even more so from Japan (after 1970) have been benefiting those two economies for decades. Large segments of other industries—above all steel, rubber, glass, plastics, and oil refining—are dependent on making and driving cars. Highway building has involved massive state participation, leading to enormous cumulative capital investments. Hitler's *Autobahnen* of the 1930s preceded Eisenhower's system of interstates by a generation (starting in 1956, the total is now just above 77,000 km), and the latter system has been far surpassed by China's National Trunk Highway System, whose total length reached 112,000 km in 2015.

Certainly the most obvious car-generated impact has been the worldwide reordering of cities through the proliferation of freeways and parking spaces and the destruction of neighborhoods. Where space allows, there has been also a rapid increase in suburbanization (in North America also in exurbanization) and changes in location and forms of shopping and services. The social impacts have been even greater. Car ownership has been an important part of *embourgeoisement,* and some affordable designs that enabled this new mass ownership enjoyed amazing longevity (Siuru 1989). The first was Ford's Model T, whose price dropped as low as $265 in 1923 and whose production lasted 19 years (McCalley 1994). Other notable models were the Austin Seven, the Morris Minor, the Citroen 2CV, the Renault 4CV, the Fiat Topolino, and, the most popular of them all, Ferdinand Porsche's Hitler-inspired Volkswagen (box 6.6).

Freedom of personal travel has had enormous effects on both residential and professional mobility. These benefits have proved to be highly addictive. Boulding's (1974) analogy of a car as a mechanical steed turning its driver into a knight with an aristocrat's mobility, looking down at pedestrian peasants (and making it almost unthinkable to rejoin them), is hardly exaggerated. In 2010 there were only 1.25 people per motor vehicle (including trucks and buses) in the United States, and the rate was 1.7 in both Germany and Japan (World Bank 2015b). This widespread addiction to on-demand mobility makes it difficult to give up the habit: after a

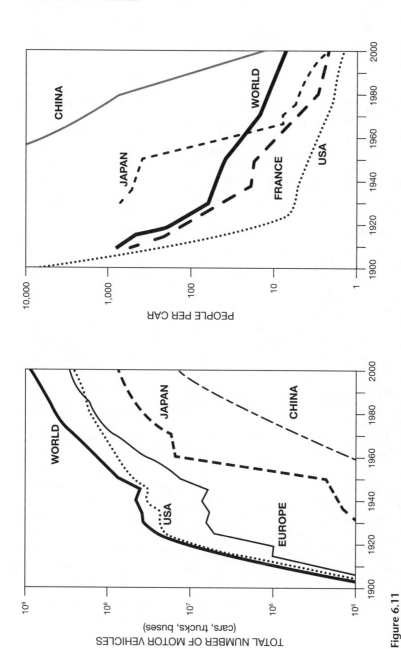

Figure 6.11

The worldwide total of road vehicles grew from about 10,000 in 1900 to more than one billion by 2010 (left). U.S. registrations were surpassed by the European total during the late 1980s, but the country still has the highest rate of ownership, about 1.25 people per vehicle in 2010 (right). Plotted from data in annual reports of the Motor Vehicle Manufacturers Association and World Bank (2015b).

Box 6.6
Volkswagen and other durable models

In terms of the aggregate production, size, and longevity (though with updated models), no car designed for the masses comes close to the one that Adolf Hitler decreed as the most suitable for his people (Nelson 1998; Patton 2004). In autumn 1933 Hitler set down the car's specifications—top speed 100 km/h, 7 L/km, capable of conveying two adults and three children, with air cooling, and at a cost below 1,000 RM—and Ferdinand Porsche (1875–1951) had the car, rather ugly and looking, at Hitler's insistence, like a beetle (*Käfer*), ready for production in 1938. War prevented any civilian production, and the Beetle's serial assembly began only in 1945, under the British Army command led by Major Ivan Hirst (1916–2000), who saved the damaged factory (Volkswagen AG 2013).

During the early years of West German *Wirtschaftswunder* (before mass ownership of Mercedeses, Audis, and BMWs), the car flooded German roads, and during the 1960s Volkswagen became the most popular import to the United States before it was displaced by Hondas and Toyotas. The production of the original Beetle stopped in Germany in 1977 but continued in Brazil until 1996 and in Mexico until 2003: the last car produced at the Puebla plant had number 21,529,464. The New Beetle, with an exterior redesigned by J. Mays and the engine in the front, was made between 1997 and 2011; since the model year 2012 the name of the latest design (A5) has reverted to Volkswagen Beetle.

The Renault 4CV, secretly designed during World War II, was the Beetle's French counterpart; more than one million cars were made between 1945 and 1961. The country's most famous basic car was the Citroen 2CV, made between 1940 and 1990: *deux cheveaux* marked just the number of cylinders; the engine actually had 29 hp (Siuru 1989). Fiat's little mouse, the Topolino, a two-seater with a wheelbase just short of 2 m, was made between 1936 and 1955, and the British Morris Minor was made between 1948 and 1971. All these models were eclipsed in popularity by Japanese designs: after relatively small exports during the 1960s and 1970s they became the global bestsellers during the 1980s.

recession-induced dip between 2009 and 2011, car sales in the United States reached near record levels of 16.5 million units by 2015.

We have gone to extraordinary lengths to preserve this privilege (and in North America we have made it easier by selling more than 90% of all vehicles on credit), and hence we cannot be surprised that Chinese and Indians want to emulate the North American experience. But like every addiction, this one exacts a high price. In 2015 the world had about 1.25 billion vehicles on the road, and in 2015 new passenger car sales reached about 73 million (Bank of Nova Scotia 2015), while traffic accidents cause annually nearly 1.3 million deaths and up to 50 million injuries (WHO 2015b), and automotive air pollution has been a key contributor to the worldwide phenomenon of seasonal (or semipermanent) photochemical smog in megacities on all continents (USEPA 2004). The life span of the average car now ranges from nearly 11 years in affluent countries to more than 15 years in low-income economies. Afterward, the steel (and copper and some rubber) is mostly recycled, but we have been willing to put up with enormous death, injury, and pollution costs.

Trucking has also had many profound socioeconomic consequences. Its first mass diffusion, in rural America after 1920, reduced the cost and sped up the movement of farm products to market. These benefits have been replicated first in Europe and Japan, and during the past two decades also in many Latin American and Asian countries. In rich countries, long-distance heavy trucking has become the backbone of food deliveries, as well as a key link in the distribution of industrial parts and manufactured goods, and its operation has benefited from the universal embrace of containers offloaded by cranes from oceangoing vessels directly onto flatbed trucks. In many rapidly growing economies trucking has obviated the construction of railways (Brazil being the best example) and opened up remote areas to commerce and development—but also to environmental destruction. In poor nations buses have been the leading means of long-distance passenger transport.

The first steamships crossed the North Atlantic no faster than the best contemporary sail ships with favorable winds. But already by the late 1840s the superiority of steam was clear, with the shortest crossing time cut to less than 10 days (fig. 6.12). By 1890 trips of less than six days were the norm, as were steel hulls. Steel did away with size restrictions: structural considerations limited the length of wooden hulls to about 100 m. Large ships of such famous lines as Cunard, Collins, or Hamburg-America became proud symbols of technical age. They were equipped with powerful engines and double-screw propellers, furnished with grand staterooms, and offered excellent service.

The opulence of these great liners contrasted with the crowding, smells, and tedium of steerage passages. By 1890 steamships carried more than half a million passengers a year to New York. By the late 1920s the total North Atlantic traffic surpassed one million passengers a year, and soon afterward the liners reached their maximum tonnages (fig. 6.12). But by 1957 airlines carried more people across the Atlantic than ships, and the introduction of regular jetliner service in the same year sealed the fate of long-distance passenger shipping: a decade later regularly scheduled transatlantic service came to an end. Commercial steamships got early boosts from the completion of the Suez Canal in 1869 and from the introduction of effective refrigeration during the 1880s. Its later growth was stimulated by the opening of the Panama Canal (1914), the deployment of large diesel engines (after 1920), and the transport of crude oil. Since the 1950s larger specialized

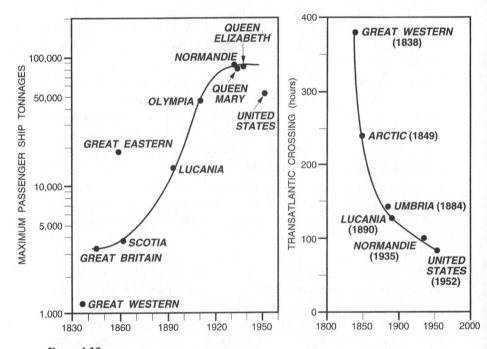

Figure 6.12
As the ships connecting Europe and North America grew in size (left) and became equipped with more powerful engines, the time needed to cross the Atlantic was cut from more than two weeks to just over three days (right). Plotted from data in Fry (1896), Croil (1898), and Stopford (2009).

ships have been needed to move not only oil but also widely traded bulky commodities (ores, lumber, grain, chemicals) and growing shipments of cars, machinery, and consumer goods.

Scheduled international air transport started with daily London-Paris flights in 1919 at speeds well below 200 km/h, and advanced to regular transoceanic links just before World War II: PanAm's Clipper reached Hong Kong from San Francisco after a six-day journey in March 1939 (fig. 6.13). The age of mass air travel came only with the introduction of jet aircraft in the late 1950s (British Comet, in service since 1952, was grounded in 1954 after three fatal disasters). The Boeing 707 (the first flight in 1957, in service since October 1958) was soon followed by the mid-range Boeing 727 (in regular service since February 1964 and produced until 1984) and the short- to mid-range Boeing 737. This smallest of all Boeing jetliners has become the bestselling airplane in history: by mid-2015 more than 8,600 planes had been delivered (compared to about 9,200 for all Airbus models). During the 1950s and 1960s McDonnell Douglas (DC-9, triple-engine DC-10), General Dynamics (Convair), Lockheed (Tristar), and Sud Aviation (Cara-velle) introduced their own jetliners, but (leaving the Russian makers aside) by the century's end only the duopoly of the American Boeing and the European Airbus consortium was left (box 6.7).

The speed and range of these planes, the proliferation of airlines and flights, and the nearly universal linking of reservation systems have made it possible to travel among virtually all major cities of the planet in a single day (fig. 6.13). By the year 2000 the maximum range of wide-body jetliners reached 15,800 km, and in 2015 the longest scheduled flights (Dallas-Sydney and Johannesburg-Atlanta) lasted nearly 17 hours, while many cit-ies are connected by frequent shuttle flights (in 2015 there were nearly 300 daily flights between Rio de Janeiro and São Paulo, nearly 200 between New York and Chicago). Moreover, the costs of flying have been steadily declin-ing in real terms, in part because of lower fuel consumption. These achieve-ments opened up new business opportunities as well as mass long-distance tourism to major cities and to subtropical and tropical beaches. They also opened up new possibilities for unprecedented migrant and refugee move-ments, for widespread drug smuggling, and for international terrorism involving aircraft hijacking.

Information and Communication

From their very conception, fossil-fueled societies have produced, stored, distributed, and used incomparably larger amounts of information than their predecessors. In East Asia and in early modern Europe, printing was an

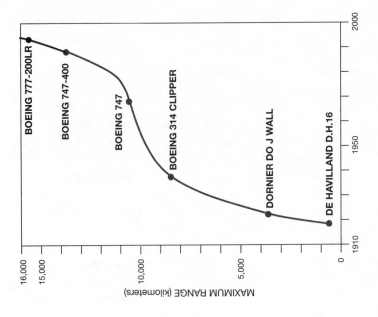

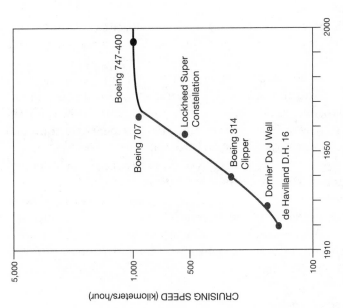

Figure 6.13

The first scheduled commercial flights (by the de Havilland D.H. 16 in 1919) averaged just over 150 km/h, and the plane had a maximum range of about 600 km (left). By the late 1950s the Boeing 707 could cruise at close to 1,000 km/h, and by the late 1990s the Boeing 777 could fly nonstop more than 15,000 km (right). The Concorde, flying at over twice the speed of sound, was a costly exception, not a precursor to a new generation of fast planes. Plotted from data in Taylor (1989) and Gunston (2002) and from technical specifications on the Boeing corporate website.

Box 6.7
Boeing and Airbus

Boeing is an old U.S. company—established by William E. Boeing (1881–1956) in 1916—and a maker of such iconic designs as the Boeing 314 Clipper and the 307 Stratoliner (both in 1938), the Boeing 707 (the first successful jetliner, in 1957), and the Boeing 747, the first wide-body plane, in 1969 (Boeing 2015). The company's latest innovation is the Boeing 787, an advanced design that uses lighter but stronger carbon fibers for 80% of the body, allowing 20% higher fuel efficiency than for the 767 (Boeing 2015). Airbus was set up in December 1970 with French and German participation, later joined by Spanish and British companies. Its first twinjet, the Airbus A300 (226 passengers), launched in October 1972, and its offerings expanded to a full range of planes, from the short-haul planes A319, 320, and 321 to the long-distance wide-body A340. In 2000 Airbus surpassed for the first time the number of planes sold by Boeing. Its greatest innovation has been the A380, a double-decker wide-body craft in service since 2007 with a maximum capacity of 853 passengers in a single class but, so far, ordered only in three-class configurations for 538 people (compared to 416 in three-class and 524 in two-class configuration for Boeing 747-400 planes).

The two companies have been in a close competition. Between 2001 and 2015 Boeing delivered 6,803 airplanes and Airbus produced 6,133 jetliners, and both companies have substantial multiyear order backlogs to supply the rising demand, particularly from Asia. Both companies have also made many cooperative agreements with aircraft and engine designers and with the suppliers of major airplane components in Europe, North America, and Asia, and both face a growing competition from below. The Canadian company Bombardier and the Brazilian Embraer have been enlarging their commuter jets: Bombardier's CRJ-900 seats 86, while Embraer's EMB-195 takes up to 122 passengers, and both of these companies, as well Russia's Sukhoi Superjet, the Commercial Aircraft Corporation of China, and Japan's Mitsubishi, are entering the lucrative market for narrow-body planes that is now served by the Boeing 737 and the Airbus A319/320.

---

established commercial activity for hundreds of years before the introduction of fossil fuels, but hand typesetting was laborious and print runs were limited by the slowness of hand-operated wooden screw presses. Iron frames sped up the work, but even advanced designs of Gutenberg's printing press could make no more than 240 impressions per hour (Johnson 1973). But even the first press powered by a steam engine—designed by Friedrich Koenig and Andreas Friedrich Bauer and sold to the *Times* in 1814—could do 1,100 impressions per hour. By 1827 that figure was 5,000, and the first rotary presses of the 1840s managed 8,000 impressions per hour; two decades later the rate was up to 25,000 (Kaufer and Carley 1993).

Mass editions of inexpensive newspaper became a quotidian reality, with news traveling faster thanks to telegraph (commercially for the first time in 1838) and less than two generations later telephone (1876), and before the century's end two new information-communication techniques had become commercial: sound recordings and replays and film. Except for printing, all these techniques were developed during the high-energy age based on fossil fuels. Except for photography and the early phonographs, none of them could function without electricity. And except for printed matter, now in retreat as many e-formats are taking its place, all these techniques have been expanding their user base and acquiring new modes of information capture, storage, recording, viewing, and sharing in the instantaneously interconnected world.

Inexpensive, reliable, and truly global telecommunication became possible only with electricity. The first century of its development was dominated by messages transmitted by wires. Decades of experiments in various countries ended with the first practical telegraph, demonstrated by William Cooke and Charles Wheatstone in 1837 (Bowers 2001). Its success depended on a reliable source of electricity, which was provided by Alessandro Volta's battery, designed in 1800. The adoption of the coding system of Samuel Morse in 1838 and the rapid extension of land lines in conjunction with railways were the most notable early developments. Undersea links (across the English Channel in 1851, across the Atlantic in 1866) and a wealth of technical innovations (including some of Edison's early inventions) combined to make the telegraph global within just two generations. By 1900 multiplex wires with automatic coding carried millions of words every day. The messages ranged from personal to diplomatic codes, and included reams of stock market quotations and business orders.

The telephone, patented by Alexander Graham Bell in 1876 just hours ahead of Elisha Grey's independent filing (Hounshell 1981), had an even faster acceptance in local and regional service (Mercer 2006). Reliable and cheap long-distance links were introduced rather slowly. The first trans-American link came only in 1915, and the transatlantic telephone cable was laid only in 1956. To be sure, radio-telephone links were available from the late 1920s, but they were neither cheap nor reliable. Great telephone monopolies provided affordable and reliable service, but they were not great innovators: the classic black rotary-dial telephone was introduced in the early 1920s and remained the only choice for the next four decades: the first electronic touch-tone phones appeared in the United States only in 1963.

Techniques for the storage, reproduction, and transmission of sound and pictures were developing concurrently with advances in telephony. Thomas Edison's 1877 phonograph was a simple hand-operated machine, as was Emile Berliner's (1851–1929) more complex gramophone in 1888 (Gronow and Saunio 1999). Electric record players took over only during the 1920s. Image making advanced rather slowly from its French beginnings, most notably in the work of J. N. Niepce and L. J. M. Daguerre during the 1820s and 1830s (Newhall 1982; Rosenblum 1997). Kodak's first inexpensive box camera came out in 1888, and developments sped up after 1890 with breakthroughs in cinematography: the first public short movies by the Lumière brothers were projected in 1895. Sound movies came in the late 1920s (the first feature film was *The Jazz Singer*, in 1927), the first color feature (after years of short color movies) came in 1935, and the invention of xerography by Chester Carlson (1906–1968) came two years later (Owen 2004).

The quest for wireless transmission started with Heinrich Hertz's (1857–1894) generation of electromagnetic waves in 1887, anticipated by James Clerk Maxwell's (1831–1879) formulation of the theory of electromagnetic radiation (Maxwell 1865; fig. 6.14). Subsequent practical progress was fast. By 1899 Guglielmo Marconi's (1874–1937) signals had crossed the English Channel, two years later the Atlantic (Hong 2001). In 1897 Ferdinand Braun (1850–1918) invented the cathode ray tube, the device that made possible both television cameras and receivers. In 1906 Lee de Forest (1873–1961) built the first triode, whose indispensability for broadcasting, long-distance telephony, and computers ended only with the invention of the transistor.

Regular radio broadcasts started in 1920. BBC offered the first scheduled television service in 1936, and RCA followed suit in 1939 (Huurdeman 2003). Mechanical calculators—starting with prescient designs by Charles Babbage and Edward Scheutz after 1820 (Lindgren 1990; Swade 1991) and culminating in the establishment of IBM in 1911—were finally left behind with the development of the first electronic computers during World War II. But these machines—the British Mark, the U.S. Harvard Mark 1, and the ENIAC—were unique, dedicated, massive (room-sized, to accommodate thousands of glass vacuum tubes) devices with no immediate commercial prospects.

This impressive concatenation of greatly improved and entirely new communication and information techniques and services was entirely

Figure 6.14
Engraved portrait of James Clerk Maxwell, based on a photograph by Fergus (Corbis). Maxwell's formulation of the theory of electromagnetism opened the way to the still unfolding exploits of modern wireless electronics that have brought inexpensive instant communication and global connectivity: the e-world of the twenty-first century rests on Maxwell's insights.

overshadowed by post–World War II developments. Their shared foundation was the rise of solid state electronics, which began with the American invention of the transistor, a miniature solid-state semiconductor device, the equivalent of a vacuum tube that can amplify and switch electronic signals. Julius Edgar Lilienfeld filed his patent for the field-effect transistor in Canada in 1925 and a year later in the United States (Lilienfeld 1930); the patent application clearly outlines the way to control and amplify the flow of current between the two terminals of a conducting solid.

But Lilienfeld did not attempt to build any device, and the first experimental success, by two Bell Labs researchers, Walter Brattain and John Bardeen, on December 16, 1947, used a germanium crystal (Bardeen and Brattain 1950). But as the Bell System Memorial site now admits, "It's perfectly clear that Bell Labs didn't invent the transistor, they re-invented it" while failing to acknowledge a great deal of pioneering research and design done since the very first decade of the twentieth century (Bell System Memorial 2011). In any case, it was not the crude point-contact device used by Brattain and Bardeen but more useful junction field-effect transistor patented in 1951 by William Shockley (1910–1989) that transformed electronic computing. In the same year Gordon K. Teal and Ernest Buehler succeeded in making larger silicon crystals and mastering improved methods for crystal pulling and silicon doping (Shockley 1964; Smil 2006).

A very important theoretical advance was made in 1948 when Claude Shannon opened the way to quantitative appraisals of the energy cost of communication (Shannon 1948). Despite the impressive progress made during the intervening years (a three orders of magnitude increase in carrying simultaneous conversation by a single cable, now no thicker than a human hair), Shannon's theoretical limits indicated that the performance could be improved by several orders of magnitude. But there was no immediate post–World War II rush to commercialize electronic computing, and Remington Rand's first UNIVAC (Universal Automatic Computer, an outgrowth of the Eckert-Mauchly ENIAC) was sold to the U.S. Census Bureau only in 1951.

The calculating speed of the new programmable machines started to rise exponentially as transistors supplanted vacuum tubes. Business use of computers in the United States finally took off only during the late 1950s, with Fairchild Semiconductor, Texas Instruments (which marketed the first silicon transistor in 1954), and IBM as the most accomplished developers of hardware and software (Ceruzzi 2003; Lécuyer and Brock 2010). In 1958–1959 Jack S. Kilby (1923–2005) at Texas Instruments and Robert Noyce (1927–1990) at Fairchild Semiconductor independently invented miniaturized circuits integrated into the body of semiconductor material (Noyce 1961; Kilby 1964). Noyce's design of a planar transistor opened the new era of solid-state electronics (box 6.8).

The U.S. military was the first customer for integrated circuits. In 1965, when the number of transistors on a microchip had doubled to 64 from 32 in the previous year, Gordon Moore predicted that this doubling would continue (Moore 1965). In 1975 he relaxed the pace to a doubling every two years (Moore 1975), and this rule, now commonly known as Moore's law, has held ever since (fig. 6.15). The world's first microprocessor-controlled commercial product was a programmable calculator by Busicom, a small Japanese company; its four-chip set was designed by the just

---

Box 6.8

Invention of integrated circuits

When working as the director of research at Fairchild Semiconductors in Santa Clara, California, Robert Noyce wrote in his lab notebook that

> it would be desirable to make multiple devices on a single piece of silicon, in order to be able to make interconnections between devices as part of the manufacturing process, and thus reduce size, weight, etc. as well as cost per active element. (Reid 2001, 13)

Noyce's 1959 patent application for a "semiconductor device-and-lead structure" showed a planar integrated circuit. It specified

> dished junctions extending to the surface of a body of extrinsic semiconductor, an insulating surface layer consisting essentially of oxide of the same semiconductor extending across the junctions, and leads in the form of vacuum-deposited or otherwise formed metal strips extending over and adherent to the insulating oxide layer for making electrical connections to and between various regions of the semiconductor body without shorting the junctions. (Noyce 1961, 1)

Noyce's patent (U.S. 2,981,877) was granted in April 1961, Kilby's (U.S. 3,138,743) only in July 1964, and lengthy interference proceeding, litigation, and appeals were settled only in 1971 when the Supreme Court ruled in Noyce's favor. By that time that was an immaterial victory because back in the summer of 1966 the two companies had agreed to share their production licenses and require other fabricators to make separate arrangements with both of them. In principle, Kilby's and Noyce's ideas were identical, but Noyce died of a heart attack in 1990, whereas Kilby lived long enough to share a Nobel Prize for Physics in the year 2000 "for his part in the invention of the integrated circuit."

---

established Intel in 1969–1970 (Augarten 1984). Busicom sold only a few large calculator models using the MCS-4 chip set before it went bankrupt in 1974; fortuitously, Intel had had the foresight to buy back the rights for the processor before that happened, and it released the world's first universal microprocessor—the 3 mm × 4 mm Intel 4004 containing 2,250 metal-oxide semiconductor transistors and priced at $200—in November 1971. With 60,000 operations per second, it was the functional equivalent of the room-sized ENIAC of 1945 (Intel 2015).

The universal deployment of these increasingly powerful microprocessors in conjunction with increasingly capacious memory devices has affected every sector of modern manufacturing, transportation, services, and communication, and the spectacular growth of these capabilities has been accompanied by steadily declining costs and improving reliability (Williams 1997; Ceruzzi 2003; Smil 2013c; Intel 2015). Microchips have

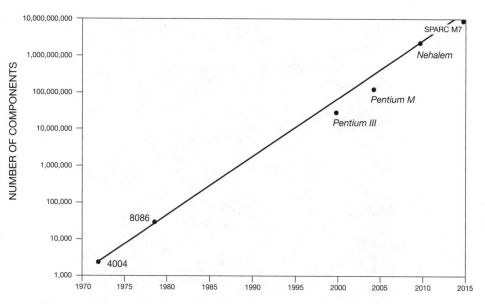

Figure 6.15

Moore's law in operation. The first commercially available microchip (Intel 4004) had 2,250 metal-oxide semiconductor transistors, while the latest designs have more than ten billion components, an increase of six orders of magnitude. Plotted from data in Smil (2006) and Intel (2015).

become the most ubiquitous complex artifacts of modern civilization: more than 200 billion of them are produced every year, and the devices can be found in products ranging from mundane household items and appliances (thermostats, ovens, furnaces, and in every electronic gadget) to the automated fabrication of complex assemblies, including the design and making of microprocessors themselves. They govern the timing of fuel ignition in automotive engines, optimize the operation of jetliner turbines, and guide rockets to place satellites on their predetermined paths.

But the most personalized impact of microprocessors has been through the mass ownership of portable electronic devices, above all cellular phones. This development was preceded by the rise of personal computers, by the surprisingly long development of the Internet, and by a period of relatively slow adoption of mobile phones. Xerox Palo Alto Research Center (PARC) invented personal computing during the 1970s by combining the processing power of microchips with a mouse, a graphical user interface, icons and pop-up menus, laser printing, text editing, spell checking, and access to file servers and printers with point-and-click actions (Smil 2006; fig. 6.16). Without these advances Steven Wozniak and Steven Jobs could not have

Figure 6.16
Utilitarian but revolutionary: the Xerox Alto desk computer, released in 1973, was the first nearly complete embodiment of the basic features characteristic of all later PCs (Wikimedia photograph).

introduced the first successful commercial PC, the Apple II with color graphics, in 1977 (Moritz 1984). IBM's PC was released in 1981, and in the United States ownership of PCs rose from two million units in 1983 to nearly 54 million units by 1990 (Stross 1996). Lighter, portable machines, laptops and tablets, matured only during the late 1990s, and Apple's iPad was introduced in 2010.

Communication using computers was first proposed in 1962 by J.C.R. Licklider, the first director of the Pentagon's Advanced Research Project Agency, and began in 1969 with ARPANET, limited to just four sites, at the Stanford Research Institute, UCLA, UCSB, and the University of Utah. In 1972 Ray Tomlinson of BBN Technologies designed programs for sending messages to other computers and chose the @ sign as the locator symbol for email addresses (Tomlinson 2002). In 1983 ARPANET converted a protocol that made it possible to communicate across a system of networks, and by 1989, when it ended its operation, it had more than 100,000 hosts. A year later Tim Berners-Lee created the hypertext-based World Wide Web at Geneva's CERN in order to organize online scientific information (Abbate 1999). The early web was not easy to navigate, but that changed rapidly with the introduction of efficient browsers, starting with Netscape in 1993.

The first major electronic advance in telephony was the possibility of inexpensive intercontinental calls, thanks to automatic dialing via geostationary satellites. This innovation resulted from a combination of microelectronic advances and powerful rocket launchers during the 1960s, and as the underlying costs declined, calls became cheaper. But the first radical change in telephony came only with the introduction of mobile phones (cell phones): first demonstrated in 1973, an expensive paid service with bulky Motorola sets was available in the United States in 1983, but ownership began to rise rapidly (with Japan and the EU ahead of the United States) only during the late 1990s. Global cell phone sales surpassed 100 million units in 1997, the year that Ericsson introduced the first smart phone.

Cell phone sales reached the billion mark by 2009, and by the end of 2015 there were 7.9 billion devices in use and the total annual shipments of mobile devices, including tablets, notebooks, and netbooks, had reached nearly 2.2 billion units, among them 1.88 billion cell phones (Gartner 2015; mobiForge 2015). This impressive and rapidly changing system of communication, entertainment, monitoring and data devices, and software requires a significant amount of energy to be embodied in highly energy-intensive electronic devices and is utterly dependent on an incessant and highly reliable electricity supply to energize the requisite infrastructures, ranging from data centers to cell towers (box 6.9).

Of special note is the enormous progress that has been made since the 1960s in designing and deploying a wide range of diagnostic, measuring, and remote sensing techniques. These advances yielded a previously unimaginable wealth of information. X-rays, discovered by W. K. Roentgen (1845–1923) in 1895, were the only such option in 1900. By 2015 these techniques ranged from ultrasound (used both in medical diagnoses and in engineering) to high-resolution imaging (MRI, CT), and from radar (developed on the eve of World War II, and now an indispensable tool in transportation and weather monitoring) to a wide range of satellite-based sensors acquiring data in various bands of the electromagnetic spectrum and

---

Box 6.9

Energy embodied in mobile phones and cars

Even a compact car weighs 10,000 times as much as a smart phone (1.4 t vs. 140 g), and hence it embodies considerably more energy. But the energy difference is far smaller than the four orders magnitude mass disparity, and aggregate accounts make for a surprising comparison. A cell phone embodies about 1 GJ of energy, whereas a typical passenger car now requires about 100 GJ to produce, only 100 times as much energy as goes into a cell phone. In 2015 worldwide sales of mobile phones came very close to 2 billion units, and hence their production consumed about 2 EJ (an equivalent of about 48 million metric tons of crude oil). About 72 million cars were sold worldwide in 2015, and their production embodied roughly 7.2 EJ—or only less than four times the total for mobile phones.

Mobile phones have very short life spans, on the average just two years, and their production now embodies globally about 1 EJ per average year of use. Passenger cars last on the average at least a decade, and their production embodies globally about 0.72 EJ per year of use—30% less than the making of mobile phones! This means that even if these approximate aggregates err in opposite directions (in reality, cars embodying more and mobile phone less energy), the two totals would still be not only of the same order of magnitude but surprisingly close. Operating energy costs are, of course, vastly different. A smart phone consumes annually just 4 kWh of electricity, less than 30 MJ during its two years of service, or just 3% of its embodied energy cost. In contrast, during its lifetime a compact car will consume four to five times as much energy (as gasoline or diesel fuel) as its embodied content. But the costs of electrifying the world's information and communications networks is rising: it claimed nearly 5% of worldwide electricity generation in 2012 and will approach 10% by 2020 (Lannoo 2013).

---

enabling much improved weather forecasting and natural resource management.

Economic Growth

To talk about energy *and* the economy is a tautology: every economic activity is fundamentally nothing but a conversion of one kind of energy to another, and monies are just a convenient (and often rather unrepresentative) proxy for valuing the energy flows. Not surprisingly, Frederick Soddy, a Nobelian physicist approaching the discipline from this perspective, argued that "the flow of energy should be the primary concern of economics" (Soddy 1933, 56). At the same time, energy flow is a poor measure of intellectual activity: education certainly embodies a great deal of energy expended on its infrastructures and employees, but brilliant ideas (which are by no means directly related to the intensity of schooling) do not require large increases of the brain's metabolic rate.

This obvious fact explains much of the recent decoupling of GDP growth from overall energy demand: we impute much higher monetary values to the nonphysical endeavors that now constitute the largest share of the economic product. In any case, energy has been of marginal concern in modern economic studies; only ecological economists have seen it as their primary focus (Ayres, Ayres, and Warr 2003; Stern 2010). And the public concern about energy and the economy has been disproportionately focused on prices in general, and on the prices of crude oil, the world's most important traded commodity, in particular.

In the West it was OPEC's two rounds of oil price increases during the 1970s—both the source of Middle Eastern consumption excesses and a threat to the region's stability—that became a particular object of critique, blamed for economic dislocations and social turmoil. But OPEC's price rise had a salutary (and long overdue) effect on the efficiency with which the countries importing OPEC oil consumed refined fuels. In 1973, after four decades of slow deterioration, the average specific fuel consumption of new American passenger cars was higher than in the early 1930s, 17.7 L/100 km versus 14.8 L/100 km, or, in American usage, 13.4 mpg versus 16 mpg (Smil 2006)—a rare example of a modern energy conversion becoming less efficient.

Higher oil prices forced the reversal, and between 1973 and 1987 the average fuel demand of new cars on the North American market was cut in half as the CAFE (Corporate Automobile Fuel Efficiency) standard fell to 8.6 L/100 km (27.5 mpg). Unfortunately, the post-1985 fall in oil prices first stopped and then even reversed (with more SUVs and pickups) this efficiency progress, and return to rationality came only in 2005. And OPEC's price rise had a beneficial effect for the global economy as it significantly reduced its average oil intensity (amount of oil used per unit of GDP). Power plants stopped burning liquid fuels; iron makers replaced injections of fuel oil to blast furnaces by powdered coal; jet engines became more efficient; and many industrial processes converted to natural gas. The results have been quite impressive. By 1985 the U.S. economy needed 37% less oil to produce a dollar of GDP than it did in 1970; by the year 2000 its oil intensity was 53% lower; and by 2014 it required 62% less crude oil to create a dollar of GDP than it did in 1970 (Smil 2015c).

And (a curiously neglected fact) Western governments have been making more money from oil than OPEC. In 2014, taxes in G7 countries accounted for about 47% of the price of a liter of oil, compared to about 39% going to the producers, with the respective national shares at 60/30 in the UK, 52/34 in Germany, and 15/61 in the United States (OPEC 2015). Moreover, to ensure a secure supply, many governments (including those of market economies) have engaged in a great deal of industry regulation, while governments in many oil-producing countries have been buying political support with heavy subsidies of energy prices (GSI 2015). Saudi subsidies claimed more than 20% of all government expenditures in 2010, and China's coal subsidies have resulted in prices fixed even below the production cost.

Growth—its origins, rate, and persistence—has been the leading concern of modern economic inquiries (Kuznets 1971; Rostow 1971; Barro 1997; Galor 2005), and hence the links between energy consumption and the increase gross economic product (either gross domestic product, GDP, for individual economies, or GWP, gross world product, for studying global trends) have received a great deal of attention (Stern 2004, 2010; World Economic Forum 2012; Ayres 2014). Traditional preindustrial economies were either largely stationary or managed to grow by a few percent per decade, and average per capita energy consumption advanced at an even slower pace: there is no shortage of testimonies from the early decades of the nineteenth century showing that the living conditions of some impoverished groups were not very different from those that had prevailed even two or three or four centuries before.

In contrast, the fossil-fueled economies have seen unprecedented rates of growth, though modified by the cyclical nature of economic expansion (van Duijn 1983; ECRI 2015) and interrupted by major internal or international conflicts. Industrializing societies of the nineteenth century saw their economies growing by 20–60% in a decade. Such growth rates meant that the output of the British economy in 1900 was nearly ten times larger than in 1800. America's GDP doubled in just 20 years, between 1880 and 1900. Japanese output during the Meiji era (1868–1912) rose 2.5 times. Economic growth during the first half of the twentieth century was affected by two world wars and the great economic crisis of the 1930s, but there had never been a period of such rapid and widespread growth of output and prosperity as between 1950 and 1973.

The steady pre-1970 decline in real crude oil prices was a critical ingredient of this unprecedented expansion. American per capita GDP, already the world's highest, rose by 60%. The West German rate more than tripled, and the Japanese rate more than sextupled. A number of the poor populous countries of Asia and Latin America also entered a phase of vigorous economic growth. OPEC's first round of oil price increases (1973–1974) temporarily stopped this growth. The second round of oil price increases, in 1979, was caused by the overthrow of the Iranian monarchy and the ascent of fundamentalist ayatollahs to power. The global economic slowdown of the early 1980s was accompanied by record inflation and high unemployment, but during the 1990s stabilized low oil prices supported another period of growth, which ended only in 2008 with the world's worst post–World War II recession, followed by a weak recovery.

Ayres, Ayres, and Warr (2003) identified the declining price of useful work as the growth engine of the U.S. economy during the twentieth century, useful work being the product of exergy (the maximum work possible in an ideal energy conversion process) and conversion efficiency. Once the historical data of economic output are normalized (with GDP values expressed in constant, inflation-adjusted monies and with the national products used to calculate GWP given in terms of purchasing power parity rather than by using official exchange rates), impressively strong long-term correlations between economic growth and energy use emerge on both global and national levels.

Between 1900 and 2000 the use of all primary energy (after subtracting processing losses and nonfuel uses of fossil fuels) rose nearly eightfold, from 44 to 382 EJ, and the GWP increased more than 18 times, from about $2 trillion to nearly $37 trillion in constant 1990 monies (Smil 2010a; Maddison Project 2013), implying an elasticity of less than 0.5. High correlations of the two variables can be found for a single country over time, but the elasticities differ: during the twentieth century, the Japanese GDP increased 52-fold and total energy use rose 50-fold (an elasticity very close to 1.0), while the multiples for the United States were, respectively, nearly 10-fold and 25-fold (an elasticity of less than 0.4), and for China nearly 13-fold and 20-fold (an elasticity of 0.6).

The expected closeness of the link between the two variables is further confirmed by very high correlations (>0.9) between averages of per capita GDP and energy supply when the set includes all the world's countries. This is clearly one of the unusually high correlations in the normally unruly realm of socioeconomic affairs, but the effect weakens considerably once we examine more homogeneous groups of countries: to become rich requires a substantial increase in energy use, but the relative energy consumption increase among affluent societies, whether measured per GDP unit or per capita, varies widely, producing very low correlations.

For example, Italy and South Korea have a very similar per capita GDP—adjusted for purchasing power, it was about $35,000 in 2014—but South Korea's per capita energy use is nearly 90% higher than Italy's. Conversely, Germany and Japan have a nearly identical annual energy consumption, about 170 GJ/capita, but in 2014 Germany's GDP was nearly 25% higher (IMF 2015; USEIA 2015d). And the rise in absolute energy consumption required to produce higher economic outputs hides an important relative decline. High-income, high-energy mature economies have a significantly lower energy intensity (energy per unit of GDP) than they had during earlier stages of their development (box 6.10, fig. 6.17).

The most important lesson to be drawn from looking at long-term trends of per capita energy use and economic growth is that respectable rates of the latter can be achieved with progressively lower use of the former. In the United States a continuing if slow population growth has brought further increases in the absolute consumption of fuels and electricity, but the average per capita use of primary energy has been flat (with only minor fluctuations) for three decades, since the mid-1980s, yet the real GDP (in chained 2009 dollars) per capita gained nearly 57%, growing from $32,218 in 1985 to $50,456 in 2014 (FRED 2015). Similarly, in both France and Japan (where population is now declining) per capita primary energy use has stabilized since the mid-1990s—yet in the following two decades the average per capita GDP increased, respectively, by about 20% and 10%.

But these outcomes must be interpreted with caution as those periods of relative energy-GDP decoupling coincided with extensive offshoring of U.S., European, and Japanese energy-intensive heavy industries and manufacturing to Asia in general and to China in particular: it would be premature to conclude that the recent experience of those three major economies is a harbinger of a widespread decoupling trend. And mainly because of China's enormous growth in pre-2014 energy demand (achieving a nearly 4.5-fold increase since 1990), the global primary energy supply had to rise nearly 60% in order to produce a 2.8-fold rise in GWP during the 25 years after 1990 (an elasticity of 0.56). Moreover, declines in electricity intensity have been much slower than the declines in overall energy intensity. Between 1990 and 2015 the global drop was just short of 20% (compared to >40% for all energy), and the U.S. decline was also 20%, but rapidly modernizing China saw no decline between 1990 and 2015.

The primary energy (and electricity) intensity of global economic growth have been declining, but, because of the size of the world economy and the continuing population growth in Asia and Africa, the coming decades will repeat, though in a modified way, the past experience as large quantities of

Box 6.10

Declining energy intensity of economic growth

Historical statistics show a steady decline in the British energy intensity following the rapid rise brought by the adoption of steam engines and railways between 1830 and1850 (Humphrey and Stanislaw 1979). Canadian and U.S. intensities followed the declining British trend with a lag of 60–70 years. The U.S. rate peaked before 1920, the Chinese maximum was reached during the late 1970s, and India's energy intensity began to decline only in the twenty-first century (Smil 2003). Between 1955 and 1973 the U.S. energy intensity was flat (fluctuating just ±2%), while the real GDP grew 2.5-fold, but then it resumed its decline, and by 2010 US it was 45% below the 1980 level.

In contrast, the Japanese energy intensity was rising until 1970, but between 1980 and 2010 it declined by 25% (USEIA 2015d), and the Chinese decline has been particularly large, almost 75% between 1980 and 2013 (China Energy Group 2014), as much a reflection of exceedingly low efficiencies of early post-Mao China as of the modernization advances since 1980. On the other hand, India, still in an earlier stage of economic development, saw only a 7% drop between 1980 and 2010. These declines stem from a combination of several factors: the declining importance of energy-intensive capital inputs that characterize earlier stages of economic development, heavily focused on basic infrastructures; improved conversion efficiencies of combustion and electricity use; and the rising shares of the service sector (retail, education, banking), where adding value requires less energy per unit of GDP than in extractive industries or manufacturing.

Major differences in the national energy intensities of otherwise similarly accomplished economies are also explained by the composition of primary energy use (somebody must produce energy-intensive metals), the efficiency of final conversions (hydroelectricity is always superior to coal), climate, and the size of the territory (Smil 2003). With the United States at 100, the relative rates in 2011 were about 60 in Japan and Germany, 70 in Sweden, 150 in Canada, and 340 in China. Interestingly, Kaufmann (1992) showed that most of the post-1950 decline in energy intensity in affluent economies resulted from shifts in the kind of energies used and the type of dominant goods and services rather than from technical advances.

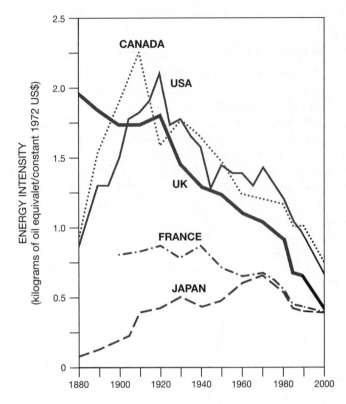

Figure 6.17
A declining energy intensity of the GDP has been a universal feature of maturing economies. Based on data in Smil 2003 and USEIA 2015d.

fuels and large additions of electricity-generating capacities will be required to energize economic growth in modernizing countries. Obviously, both the initiation and the maintenance of strong economic growth are matters of complex, interdependent inputs. They require technical improvements and responsive institutional arrangements, most notably sound banking and legal systems. Appropriate government policies, good educational systems, and a high level of competitiveness are also essential. But if today's low-income countries are to move from poverty to an incipient affluence (replicating China's post-1990 economic trajectory), then none of those factors could make a difference without the rising consumption of fuels and electricity: a decoupling of economic growth and energy consumption during early stages of modern economic development would defy the laws of thermodynamics.

Consequences and Concerns

The negative consequences of high energy use by modern societies range from obvious physical manifestations to gradual changes whose undesirable outcomes become apparent only after many generations. In the first category is an abundant food supply fostering indefensibly high food waste and contributing to unprecedented rates of overweight (a body mass index between 25 and 30) and obesity (body mass index >30). This trend toward heavier bodies is further reinforced by reduced energy expenditures, by more sedentary lifestyles resulting from mass replacement of muscle exertion by machines, and by the ubiquitous use of cars even for short trips that used to be made on foot. By 2012, 69% of the U.S. population was overweight or obese, up from 33% during the 1950s (CDC 2015), a clear proof that those conditions have been acquired through the combination of overeating and reduced physical activity.

The United States is hardly the only country with increasing shares of overweight and obese population (the rates are even higher in Saudi Arabia, and some of the fastest increases in excess weight are now found among Chinese children), but the trend is not (yet?) global: many European populations and most of the sub-Saharan Africa populations still have appropriate body masses. In any case, my intent is not to focus only on the negative impacts of intensive use of energy. Every one of the five fundamental global consequences of modern energy use I will examine has brought many welcome improvements along with effects whose worrisome impacts can be seen on scales ranging from local to global.

Continuing urbanization—since 2007, more than half of humanity has been living in cities—has been a major source of innovation. It has improved the physical quality of life and offers unprecedented opportunities for education and cultural exploits even as it has caused harmful levels of air and water pollution, led to excessive crowding, and created appalling living conditions for the poorest urban residents. High-energy societies enjoy a much higher standard of living than their traditional predecessors, and these gains have led to expectations of continued improvements: but because of persevering (and often deep) economic inequalities, these benefits have been unevenly distributed; moreover, there is no guarantee that further gains, requiring further deficit spending, will continue as populations age.

Energy prices, trade in fuels and electricity, and the security of energy supplies have become important political factors in both energy-importing and energy-exporting countries; in particular, periods of high and low oil prices have had major consequences for economies heavily dependent on hydrocarbon exports. The increased destructiveness of weapons and the increased risks of a nuclear conflict with truly global environmental and economic consequences have been accompanied by a widespread recognition of the futility of thermonuclear war and by steps to reduce the possibilities of such conflicts. And the massive combustion of fossil fuels has brought many negative environmental impacts, above all the risk of rapid global warming, and it will be very challenging to mitigate this threat.

Urbanization

Cities, even large cities, have a long history (Mumford 1961; Chandler 1987). Rome of the first century CE housed more than half a million people. Harun ar-Rashid's early ninth-century Baghdad had 700,000 people, and contemporary Changan (the capital of the Tang dynasty) had about 800,000 inhabitants. A thousand years later Beijing, the capital of the Qing dynasty, topped one million, and in 1800 there were about 50 cities above 100,000. But even in Europe no more than 10% of people lived in cities in 1800. The subsequent rapid increases in both the population of the world's largest cities and the overall shares of city dwellers would have been impossible without fossil fuels. Traditional societies could support only a small number of large cities because their energies had to come from croplands and woodlands that were at least 50 times and commonly about 100 times larger than the size of the settlement itself (box 6.11).

Modern cities use fuels with much higher efficiency, but their high concentrations of housing, factories, and transport push their power density to 15 W/m^2 in sprawling, warm-climate places and, in industrial cities in colder climates, up to 150 W/m^2 of their area. However, both coals and crude oils supplying these needs are extracted with power densities ranging usually between 1,000 and 10,000 W/m^2 (Smil 2015b). This means that an industrial city needs to rely on a coalfield or oil field whose size is no more than one-seventh and as little as 1/1,000th of its built-up area, and on new powerful prime movers to transport fuels from their basically punctiform places of extraction to urban users. While traditional cities had to be supported by the concentration of diffuse energy flows harvested over large areas, modern cities are supplied by the diffusion of fossil energies extracted in concentrated fashion from relatively small areas.

As far as food is concerned, a modern city of 500,000 people consuming daily 11 MJ/capita (with one-third coming from animal foods requiring, on the average, four times their energy value in feed) needs only about 70,000 ha to grow the crops, even when their mean yield would be just 4 t/ha. This

---

Box 6.11
Power densities of traditional urban energy supply and use

With average per capita food intakes of about 9 MJ/day originating, as preindustrial diets usually did, overwhelmingly (90%) from plant foods, and with typical grain yields of just 750 kg/ha, a traditional city of 500,000 people would have needed about 150,000 ha of cropland. In a colder climate, annual fuel (wood and charcoal) needs would have been about 2 t/capita. If supplied on a sustainable basis from forests or from fuelwood groves with annual yields of 10 t/ha, around 100,000 ha would have been needed to fuel the city. A densely populated city of that size occupied as little as 2,500 ha and had to rely on an area about 100 times its size for its food and fuel.

In terms of average power densities, this example implies about 25 W/m^2 for total energy consumption and 0.25 W/m^2 for the supply. The actual range of power densities was fairly large. Depending on their food intakes, cooking and heating practices, energy requirements for small manufactures, and combustion efficiencies, the total energy consumption of preindustrial cities was between 5 and 30 W/m^2 of their area. The sustainable production of fuel from nearby forests and woodlots yielded anywhere between 0.1 and 1 W/m^2. Consequently, cities had to rely on cropped and wooded areas 50–150 times larger than their own size—and the absence of powerful and inexpensive prime movers limited the capacity to transport food and fuel from distant regions, putting pressure on the plant resources of the surrounding areas (Smil 2015b).

---

would be less than half the total in the traditional city example, and fossil fuels and electricity also make large-scale, long-distance food imports affordable. And only electricity and liquid transportation fuels have made it possible to pump drinking water, remove and treat sewage and garbage, and meet the transportation and communication needs of megacities (cities with more than 10 million people). All modern cities are creations of fossil energy flows converted with high power densities, but megacities make exceptionally high claims: a survey by Kennedy and co-workers (2015) concluded that in 2011 the world's 27 megacities (with less than 7% of the global population) consumed 9% of all electricity and 10% of all gasoline.

The rise of fossil-fueled (initially just coal-fueled) cities was rapid. In 1800 only one of the world's ten largest cities, London (number two), was in a country whose energy use was dominated by coal. A century later nine out of ten were in that category: London, New York, Paris, Berlin, Chicago,

Vienna, Saint Petersburg, Philadelphia, and Manchester, while Tokyo was the capital of a country where biomass fuels provided still about half of all primary energy (Smil 2010a). The worldwide share of urban population in 1900 was only about 15%—but it was far higher in the world's three largest coal producers. The rate was over 70% in UK, approaching 50% in Germany, and nearly 40% in the United States. The subsequent continuation of urban growth has also brought a remarkable increase in the total number of very large cities. By 2015 nearly 550 urban agglomerations surpassed one million inhabitants compared to 13 in 1900 and only two, Beijing and Greater London, in 1800 (City Population 2015).

Fossil fuels also energized the push and pull forces of migration: urban growth has been driven by the push of agricultural mechanization and by the pull of industrialization. Urbanization and industrialization are not, of course, synonymous, but the two processes have been closely tied by many mutually amplifying links. Most notably, technical innovation in Europe and North America had overwhelmingly urban origins, and cities continue to be the fonts of innovation (Bairoch 1988; Wolfe and Bramwell 2008). Bettencourt and West (2010) concluded that as the population of a city doubles. economic productivity goes up by an average of 130%, with both total and per capita productivity rising, and Pan and co-workers (2013) attributed this result largely to "superlinear scaling" as increases in urban population density give residents greater opportunity for face-to-face interaction.

The massive shift of urban jobs into service sectors is largely a post–World War II development. By 2015 these transfers had brought urban populations above 75% of the total not only in nearly all Western nations but also in Brazil and Mexico (respectively about 90% and 80%). Only in many African and Asian countries do urban shares of the population remain below 50%, with India at 35% and Nigeria at 47%, but China is at 55%. China's relatively low figure has been heavily influenced by decades of tightly controlled migration in Maoist China, with rapid urbanization beginning only in the 1990s. The economic, environmental, and social effects of these great human translocations have been among the most avidly studied phenomena of modern history. The misery, deprivation, filth, and disease common in rapidly growing nineteenth-century cities spawned a particularly vast literature. These writings ranged from primarily descriptive (Kay 1832) to largely indignant (Engels 1845) and from a series of parliamentary hearings to bestselling novels (Dickens 1854; Gaskell 1855).

Similar realities—minus the threat of most contagious diseases, now eliminated by inoculation—can be seen today in many Asian, African, or Latin American cities. But people are still moving in. Now as earlier, they are often leaving conditions that, on balance, were even worse, a fact commonly neglected both by the original reformist writings and by subsequent debates about the disadvantages of urbanization. Now as earlier, one must weigh the dismal state of urban environments—aesthetic affronts, air and water pollution, noise, crowding, dismal living conditions in slums—against their often no less objectionable rural counterparts.

Common rural environmental burdens include very high concentrations of indoor air pollutants (particularly fine particulate matter) from unvented biomass combustion, inadequate heating in colder climates, unsafe water supplies, poor personal hygiene, dilapidated, overcrowded housing, and minimal or no opportunities to see the children properly educated. Moreover, the drudgery of field labor in the open is seldom preferable even to unskilled industrial work in a factory. In general, typical factory tasks require lower energy expenditures than does common farm work, and in a surprisingly short time after the beginning of mass urban industrial employment the duration of factory work became reasonably regulated.

Later came progressively higher wages, in combination with such benefits as health insurance and pension plans. Together with better educational opportunities these changes led to appreciable improvements in typical standards of living. This led eventually to the emergence of a substantial urban middle class in all largely laissez-faire economies. The appeal of this great, although now certainly tarnished, Western accomplishment is felt strongly throughout the industrializing world. And it was undoubtedly an important factor in the demise of Communist regimes, which proved slow to deliver similar benefits. And there is no doubt about the consequence of urbanization for energy consumption; living in cities requires substantial increases in the per capita provision of energy even in the absence of heavy industries or large ports: the fossil fuels and electricity required to sustain a person who moved to one of Asia's new growing cities can be easily an order of magnitude higher than the meager amounts of biomass fuels used in the village of her birth to cook and (if need be) to heat a room.

Quality of Life

Rising energy consumption has been exerting usually gradual (but in some instances, as in post-1990 China, fairly abrupt) and largely desirable effects on the average quality of life—a term broader than standard of living as it also encompasses such key intangible variables as education and personal freedoms. During the decades of rapid post–World War II economic growth, many previously poor countries moved to the intermediate energy consumption category as their inhabitants improved their overall quality of life (though often at the price of concomitant environmental degradation), but the distribution of global energy use remains extremely skewed. In 1950 only about 250 million people, or one-tenth of the global population, living in the world's most affluent economies consumed more than 2 t of oil equivalent (84 GJ) a year per capita—yet they claimed 60% of the world's primary energy (excluding traditional biomass). By the year 2000, such populations numbered nearly a quarter of all mankind and claimed nearly three quarters of all fossil fuels and electricity. In contrast, the poorest quarter of humanity used less than 5% of all commercial energies (fig. 6.18).

By 2015, thanks to China's rapid economic growth, the share of the global population consuming more than 2 t of oil equivalents jumped to 40%, the greatest equalization advance in history. Stunning as they are, these averages do not capture the real differences in the average quality of life because poor countries devote a much smaller share of their total energy

Figure 6.18
Kibera, one of Nairobi's largest slums (Corbis). Kenya's per capita use of modern energies averages about 20 GJ/year, but slum dwellers in Africa and Asia consume as little as 5 GJ/year, or less than 2% of the U.S. mean.

consumption to private household and transportation uses and convert those energies with lower efficiency. The real difference in typical direct per capita energy use among the richest and the poorest quarters of the mankind is thus closer to being 40-fold rather than "just" 20-fold. This enormous disparity is one of the few main reasons for the chronic gap in economic achievements and in the prevailing quality of life. In turn, these inequalities are a major source of persistent global political instability.

Those countries that have made it into the intermediate consumption category have gone through similar stages of improvements but at a very different pace: what took the early industrializers of Western Europe two or even three generations has been recently accomplished in South Korea and China in a single generation of compressed development (an advantage of determined late-starters). In the early stages of economic growth, these benefits are rather limited because fuels and electricity are overwhelmingly channeled into building up an industrial base. The slowly increasing acquisition of household and personal goods and better basic diets have been the first signs of improvement, starting in the cities and gradually diffusing to the countryside.

Among the first gains are a greater variety and better quality of basic cookware, dishes, and utensils; more, and usually more colorful, pieces of clothing; better shoes; better personal hygiene (more frequent washing and laundering); purchases of additional pieces of furniture; purchases of small gifts for special occasions; and pictures (starting with cheap reproductions) on walls. In North America and Europe of the early twentieth century possession of an increasing range of electrical appliances came during the next stage of *embourgeoisement,* but the low cost of new electric (air conditioning, microwave ovens, TVs) and electronic appliances and devices (above all mobile phones) means that in many Asian and in some African countries, families acquired them even before they owned other better household items.

The next stage sees further improvements in the variety and quality of the food supply and better health care, and the progress starts spilling into the countryside. The educational level of urban populations begins to rise, and there are increasing signs of incipient affluence, including car ownership, new house comforts, and travel abroad for people in higher income groups. Again, some of these gains have been recently conflated or inverted, particularly in Asia. Eventually comes the stage of mass consumption with its many physical comforts and frequent ostentatious displays. Longer periods of schooling, high personal mobility, and growing expenditures on leisure and health are part of this change.

Correlations of this sequence with average per capita energy consumption have been unmistakable, but what is usually compared—average per capita consumption calculated by aggregating a nation's primary energy supply and dividing it by the population total—is not the best variable. The per capita average consumption of the total primary energy supply tells us nothing either about the consumption breakdown (the military may claim a disproportionately large amount, as it did in the USSR and as it does in North Korea and Pakistan) or about the typical (or average) efficiency of energy conversions (higher, and hence delivering more final services per unit of gross energy, in Japan than in India). Better insights might come from comparing average rates of residential energy consumption, but that tack, too, is hardly perfect: fuels and electricity consumed by households will count, but considerable indirect energy inputs (required to build houses or to manufacture cars, household appliances, electronics, and furniture) are excluded.

Keeping this mind, and also realizing that national peculiarities (from climatic to economic singularities) preclude any simple classification, the relationship between energy use and quality of life can be divided into three basic categories. No country whose annual primary commercial energy consumption (leaving aside traditional biofuels) averages less than 5 GJ/capita (that is, about 120 kg of oil equivalent) can guarantee even basic necessities to all its inhabitants. In 2010 Ethiopia was still well below that minimum, Bangladesh barely above it; China was there before 1950, as were large parts of Western Europe before 1800.

As the rate of commercial energy use approaches 1 t of oil equivalent (42 GJ), industrialization advances, incomes rise, and the quality of life improves noticeably. China of the 1980s, Japan of the 1930s and again of the 1950s, and Western Europe and the United States between 1870 and 1890 are all examples of this stage of development. Incipient affluence requires, even with fairly efficient energy use, at least 2 t of oil equivalent (84 GJ) per capita per year. France made it during the 1960s, Japan during the 1970s. China reached that level by 2012, but its rate is not fully comparable with the Western rates because too much of its energy use still goes into industry (almost 30% in 2013), too little for private discretionary energy use (IEA 2015a).

But both the French and the Chinese gains illustrate the speed of recent changes. The French census of 1954 revealed the striking deficiencies in housing: less than 60% of households had running water, only 25% had an indoor toilet, and only 10% had a bathroom and central heating (Prost 1991). By the mid-1970s refrigerators were in almost 90% of households, toilets in 75%, 70% had bathrooms, and about 60% enjoyed central heating and washing machines. By 1990 all these possessions became virtually universal, and 75% of all families also owned a car, compared to fewer than 30% in 1960. Such growing affluence had to be reflected in a rising use of energy. Between 1950 and 1960 the average French per capita energy consumption rose by about 25%, but between 1960 and 1974 it soared by over 80%; and while between 1950 and 1990 the per capita supply of all fuels more than doubled, gasoline consumption rose nearly sixfold and electricity use went up more than eightfold (Smil 2003).

Even faster advances have taken place in China. In 1980, when the economic reforms started (four years after Mao Zedong's death), per capita energy consumption averaged about 19 GJ; by 2000 it was nearly 35 GJ; in 2010, after quadrupling in three decades, it was roughly 75 GJ; and in 2015 it was just above 90 GJ (Smil 1976; China Energy Group 2015), a level comparable to the Spanish mean during the early 1980s. Moreover, disproportionate shares of these gains have been used in construction. Nothing indicates this better than this fact: while the US consumption of cement added up to about 4.5 Gt during the entire twentieth century, China emplaced more of it (4.9 Gt) in its new construction projects in just the three years of 2008–2010 (Smil 2014b). No wonder that the country now has the world's largest modern networks of high-speed railways and interprovincial freeways.

No other form of energy has had a more wide-ranging impact on rising quality of life than the provision of affordable electricity: on the personal level the effects have been pervasive and life-spanning (premature babies are kept in incubators, vaccines to inoculate them are kept in refrigerators, dangerous illness are diagnosed by noninvasive techniques in time to be treated, the critically ill are hooked up to electronic monitors). But one of electricity's most consequential social impacts has been to transform many chores of household work and hence to disproportionately benefit women. This change has been, even in the Western world, fairly recent.

For generations, a rising energy consumption made little difference for everyday household work. Indeed, it could make it worse. As the standards of hygiene and social expectations rose with better education, women's work in Western countries often got harder. No matter if it was washing, cooking, and cleaning in cramped English apartments (Spring-Rice 1939) or doing daily chores in American farmhouses, women's work was still exceedingly hard during the 1930s. Electricity was the eventual liberator. Regardless of the availability of other energy forms, it was only the introduction of electricity that did away with exhausting and often dangerous labor (Caro 1982; box 6.12).

Many electric appliances were available already by 1900: during the 1890s General Electric was selling electric irons, fans, and an immersion water heater coil that could boil a pint of water in 12 minutes (Electricity Council 1973). The high cost of these appliances, limited house wiring, and slow progress in rural electrification delayed their widespread adoption, both in Europe and in North America, until the 1930s. Refrigeration has been a more important innovation than gas or electric cooking (Pentzer 1966). The first home refrigerators were marketed by Kelvinator Company in 1914. American ownership rose sharply only during the 1940s, and refrigerators became common in Europe only after 1960. Their importance has increased with the growing reliance on fast food. Refrigeration now accounts for up to 10% of all electricity used in the households of rich nations.

Electricity's conquest of household services continues to bring further time and labor savings in rich countries. Self-cleaning ovens, food processors and microwave cooking (developed in 1945, but introduced in small household models only during the late 1960s) have become common throughout the rich world. Refrigerator, washing machine, and microwave

---

Box 6.12
Importance of electricity for easing housework

The liberating effects of electricity are unforgettably illustrated in Robert Caro's (1982) first volume of Lyndon Johnson's biography. As Caro points out, it was not the shortage of energy that made life in Texas Hill County so hard (households had plenty of wood and kerosene) but the absence of electricity. In a moving, almost physically painful, account Caro describes the drudgery, and danger, of ironing with heavy wedges of metal heated on wood stoves, the endless pumping and carrying of water for cooking, washing, and animals, the grinding of feed, and sawing wood. These burdens, which fell largely on women, were much harder than the typical labor requirements in poor countries as the Hill County farmers of the 1930s strove to maintain a much higher standard of life and run much larger farming operations than peasants in Asia or Latin America. For example, the water needs for a family of five came to nearly 300 t/year, and to supply them required an equivalent of more than 60 eight-hour days and walking about 2,500 km. Not surprisingly, nothing could have been as revolutionary in the life of these people as the extension of transmission lines.

ownership has also approached saturation levels among better-off segments of Asian and Latin American populations, and they also have a high ownership of air conditioning units. Patented first by Willis Carrier (1876–1950) in 1902, air conditioning was limited for decades to industrial applications. The first units scaled down for household use came during the 1950s in the United States, and their widespread adoption opened up the American Sun Belt to mass migration from northern states and increased the appeal of subtropical and tropical tourist destinations (Basile 2014). Household air conditioners are now also used widely in urban areas of hot-weather countries, most of them being single-room wall units (fig. 6.19).

Modern societies have elevated economic growth, and hence rising energy use, to the level of unquestioned desiderata, implicitly assuming

Figure 6.19
A Shanghai high-rise apartment building with air conditioners for virtually every room (Corbis).

that using more will always have its rewards. But economic growth and rising energy use should be seen only as the means of securing a better quality of life, a concept that includes not only the satisfaction of basic physical needs (health, nutrition) but also the development of the human intellect (ranging from basic education to individual freedoms). Such an inherently multidimensional concept cannot be contracted into a single representative indicator, but it turns out that a few variables serve as its sensitive markers.

Infant mortality (deaths/1,000 live births) and life expectancy at birth are two obvious and unambiguous indicators of the physical quality of life. Infant mortality is an excellent proxy for conditions ranging from disposable income and quality of housing to the adequacy of nutrition, level of education, and a state's investment in health care: very few babies die in countries where families live in good housing and where well-educated parents (themselves well nourished) feed them properly and have access to medical care. And, naturally, life expectancy quantifies the long-term effects of these critical factors. Education and literacy data are not as revealing: enrollment ratios tell us about access but not about quality, and detailed achievement studies (such as the OECD's Programme for International Student Assessment, or PISA) are not available for most countries. Another option is to use the UNDP's Human Development Index (HDI), which combines life expectancy at birth, adult literacy, combined educational enrolment, and per capita GDP.

Comparing these measures with average energy use leads to some important conclusions. Some societies have been able to secure adequate diets, basic health care and schooling, and a decent quality of life with an annual energy use as low as 40–50 GJ/capita. Relatively low infant mortalities, below 20/1,000 newborns; relatively high female life expectancies, above 75 years; and an HDI above 0.8 could be achieved with 60–65 GJ/capita, while the world's top rates (infant mortality below 10/1,000 newborns, female life expectancies above 80, HDI > 0.9) require at least 110 GJ/capita. There is no discernible improvement in fundamental quality of life above that level.

Energy use is thus related to quality of life in a fairly linear manner only during the lower stages of development (going from quality of life in Niger to quality of life in Malaysia). Plotted values show distinct inflections of the best-fit lines at between 50 and 70 GJ/capita, followed by diminishing returns, topped by a plateau above (depending on the studied quality-of-life variable) 100–120 GJ/capita (fig. 6.20). This means that the effect of energy consumption on improving quality of life—measured by variables that

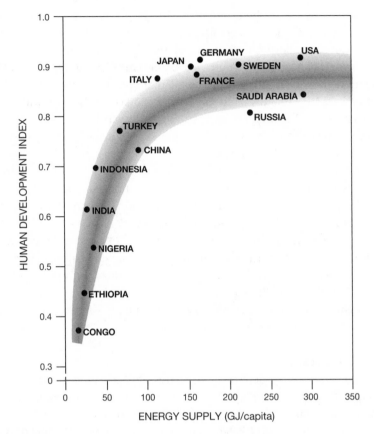

Figure 6.20
Average per capita energy consumption and the human development index in 2010.
Plotted from data in UNDP (2015) and World Bank (2015a).

truly matter, not by the ownership of yachts—reaches a saturation level
well below the rates of energy use prevailing in affluent countries, with the
leading EU economies and Japan at about 150 GJ/capita, Australia at 230
GJ/capita, the United States at 300 GJ/capita, and Canada at about 385 GJ/
capita in 2015 (BP 2015). Additional increases in discretionary energy
use go into ostentatious housing (as average family size has declined, the
average size of U.S. houses has more than doubled since the 1950s), the
ownership of multiple expensive vehicles, and frequent flying.

More remarkably, America's high energy use has been accompanied
by quality-of-life indicators that are inferior not only when compared with
the performance of leading EU countries or Japan (whose energy use is only

half the U.S. rate) but when compared with the performance of many countries with intermediate energy use. In 2013 the United States, with 6.6 of every 1,000 live-born babies dying in the first year of life, ranked 31st worldwide, below not only France (3.8), Germany (3.5), and Japan (2.6) but also more than twice as high as Greece's infant mortality (CDC 2015). Even worse, in 2013 America's life expectancy ranked 36th worldwide, with an average of 79.8 years for both sexes, which was hardly better than in Castro's Cuba (79.4) and behind the life expectancy of Greece, Portugal, and South Korea (WHO 2015a).

The educational achievements of students in OECD countries are regularly assessed by PISA,, and the latest results show America's 15-year-old ranking just below that of Russia, Slovakia, and Spain and far lower than that German, Canadian, or Japanese teenagers (PISA 2015). In science, U.S. children were just below the mean OECD score (497 vs. 501); in reading they were barely above the mean (498 vs. 496), and far behind all populous affluent Western nations. PISA, much like any such study, has its weaknesses, but the large differences in relative rankings are clear: there is not the slightest indication that America's high energy use has any beneficial effect on the country's educational achievements.

Political Implications

The dependence of modern societies on incessant, reliable, and inexpensive supplies of fossil fuels and electricity (delivered at required, and now invariably massive, rates) has generated a multitude of political concerns and responses, domestic and foreign. Perhaps the most universal concern is the concentration of decision-making power resulting from higher levels of integration, be it in government, business, or the military. As Adams (1975, 120–121) noted, when "more energetic processes and forms enter a society, control over them becomes disproportionately concentrated in the hands of a few, so that fewer independent decisions are responsible for greater releases of energy."

But far greater perils arise when these concentrated controls become superconcentrated in a single individual who decides to use them in an aggressive and destructive manner. Their misdirection can result in enormous human suffering, the prodigious waste of labor and resources, damage to the environment, and the destruction of a cultural heritage. Examples of such excessive concentration of control to unleash destructive forces have been a recurrent phenomenon in history; if measured only in human casualties, then the decisions made by the Spanish kings of the sixteenth century, by Napoleon Bonaparte (1769–1821), by Kaiser Wilhelm II

(1859–1941), or by Adolf Hitler (1889–1945) resulted in millions of deaths. The Spanish *conquista* of the Americas had eventually led, directly (through battle death and enslavement) and indirectly (through infectious diseases and famines) to the deaths of tens of millions (López 2014); Napoleon's serial aggression cost at least 2.5 million and as many as 5 million lives (Gates 2011); Prussian aggression was the proximate cause of more than 17 million World War I deaths; and the total death toll in World War II, military and civilian, approached 50 million (War Chronicle 2015).

But the unchallenged decisions of the two Communist dictators who could convert their manias into terrible realities with greater flows of fossil fuels and electricity are the unsurpassed epitomes of the perils of concentrated control. In 1953, the year of Stalin's death, the USSR's energy use was more than 25 times the total in 1921 when the country emerged from its civil war (Clarke and Dubravko 1983). Yet the generalissimo's paranoia led to the deaths of tens of millions in massive purges, the resettlement of entire populations (Crimean Tatars, Volga Germans, Chechens), the gulag empire, and the economic prostration of the world's potentially richest nation; the total number of deaths will be never tallied accurately, but it is at least on the order of 15–20 million (Conquest 2007).

Similarly, on Mao Zedong's death in 1976, China's energy production was more than 20 times the 1949 total (Smil 1988). But the Great Helmsman's delusions brought successive waves of death in the Great Leap Forward, followed by the worst famine in human history—between 1959 and 1961 more than 30 million Chinese died (Yang 2012)—and then the destruction of the Cultural Revolution. Again, no accurate total will be ever known, but the total of 1949–1976 deaths could be close to 50 million (Dikötter 2010). And while the probability of the ultimate threat—a thermonuclear war between great powers—has been reduced, thanks to the reduction of American and Russian warhead arsenals, its possibility remains, and the decision to launch would be made, on either side, by a very small group of people.

There has been no better example of the global political and economic consequences of concentrated controls of energy flows than the decisions made by the Organization of Petroleum Exporting Countries since 1973. Given the importance of crude oil in modern economies and the dominance of the global export market by a few Middle Eastern countries, it is inevitable that any decisions of a few individuals, particularly those in Saudi Arabia whose enormous oil-production capacity dominated OPEC's directions, will have profound consequences for global prosperity. OPEC's dissatisfaction with low royalties and the ensuing quintupling of world oil prices in 1973–1974, and their further near quadrupling in 1979–1980, ushered in, and then further deepened, a period of worldwide economic dislocation marked by high inflation and significantly reduced economic growth (Smil 1987; Yergin 2008).

In response, all major Western importers and Japan set up emergency energy-sharing agreements coordinated by the International Energy Agency, mandated the establishment of strategic petroleum reserves (some countries had also promoted closer bilateral ties with OPEC nations), and subsidized the quest for domestic fuel self-sufficiency by promoting alternative sources of energy. France's development of nuclear electricity and Japan's energy conservation effort have been especially remarkable and effective. But China's rapid economic rise—the country became a net oil importer in 1994—and the declining output of traditional oil fields, whether in Alaska or in the North Sea, were key reasons behind another rise in world oil price to a record level of about $145/bbl in July 2008, a run-up that ended only with the economic crisis in the fall of 2008 and with the oil price barely above $30/bbl in December 2008.

As the economies recovered and the Chinese demand continued to rise, oil prices rose once again above $100/bbl in July 2014, but then the falling demand and rising supply (mainly owing to the reemergence of the United States as the world's largest producer, thanks to rapid increases in shale oil production through hydraulic fracturing) brought a deep reversal. But this time there was a key difference: in order to protect the country's global market share, Saudi leaders decided to keep producing at maximum output rather than, as in the past, cutting output and propping up the price. Once again, the decisions made by a few men has worldwide consequences for the political stability of countries heavily dependent on oil exports, as well as for all major non-OPEC oil producers, including the United States and Canada.

Falling oil prices have once again brought expectations of OPEC's near demise—but the peculiarities of the highly uneven distribution of crude oil reserves (a leading strategic concern of the twentieth century that has not lost its importance in the twenty-first century) remain in favor of the Middle Eastern producers. The Persian Gulf basin is an unparalleled singularity: it has 12 of the world's 15 largest oil fields, and in 2015 it contained about 65% of the world's reserves of liquid oil (BP 2015). These riches explain the lasting interest in the region's stability. This desire is immensely complicated by the near chronic disarray of the area, which is made up of artificial states separated by arbitrary borders cutting across ancient ethnic group and containing complex religious enmities.

Notable post–World War II outside involvements in the region started with the Soviet attempt to take over northern Iran (1945–1946). Americans had landed twice in Lebanon, in 1958 and 1982, when their resolve was broken by a single terrorist bombing of the Beirut barracks in 1983 (Hammel 1985). Western countries armed Iran heavily (before 1979, during the last decade of Shah Reza Pahlevi's reign) and also Saudi Arabia, and the Soviets did the same with Egypt, Syria, and Iraq. The Western tilt (weapons, intelligence, and credit) benefited Iraq during the Iraq-Iran War (1980–1988). The pattern of intervention culminated in the Desert Shield and Desert Storm Operations of 1990–1991, a massive response by the U.S.-led, UN-sanctioned alliance assembled to reverse the Iraqi invasion of Kuwait (CMI 2010).

By that move, Iraq doubled the oil reserves under its control, pushing them to about 20% of the global total. The Iraqi advance seriously threatened the nearby Saudi oil fields, and perhaps even the very existence of the monarchy, which controlled one quarter of the world's oil reserves. But after the swift defeat, Saddam Hussein remained in power, and after the events of 9/11, fears of further aggression (misplaced, as proven later when no weapons of mass destruction were found in Iraq) led to the U.S. occupation of Iraq in March 2003, which was followed by years of internal violence and the eventual loss of part of the country to the so-called Islamic State. But later in this chapter I will argue, agreeing with Lesser (1991), that resource-related objectives, seemingly so paramount in the Middle Eastern conflicts, have historically been determined by broader strategic aims, not vice versa. And the failure of the Arab OPEC nations to turn oil into a political weapon (enacting an oil embargo against the United States and the Netherlands in the wake of the October 1973 Arab-Israeli Yom Kippur War) was not the first instance of using the energy supply to carry an ideological message.

The symbolic power of electric light was exploited by such diverse actors as large U.S. companies and Germany's Nazi party. American industrialists displayed the power of light for the first time during the 1894 Columbian Exposition in Chicago, and then by flooding the downtowns of large cities with "White Ways" (Nye 1992). The Nazis used walls of light to awe the participants at massed party rallies of the 1930s (Speer 1970). Electrification became the embodiment of such disparate political ideals as Lenin's quest for a Communist state form and Franklin Roosevelt's New Deal. Lenin summarized his goal in a terse slogan, "Communism equals the Soviet power plus electrification," and the Soviet preference for building giant hydroelectric projects was kept alive after the USSR's demise in post-Mao China.

Roosevelt used federal involvement in building dams and electrifying the countryside as a means of economic recovery, some of it in the country's most backward regions (Lilienthal 1944).

Weapons and Wars

Weapons production has become a leading industrial activity, one now heavily supported by advanced research, and all major economies have also become large-scale exporters of armaments. Only a fraction of these expenditures could be justified by real security needs, and waste and misallocation of investments and skilled manpower—most notably the development of weapons irrelevant for new forms of warfare (massed tank warfare is hardly the best response to jihadi terrorism)—have marked the history of modern weapon procurement. Not surprisingly, many technical advances brought about by new fuels and new prime movers were rapidly adapted for destructive uses. First they increased the power and the effectiveness of existing techniques. Later they made it possible to design new classes of weapons capable of unprecedented reach, speed, and destruction.

The culmination of these efforts came with the construction of enormous nuclear arsenals and with the deployment of intercontinental ballistic missiles capable of reaching any target on Earth. Accelerated destructiveness of modern weapons is well illustrated by contrasting the typical mid-nineteenth-century and the mid-twentieth-century weapons with their predecessors half a century before. The two principal classes of weapons used during the American Civil War (1861–1865), infantry muskets and 12-pound guns (both muzzle-loading with smooth bores), would have been quite familiar to the veterans of the Napoleonic Wars (Mitchell 1931). In contrast, among the weapons that dominated the battlefields of World War II—tanks, fighter and bomber planes, aircraft carriers, and submarines—only the last ones existed, and just in early experimental stages, during the 1890s. A revealing way to illustrate the energetic dimension of these developments is to compare the actual kinetic and explosive power of commonly used weapons.

As the basis for the first kind of comparison, it is useful to recall (as shown in chapter 4) that the kinetic energy of the two most common handheld weapons of the preindustrial era, arrows (shot from bows) and swords, was merely on the order of 10^1 J (mostly between 15 and 75 J), and that an arrow released from a heavy crossbow may hit a target with 100 J of kinetic energy. In contrast, bullets shot from the muzzles of muskets and rifles would have kinetic energies on the order of 10^3 J (10 to 100 times higher), while the shells fired from modern guns (including those mounted on tanks) rate at 10^6 J. The calculations for half a dozen specific weapons are shown in box 6.13: the values for gun shells are only the kinetic energies of projectiles and exclude the energies of the explosives they may or may not carry.

Rockets and missiles, propelled by solid or liquid fuels, cause most of their damage by targeted explosion of their warheads, not by their kinetic energy, but when the first (unguided) World War II German V-1 missiles failed to explode, the kinetic energy of their impact was 15–18 MJ. And the most famous recent example using an object with high kinetic energy to inflict extraordinary damage was the steering of large Boeing aircraft (767 and 757) into the World Trade Center skyscrapers by jihadi hijackers on September 11, 2001. The towers were actually designed to absorb a jetliner impact, but only of a slow-flying (80 m/s) Boeing 707 that might get lost on its approach to Newark, La Guardia, or JFK airport. The Boeing 767–200 is only about 15% heavier than was the 707, but the plane hit the tower at no less than 200 m/s, and hence its kinetic energy was more than seven times higher (about 3.5 GJ vs. roughly 480 MJ).

Even so, the structures were not brought down by the impact as the airplanes acted much as bullets hitting a massive tree: they could not push the massive structure, but they penetrated it by first destroying exterior columns. Karim and Fatt (2005) showed that 46% of the initial kinetic energy of the aircraft was used to damage the exterior columns, and that they would not have been destroyed if they had had a minimum thickness of 20 mm. The collapse of the towers was thus caused by the burning of fuel (more than 50 t of kerosene, or 2 TJ) and the building's interior combustibles, which caused thermal weakening of structural steel and nonuniform heating of the long floor joists, which precipitated the staggered floor

---

Box 6.13
Kinetic energy of projectiles propelled by explosives

Weapon	Projectile	Kinetic energy (J)
Civil war musket	Bullet	1×10^3
Assault rifle (M16)	Bullet	2×10^3
Eighteenth-century cannon	Iron ball	300×10^3
World War I artillery gun	Shrapnel shell	1×10^6
World War II heavy AA gun	High-explosive shell	6×10^6
M1A1 Abrams tank	Depleted U shell	6×10^6

collapse and led to the free-fall speed, as the towers fell in only about 10 s (Eagar and Musso 2001).

The explosive power of modern weapons began to rise with the invention of compounds more powerful than gunpowder: they, too, are self-oxidizing, but their high detonation velocities create a shock wave. This new class of chemicals was prepared by the nitration of such organic compounds as cellulose, glycerine, phenol, and toluene (Urbanski 1967). Ascanio Sobrero prepared nitroglycerin in 1846 and J. F. E. Schultze introduced nitrocellulose in 1865, but nitroglycerin's practical use was made possible only by Alfred Nobel's two inventions: mixing the compound with diatomaceous earth (an inert porous substance) to create dynamite and the introduction of a practical detonator, the Nobel igniter (Fant 2014).

Depending on the composition, gunpowder's detonation velocity can be only a few hundred m/s, while dynamite's is up to 6,800 m/s. Trinitrotoluene (TNT) was synthesized by Joseph Wilbrand in 1863 and was used as an explosive (detonation velocity of 6,700 m/s) by the end of the nineteenth century, while the most powerful prenuclear explosive, cyclonite (cyclotrimethylenetrinitramine or RDX, Royal Demolition eXplosive, detonation velocity of 8,800 m/s), was first made by Hans Henning in 1899. These explosives have been used ever since in gun shells, mines, torpedoes, and bombs, and in recent decades also strapped to the bodies of suicide bombers. But many terrorist attacks using car and truck bombs have been made just with a mixture of a common fertilizer (ammonium nitrate) and fuel oil: ANFO consists 94% of NH_4NO_3 (as an oxidizing agent) and 6% of fuel oil, both readily available ingredients whose effect is the result of the mass of the explosive used, not of any extraordinary detonation velocity (box 6.14).

The combination of better propellants and high-quality steels increased the range of field and naval guns from less than 2 km during the 1860s to

---

Box 6.14
Kinetic energy of explosive devices

Explosive device	Explosive	Kinetic energy (J)
Hand grenade	TNT	2×10^6
Suicide bomber with a belt	RDX	100×10^6
World War II gun shrapnel	TNT	600×10^6
Truck bomb (500 kg)	ANFO	2×10^9

over 30 km by 1900. The combination of long-range guns, heavy armor, and steam turbines for naval propulsion made it possible to build new heavy battleships: the HMS *Dreadnought*, launched in 1906, was their prototype (Blyth, Lambert, and Ruger 2011). The ship was powered by steam turbines (introduced by the Royal Navy in 1898), as were all of the largest passenger ships of the pre–World War I years, starting in 1907 with the *Mauretania* and *Lusitania*, and as are today the U.S. nuclear aircraft carriers of the Nimitz class (Smil 2005). Other notable pre–World War I destructive innovations included machine guns, submarines, and the first prototypes of military planes. The horrible trench stalemates of World War I were sustained by the massive deployment of heavy field guns, machine guns, and mortar launchers. Neither poisonous gases (first used in 1915) nor the first extensive use of fighter planes and tanks (in 1916, heavily only in 1918) could break the hold of that massive firepower deployed in frontal attacks (Bishop 2014).

The interwar years saw the rapid development of tanks and fighter and bomber planes. All-metal bodies replaced the early wood-canvas-wire construction, and the first purpose-built aircraft carriers came in 1922 (Polmar 2006). These weapons launched the aggression of World War II. Early German successes were largely a matter of rapid tank-led penetrations, and Japan's surprise attack on Pearl Harbor on December 7, 1941, could be made only with long-range fighters (the Mitsubishi A6M2 Zero, range of 1,867 km) and bombers (the Aichi 3A2, range of 1,407 km, and the Nakajima B5N2, range of 1,093 km) launched from a large carrier force (Hoyt 2000; National Geographic Society 2001; Smith 2015).

The same classes of weapons were essential in defeating the Axis powers. First it was a combination of excellent fighter planes (Supermarine Spitfires and Hawker Hurricanes) and radar during the Battle of Britain in August and September 1940 (Collier 1962; Hough and Richards 2007). Then came America's effective use of carrier planes (starting with the pivotal Battle of Midway in 1942), and the crushing Soviet tank superiority (model T-42) during the Red Army's westward thrust. The postwar arms race began during the war with the development of jet propulsion, the firing of German ballistic missiles (the V-2 was first used in 1944), and the explosion of the first nuclear bombs, the Trinity, New Mexico, test on July 11, the Hiroshima bombing on August 6, 1945, and the Nagasaki bombing three days later. The total energy released by these first nuclear bombs was orders of magnitude above that of any previous explosive weapons—but also orders of magnitude below those of subsequent hydrogen bomb designs.

The first modern field gun, the French *canon 75 mm modèle 1897*, fired shells filled with nearly 700 g of picric acid, whose explosive energy reached 2.6 MJ (Benoît 1996). Perhaps the best-known gun of World War II was the German anti-aircraft FlaK *(Flugzeugabwehrkanone)* 18, whose variant was also used in Tiger tanks (Hogg 1997); it fired shrapnel shells whose explosive energy was 4 MJ. But the most powerful explosives of World War II were the massive bombs that were dropped on cities. The most powerful bomb carried by the Flying Fortress (Boeing B-17) had an explosive energy of 3.8 GJ. But the greatest damage was done by dropping incendiary bombs on Tokyo on March 9–10, 1945 (box 6.15, fig. 6.21).

The Hiroshima bomb released 63 TJ of energy, about half of it as the blast and 35% as thermal radiation (Malik 1985). These two effects caused a large number of instant deaths, while ionizing radiation caused both instant and delayed casualties. The bomb exploded at 8:15 a.m. on August 7, 1945, about 580 m above ground; the temperature at the point of explosion was several million degrees Centigrade, compared to 5,000°C for conventional

---

Box 6.15
Firebombing of Tokyo, March 9–10, 1945

The raid, the largest of its kind ever, involved 334 B-29 bombers that offloaded their bombs at low (about 600–750 m) altitude (Caidin 1960; Hoyt 2000). Most of these were large, 230 kg cluster bombs, each releasing 39 M-69 incendiary bombs filled with napalm, a mixture of polystyrene, benzene, and gasoline (Mushrush et al. 2000); simple 45 kg jelled-gasoline and phosphorus bombs were also used. About 1,500 t of incendiary compounds were dropped on the city, and their total energy content (assuming an average density of napalm of 42.8 GJ/t) amounted to about 60 TJ, nearly as much as the power of the Hiroshima bomb.

But energy released by burning napalm was only a small fraction of the total released by the city's incinerated wooden buildings. According to the Tokyo Metropolitan Police Department, the fire destroyed 286,358 buildings and structures (U.S. Strategic Bombing Survey 1947), and conservative assumptions (250,000 wooden buildings, just 4 t of wood per building, 18 GJ/t of dry lumber) result in in some 18 PJ of energy released from the combustion of the city's wooden housing, two orders of magnitude (300 times) larger than the energy of incendiary bombs. The destroyed area amounted to about 4,100 ha, and at least 100,000 people died. For comparison, the totally destroyed area in Hiroshima was about 800 ha, and the best estimate of immediate deaths was 66,000.

Figure 6.21
Aftermath of Tokyo bombing of March 1945 (Corbis).

explosives. The fireball expanded to its maximum size of 250 m in one second, the highest blast velocity at the hypocenter was 440 m/s, and the maximum pressure reached was 3.5 kg/cm^2 (Committee for the Compilation of Materials 1991). The Nagasaki bomb released about 92 TJ.

These weapons appear minuscule compared to the most powerful thermonuclear bomb, tested by the USSR over Novaya Zemlya on October 30, 1961: the *tsar bomba* released 209 PJ of energy (Khalturin et al. 2005). Less than 15 months later Nikita Khrushchev revealed that Soviet scientists had built a bomb that was twice as powerful. Comparisons of explosive powers are usually done not in joules but in units of TNT equivalents (1 t TNT = 4.184 GJ): the Hiroshima bomb was equivalent to 15 kt TNT, the *tsar bomba* to 50 Mt TNT. Typical warheads on intercontinental missiles have a power of between 100 kt and 1 Mt, but up to 10 of them can be carried by such missiles as the U.S. submarine-launched Poseidon or the Russian SS-11. To emphasize the magnitudes of energy release I do not use scientific notation (exponents) in the staggering ladder of maximum destructivity of explosive weapons (box 6.16).

Box 6.16
Maximum energy of explosive weapons

Year	Weapon	Energy (J)
1900	Picrite-filled shell from French 75 mm modèle 1897 gun	2,600,000
1940	Amatol/TNT-filled shrapnel from German 88 mm FlaK	4,000,000
1944	The largest bomb carried by the Boeing B-17	3,800,000,000
1945	Hiroshima bomb	63,000,000,000,000
1945	Nagasaki bomb	92,400,000,000,000
1961	Soviet *tsar bomba* tested in 1961	209,000,000,000,000,000

The two nuclear superpowers eventually amassed about 5,000 strategic nuclear warheads (and an arsenal of more than 15,000 other nuclear warheads on shorter-range missiles) with an aggregate destructive energy of about 20 EJ. This was irrational overkill. As Victor Weisskopf (1983, 25) noted, "Nuclear weapons are not weapons of war. The only purpose they can possibly have is to deter their use by the other side, and for that purpose far fewer are good enough." And yet this excess actually served the West well as a mighty deterrent that prevented an obviously unwinnable global thermonuclear war.

But the development of nuclear bombs imposed a significant drain on national treasuries as it required enormous investment and very large amounts of energy, mostly for separating the fissile isotope of uranium (Kesaris 1977; WNA 2015a). Gaseous diffusion required about 9 GJ/SWU (separative work unit), but modern gas centrifuge plants need only 180 MJ/ SWU, and with 227 SWU needed to produce a kilogram of weapons-grade uranium, the latter rate works to or about 41 GJ/kg. And the triad of means amassed to deliver nuclear warheads—long-distance bombers, intercontinental ballistic missiles, and nuclear submarines—also consisted of prime movers (jet and rocket engines) and structures whose production and operation were highly energy-intensive.

The production of conventional weapons also requires energy-intensive materials, and their deployment is energized by secondary fossil fuels (gasoline, kerosene, diesel fuel) and the electricity used to power the machines that carry them and to equip and to provision the soldiers who operate them. While ordinary steel could be made from iron ore and pig iron with as little as 20 MJ/kg, specialty steels used in heavy armored equipment require 40–50 MJ/kg, and the use of depleted uranium (for armor-piercing shells and enhanced armor protection) is even more energy-intensive. Aluminum and titanium (and their alloys), the principal materials used to build modern aircraft, embody respectively between 170 and 250 MJ/kg (aluminum) and 450 MJ/kg (titanium), while lighter and stronger composite fibers require typically between 100 and 150 MJ/kg.

Such powerful modern war machines are obviously designed for optimized combat performance, not for minimized energy consumption, and they are extraordinarily energy-intensive. For example, America's 60 t M1/A1 Abrams main battle tank is powered by a 1.1 MW AGT-1500 Honeywell gas turbine and consumes (depending on mission, terrain, and weather) 400–800 L/100 km (*Army Technology* 2015). By comparison, a large Mercedes S600 needs about 15 L/100 km and a Honda Civic sips just 8 L/100 km. And flying at supersonic speeds (up to 1.6–1.8 Mach) such highly maneuverable combat aircraft as the F-16, Lockheed's Fighting Falcon, and F/A-18, the McDonnell Douglas Hornet, needs so much aviation fuel that their extended missions are possible only with in-flight refueling from large tanker planes, such as the KC-10, the KC-135, and the Boeing 767.

Another feature of modern warfare that demands high energy inputs is the use of weapons in massive configurations. The most concentrated tank attack during 1918 involved almost 600 machines (at that time relatively light models), but nearly 8,000 tanks, 11,000 airplanes, and more than 50,000 guns and rocket launchers were deployed by the Red Army during its final assault on Berlin in April 1945 (Ziemke 1968). As an example of the intensity of modern airfare, during the Gulf War (Operation Desert Storm, January–April 1991) and the months preceding it (Operation Desert Shield, August 1990–January 1991), some 1,300 aircraft flew more than 116,000 sorties (Gulflink 1991).

Yet another phenomenon that has greatly contributed to overall energy costs has been is the necessity to ramp up the mass production of military equipment in very short periods of time. The two world wars offer the best examples. In August 1914 Britain had only 154 military airplanes, but four years later the country's aircraft factories were employing about 350,000 people and producing 30,000 airplanes a year (Taylor 1989). When the United States declared war on Germany in April 1917 it had fewer than 300 second-rate planes, none able to carry machine guns or bombs, but three months later the U.S. Congress approved an unprecedented appropriation of $640 million (almost U.S. $12 billion in 2015 monies) to build 22,500

Liberty engines for new fighters (Dempsey 2015). And the American industrial acceleration during World War II was even more impressive.

During the last quarter of 1940 only 514 planes were delivered to the U.S. Army Air Force. In 1941 the total reached 8,723, in 1942 it was 26,448, in 1943 the total surpassed 45,000, and 1944 saw American factories complete 51,547 new planes (Holley 1964). America's aircraft production was the largest manufacturing sector of the wartime economy: it employed two million workers, required nearly a quarter of all wartime expenditures, and produced in total 295,959 airplanes, compared to 117,479 British, 111,784 German, and 68,057 Japanese airplanes (Army Air Forces 1945; Yenne 2006). Ultimately, Allied victories were to the result of their superiority in harnessing destructive energy. By 1944 the United States, the USSR, the UK, and Canada were making three times as much combat munitions as Germany and Japan (Goldsmith 1946). The increasing destructiveness of weapons and the more concentrated delivery of explosives can be illustrated by comparing both discrete events and overall conflict casualties (box 6.17).

Box 6.17
Casualties of modern wars

Combat casualties during the Battle of the Somme (July–November 19, 1916) totaled 1.043 million. Those during the Battle of Stalingrad (August 23, 1942–February 2, 1943) surpassed 2.1 million (Beevor 1998). Battle death rates—expressed as fatalities per 1,000 men of armed forces fielded at the beginning of a conflict—were below 200 during the first two modern wars involving major powers (the Crimean War of 1853–1856 and the Franco-Prussian War of 1870–1871); they surpassed 1,500 during World War I and 2,000 during World War II, and were above 4,000 for Russia (Singer and Small 1972). Germany lost about 27,000 combatants per million people during World War I but more than 44,000 during World War II.

The civilian casualties of modern warfare grew even faster. During World War II they reached about 40 million, more than 70% of the 55 million total casualties. The bombing of large cities produced huge losses within days or just hours (Kloss 1963; Levine 1992). The total German bombing casualties reached nearly 600,000 dead and almost 900,000 wounded. About 100,000 people died during nighttime raids by B-29 bombers, which leveled about 83 km^2 of Japan's four principal cities between March 10 and 20, 1945. The effects of the firebombing of Tokyo and of the nuclear attack on Hiroshima have already been described (see box 6.15).

Calculating the energy cost of major armed conflicts requires important arbitrary delimitations of what should be included in such totals. After all, societies in mortal danger do not operate two separate civilian and military sectors, for wartime economic mobilization affects nearly all activities. Available summations put the total U.S. cost of major twentieth-century conflicts at about $334 billion for World War I, $4.1 trillion for World War II, and $748 billion for the Vietnam War (1964–1972), all expressed in constant 2011 dollars (Daggett 2010). Expressing those costs in current monies and multiplying those totals by adjusted averages of prevailing energy intensities of the country's GDP would amount to defensible approximations of the minimum energy costs of those conflicts.

Adjustments are required because the wartime industrial production and transportation consumed more energy per unit of their output than did the average unit of GDP. As approximations, I chose the respective multiples of 1.5, 2, and 3 for the three conflicts. As a result, participation in World War I required about 15% of the total U.S. energy consumption in 1917 and 1918, and it averaged about 40% during World War II, but it was no more than 4% for the years of the Vietnam War. Peak shares were obviously higher, ranging from 54% for the United States in 1944 to 76% for the USSR in 1942 and a similar share for Germany in 1943.

There is no obvious correlation between overall energy use and success in waging modern acts of aggression (or preventing them). The clearest case for a positive correlation between energy outlay and a fairly swift victory is the U.S. mobilization for World War II, energized by a 46% increase in the total use of primary energy between 1939 and 1944. But in conventional sense America was even more dominant during the Vietnam War—the amount of explosives used was three times as much as all bombs dropped by the U.S. Air Force during World War II on Germany and Japan, and the United States had state-of-the art jet fighters, bombers, helicopters, aircraft carriers, and defoliants—yet it could not, for a variety of political and strategic reasons, translate that dominance into another victory.

And, of course, the absence of any correlation between energies expended and results achieved is most obviously illustrated by terrorist attacks. Completely reversing the Cold War paradigm, in which weapons were extremely expensive to produce and carefully guarded by states, terrorists use weapons that are cheap and widely available. A few hundred kilograms of ANFO (ammonium nitrate/fuel oil) for a truck bomb, a few tens of kilograms for a car bomb, or just a few kilograms of high explosives (often spiked with metal bits) fastened to the bodies of suicide bombers suffice to cause scores or even hundreds of deaths (in 1983 two truck bombs killed 307 people, mostly U.S. servicemen, in their Beirut barracks) and many more injuries, and to terrorize the targeted population.

The 19 hijackers of 9/11 had no weapons other than a few box cutters, and the entire operation, including flight lessons, cost less than $500,000 (bin Laden 2004, 3)—while even the narrowest estimate of the monetary burden (New York City's comptroller report, issued a year after the attack) put the city's direct cost at $95 billion, including about $22 billion to replace the buildings and infrastructure and $17 billion in lost wages (Thompson 2002). A national perspective evaluating lost GDP, the decline of stock values, losses by the airline and tourist industries, higher insurance and shipping rates, and increased security and defense spending put the cost at more than $500 billion (Looney 2002). Adding even a partial cost of the subsequent invasion and occupation of Iraq would raise the total well above a trillion dollars, and as the experience since the attack has shown, there is no easy military solution, as both the classic powerful weapons and the latest smart machines are of a limited use against fanatically motivated individuals or groups willing to die in suicide attacks.

There is no doubt that the mutually assured destruction (MAD) concept has been the main reason why the two nuclear superpowers have not fought a thermonuclear war, but at the same time, the magnitude of the nuclear stockpiles amassed by the two adversaries, and hence their embedded energy cost, has gone far beyond any rationally defensible deterrent level. Every step in developing, deploying, safeguarding, and maintaining nuclear warheads and their carriers (intercontinental bombers and ballistic missiles, nuclear-powered submarines) is energy-intensive. An order-of-magnitude estimate is that at least 5% of all U.S. and Soviet commercial energy that was consumed between 1950 and 1990 was claimed by developing and amassing these weapons and the means of their delivery (Smil 2004).

But even if the burden was twice as high, it might be argued the cost has been acceptable compared to the toll of a thermonuclear exchange that would have, even in the case of a limited exchange, resulted in tens of millions of casualties from the direct effects of blast, fire, and ionizing radiation (Solomon and Marston 1986). A thermonuclear exchange between the United States and the USSR limited to targeting strategic facilities would have caused at least 27 million and up to 59 million deaths during the late 1980s (von Hippel et al. 1988). Such a prospect has acted as a very powerful deterrent from launching, and after the 1960s even seriously contemplating, the first attack.

Unfortunately, the cost attributable to nuclear weapons would not cease even with their instant abolishment: their disarming and expensive safeguarding and the cleanup of contaminated production sites would continue for decades to come, and the estimated U.S. costs of these operations have been rising. It would be even more costly to clean up the more severely contaminated nuclear weapons sites in the countries of the former USSR. Fortunately, the costs of nuclear warhead decommissioning can be much reduced by reusing the recovered fissile material for electricity generation (WNA 2014).

Highly enriched uranium (HEU, containing at least 20% and up to 90% U-235) is blended down with depleted uranium (mostly U-238), natural uranium (0.7% U-235), or partially enriched uranium to produce low-enriched uranium (<5% U-235) used for power reactors. According to a 1993 agreement between the United States and Russia (megatons for megawatts), Russia converted 500 t of HEU from its warheads and strategic stockpiles (equivalent to around 20,000 nuclear bombs) to reactor-ready fuel (averaging about 4.4% U-235) and sold it to power the U.S. civilian reactors.

I cannot leave this section on energy and war without making a few remarks about energy as the casus belli. The belief in this link has been all too common, its latest iteration being the U.S. invasion of Iraq in 2003, done, we are assured, to get at Iraqi oil. And for historians, the most often cited example of the link is the Japanese attack on the United States in December 1941. Roosevelt's administration first abrogated the 1911 Treaty of Commerce and Navigation (in January 1940), then it stopped licensing for exports of aviation gasoline and machine tools (in July 1940) and followed that by a ban on exporting scrap iron and steel (in September 1940). That, according to a still far from abandoned Japanese justification, led the country with little choice but to attack the United States in order to have a free hand to assault Southeast Asia with its Sumatran and Burmese oil fields.

But Pearl Harbor was preceded by nearly a decade of expansive Japanese militarism, beginning with the 1933 conquest of Manchuria and escalating with the 1937 attack on China: Japan could have had continued access to U.S. oil if it had abandoned its aggressive China policy (Ienaga 1978). Not surprisingly, Marius Jansen, one of the leading historians of modern Japan, wrote about the peculiarly self-inflicted nature of the entire confrontation with the United States (Jansen 2000). And who would claim that Hitler's serial aggression—against Czechoslovakia (in 1938 and 1939), Poland (1939), Western Europe (beginning in 1939), and the USSR (1941)— and his genocidal war against Jews were motivated by a quest for energy resources?

Neither were there any energy-related motives for the Korean War (started on Stalin's orders), for the conflict in Vietnam (the French fighting the Communist guerrillas until 1954, the United States between 1964 and 1972), the Soviet occupation of Afghanistan (1979–1989), the U.S. war against the Taliban (launched in October 2001)—or for late twentieth-century cross-border conflicts (China-India, several rounds between India-Pakistani, Eritrea-Ethiopia, and many more) and civil wars (Angola, Uganda, Sri Lanka, Colombia). And while Nigeria's war with the secessionist Biafra (1967–1970) and Sudan's endless civil war (now transformed into the Sudan–South Sudan conflict and tribal warfare within South Sudan) had a clear oil component, both stemmed primarily from religious and ethnic enmities, and the Sudanese conflict began in 1956, decades before any oil discoveries.

Finally, we are left with the two wars in which oil has been widely seen as the real cause. The Iraqi invasion of Kuwait in August 1990 doubled the conventional crude oil reserves under Saddam Hussein's control and threatened the nearby Saudi giant oil fields (Safania, Zuluf, Marjan, and Manifa, on- and offshore just south of Kuwait) and the survival of the monarchy. But there was more at stake than oil, including the Iraqi quest for nuclear and other nonconventional weapons (in 1990 nobody doubted that) and the risks of another Arab-Israeli war (the Iraqi missile attacks on Israel were designed to provoke such a conflict). And if control of oil resources was the primary objective of the 1991 Gulf War, why was the victorious army ordered to stop its uncheckable progress, and why it did not occupy at least Iraq's richest southern oil fields?

What have been the results of the 2003 U.S. invasion of Iraq? American imports of Iraqi oil had actually peaked in 2001 when Saddam Hussein was still in control, at about 41 Mt, after the invasion they kept on declining steadily, and in 2015 they totaled less than 12 Mt, not even 3% of all U.S. imports (USEIA 2016b)—and those, of course, have been diminishing steadily as hydraulic fracturing has made the country once again the world's largest producer of crude oil and natural gas liquids (BP 2016). The verdict is simple: the United States does not need Iraqi oil, East Asia has been its largest buyer—so did the United States go into Iraq to secure Chinese oil supplies? Even the case seen by many as a clear-cut demonstration of energy-driven war is anything but! The conclusion is clear: broader strategic aims, whether well justified or misplaced, and not a quest for resources have led America into its post–World War II conflicts.

Environmental Changes

The provision and the use of fossil fuels and electricity are the largest causes of anthropogenic pollution of the atmosphere and greenhouse gas emissions and are leading contributors to water pollution and land use changes. The combustion of all fossil fuels entails, of course, rapid oxidation of their carbon, which produces increasing emissions of CO_2, while methane (CH_4), a more potent greenhouse gas, is released during the production and transportation of natural gas; small volumes of nitrous oxide (N_2O) are also released from fossil fuel combustion. The combustion of coal used to be a large source of particulate matter and sulfur and nitrogen oxides (SO_x and NO_x), but the stationary emissions of these gases are now largely controlled by electrostatic precipitators, desulfurization, and NO_x-removal processes (Smil 2008a). Even so, emissions from coal combustion continue to have significant health impacts (Lockwood 2012).

Water pollution arises mainly from accidental oil spills (from pipelines, rail cars, barges and tankers, refineries) and acid mine drainage. Major land use changes are caused by surface coal mining, by reservoirs created by major hydroelectric dams, by right-of-way corridors for high-voltage transmission lines, by building extensive storage, refining, and distribution facilities for liquid fuels, and, most recently, by construction of large wind and solar farms. Indirectly, fuels and electricity are responsible for many more pollution flows and ecosystemic degradations. The most notable ones arise from industrial production (above all from ferrous metallurgy and chemical syntheses), agricultural chemicals, urbanization, and transportation. These impacts have been increasing in both extent and intensity and affecting the environment on scales ranging from local to regional. Their costs have been forcing all major economies to devote growing attention to environmental management.

By the 1960s one of these degradations, acid deposition in Central and Western Europe and in eastern North America, created mostly by emissions of SO_x and NO_x from large coal-fired power plants but also by automotive emissions, reached semicontinental scale and until the mid-1980s was widely seen as the most pressing environmental problem facing affluent countries (Smil 1985, 1997). A combination of actions—the switch to low-sulfur coal and to sulfur-free natural gas in electricity generation, the use of cleaner gasoline and diesel and more efficient car engines, and the installation of flue gas desulfurization at major pollution sources—not only arrested the acidification process but by 1990 had reversed it, and precipitation in Europe and North America became less acid (Smil 1997). But the problem has reoccurred since 1990 in East Asia following China's large post-1980 increase in coal combustion.

The partial destruction of the ozone layer above Antarctica and the surrounding ocean briefly assumed the top spot among environmental concerns associated with energy use. The possibility of reduced concentrations of stratospheric ozone protecting the planet from excessive ultraviolet radiation was accurately foreseen in 1974, and the phenomenon was first measured above Antarctica in 1985 (Rowland 1989). Ozone loss has been caused largely by releases of chlorofluorocarbons (CFCs, used mostly as refrigerants), but an effective international treaty, the Montreal Protocol, signed in 1987, and a switch to less harmful compounds soon eased the worries (Andersen and Sarma 2002).

The threat to stratospheric ozone was only the first of several new concerns about the global consequences of environmental change (Turner et al. 1990; McNeill 2001; Freedman 2014). Prominent worries have ranged from the loss of global biodiversity to plastic accumulation in the oceans, but one global environmental concern has been paramount since the late 1980s: anthropogenic emissions of greenhouses gases causing relatively rapid climate change, above all tropospheric warming and ocean acidification and sea-level rise. The behavior of greenhouse gases and their likely warming effect were fairly well understood by the end of the nineteenth century (Smil 1997). The leading anthropogenic contributor is CO_2, the end-product of the efficient combustion of all fossil and biomass fuels, and the destruction of forests (above all in wet tropics) and grasslands has been the second most important source of CO_2 emissions (IPCC 2015).

Since 1850, when it was just 54 Mt C (multiply by 3.667 to convert to CO_2), the global anthropogenic generation of CO_2 has been rising exponentially with the increasing consumption of fossil fuels: as already noted, by 1900 it had risen to 534 Mt C and in 2010 it surpassed 9 Gt C (Boden and Andres 2015). In 1957 Hans Suess and Roger Revelle concluded that

> human beings are now carrying out a large scale geophysical experiment of a kind that could not have happened in the past nor be reproduced in the future. Within a few centuries we are returning to the atmosphere and oceans the concentrated organic carbon stored in sedimentary rocks over hundreds of millions of years. (Revelle and Suess 1957, 19)

The first systematic measurements of rising background CO_2 levels, organized by Charles Keeling (1928–2005), began in 1958 near the summit of Mauna Loa in Hawaii and at the South Pole (Keeling 1998). Mauna Loa concentrations have been used as the global marker of rising tropospheric

CO_2: they averaged almost 316 ppm in 1959, surpassed 350 ppm in 1988, and were 398.55 ppm in 2014 (NOAA 2015; fig. 6.22). Other greenhouse gases are emitted by human activities in much smaller volumes than CO_2, but because their molecules absorb relatively more of the outgoing infrared radiation (methane 86 times as much over 20 years, nitrous oxides 268 times as much as CO_2), their combined contribution now accounts for about 35% of anthropogenic radiative forcing (box 6.18).

The consensus position is that, to avoid the worst consequences of global warming, the average temperature rise should be limited to less than 2°C, but this would require immediate and substantial curtailment of fossil fuel combustion and a rapid transition to noncarbon sources of energy—not an impossible but a highly unlikely development, given the dominance of fossil fuel in the global energy system and the enormous energy requirements of low-income societies: some of those large new needs can come from renewable electricity generation, but there is no affordable, mass-scale alternative available for transportation fuels, feedstocks (ammonia, plastics), or iron ore smelting.

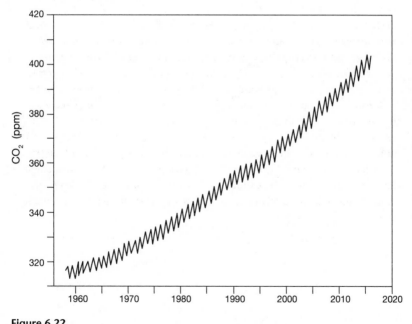

Figure 6.22
Atmospheric CO_2 measured at Mauna Loa Observatory in Hawaii (NOAA 2015).

Box 6.18
Greenhouse gases and rising tropospheric temperature

In 2014 the global rate of anthropogenic radiative forcing (capacity of green-house gases to affect the planet's energy balance) reached 2.936 W/m^2, with CO_2 contributing 65% (Butler and Montzka 2015). As for the sources, fossil fuels account for more than 60%, land use changes (mainly deforestation) for 10%, and methane emissions (mainly from livestock) for about 20%. The globally averaged surface temperature increase (combined data for ocean and land) shows a linear rise of 0.85°C (0.65–1.06°C) between 1880 and 2012 (IPCC 2015). Uncertainties regarding the future level of global emissions and the complexity of atmospheric, hydrospheric, and biospheric processes and interactions governing the global carbon cycle make it impossible to construct reliable models forecasting temperature and sea-level rises for the year 2100. The latest consensus assessment shows that (depending largely on future emission rates) by the end of the twenty-first century (2081–2100), the average global temperature will be at least 0.3–1.7°C higher than during 1986–2005, but that it may rise by as much as 2.6–4.8°C (IPCC 2015).

In any case, the Arctic region will continue warming more rapidly. Obviously, the lower rates would make it easier to adapt, while the highest likely increases would pose many serious problems. The multitude of changes attributable to global warming range from new precipitation patterns, coastal flooding, and shifts in ecosystem boundaries to the spread of warm-climate, vector-borne diseases. Changes in plant productivity, loss of near-shore real estate, sectoral unemployment, and large-scale migration from affected regions would be the key economic consequences. There is no easy technical fix (such as CO_2 capture from the air or CO_2 storage underground, both required to handle, at an affordable cost, more than 10 Gt CO_2/year in order to be effective) for anthropogenic greenhouse gas emissions. The only poten-tially successful approach to deal with these changes is through unprecedented international cooperation. Unintentionally, this worrisome challenge also offers a fundamental motivation for a new departure in managing human affairs.

7 Energy in World History

All natural processes and all human actions are, in the most fundamental physical sense, transformations of energy. Civilization's advances can be seen as a quest for higher energy use required to produce increased food harvests, to mobilize a greater output and variety of materials, to produce more, and more diverse, goods, to enable higher mobility, and to create access to a virtually unlimited amount of information. These accomplishments have resulted in larger populations organized with greater social complexity into nation-states and supranational collectives, and enjoying a higher quality of life. Outlining the milestones of this history in terms of dominant energy sources and leading prime movers is, as I hope this book demonstrates, fairly straightforward. Nor is it difficult to recount the most important socioeconomic consequences of these technical changes.

What is much more challenging is to find a sensible balance between seeing history through the prism of energy imperatives and paying proper attention to the multitude of nonenergy factors that have always initiated, controlled, shaped, and transformed human use of energy. Even more fundamentally, it is also necessary to note the basic paradox of energy's role in life's evolution in general, and in human history in particular. All living systems are sustained by incessant imports of energy, and this dependence necessarily introduces a number of fundamental constraints. But these life-sustaining energy flows cannot explain either the very existence of organisms or the particular complexities of their organization.

Grand Patterns of Energy Use

The long-term relationship between human accomplishments and dominant energy sources and changing prime movers is perhaps best revealed when viewed in terms of energy eras and transitions. This approach must eschew rigid periodization (as some transitions unfolded very slowly) and it must recognize that generalizations regarding specific periods must take into account differences in the onset and pace of key underlying processes: perhaps the best recent example is China's exceptionally rapid post-1990 development, accomplishing in a single generation what took many nations during the earlier stages of industrialization three generations. There are also many national and regional particularities driving and shaping such complex changes.

The most obvious uniformities dictated by specific energy eras could be seen in activities pertaining to the extraction, conversion, and distribution of energies. Human muscles and harnessed oxen put very similar limits on the extent of land that could be planted or harvested in a day, be it in Punjab or in Picardy; the charcoal yield from traditional piles in Tohoku (northern Honshu) differed little from that in Yorkshire (northern England). In modern global civilization these commonalities have become absolute identities: the same energy sources and the same prime movers are now managed, extracted, and converted worldwide with the same processes and machines, and most of them are often produced or deployed by a small number of globally dominant companies.

Examples of such global companies include Schlumberger, Halliburton, Saipem, Transocean, and Baker for oil field services; Caterpillar, Komatsu, Volvo, Hitachi, and Liebherr for heavy construction machinery; General Electric, Siemens, Alstom, Weir Allen, and Elliott for large steam turbines; and Boeing and Airbus for large jet airliners. As the reach of services and products offered by these firms became truly global, former international differences in performance and reliability have been greatly reduced or even entirely eliminated, and in some cases the late starters now have higher shares of advanced techniques than do the pioneering industrializers. And despite large differences in cultural and political settings there is also a surprisingly wide scope for generalizing about the socioeconomic consequences of these fundamental energetic changes.

Because the most rewarding exploitation of identical energy sources and prime movers requires the same techniques, this uniformity also imposes many identical, or very similar, imprints not only on crop cultivation (leading to the dominance of a few commercial crops and the mass production of animal foods), industrial activities (entailing specialization, concentration, and automation), the organization of cities (leading to the rise of downtown business districts, suburbanization, and subsequently the desirability of green spaces), and transportation arrangements (in large cities manifesting as the need for subways, suburban trains, commuting by car, and taxi fleets) but also on consumption patterns, leisure activities, and intangible aspirations.

In every mature high-energy society, and in the urban areas of many still relatively fast-growing economies, TVs, refrigerators, and washing machines are owned by more than 90% of households, and other items with high rates of ownership range from personal electronic devices to air conditioning units and passenger vehicles. Globally shared food consumption trends include internationalization of tastes (tikka masala being the most popular food item in England, kare raisu in Japan), the popularization of fast food, and the year-round availability of seasonal fruit and vegetables, a convenience bought at a significant energy cost of intercontinental shipments in refrigerated containers and airborne deliveries. Among the now universal leisure activities are flights to warm beaches, visits to theme parks (American Disneylands are now in France, China, Hong Kong, and Japan), and traveling on cruise ships (cruising, formerly a European and American pastime, is now experiencing its fastest growth in Asia). And, taking this a step further, shared energy foundations eventually affect many intangible aspirations, especially for advanced (and elite) education.

But what recurs again and again is the enormous gap between the low-income societies (whose energetic foundations are an amalgam of traditional biomass fuels and animate prime movers and increasing shares of fossil fuels and electricity) and high-energy (industrialized or postindustrial) countries whose per capita consumption of fossil fuels and electricity has reached, or closely approached, saturation levels. This gap can be seen at every level, when looking at the overall economic output or at the average standard of living, at labor productivities or access to education. And this gap is becoming less a matter of international disparities and more a divide based on privilege (access, education, opportunity), the reality best illustrated by the affluent class in China and India. In 2013 one branch of China's Sports Car Club required its members to own a car better than the $440,000 Porsche Carrera GT (Taylor 2013), while Asia's most expensive private residential building, Mukesh Ambani's 27-story $2 billion skyscraper in downtown Mumbai, has an unimpeded view of sprawling slums.

Energy Eras and Transitions

Any realistic periodization of human energy use must take into account both the dominant fuels and the leading prime movers. This need disqualifies the two conceptually appealing divisions of history into just two distinct energy eras. Animate versus inanimate contrasts the traditional societies, in which human and animal muscles were the dominant prime movers, with modern civilization, dependent on fuel- and electricity-powered machines. But this division misleads both about the past and the present. In a number of old high cultures two classes of inanimate prime movers, waterwheels and windmills, were making critical differences centuries before the advent of modern machines.

And the rise of the West owes a great deal to a powerful combination of two inanimate prime movers: to effective harnessing of wind and to the adoption of gunpowder, embodied by oceangoing sail ships equipped with heavy guns (McNeill 1989). Moreover, the cleavage between animate and inanimate prime movers has been fully accomplished only among the richest fifth of the humanity. Substantial reliance on heavy human and animal labor is still the norm in the poorest rural areas of Africa and Asia, and exhaustive (and often risky) manual tasks are performed daily by hundreds of millions of workers in many extractive, processing, and manufacturing industries of low-income countries (ranging from crushing stones to make gravel to dismantling old oil tankers).

The second simplification, the use of renewable versus nonrenewable energy sources, captures the basic dichotomy between the millennia dominated by animate prime movers and biomass fuels and the more recent past, heavily dependent on fossil fuels and electricity. Once again, actual developments have been more complex. Biomass supply in wooden-era societies was not a matter of assured renewability: excessive tree cutting followed by destructive soil erosion on vulnerable slope lands destroyed the conditions for sustainable forest growth over large areas of the Old World, especially around the Mediterranean Sea and in North China. And, in today's fossil fuel–dominated world, water power, a renewable resource, generates roughly one-sixth of all electricity, while (as just noted) most farmers in poor countries still rely on human and animal labor for field work and for the maintenance of irrigation systems.

Clear divisions into specific energy eras are unrealistic not only because of obvious national and regional differences at the time of innovation and the widespread adoption of new fuels and prime movers but also because of the evolutionary nature of energy transitions (Melosi 1982; Smil 2010a). Established sources and prime movers can be surprisingly persistent, and new supplies or techniques may become dominant only after long periods of gradual diffusion. A combination of functionality, accessibility, and cost explains most of this inertia. As long as the established sources or prime movers work well within established settings, are readily available, and are profitable, their substitutes, even those with some clearly superior attributes, will advance only slowly. Economists might see these realities as examples of lock-in or path dependence as conceptualized by David (1985), who based his argument on a takedown of the QWERTY keyboard (as opposed to a supposedly superior Dvorak layout).

But we do not need any new questionable labels to describe what is a very common process of slow, evolutionary progression noticeable in organismic evolution and personal decision making, as well as in technical advances and economic management. Examples from energy history abound. Roman water mills were first used during the first century BCE, but they became really widespread only about 500 years later. Even then their use was almost completely limited to grain grinding. As Finley (1965) noted, freeing slaves and animals from their drudgery was not a powerful enough incentive for the rapid introduction of water mills. By the end of the sixteenth century circumnavigation of the Earth by sail ships had become almost commonplace—but in 1571 in the battle of Lepanto each side used more than 200 galleys, in 1588 the Spanish Armada set to invade England still had four large galleys and four galleasses manned by more than 2,000 convicted oarsmen, and heavily gunned Swedish galleys were used to destroy most of the Russian fleet at Svensksund in 1790 (Martin and Parker 1988; Parker 1996).

Draft animals, water power, and steam engines coexisted in industrializing Europe and North America for more than a century. In the wood-rich United States, coal surpassed fuelwood combustion, and coke became more important than charcoal, only during the 1880s (Smil 2010a). Mechanical power in farming topped horse and mule power only during the late 1920s, there were still millions of mules in the U.S. South in the early 1950s, and the U.S. Department of Agriculture stopped counting working animals only in 1963. And during World War II the mass-produced Liberty (EC2) class ships, the dominant U.S. cargo carriers, were not powered by new, efficient diesel engines but by well-proven three-cylinder steam engines supplied by oil-fired boilers (Elphick 2001).

Only suggestive approximations are possible in charting long-term patterns of prime mover deployment in the Old World's preindustrial societies. Their most remarkable feature is the long dominance of human labor (fig. 7.1). Human muscles were the only source of mechanical energy from the beginning of hominin evolution until the domestication of draft animals, which started only about 10,000 years ago. Human power was increased by using a growing number of better tools, while animal work throughout the Old World remained limited for millennia by poor harnessing and inadequate feeding, and draft beasts were absent in the Americas and Oceania. Human muscles thus remained indispensable prime movers in all preindustrial societies.

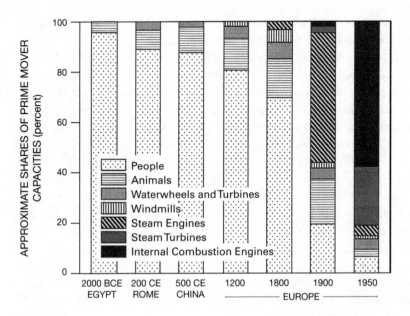

Figure 7.1
The prolonged dominance of human labor, the slow diffusion of water- and wind-driven machines, and the rapid post-1800 adoption of engines and turbines are the three most remarkable features in the history of prime movers. Approximate ratios are estimated and calculated from a wide variety of sources cited in this book.

A remarkable dichotomy characterized the use of human labor in all ancient civilizations. In contrast to its massed deployment to accomplish remarkable feats of heavy construction, old high cultures, whether based on slave, corvée, or mostly free labor, never took the steps to the really large-scale manufacture of goods. The atomization of production remained the norm (Christ 1984). Han Chinese mastered some potentially large-scale production methods. Perhaps most notably, they perfected the casting of iron suitable for mass-producing virtually identical multiple pieces of small metal articles from a single pouring (Hua 1983). But the largest discovered Han kiln was just 3 m wide and less than 8 m long. Outside Europe and North America, relatively small-scale, artisanal manufactures remained the norm until the twentieth century. The lack of inexpensive land transportation was obviously a major factor militating against mass production.

The costs of distribution beyond a relatively small radius would have surpassed any economies of scale gained by centralized manufacturing. And many ancient construction projects did not really require extraordinarily massive labor inputs either. Several hundred to a few thousands corvée laborers working for only two to five months every year could erect enormous religious structures or defensive walls, dig long irrigation and transportation canals, and build extensive dikes over a period of just 20–50 years. But many stupendous projects were under construction for much longer periods. Ceylon's Kalawewa irrigation system took about 1,400 years to build (Leach 1959). Piecemeal construction and repairs of China's Great Wall extended over an even longer period of time (Waldron 1990). And a century or two was not an exceptionally long time to finish a cathedral.

The first inanimate prime movers begun to make notable differences in some parts of Europe and Asia only after 200 CE (water mills) and 900 CE (windmills). Gradual improvements of these devices replaced and sped up many tiresome, repetitive tasks, but the substitution for animate labor was slow and uneven (fig. 7.2). In any case, except for water pumping, waterwheels and windmills could do little to ease field tasks. That is why Fouquet's (2008) approximate calculations for England show human

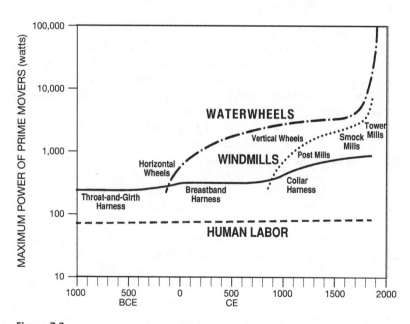

Figure 7.2
The average unit power of traditional prime movers remained limited even after the adoption of larger waterwheels at the beginning of the early modern era. The change came only with steam engines of the nineteenth century. Maximum capacities are plotted from prime mover–specific sources cited in this book.

and animal exertion accounting for 85% of all power in 1500 and still 87% in 1800 (when water and wind delivered about 12%)—but only 27% by 1900: by that time steam had taken over in industries. But even during the steam engine era animate labor remained indispensable for extracting and distributing fossil fuels and in countless manufacturing tasks; and in farming it dominated field work throughout the nineteenth century (box 7.1).

But long before the maximum power of working animals was itself tripled (by strong horses with collar harnesses), waterwheels had become the most powerful prime movers. Their subsequent development was slow: the first tenfold increase in highest capacities took about 1,000 years, the second one about 800. Their peak unit power was finally surpassed by steam engines of the late eighteenth century, but their dominance ended only with the introduction and growth of internal combustion engines and

---

Box 7.1
Persistence of animate power

In the Americas, horse, mule, and ox teams also converted most of the currently cultivated land by plowing up the extensive grasslands of the U.S. Great Plains, the Canadian prairies, the Brazilian cerrado, and the Argentinian pampas during the closing decades of the nineteenth century and at the very beginning of the twentieth century. Only by 1963, when America's tractor power was nearly 12 times the record draft animal capacity of 1920, did the U.S. Department of Agriculture stop counting draft animals. In late dynastic and early republican China, the contribution of windmills, water mills, and steam remained marginal in comparison with that of human labor, whose aggregate power also greatly surpassed that of draft animals. My best estimate was that even by 1970, China's human labor contributed about 200 PJ of useful energy, compared to just over 90 PJ for the country's draft animals (Smil 1976).

The dominance of human muscles limited the most commonly deployed single labor units to 60–100 W of sustained (daylong) useful labor. This means that in all but a few exceptional circumstances, the highest power concentrations of human labor under a single command (hundreds to thousands of laborers at construction sites) reached no more than 10,000–100,000 W during sustained effort, although brief peaks were multiples of those rates. A traditional master architect or a canal builder thus controlled energy flows equivalent to no more than those produced by a single engine powering today's small earth-moving machines.

steam turbines, both of which entered use during the 1880s, became dominant by the 1920s, and remain the leading mobile and stationary prime movers, respectively, of the early twenty-first century.

Despite some important continental and regional differences, typical levels of fuel consumption and the prevailing modes of prime mover use in old high cultures were fairly similar. If there is an ancient society to be singled out for its notable advances in fuel use and prime mover development, it must be Han China (207 BCE–220 CE). Its innovations were adopted elsewhere only centuries, or even more than a millennium, later. The most notable contributions of the Han Chinese were the use of coal in iron making, drilling for natural gas, the making of steel from cast iron, the widespread use of curved moldboard iron plows, the beginning use of collar harness, and use of the multitube seed drill. There was no similar cluster of such key advances for more than a millennium.

Early Islam brought innovative designs for water-raising machines and windmills, and the realm's maritime trade benefited from an effective use of triangular sails. But the Islamic world did not introduce any radical innovations in fuel use, metallurgy, or animal harnessing. Only medieval Europe, borrowing eclectically from earlier Chinese, Indian, and Muslim accomplishments, began to innovate in a number of critical ways. What really set European medieval societies apart in terms of energy use was their rising reliance on the kinetic energies of water and wind. These flows were harnessed by increasingly more complex machines, providing unprecedented concentrations of power for scores of applications. By the time of the first great Gothic cathedrals the largest waterwheels rated close to 5 kW, an equivalent of more than three score men. Long before the Renaissance some European regions became critically dependent on water and wind, first for their grain milling, then for cloth fulling and iron metallurgy; and this dependence also contributed to the sharpening and diffusion of many mechanical skills.

Late medieval and early modern Europe was thus a place of broadening innovation, but, as attested by reports of contemporary European travelers admiring the riches of the Heavenly Empire, the overall technical prowess of contemporary China was certainly more impressive. Those travelers could not know how soon the reverse would be true. By the end of the fifteenth century Europe was on a road of accelerating innovation and expansion while the elaborate Chinese civilization was about to start its long and deep technical and social involution. Western technical superiority did not take very long to transform European societies and extend their reach to other continents.

By 1700 the Chinese and European levels of typical energy use, and hence of average material affluence, were still broadly similar. By the mid-eighteenth century the real incomes of building workers in China were similar to those in less developed parts of Europe but lagged behind the continent's leading economies (Allen et al. 2011). Then the Western advances gathered speed. In the energy realm they were demonstrated by the combination of rising crop yields, new coke-based iron metallurgy, better navigation, new weapon designs, a keenness for trade, and the pursuit of experimentation. Pomeranz (2002) argued that this takeoff had less to do with institutions, attitudes, or demography in the core economic regions of Europe and China than with the fortuitous location of coal, and with the very different relationships between these cores and their respective peripheries, as well as with the process of invention itself.

Others saw the foundations of this success going back to the Middle Ages. Christianity's favorable effect on technical advances in general (including a critical concept of the dignity of manual labor), and medieval monasticism's quest for self-sufficiency in particular, were important ingredients of this process (White 1978; Basalla 1988). Even Ovitt (1987), who questions the importance of these links, acknowledges that the monastic tradition, by upholding the fundamental dignity and spiritual usefulness of labor, was a positive factor. In any case, by 1850 the most economically advanced parts of China and Europe belonged to two different worlds, and by 1900 they were separated by an enormous performance gap: Western European energy use was at least four times the Chinese mean.

The period of very rapid advances after 1700 was ushered in by ingenious practical innovators. But its greatest successes during the nineteenth century were driven by close feedbacks between the growth of scientific knowledge and the design and commercialization of new inventions (Rosenberg and Birdzell 1986; Mokyr 2002; Smil 2005). The energy foundations of nineteenth-century advances included the development of steam engines and their widespread adoption as both stationary and mobile prime movers, iron smelting with coke, the large-scale production of steel, and the introduction of internal combustion engines and of electricity generation. The extent and rapidity of these changes came from a novel combination of these energy innovations with new chemical syntheses and with better modes of organizing factory production. Aggressive development of new ways of transport and telecommunication was also essential, both for boosting production and for promoting national and international trade.

By 1900, the accumulation of technical and organizational innovations gave the West, now including the new power of the United States, command over an unprecedented share of global energy. With only 30% of the world's population the Western nations consumed about 95% of fossil fuels. During the twentieth century the Western world increased its total energy use nearly 15 times. Inevitably, its share of global energy use declined, but by the end of the twentieth century the West (the EU and North America), with less than 15% of the global population, consumed nearly 50% of all primary commercial energy. Europe and North America remain the dominant consumers of fuels and electricity in per capita terms and have retained technical leadership. China's rapid economic growth changed the absolute ranking: the country became the world's largest energy consumer in 2010, by 2015 it was about 32% ahead of the United States, but its per capita energy use was only a third of the U.S. mean (BP 2016).

Only rough approximations are possible in presenting long-term patterns of the Old World's primary energy consumption (fig. 7.3). In the UK, coal had displaced wood already during the seventeenth century; in France and Germany wood receded rapidly in importance only after 1850; and in Russia, Italy, and Spain biomass energies remained dominant into the twentieth century (Gales et al. 2007; Smil 2010a). Once the basic energy statistics become available, it is possible to quantify the transitions and to discern long substitution waves (Smil 2010a; Kander, Malanima, and Warde 2013). In global terms this can be done with fair accuracy since the middle of the nineteenth century (fig. 7.3). Substitution rates have been slow, but, considering the variety of intervening factors, they have been surprisingly similar.

My reconstruction of global energy transitions shows coal (replacing wood) reaching 5% of the global market around 1840, 10% by 1855, 15% by 1865, 20% by 1870, 25% by 1875, 33% by 1885, 40% by 1895, and 50% by 1900 (Smil 2010a). The sequence of years needed to reach these milestones was 15–25–30–35–45–55–60. The intervals for oil replacing coal, with 5% of the global supply reached in 1915, were virtually identical: 15–20–35–40–50–60 (oil will never reach 50%, and its share has been declining). Natural gas reached 5% of the global primary supply by 1930 and 25% of it after 55 years, taking significantly longer to reach that share than coal or oil.

The similar progress of three global transitions—it takes two or three generations, or 50–75, years for a new resource to capture a large share of the global energy market—is remarkable because the three fuels require

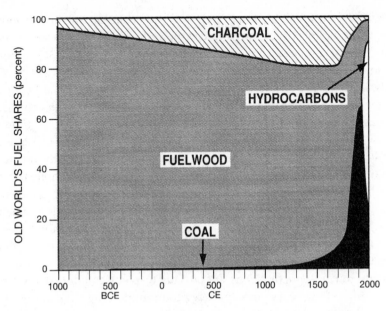

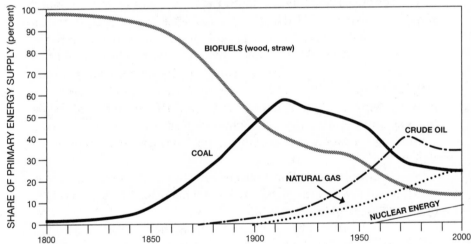

Figure 7.3
Approximate estimates chart the shares contributed by major fuels to the Old World's
primary energy supply during the past 3,000 years (top). Reasonably accurate (except
for the consumption of traditional biomass fuels) post-1850 statistics reveal successive waves of slow energy transitions (bottom): by 2010 crude oil was the leading
fossil fuel, but coal and natural gas were not far behind. Plotted from data in UNO
(1956) and Smil (2010a).

different production, distribution, and conversion techniques and because the scales of substitutions have been so different: going from 10% to 20% for coal required increasing the fuel's annual output by less than 4 EJ, whereas going from 10% to 20% of natural gas needed roughly an additional 55 EJ/year (Smil 2010a). The two most important factors explaining the similarities in the pace of transitions are the prerequisites for enormous infrastructural investment and the inertia of massively embedded energy systems.

Although the sequence of three substitutions does not mean that the fourth transition, now in its earliest stage (with fossil fuels being replaced by new conversions of renewable energy flows), will proceed at a similar pace, the odds are highly in favor of another protracted process. In 2015 the two new renewable ways of electricity generation, solar (at 0.4%) and wind (at 1.4%), were still below 2% of the world's primary energy supply (BP 2016). Two early breakthroughs would accelerate the shift: swift construction of new nuclear plants based on the best available designs, and the availability of new, inexpensive ways to store wind and solar electricity on massive scales. And even then we would still face the challenges of replacing billions of tonnes of high energy-density liquid fuels in transportation, and producing pig iron, cement, plastics, and ammonia without any fossil carbon.

Long-Term Trends and Falling Costs
Secular transitions to more powerful prime movers can be traced quite accurately in terms of both typical and maximum capacities (fig. 7.4). The power envelope connecting the peak prime mover capacities moved from roughly 100 W of sustained human labor to about 300–400 W for draft animals sometime during the third millennium BCE; then the line rose to about 5,000 W (5 kW) for horizontal waterwheels by the end of the first millennium of the Common Era. By 1800 it had surpassed 100,000 W (100 kW) in steam engines, and they remained by far the most powerful units until the middle of the nineteenth century, when water turbines gained a short-lived primacy between 1850 and 1910 (reaching 10 MW). Afterward steam turbines have been the most powerful single-unit prime movers, reaching a plateau at more than 1,000,000,000 W (1 GW) in the largest units installed after 1960.

A different perspective is gained by looking at total prime mover capacities. After 1700 the basic global pattern can be reasonably approximated, and accurate historical statistics make the retrospective easy for the United States (fig. 7.5). In 1850 animate labor still accounted for more than 80% of

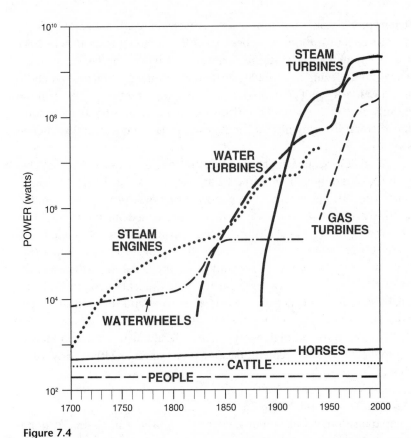

Figure 7.4
Maximum capacities of prime movers predating 1700 and those introduced during the past three centuries. The largest turbogenerators are now six orders of magnitude (nearly two million times) more powerful than heavy draft horses, the most powerful animate prime movers. Waterwheel ratings were surpassed by steam engines before 1750, by 1850 water turbines had become briefly the most powerful prime movers, and steam turbines have been the most powerful prime movers ever since the second decade of the twentieth century. Plotted from data cited in the sections concerning specific prime movers.

the world's prime mover capacity. Half a century later its share was about 60%, with steam engines supplying about one-third. By the year 2000 all but a small fraction of the world's available power was installed in internal combustion engines and electricity generators. U.S. prime mover substitutions predated these global changes. Of course, internal combustion engines (be they in vehicles, tractors, combines, or pumps) are rarely deployed in such a sustained manner as electricity generators. Automobiles and farm machines usually operate less than 500 hours a year, compared to over

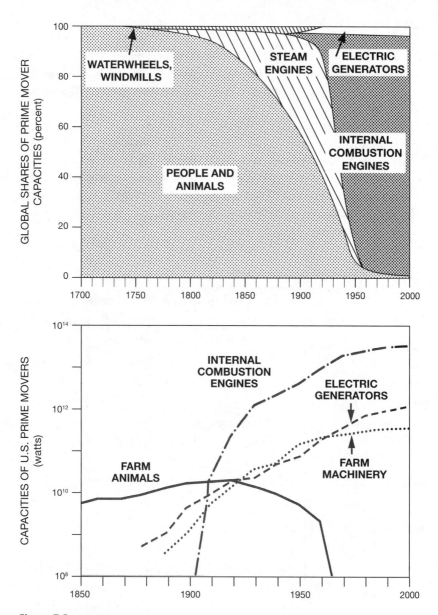

Figure 7.5

Global prime mover shares in 1700 were only marginally different from those of 500 or even 1,000 years ago. In contrast, by 1950 all but a very small fraction of the world's available power was installed in internal combustion engines (mostly in passenger cars) and in steam and water turbines (top). Disaggregated U.S. statistics (bottom) show this rapid transformation with greater detail and accuracy. The global rates were estimated and plotted from data in UNO (1956), Smil (2010a), and Palgrave Macmillan (2013); the bottom graph was plotted from data in USBC (1975) and from subsequent issues of *The Statistical Abstract of the United States*.

5,000 hours for turbogenerators. Consequently, in terms of actual energy production, the global ratio between internal combustion engines and electricity generators is now about 2:1.

Two important general trends have accompanied the growth of unit power of inanimate prime movers and the accumulation of their total capacity: their mass/power ratios have declined (producing more power from smaller units), and their conversion efficiencies have increased (producing more useful work from the same amount of initial energy input). The first trend has brought progressively lighter and hence more versatile fuel converters (fig. 7.6). The earliest steam engines, while much more powerful than horses, were exceedingly heavy because their mass/power

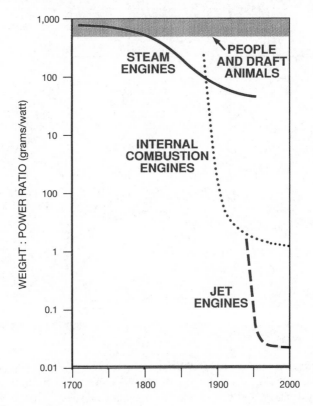

Figure 7.6

Every new inanimate energy converter became eventually much lighter and more efficient. The steady decline in the weight/power ratio of leading prime movers means that the best internal combustion engines now weigh less than 1/1,000 of what the equally powerful draft animals or early steam engines weighed. Plotted from data cited throughout this book.

ratio was of the same order of magnitude as that of draft animals. More than two centuries of subsequent development lowered the mass/power ratio of steam engines to about one-tenth of initial values, still too high for the engines to serve on roads or to power flight.

The mass/power ratio of internal combustion engines (first gasoline engines, then diesels) declined by two orders of magnitude in less than 50 years after the first commercial models, horizontal engines fueled by coal gas, were introduced during the 1860s. That precipitous drop opened the way to the affordable mechanization of road transport (cars, buses, trucks) and made aviation possible. Starting in the 1930s (both for stationary use and in flight), gas turbines carried these improvements by almost another two orders of magnitude, making speedy jet-powered air travel possible, starting in 1958 and on a mass scale after the introduction of wide-body jets, with the Boeing 747 leading the way in 1969. Concurrently, gas turbines have emerged as a leading choice for flexible and clean electricity generation.

The efficiencies of prime movers are limited by fundamental thermodynamic considerations. Technical advances have been narrowing the gaps between best performances and theoretical maxima. The efficiencies of steam-driven machines rose from a fraction of a percent for Savery's primitive engines to just over 40% for large turbogenerators of the early twenty-first century. Only marginal improvements are now possible for turbogenerators, whether steam- or water-driven, but combined-cycle gas turbines can reach efficiencies of 60%. Similarly, the best combustors now perform close to theoretical limits. Both large power plant boilers and household natural gas furnaces may be up to 97% efficient. In contrast, the everyday performances of internal combustion engines, the prime movers with the largest aggregate installed power, are still very low. Poorly maintained car engines often perform at less than a third of their rated maxima. Improvements in lighting efficiency have been even more impressive (box 7.2).

More powerful yet more efficient and lighter mechanical prime movers have increased the typical speeds of long-distance travel by more than tenfold on land and on water, and made flying possible (fig. 7.7). In 1800 horse-drawn coaches usually covered less than 10 km/h, while heavy freight wagons moved at half that speed. In the year 2000 highway traffic could flow at speeds above 100 km/h and high-speed passenger trains ran at speeds approaching, or even surpassing, 300 km/h, while the standard cruising speed for jet planes is 880–920 km/h at about 11 km above the

---

Box 7.2
Efficiency and efficacy of lighting

Candles convert as little as 0.01% and no more than 0.04% of the chemical energy of the burning wax, tallow, or paraffin into light. Edison's first light bulbs, which used oval loops of carbonized paper secured by platinum clamps to platinum wires sealed through the glass, converted 0.2%, an order of magnitude better than candles but no better than contemporary gas lights (0.15–0.3%). Osmium filaments, introduced in 1898, converted nearly 0.6% of electric energy into light. This rate was more than doubled after 1905 with tungsten filaments in a vacuum, and then doubled again with inert gas in bulbs. In 1939 the first fluorescent light pushed the efficiency above 7%, and the rates rose well above 10% after World War II (Smil 2006).

But the best way to appreciate these gains is in terms of luminous efficacy. This ratio of luminous and radiant flux (expressed in lumens per watt, lm/W) measures the efficiency with which a source of radiant energy produces visible light, and its maximum is 683 lm/W. Here are the ascending luminous efficacies, all in lm/W (Rea 2000): candle, 0.3; gas light, 1–2; early incandescent light bulbs, less than 5; modern incandescent lights, 10–15; fluorescent lights, up to 100. Low-pressure sodium lamps are currently the most efficient commercial source of light (with maxima just above 200 lm/W), but their yellowish light is used only for street lighting. Light-emitting diodes, suitable for any indoor applications, already deliver close to 100 lm/W, and soon they will go above 150 lm/W (USDOE 2013).

---

ground. Increasing speeds have been accompanied by growing capacities and ranges in transporting both goods and people.

On land, this mechanical evolution has recently peaked with multi-axle trucks, unit trains (carrying up to 10,000 t of bulk materials), and fast electric passenger trains (for up to 1,000 people). Supertankers move up to 500,000 t of crude oil, while the largest passenger planes, the Boeing 747 and the Airbus 380, carry about 500 people and the largest cargo plane, the Antonov 225, can lift 250 t. The range increase has been equally impressive: the greatest distance that can be covered by a passenger car without refueling is now just over 2,600 km—the record made in 2012 with the diesel-powered Volkswagen Passat TDI (Quick 2012)—and the Boeing 777-200LR can fly more than 17,500 km.

The increased speed and range of passenger and goods transportation had its destructive counterpart in the increased speed, range, and effective power of projectiles released by weapons. The killing range of spears was

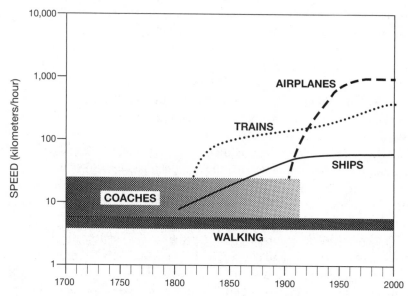

Figure 7.7

Maximum speeds of passenger transportation rose from less than 20 km/h for coach-es of the pre-railway era to well over 100 km/h after just a few decades of better loco-motive designs. Modern rapid trains commonly travel at 200–300 km/h and jetliners cruise at speeds just above 900 km/h. Plotted from data in numerous references cited in the book's sections dealing with transportation.

just a few tens of meters; an expertly fielded spear-thrower could increase this distance to more than 60 m. Good composite bows delivered piercing arrows up to 500–700 m. This was also the range for the more powerful crossbows. Various catapults could throw stones of 20–150 kg 200–500 m. The reach rose rapidly when muscles were replaced by gunpowder. Just before 1500 the heaviest cannons could fire 140 kg iron balls about 1,400 m and lighter stone balls twice as far (Egg et al. 1971).

By the beginning of the twentieth century, when the ranges of big field pieces had reached several tens of kilometers, guns lost their primacy in the long-range delivery of destruction to bombers. Bomber ranges surpassed 6,000 km, with the delivery of up to 9 t of bombs, by the end of World War II, and bombers in turn were surpassed by ballistic missiles (Spinardi 2008). Since the early 1960s these missiles could deliver more powerful nuclear bombs with greater accuracy either from land-based silos or from subma-rines to any place on Earth. The increase in range from the ancient Old World's compound bow to a late twentieth-century ballistic missile has been about 30,000-fold, while a missile has a destructive power 16 orders of magnitude greater than an arrow's.

Long-term consumption trends, both in absolute and in relative terms, have been no less impressive. On the global scale, total primary energy flows, including traditional biomass fuels, reached 20 EJ in 1800, nearly 45 EJ in 1900, 100 EJ in 1950, just over 380 EJ in the year 2000, and more than 550 EJ in 2015. This means that the annualized power rose from about 650 GW in 1800 to 12.2 TW in 2000, a nearly 20-fold rise in two centuries, and by 2015 it had increased by more than 40%, to about 17.5 TW. The increase in fossil fuel extraction between 1800 and 2000 was nearly 900-fold, from less than 0.4 EJ to more than 300 EJ. Rising energy use has profoundly changed both the absolute and relative levels of typical per capita consumption.

The energy needs of foraging societies were dominated by the provision of food, basic clothing, and temporary shelters. Ancient high cultures channeled slowly rising energy use into permanent shelters, a greater variety of cultivated and processed food, better clothing, transportation, and a variety of manufactures (with charcoal the dominant source supplying the heat needed for smelting ores and firing bricks). Early industrial societies—with larger numbers of domestic animals, with the kinetic energy of waterwheels and windmills, and with a rising extraction of coal—easily doubled the per capita energy use that prevailed in the high Middle Ages.

Initially, most of that increase went into new manufactures, construction, and transportation (including extensive infrastructural developments), but the eventual rise of the discretionary private use of energies is not captured by the standard reporting of sectoral energy use: for example, International Energy Agency statistics show that in 2013, only 12% of the U.S. primary energy use was residential, while the U.S. Energy Information Agency put that share (including all electricity and losses in its generation) at about 22%, and the actual share (including large proportions of energy use classed under commercial and transportation categories) is well over 30%.

The U.S. per capita energy supply was already fairly high in 1900 and hence its rate during the first decade of the twenty-first century was "only" 2.5 times higher (330 vs. 132 GJ/capita), while Japan's per capita consumption between 1900 and 2015 rose 15-fold, and the multiple was nearly 10 in China. Owing to the steady rise of average conversion efficiencies, increases in per capita consumption of useful energy have been far higher: depending on the country, they were at least fourfold and up to 50-fold during the twentieth century. With an overall energy efficiency no higher than 20%, the United States consumed no more than about 25 GJ/t of useful energy in 1900, but by the year 2000, with an average efficiency of 40%, the rate was about 150 GJ/capita, a sevenfold rise in a century. My best calculation for China shows the rise from 0.3 GJ/capita of useful energies in 1950 to about 15 GJ in 2000, a 50-fold increase in just two generations.

Fouquet's (2008) British data illustrate these useful gains for major energy consumption categories for 250 years between 1750 and 2000. For all industrial power (in 1750 provided by animate labor, waterwheels, wind mills, and a few steam engines; in 2000 delivered mostly by electric motors and internal combustion engines) the multiple was 13 in 250 years; for space heating it was 14; for all passenger transport (in 1750 horses, carts, coaches, barges, and sail ships; in 2000 vehicles and ships powered by internal combustion engines and mostly jet-powered airplanes) it was nearly 900; and (as already noted) lighting gains take the top place, with the average Briton consuming about 11,000 times more light in 2000 than in 1750.

These multiples tracing the gains of useful energy services are the most revealing energy metric as they explain large gains in productive capacity, quality of life, unprecedented mobility, and (if a sapient extraterrestrial were to take a look) so much light that nighttime satellite images show large regions of Europe, North America, and Asia as continuous patches of brilliance. But higher energy efficiencies have been swamped by the combination of growing demand and larger populations, and although the global economy has become relatively less energy-intensive, its aggregate energy use has been increasing, and only some of the most advanced economies have shown saturation of average per capita energy demand during the recent three decades.

Concurrently, energy used to provide the physical necessities of life has become a steadily smaller part of the rising consumption, and the production of an enormous variety of goods, the provision of countless services, and transportation and leisure activities now consume the bulk of fuels and electricity in all affluent countries; the same pattern applies to increasing numbers of affluent urbanites in all populous modernizing countries, above all in China, India, and Brazil. And long-run efficiency gains have been the most important reason for substantial declines in energy prices (compared in real, inflation-adjusted, terms).

Kander (2013) showed that during the twentieth century, real Western European energy prices declined by 75%, ranging from an 80% decline in the UK to a 33% decline in Italy. Some of the most interesting long-term trends (properly compared in constant monies, or per unit of specific performance or delivered service) were presented by Fouquet (2008), who took advantage of English price data, some going back as far as the Middle Ages. Between the years 1500 and 2000 the cost of domestic heating fell by nearly 90%, the cost of industrial power by 92%, the cost of freight transport on land by 95%, and the cost of ocean freight transport by 98%. But by far the most impressive decline was for lighting.

The falling cost of fuels used to generate light directly or via electricity and the rising efficiency of lighting devices have combined to produce the secular drop in the cost of lighting services (monies/lumens) that is unequaled by any other kind of energy conversion. In the year 2000 a lumen of light in Britain cost merely 0.01% of what it did in 1500 and about 1% of what it did in 1900 (Fouquet 2008). And Nordhaus (1998) calculated that by the end of the twentieth century the cost of illumination in the United States was four orders of magnitude lower (the actual fraction ratio was about 0.0003) than it was in 1800. Real electricity prices declined by 97–98% during the twentieth century both in Europe and in North America (Kander 2013), and when this decline is combined with a concurrent fivefold rise in average per capita disposable income and up to an order of magnitude increase in conversion efficiency, it means that in the year 2000 a unit of electricity service in the United States was at least 200 times, and up to 600 times, more affordable than in 1900 (Smil 2008a). And since the year 2000 total energy expenditures of an average American family have been just 4-5% of its disposable income, an extraordinary bargain considering the typical housing size and the intensity of transportation (USEIA 2014).

All of these secular price declines portray indisputable trends, but at the same time, it must be kept in mind that virtually all of these trajectories would look different if energy prices had fully reflected a variety of externalities, including the environmental and health impacts associated with fuel extraction, transportation, processing, and combustion, and the various ways of electricity generation. That was never the case anywhere. Some externalities, including the capture of particulate matter and flue gas desulfurization, have been largely internalized, while others remain ignored: most notably, no fossil fuel has borne the eventual cost of CO_2-driven global warming. In addition, most energy prices—no matter whether in so called free-market economies or in states with heavy dirigiste economic policies, whether in high- or low-income countries—have been subsidized, often heavily, mostly by ignoring externalities, by setting low tax rates, and by other preferential treatments (box 7.3).

Box 7.3
Energy subsidies

The International Monetary Fund (IMF 2015) more than doubled its original 2011 estimate of $2.0 trillion of global energy subsidies, to $4.2 trillion, and put the 2015 total at $5.3 trillion, or about 6.5% of the world economic product. Most of these subsidies stem from undercharging for domestic environmental and health burdens and other externalities (including traffic congestion and accidents). China, with its massive coal combustion, has been the leading subsidizer in absolute terms (about $2.27 trillion in 2015); Ukraine's subsidies amounted to 60% of the country's GDP; and Qatar's per capita subsidies ranked first, with about $6,000 for every inhabitant. A new wave of energy subsidies has been used to establish and then to expand solar and wind generation, the two leading ways of renewable electricity production, and fermentation of carbohydrate crops to produce automotive ethanol (Charles and Wooders 2011; Alberici et al. 2014; USEIA 2015c).

What Has Not Changed?

Given the fundamental nature of energy-driven developments, this is a fair question to ask—and the obvious simple answer must be that the adoption and diffusion of new energy sources and new prime movers have been the fundamental physical reasons for economic, social, and environmental change and that they have transformed virtually every facet of modern societies: the process has always been with us, but its pace has been accelerating. Prehistoric changes brought about by better tools, the mastery of fire, and better hunting strategies were very slow, unfolding over tens of thousands of years. The subsequent adoption and intensification of permanent farming lasted for millennia. Its most important consequence was a large increase in population densities, leading to social stratification, occupational specialization, and incipient urbanization. High-energy societies created by the rising consumption of fossil fuels became the very epitomes of change, leading to a widespread obsession with the need for constant innovation.

The densities of foraging populations spanned a wide range, but, with the exception of some maritime cultures, they never surpassed one person per square kilometer. Even the least productive shifting farming raised this rate at least ten times. Permanent cropping resulted in yet another tenfold increase. Intensification of traditional farming required higher energy inputs. As long as animate labor remained the sole prime mover of field work, the share of the population engaged in crop cultivation and animal husbandry had to remain very high, more than 80%, commonly over 90%. The net energy returns of intensive farming involving irrigation, terracing, multicropping, crop rotation, and fertilization were generally lower than those in extensive agriculture but allowed unprecedented population densities.

The most intensive traditional farming—most notably Asia's year-round multicropping, sustaining largely vegetarian diets—could commonly support more than five people per hectare of cultivated land. Such densities led to gradual urbanization, but the growth of cities, extensive trade, and the effective integration of expanding empires were restricted above all by the slow speeds and low capacities of land transport. But in maritime societies they were aided by the improving capabilities of sail ships, used both for lucrative intercontinental trade and for the long-distance projection of power.

In contrast to the slow, cumulative transformations of traditional societies, the socioeconomic consequences of fossil fuel-based industrialization were nearly instantaneous. The substitution of fossil fuels for biomass fuels and the later replacement of animate energies by electricity and internal combustion engines created a new world within just a few generations (Smil 2005). The American experience was the extreme example of these compressed changes. More than in any other modern nation, the power and influence of the United States have been created by its extraordinarily high use of energy (Schurr and Netschert 1960; Jones 1971; Jones 2014; Smil 2014b). In 1850 the country was an overwhelmingly rural, wood-fueled society of marginal global import. A century later—after more than tripling its per capita consumption of useful energy and becoming both the world's largest producer and consumer of fossil fuels and the leading technical innovator able to translate these advantage into a complete victory in World War II—it was both an economic and military superpower and the world's leading technical innovator.

The most obvious physical transformations of the new fossil-fueled world have been created by the intertwined processes of industrialization and urbanization. On the most fundamental level, they released hundreds of millions of people from hard physical labor and brought a growing supply and greater variety of food and better housing conditions. The combination of more productive farming and new labor opportunities in expanding industries led to mass migration from villages and sustained rapid urbanization on all continents. In turn, this change has had an enormous positive feedback on the global use of energy. The infrastructural requirements of urban life increase the average per capita energy consumption much above rural means even if the cities are not highly industrialized. These relatively high-energy-density needs could not be supported without cheap means of long-distance transportation of food and fuel, and later without the transmission of electricity.

The mechanization of mass factory production energized by fossil fuels and electricity has enabled the mass production of common goods, bringing their greater variety and improved quality at affordable prices. It introduced new materials (metals, plastics, composites) and greatly intensified trade, transport, and telecommunication, all now at truly global levels and accessible to all individuals with reasonable disposable incomes (with crowding and commodification of experiences as inevitable consequences, evident as armies of tourists besiege every notable architectural or scenic site for a perfunctory look and a bunch of selfies on a stick).

These developments have also accelerated every facet of social change. They broke the traditional circle of limited social and economic horizons, in the first instance (leaving aside the inevitable inverse relationship between the quantity of communication and its quality) all the way to billions of "social media" users, in the second instance to often counterproductive offshoring and subcontracting of industrial activities (owing to inherently higher transportation costs and the loss of proper quality control). They have improved health and prolonged lives, nearly universal benefits (whose obverse is the burden of dealing with aging populations). They spread both basic literacy and higher education (although the mass granting of all university degrees has diminished their worth) and allowed a modicum of affluence for a rising share of the world's population. They have made more space for democracy and human rights (but they certainly did not make the world truly more democratic).

Electricity must be singled out for its many unique roles. Reliance on this most flexible and most convenient form of energy has rapidly developed into an all-encompassing dependence. Without electricity, modern societies could not farm or eat the way they do: electricity powers compressors in both ammonia plants and domestic refrigerators. They could not prevent disease (now controlled with refrigerated vaccines) and take care of the sick (with diagnoses dependent on electricity-powered machines, from venerable x-ray machines to the latest MRI, and with extensive monitoring in intensive care units), control their transportation networks, or handle their enormous volume of information (with data centers becoming some of the largest point consumers of electricity) or urban sewage.

And, of course, without electricity modern societies could not operate and manage their industries to mass-produce a growing range of higher-quality yet more affordable goods. This production has erased most of the ancient division between an admirable variety of refined luxury goods produced in small numbers for the richest few and the limited assortment of generally available crude manufactures. A rising share of this advancing output has found its way onto the world market. In 2015 foreign trade accounted for about 25% of the gross world economic product, compared to less than 5% in 1900 (World Bank 2015c). This trend has accelerated with faster and more reliable methods of transportation and with instant electronic telecommunication. Fossil fuels and electricity have moved the world from a mosaic of economic autarchies and limited cultural horizons to an increasingly interdependent whole.

No less profound transformations of the fossil-fuel era have included new structures of social relations. Perhaps the most important one has been a new system of distributing wealth. The change from status to contract has led to greater personal and political independence. This transformation has brought into existence new work regimens (typically, fixed working hours and multilayered organizational hierarchies) and new social groupings with special interests (labor unions, management, investors). Almost from its beginning it also introduced new national challenges, above all the need to cope with the extremes of rapid regional industrial growth and chronic economic decline. This disparity continues to plague even the richest nations. New tensions in international relations have been caused by trade barriers, subsidies, tariffs, and foreign ownership.

The introduction of new sources of primary energy and new prime movers has also had a profound impact on economic growth and technical innovation cycles. Substantial investment is needed to develop the extensive infrastructure needed to extract (or to harness) new energy sources, to transport (or transmit) fuels and electricity, to process fuels, and to mass-manufacture new prime movers. In turn, the introduction of these new sources and prime movers elicits clusters of gradual improvements and fundamental technical innovations. Schumpeter's (1939) classic account of business cycles in industrializing Western countries showed the unmistakable correlation between new energy sources and prime movers, on the one hand, and accelerated investment on the other (box 7.4, fig. 7.8).

Subsequent extensions of these long cycles work quite well. The postwar economic upswing was associated with the global substitution of hydrocarbons for coal, with the worldwide rise in electricity generation (including by nuclear fission), and with mass car ownership and extensive energy

Box 7.4

Business cycles and energy

The first well-documented upswing (1787–1814) coincides with the spreading extraction of coal and with the initial introduction of stationary steam engines. The second expansion wave (1843–1869) was clearly driven by the diffusion of mobile steam engines (railroads and steamships) and by advances in iron metallurgy. The third upswing (1898–1924) was influenced by the rise of commercial electricity generation and by the rapid replacement of mechanical drives by electric motors in factory production. The center points of these upswings are about 55 years apart. Fascinating post-1945 research brought a great deal of confirmation about the existence of approximately 50-year pulsations in human affairs (Marchetti 1986), and about the recurrence of such long waves in economic life and in technical inventions in particular (van Duijn 1983; Vasko, Ayers, and Fontvieille 1990; Allianz 2010; Bernard et al. 2013).

These studies indicate that the initial stages of adopting new primary energies correlate significantly with the starts of major innovation waves. The history of energy innovations also strongly confirms a still contentious proposition that economic depressions act as triggers of innovative activity. The center points of the three temporal innovation clusters identified by Mensch (1979) fall almost perfectly into the midpoints of Schumpeterian downswings. The first cluster, peaking in 1828, is clearly associated with the deployment of stationary and mobile steam engines, the substitution of coke for charcoal, and the generation of coal-derived gas. The second one, peaking in 1880, includes the revolutionary innovations of electricity generation, electric light, telephone, steam turbine, the electrolytic production of aluminum, and the internal combustion engine. The third one, clustered around 1937, includes the gas turbine, jet engine, fluorescent lights, radar, and nuclear energy.

subsidies in agriculture. This expansion was arrested by OPEC's quintupling of oil prices in 1973. The latest wave of innovations has included a variety of highly efficient industrial and household energy converters and progress in photovoltaics. The rapid diffusion of microchips, advances in computing, a greater use of optical fibers, the introduction of new materials and new ways of industrial production, and ubiquitous automation and robotization will have even greater energy implications.

The economic consequences of the world's prodigious energy use are also reflected in the roster of the world's largest companies (Forbes 2015). In 2015, five of the top 20 nonfinancial multinational corporations were oil

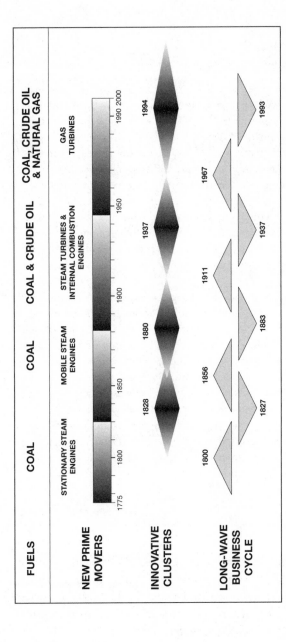

Figure 7.8

Comparison of the onsets of major energy eras (identified by principal fuels and prime movers) with innovation clusters according to Mensch (1979) and with Schumpeter's (1939) long waves of Western business cycles. I have extended both waves to the year 2000.

companies—EXXON, PetroChina, Royal Dutch Shell, Chevron, and Sino-pec—and three were car and truck makers, Toyota, Volkswagen, and Daimler. The intensification of production has been enabled by a reliable supply of affordable energies and has promoted the economies of scale evident in industrial concentration. Virtually every sector offers good examples of this process. In 1900 the United States had about 200 car manufacturing companies and France had more than 600 (Byrn 1900). By the year 2000 there were only three American firms, GM, Ford, and Chrysler, and two French firms, Renault and Citroen-Peugeot. The number of British beer breweries fell from more than 6,000 in 1900 to just 142 by 1980 (Mark 1985). But in a number of industries (including beer microbreweries) a reverse movement has been under way since the 1970s. This change is largely attributable to a combination of better communication, faster deliveries, and opportunities to supply specialized demands.

In personal terms, by far the most important consequence of the high-energy era has been the unprecedented degree of affluence and the improving quality of life. Most fundamentally, this achievement is based on an abundant and varied supply of food. People in rich countries have average per capita availabilities much above any realistic needs. Persisting malnutrition, even hunger, in the midst of this surplus (in 2015 about 45 million Americans were receiving food stamps) is a matter of distributional inequalities. In physical terms the rising affluence has manifested most convincingly in drastically lower infant mortalities and longer life expectancies. Intellectually, it has been reflected in higher literacy rates, longer years of schooling, and easier access to a greater variety of information.

Another important ingredient of this affluence has been the use of energy to save time. These uses include much more than a widespread preference for more energy-intensive but faster private cars as opposed to the use of public transportation. Refrigeration (obviating daily food shopping), electric and gas ranges, microwave ovens and food processors (simplifying and accelerating the cooking or reheating of food), and central heating (eliminating the need for repeated kindling of fires and stocking up on fuels) have all been excellent time-saving techniques now universally adopted throughout the affluent world. In turn, the time gained by these energy investments is increasingly used for leisure trips and pastimes often requiring further considerable energy inputs.

But one fundamental reality has not changed: all of these clear and impressive historical trends tracing the rise of new sources, new superior performances, and efficiency gains do not mean that humanity has been using energy in a progressively more rational manner. Urban car driving, preferred by many because of its supposedly faster speed, is a perfect example of an irrational energy use. After taking into account the time needed to earn monies for buying (or leasing) the car and to fuel it, maintain it, and insure it, the average speed of U.S. car travel amounted to less than 8 km/h in the early 1970s (Illich 1974)—and, with more congestion, by the early 2000s the speed was no higher than 5 km/h, comparable to speeds achieved before 1900 with horse-drawn omnibuses or by simply walking. In addition, with well-to-wheel efficiencies well below 10%, cars remain a leading source of environmental pollution; as already noted, they also exact a considerable death and injury toll (WHO 2015b).

Fuels, electricity, and lighter, more reliable, more flexible, and more efficient converters are too often deployed in a wasteful manner, causing environmental problems while producing ephemeral personal satisfaction (or at least a claim of it) as their only positive payoff. As Rose (1974, 359) concluded, "So far, increasingly large amounts of energy have been used to turn resources into junk, from which activity we derive ephemeral benefit and pleasure; the track record is not too good." But there is nothing new about unproductive energy uses, and they could be seen as wasteful only if human societies were driven strictly by one overarching goal, to minimize the use of energy devoted only to tasks or processes directly relevant to species survival.

But as soon as our mastery of the physical world began to yield modest energy surpluses, human ingenuity used them to create a man-made world of diversity and (for some) leisure, even though more energy could be used to secure basic physical needs. A weight-bearing column could be just a simple smooth stone cylinder or an elongated prism; there was never any structural or functional need for the three orders of ancient Greek architecture (the Doric, Ionic, and Corinthian). A rich dinner was not enough: Roman feasts had to go on for days. This quest for distinction, novelty, variety, and diversity reached a new (relative) ubiquity during the Renaissance and early modern (1500–1800) eras, but even at that time its most captivating displays were still few and overwhelmingly designed for public consumption and for posterity.

Moreover, it is easy to conclude that the monumental structures of premodern societies were not simply wasted repositories of scarce resources. Norenzayan (2013) has argued that the belief in judgmental deities ("big gods") was critical to foster the cooperation needed to build and sustain complex societies and monumental structures. As material expressions of such beliefs they thus contributed to social cohesion and encouraged awe, respect, humility, contemplation, and charity. In any case, the posterity intent has often worked to perfection, as attested by the numbers of visitors who each year travel to admire Rome's Saint Peter's Basilica or Agra's Taj Mahal (fig. 7.9). In comparison, does not the label of wasteful energy diversions apply much more readily to the extravagant, and mostly uninspiring, structures we build to manage monies or to watch modern gladiators kicking, throwing, or hitting assorted balls?

More important, modern societies have carried this quest for variety, leisure pastimes, ostentatious consumption, and differentiation through ownership and variety to ludicrous levels and have done so on an unprecedented scale. There are now hundreds of millions of people whose annual discretionary spending on nonessential items (including a rising share of luxury products) far surpasses the average income of a Western family a century ago. Examples of these extravaganzas abound. Family size in affluent countries keeps shrinking, but the average size of U.S. custom-built houses has passed 500 m^2; shipbuilders have waiting lists for yachts with helicopter pads; many cars on the market have so much excess power that they can be never fully tested on any public road: the Koenigsegg Regera

Figure 7.9
Saint Peter's Basilica completed in 1626 (Corbis).

engine rates 1.316 MW, while the Lamborghini and the top Mercedes-Benzes come with "only" 1.176 MW, the latter value being equal to nearly 1,600 hp, or 11 times the power of the small car, a Honda Civic, that I drive.

On a more mundane level, tens of millions of people annually take intercontinental flights to generic beaches in order to acquire skin cancer faster; the shrinking cohort of classical music aficionados has more than 100 recordings of Vivaldi's *Quattro Stagioni* to choose from; there are more than 500 varieties of breakfast cereals and more than 700 models of passenger cars. Such excessive diversity results in a considerable misallocation of energies, but there appears to be no end to it: electronic access to the global selection of consumer goods has already multiplied the choice available for Internet orders, and the customized production of many consumer items (using individualized adjustments of computer designs and additive manufacturing) would raise it to yet another level of excess. The same is true of speed: do we really need a piece of ephemeral junk made in China delivered within a few hours after an order was placed on a computer? And (coming soon) by a drone, no less!

But regardless of the indicators used, those kinds of wasteful, unproductive, and excessive final energy use are still in the global minority. When looking at average per capita energy supply, then only about one-fifth of the world's 200 countries have accomplished the transition to mature, affluent industrial societies supported by the high consumption of energy (>120 GJ/capita), and the share is even lower in population terms, about 18% (1.3 billion among 7.3 billion in 2015). Adding rich households in such low- and middle-income countries as China, India, Indonesia, and Brazil would raise the population share only marginally, to about 20%. For example, China now has the world's fourth-largest number of affluent families (following the United States, Japan, and the UK), but that still worked out to fewer than five million of such families in 2015 (Atsmon and Dixit 2009; Xie and Jin 2015).

Consequently, the global diffusion of the Western miracle of rapid technical innovation has resulted in a worrisome global split, in an unprecedented level of economic inequality among nations. By 2015 the richest 10% of humanity (living in 25 nations) claimed about 35% of the world's energy. In personal terms, this meant that a week's worth of per capita energy use in the United States is equivalent to the total annual primary energy consumption of an average Nigerian and two years of the annual energy supply for an average Ugandan. Conversely, the poorest 5% of humanity (living in 15 African countries) consumes no more than 0.2% of the world's primary commercial energy supply.

These disparities have no easy remedies, and narrowing the gap takes time, even with an extraordinarily rapid economic growth: during 35 years of rapid modernization, from 1980 to 2015, China nearly quintupled its average per capita energy consumption to reach just over 90 GJ/capita. In the process it paid major environmental and health costs and strained the global energy trade, but it is still 20–25% below the comfortable supply rate. Most fundamentally, even if the requisite resources were readily available, the environmental consequences of raising the rest of the world to the West's level of primary energy consumption would be unacceptable. Already concerns about biospheric integrity have become major considerations in contemplating the future of high-energy civilization. They range from preservation of biodiversity to rapid anthropogenic climate change.

Between Determinism and Choice

Many historical developments are the result of a limited array of outcomes that follow using particular energies in certain ways. A reliance on different primary energies leaves distinct imprints on everyday work and leisure. Life spent breaking clods with heavy hoes, transplanting dripping seedlings, grasping handfuls of stems and cutting them with a sickle, husbanding the straw for cooking, and hand-pounding the grain (all still quite common in the late nineteenth-century Chinese countryside) creates a different world from one in which teams of strong horses pull curved moldboard plows, mechanical seeders, and harvesters, in which a fuelwood grove yields plenty of wood for large stoves and flour is ground by steam-powered mills (all common in the late nineteenth-century United States).

Similarly, a reliance on different prime movers determines a different scope and tempo of everyday affairs. The laborious harnessing of horses with bits, nosebands, head crowns, hames, collars, backbands, and traces, the clatter of horseshoes, the jolting of poorly sprung coaches, the muzzles of resting animals stuck in feedbags, the sweeping of horse manure from city streets and carting it away to nourish suburban gardens—these images evoke a pace of life profoundly unlike that dominated by turns of the ignition key, the swishing of radial tires, the smooth and swift ride of sedans and SUVs, networks of filling stations, and the easy availability of vegetables and fruits carried by processions of heavy trucks from California or Spain—or in refrigerated containers and in cargo holds of jetliners from other continents.

Using energy as a principal analytical concept of human history is thus an obvious, profitable, and desirable choice. But we should not look at it as the principal explanatory factor. The explanatory power of an energy approach to history must not be exaggerated. Highly generalized claims lead to indefensible conclusions. To generalize, across millennia, that higher socioeconomic complexity requires higher and more efficiently used inputs of energy is to describe indisputable reality. To conclude that every refinement of energy flows brought a refinement in cultural mechanisms, as Fox (1988) did, is to ignore a mass of contradictory historical evidence.

The only rewarding and revealing way to assess energy's importance in human history is neither to succumb to the simplistic, deterministic explanations buttressed by recitals of countless energy imperatives nor to belittle it by reducing it to a marginal role compared to many other history-shaping factors, be they climatic changes and epidemics or human whims and passions. Energy conversion is always necessary to get anything accomplished, but none of the extrasomatic conversions initiated and controlled by people is predestined, and only a few of them arise simply from chaos or accident. This dichotomy is as important for interpreting the past as it is for understanding future possibilities: they too are not predestined, but their scope is definitely restricted, and energy flows impose the most fundamental limits.

Imperatives of Energy Needs and Uses

Energy's essential role in the physical world and in the maintenance of life is necessarily reflected in evolutionary and historical developments. The prehistoric development of human societies and the growing complexity of high civilizations have been marked by countless energy imperatives. The most fundamental physical limit is, of course, the incoming solar radiation. This flow keeps the planet's temperatures within the range suitable for carbon-based life and powers the planet's atmospheric circulation and water cycle. Temperature, precipitation, and the availability of nutrients are the key determinants of plant productivity, but only a part of the newly synthesized biomass is digestible. These realities shaped the basic existential modes, population densities, and social complexities of all foraging societies. In an overwhelming number of cases these groups had to be omnivorous. Most of their food energy had to come from collecting abundant seeds (combining starches with proteins and oils) and tubers (rich in carbohydrates).

Concentrated abundance and the easy accessibility of collectible plants mattered more than their overall biomass and variety. Grasslands and woodlands offered a better existence than dense forests. The killing of large (meaty and fatty) mammals brought high net energy returns, while the hunting of smaller species was almost always a much less energy-rewarding pursuit than plant collecting. Lipids were the most desirable nutrients commonly in short supply. Their high energy density provided the satisfactory feeling of satiety. These energy imperatives dictated collecting and hunting strategies and contributed to the emergence of social complexity.

As long as human, and later animal, muscles remained the only prime movers, all labor rates were determined by metabolic imperatives: by the digestion rates of food and feed, by the basal metabolic and growth requirements of homeothermic bodies, and by the mechanical efficiency of muscles. The sustained delivery of human power by adults could not be higher than about 100 W. The efficiency of converting food to mechanical energy could not exceed 20–25%. Only a massing of people or draft animals could overcome these limits, and, as attested by prehistoric and ancient monumental structures, such feats, requiring effective, coordinated control, were repeatedly accomplished by societies as different as the menhir builders of Ireland and Brittany, the Egyptians of the early dynasties, and a small population on Easter Island.

Aggression powered by human muscles had to be discharged either in hand-to-hand combat or by an attack launched stealthily from no farther away than a couple hundred meters. For millennia, killing had to be done at fairly close quarters. Human anatomy makes it impossible for an archer to exert the maximum force when one extended arm and one flexed arm are separated by more than about 70 cm. This limits a draw, and hence the arrow's range. Catapults, tensioned by many hands, boosted the mass of projectiles, but they did not increase the range of attack. Face-to-face combat ensued eventually, with individual outcomes depending heavily on skill, experience, and chance.

The shift from foraging to farming was driven by a combination of energy-related (that is, primarily nutritional) and social factors, but the subsequent intensification of sedentary farming can be explained as a clear energy imperative (Boserup 1965, 1976). When an existing way of food production approaches the physical limits of its performance, the population can either stabilize its size (by controlling births or by outmigration) or adopt a more productive system of food production. The onset and duration of successive intensification steps have varied greatly around the world, but in order to harness more of the site's photosynthetic potential, every advance demanded higher energy inputs. In return, greater harvests were able to support a higher population density.

Farming intensification also required a higher indirect energy investment in breeding and feeding draft animals, in making, bartering, or buying increasingly more complex tools and implements, and in such long-term infrastructural projects as terracing and the building of irrigation canals, impoundments, granaries, and roads. In turn, this intensification led to a growing reliance on sources of energy other than human muscle. Plowing in heavier soils is either enormously taxing or outright impossible without draft animals. Manual milling of grains is so labor-intensive that animals, and later also water and wind power, were necessary to process concentrated harvests. Long-distance deliveries of grain to cities had to rely largely on animal power, sometime also on wind. Manufacturing more durable and more efficient iron tools and implements consumed charcoal for smelting the ores.

A number of particular energy imperatives shaped the world of traditional agricultures. Where no extensive grazing lands were available and where all arable land had to be used to grow food, human energy needs set the limits to the production of feed and hence the number of large draft animals. In all other cases (excepting the Americas and Australia) draft animals were increasingly used, but often they were fed almost exclusively by grazing and with crop residues. As the per capita availability of grain rose, enough cultivable land could be set aside to grow good-quality animal feed: in the United States its supply eventually claimed nearly 25% of all farmland, whereas in the densely populated lowlands of traditional Asia it was usually below 5%.

Intensity of cultivation, that is, the energy density of food production, also made its impact. There was hardly any space for large grazing animals in intensive (on sloping land, terraced) wet-field agriculture producing Asian rice, and water buffaloes often fed on streamside grasses or even on submerged aquatic plants. In contrast, the presence of large numbers of cattle, horses and other domestic animals in land-rich regions of Europe (and nineteenth-century North America) influenced both the density of rural populations and the organization of settlements. Plenty of space had to be devoted to barns and stables and to the storage of manure before it was recycled to fields.

The two agroecosystems embodying these extremes have been the traditional rice-dominated multicropping of China south of the Yangzi and Western European mixed farming with its heavy reliance on animals both for food (dairy and meat) and draft. Pre-Columbian agricultures were shaped by different energy imperatives. All else being equal, corn, a C_4 crop, yields more than other dryland cereals (wheat, barley, rye, all C_3 crops), and this advantage increases further when corn is intercropped with a leguminous species: corn and beans were staples of pre-Columbian farming everywhere in the Americas with the exception of the high-altitude Andean regions, where potatoes and quinoa prevailed. In addition, the absence of domestic animals in the Americas left more energy and time for other pursuits.

Energy imperatives shaping nonagricultural activities and structures of traditional societies ranged from locational limits to the challenges of effective administration. Metal smelting and forging on larger scales were made possible only by the use of water power, restricting the location of furnaces and forges to mountainous areas even if a cheap transport of ore and charcoal was possible. In reality, the power of draft animals and poor roads greatly limited the range of profitable land-borne movements of bulky materials; streams were preferred, and canals were built. Inefficient methods of charcoaling (<20% of wood energy converted into the smokeless fuel) caused extensive deforestation.

The administration of distant territories and trading and military ventures were made difficult not only by the slowness but also by the unreliability of both land and sea travel. Sailings from Rome to Egypt, the empire's largest producer of surplus grain, could take as little as a week—and as long as three months or more (Duncan-Jones 1990). Much of the Spanish Armada's failure to land in England in 1588 can be ascribed to wind, either the lack of it or wind blowing from an unwelcome quarter (Martin and Parker 1988). And as late as 1800 English ships had to wait, sometimes for weeks, for the right wind to take them into Plymouth Sound (Chatterton 1926).

Energy imperatives had a profound influence on national and regional fortunes during modern energy transitions. Countries and locales with a relatively easy access to fuels that could be produced and distributed with less energy than the previously dominant source enjoyed faster economic growth, with its welcome correlates of greater prosperity and higher quality of life. The earliest nationwide example of this advantage is the heavy Dutch reliance on peat, which ushered in the republic's Golden Age during the seventeenth century. Although Unger (1984) questioned De Zeeuw's (1978) estimates of high annual peat extraction there is no doubt that this young fossil fuel was at that time the country's most important source of prime energy. Only a few generations later an even more far-reaching advantage of this kind was demonstrated with the virtually complete replacement of English wood and charcoal by bituminous coal and coke (King 2011). After 1870 that experience was greatly surpassed by the rise of the American economy, fueled initially by excellent coals and later by hydrocarbons.

Undoubtedly, the successive positions of economic leadership and international influence of the Dutch Republic, Great Britain, and the United

States have been closely related to their early exploitation of fuels extractable with a smaller amount of investment per unit of useful energy (that is, with a higher net energy return). The dominance of fossil fuels and electricity has created an unprecedented degree of technical and, by a gradual extension, also economic and social uniformity (fig. 7.10). Even a basic list of the universal infrastructures of a high-energy civilization is a lengthy one: coal mines, oil and gas fields, thermal power stations, hydroelectric dams, pipeline networks, ports, refineries, iron and steel mills, aluminum smelters, fertilizer plants, countless processing, chemical, and manufacturing enterprises, railroads, multilane highways, airports, skyscraper-dominated downtowns, an extensive suburbia.

As they perform the same functions, their outward physical appearance must be identical or very similar, and the construction and management of many of their components have increasingly originated with a relatively small number of companies supplying the global market with critical machines, processes, and know-how. The two most obviously worrisome consequences of the dependence on high energy flows is the restriction of

Figure 7.10
Brazil's largest megalopolis, São Paulo, photographed in 2013. Megacities are the foremost examples of global uniformity imposed by high levels of fuel and electricity use (Corbis).

choices (that is, the impossibility of abandoning existing practices without causing numerous massive dislocations) and degradation of the environment. The first phenomenon is perhaps best illustrated by the impossibility of cutting off high-energy subsidies in modern farming without profoundly transforming the entire society.

For example, replacing the existing American field machinery by draft animals would require horse and mule stock at least ten times as large as its record numbers from the early twentieth century. Some 300 Mha, or twice the total area of U.S. arable land, would be needed just to feed the animals, and masses of urbanites would have to leave the cities for farms. And affluent countries are not the only ones that cannot go back to traditional farming without transforming the whole according to the preindustrial image: because of the world's highest intensity of fertilization and irrigation, China's dependence on fossil energies in food production is even higher.

Restriction of choices is a paradoxical result of a world dominated by what Jacques Ellul (1912–1994) called, simply and all-encompassingly, *la technique*, "the totality of methods rationally arrived at and having absolute efficiency (for a given stage of development) in every field of human activity" (Ellul 1954, xxv). This world gives us unprecedented benefits and almost magical freedoms, but in return, modern societies must not only adapt to it but submit to its rules and strictures. Every person now depends on these techniques, but no single individual understands them in their totality; we just follow their dictates in everyday life.

The consequences go beyond ignorant obedience as the spreading power of techniques has already made a large part of humanity irrelevant to production processes, and only a small part of the labor force is now required (with an increasing help of computers) to design and produce items destined for global mass consumption. As a result, many more people are now commonly employed in selling a product rather than in designing, improving, and making it. When ranked by the size of their labor force, in 1960 11 out of America's 15 largest companies (led by GM, Ford, GE, and United States Steel) were producers of goods employing more than 2.1 million workers; by 2010 just two makers of goods, HP and GE, employing about 600,000 people, were among the top 15, and the group is now dominated by retailers and service-providing firms (Walmart, UPS, McDonald's, Yum, Target).

The next logical step is to see this reality as part of the process leading to the eventual displacement of carbon-based life by machines (Wesley 1974). Evolutionary parallels between the two entities are intriguing.

Machines are thermodynamically alive, and their diffusion conforms to natural selection: failures do not reproduce, new species proliferate, and they tend toward maximum supportable mass; successive generations are also progressively more efficient (recall all those impressively lower mass/power ratios!), more mobile, and have longer life spans. These parallels can be dismissed as merely intriguing biomorphizations, but the ascendance of machines has been an undeniable fact.

They have already replaced enormous areas of natural ecosystems with the infrastructures needed for their making, motion, and storage (mines, railroads, roads, factories, parking lots); man's time has been increasingly spent by serving them; their waste products have caused extensive degradation of soils, waters, and the atmosphere; and the global mass of automobiles alone is already much higher than that of all of humanity. The finiteness of fossil fuel resources may do little to stop the ascent of machines. In the near term they can adapt by becoming more efficient; in the long term they can rely on renewable flows.

In any case, only a fundamental misinterpretation of clear geological evidence can see in the rising use of fossil fuels a cause for concern about their early exhaustion. Fossil fuel reserves are that small part of the resource base whose spatial distribution and recovery costs (at current prices and with existing techniques) are known in enough detail to justify their commercial exploitation. As we recover higher shares of originally available resources, the best measure of their availability is the cost of producing a marginal unit of a mineral. This approach takes into account improvements in our exploitation techniques and our ability to pay the price of recovery. The impressive post-2005 rise in the recovery of crude oil from America's abundant shale deposits—relying on a combination of horizontal drilling and hydraulic fracturing (Smil 2015a) and making the country, once again, the world's largest producer of oil and gas—illustrates enormous opportunities that remain to be fully exploited.

Resource exhaustion is thus not a matter of actual physical depletion but rather a burden of eventually insupportable cost increases. Some notable exceptions aside (such as the rapid Dutch exit from coal mining after the discovery of the supergiant Groningen natural gas field), there are no sudden ends, only prolonged declines, slow exits, and gradual shifts onto new supply planes (British coal mining being a perfect illustration of this process). This understanding is critical in appraising the rise and prospects of fossil-fueled civilization. That fossil fuel resources are finite does not imply any fixed dates for the physical exhaustion of coals or hydrocarbons, nor does it mean the early onset of unbearably rising real costs of recovering these resources and hence the necessity of a rapid transition to a post–fossil fuel era.

Reserve estimates and resource appraisals are not enough for speculating about the future of fossil fuels. Global demand and efficiency of use are no less important: demand (driven by the combination of economic and population growth) may be predictably increasing, but it is also highly modifiable, and energy conversion efficiencies, even after generations of improvements, remain greatly improvable. Consequently, it is not worries about an early exhaustion of fossil fuels—most prominently expressed by the advocates of imminent peak oil (Deffeyes 2001)—but rather the impact on the habitability of the biosphere (above all through global climate change) that is the most important near- and long-term concern resulting from the world's dependence on coals and hydrocarbons.

The Importance of Controls

Past adoptions of new energy sources and new prime movers could never have had such far-reaching consequences without introducing and perfecting new modes of harnessing those energies and controlling their conversion to supply required energy services (heat, light, motion) at desirable rates. These controls or triggers can open previously shut gates and release new flows of energy—or they can increase the overall working rates of established processes, or make them more reliable or more efficient. They can be simple mechanical devices (waterwheels) as well as sophisticated arrangements that themselves require considerable energy inputs: microprocessors in modern cars are an excellent example of this category. Or they can just be better sets of managerial procedures, newly identified and developed markets, or fundamental political or economic decisions.

No matter how strong and capable, a horse becomes an effectively controlled prime mover only when a bit is inserted in its mouth and connected to reins held in a rider's hand; it can pull a battle chariot only with a good light harness; it can be used in armored combat only when saddled and fitted with stirrups; it can provide heavy draft only when shod and harnessed with a comfortable collar; and it can be a part of an efficient team only when the unequal pulls of differently sized animals are equalized by swingletrees.

The absence of proper controls could mar the performance of otherwise admirable prime movers. The remedy has been sometimes very slow in coming. Perhaps the best illustration of such a failure was the inability to determine longitude. By the beginning of the eighteenth century, fully rigged sail ships were efficient converters of wind energy and Europe's potent empire-building tools, but their captains still could not find their longitude. As summed up in a petition presented by English captains and merchants to the Parliament in 1714, too many ships were delayed and many lost. Because Earth's rotation amounts to about 460 m/s on the equator, finding the longitude requires chronometers that would lose no more than a fraction of a second a week if a ship was to be positioned with an error smaller than a couple of kilometers after a journey of two to three months. In 1714 an act of the British Parliament offered up to £20,000 for such an achievement. The reward was finally given to John Harrison (1693–1776) in 1773 (Sobel 1995).

As far as fuels are concerned, history would have taken a different course if coal had been used merely as a substitute for wood in open fireplaces, or if crude oil use had remained limited to kerosene for lighting. In most cases it has not been the access to abundant energy resources or to particular prime movers that made the long-term difference. Decisive factors were rather the quest for innovation and the commitment to deploying and perfecting new resources and techniques and finding new uses. The combination of these factors has determined the energy efficiencies of whole economies or particular processes, as well as the safety and acceptability of new conversion techniques. Examples of these sometimes striking but often subtle contributions can be found in all energy eras and for all fuels and prime movers.

In a narrow technical sense, the most important class of controls includes feedback devices and systems (Doyle, Francis, and Tannenbaum 1990; Åström and Murray 2009). They transfer the information about a particular process back to the controlling mechanism, which can then adjust the operation. Early modern Europe gained a decisive leadership in the development of these feedbacks. The earliest applications included thermostats (for the first time around 1620 by a Dutch engineer, Cornelis Drebbel), the automatic adjustment of windmills by fantails (patented in 1745 by an English blacksmith, Edmund Lee), floats in domestic cisterns and in steam boilers (1746–1758), and James Watt's famous centrifugal governor regulating the power of steam engines (1789). Today's most numerous example in this category is microprocessors controlling the operation of car and jet engines.

An indispensable class of controls embraces the instructions that make it possible to duplicate production and management processes and to turn out standardized goods and services. The rapid development of printing in early modern Europe made an immense contribution in this regard. By 1500 more than 40,000 different books or editions had been published in

Western Europe in more than 15 million copies (Johnson 1973). The introduction of detailed copper engraving during the sixteenth century and the contemporaneous development of various map projections were other notable early advances. Another outstanding innovation in this class was a punched card device invented by Joseph Marie Jacquard (1752–1834) in 1801 to control loom operations. Before 1900 punched cards were employed on Herman Hollerith's (1860–1929) machines used to process census data (Lubar 1992). After 1940, punched input controlled first electromechanical and then electronic computers, all now supplanted by electronic data storage.

Until the end of the nineteenth century the new controls remained largely mechanical. During the twentieth century advances driven by applied mathematics and physics, above all the emergence and widespread applications of transistors, integrated circuits, and microprocessors, created a vast new field of increasingly sophisticated automatic control using electrical and electronic devices. Critical innovations ranged from the widespread deployment of radar (in fire control, bombing, missile guidance, and autopilot navigation) to myriads of microchip-based controls in computing, consumer electronics, and industrial processes.

In a wider sense, certainly the most basic control consideration is what societies do with their energy sources and prime movers. How do they apportion them among productive uses and discretionary personal consumption? What balance, if any, do they want to strike among the contradictory pulls of autarchy and an extensive reliance on foreign trade? How open do they want to be to both goods and ideas? What price are they willing to pay for military spending? How much central control do they wish to exercise? In all of these respects, cultural, religious, ideological, and political constraints, impulses, and leanings have been decisive. Again, numerous examples can be selected from all energy eras. Two momentous contrasts are especially revealing: the first one between the Western and the Chinese nautical voyages of discovery, the other one between the Russian and the Japanese approaches to economic modernization.

Transoceanic voyages required not only sails capable of taking the ship closer into the wind but also stronger hulls, good sternpost rudders, and reliable navigation devices. The Chinese originated most of these advances and combined them in their great Ming fleets. Just over a century after Marco Polo's return from China these fleets penetrated farther westward from China than any Europeans did eastward at that time. Between 1405 and 1433 they repeatedly sailed in Southeast Asian waters and in the Indian Ocean and visited the coast of East Africa (Needham et al. 1971). Then a sudden involution of the centrally governed empire made further sailings impossible.

In contrast, late medieval European ships started out as decidedly inferior prime movers, but the enterprise was sustained by a combination of inquisitive, aggressive, and zealous attitudes of Spanish and Portuguese rulers and navigators. By the seventeenth century English and Dutch mercantile ventures had become prominent: the East India Company was set up in London in 1600; the Dutch VOC (Vereenigde Oost-Indische Compagnie) was chartered in 1602 (Keay 2010; Gaastra 2007). Between 1602 and 1796 VOC ships made nearly 4,800 trips to the East Indies, and the East India Company ruled large parts of the subcontinent between 1757 and 1858. This amalgam of economic, religious, and political aspirations resulted in the eventual European domination of seas and establishment of extensive empires.

A comparison of Russia's and Japan's post-1945 economic fortunes contrasts quantity versus quality, autarchy versus commerce, and the role of the state as the sole arbiter versus the state as the leading catalyst of modernization. Japan's minor coal resources, limited hydroelectric potential, and virtual absence of hydrocarbons forced it to be a major energy importer, and in order to reduce its vulnerability to high fuel prices and import interruptions it became one of the world's most efficient users of energy (Nagata 2014). Its governing bureaucracies have promoted state cooperation with industries, technical innovation, and value-added exports.

In contrast, thanks to its exceedingly rich mineral patrimony in European Russia, Siberia, and Central Asia, the Soviet Union became not only self-sufficient in all forms of energy but a major exporter of fuels. But generations of rigid central planning, autarchic Stalinist five-year plans that continued long after the dictator's death, and excessive militarization of the economy made the country the least efficient user of energy in the industrialized world: during the last years before its collapse the USSR was by far the world's largest producer of both crude oil (extracting 1.66 times as much as Saudi Arabia) and natural gas (nearly 1.5 times as much as the United States), but its per capita GDP was only about 10% of the U.S. total (Kushnirs 2015).

Decisive controls of critical energy flows have been often beyond human influence, or they have been usurped, or at least heavily influenced, by parasitic diversions. McNeill (1980) conceptualized these notions in his dual treatment of micro- and macroparasitism. Microparasites—bacteria, fungi, insects—forestall human efforts to secure sufficient food energy. They damage or destroy crops and domestic animals, or they prevent the efficient use of digested nutrients by directly invading human bodies, and modern societies have had to devote a great deal of energy to limit their spread in fields and among populations by, above all, resorting to pesticides and antibiotics.

Macroparasitism has assumed a variety of social controls of energy flows relying both on coercion—ranging from slavery and corvée labor to military conquest—and on complex (and partly voluntary) relationships among unequal groups of people. Special interest groups have become certainly the most important macroparasites in modern affluent countries. They range from various professional associations and labor unions with restricted entries to oligopolistic industrial cartels and lobbies. By shaping, and often vetoing or disabling, government policies and fixing prices, these groups work against the optimized use of all resources, and inevitably they have a notable effect on the development of energy resources and on the efficiency of their use. They have been behind the heavy subsidies that for decades have gone to producers of various fossil fuels and nuclear electricity, and now they are behind the new subsidies promoting the construction of solar photovoltaics and wind turbines (see box 7.3).

Olson (1982) aptly called such groupings distributional coalitions, and pointed out that stable societies will gradually acquire a greater number of such alliances. Acceptance of this argument helps to explain the British and U.S. industrial decline, as well as the post–World War II success of Germany and Japan. Organizations created by the victorious powers after 1945 in the two defeated nations were much more inclusive, and their energy performance supports this view: Japanese and German economies are clearly less energy-intensive than the British and American ones. This is true not only in aggregate but also in virtually every major sectoral comparison.

On the other hand, the actions of some special groups have opened new energy gates and improved the efficiency of conversions. The skills of nineteenth-century British emigrants were higher than the migrants' proportion in the native population. Their influx into the United States was obviously an important trigger of higher energy flows (Adams 1982) but eventually also of notable energy savings. A similar process has been under way since the 1990s with the large-scale immigration of Indian engineers to electronics and Internet companies in general and to Silicon Valley in particular (Bapat 2012). And, forced to compete globally, multinational companies strive to lower the energy intensity of their production, diffusing new techniques and fostering higher energy conversion efficiencies worldwide.

The Limits of Energy Explanations

Most historians have had no use for energy as an essential explanatory variable. Even Fernand Braudel (1902–1985), known for his insistence on the importance of the material world and economic factors, does not mention energy in any of its forms in his lengthy definition of a civilization:

> A civilization is first of all a space, a "cultural area," as the anthropologists would say, a locus. Within the locus … you must picture a great variety of "goods," of cultural characteristics, ranging from the form of its houses, the material of which they are built, their roofing, to skills like feathering arrows, to a dialect or a group of dialects, to tastes in cooking, to a particular technology, a structure of beliefs, a way of making love, and even to the compass, paper, and the printing press. (Braudel 1982, 202)

As if the materials, houses, arrows, and printing presses came about ex nihilo, without any expenditures of energies! This omission is indefensible if one seeks to understand all of the fundamental factors shaping history— but it is justifiable if one notes that the kinds of energy sources and prime movers and levels of energy use do not determine the aspirations and achievements of human societies. There are undeniable natural reasons for this reality. Of course, energy conversions are absolutely essential for the survival and evolution of all organisms—but their modification and differential utilization are governed by properties intrinsic to the organisms.

As fundamental as the laws of thermodynamics are, energy is not the only determinant of the biosphere's evolution, or of life in general and human actions in particular: evolution is inevitably entropic, but there are other inputs that cannot be either substituted or recycled. The Earth suffused with radiation could not host carbon-based life without an adequate availability of elements indispensable for biochemical conversions, including phosphorus in ATP, nitrogen and sulfur in proteins, cobalt and molybdenum in enzymes, silicon in plant stems, or calcium in animal shells and bones. Epigenetic information channels energy into maintenance, growth and differentiation, and reproduction; these irreversible transformations dissipate both matter and energy and are affected by the availability of land, water, and nutrients and by the need to cope with interspecies competition and predation.

Energy flows limit but do not determine the biospheric organization on any scale. As Brooks and Wiley (1986, 37–38) put it,

> Energy flows do not provide an explanation for why there are organisms, why organisms vary, or why there are different species. … It is an organism's intrinsic properties that determine how energy will flow, not the opposite. If the flow of energy were deterministic for biological systems, it would be impossible for anything

living to starve to death. ... We suggest that organisms are physical systems with genetically and epigenetically determined individual characteristics, which utilize energy that is flowing through the environment in a relatively stochastic manner.

But these fundamental realities do not justify ignoring energy's role in history; rather, they argue for its properly qualified inclusion. In modern, complex human societies energy use is clearly much more a matter of desires and displays rather than of mere physical needs. The amount of energy at a society's disposal puts clear limits on the overall scope of action, but it may tell us little about the group's basic economic accomplishments or its ethos. Dominant fuels and prime movers are among the most important factors shaping a society, but they do not determine the particulars of its successes or failures. This is especially clear when one examines the energy-civilization equation. This concept, so pervasive in modern society, equates high energy use with a high level of civilization: we need only recall Ostwald's work, or Fox's conclusion that "a refinement in cultural mechanisms has occurred with every refinement of energy flux coupling" (Fox 1988, 166).

The genesis of the link is not surprising. Only the rising consumption of fossil energies has been able to satisfy so many material desires on such a large scale. Greater possessions and comforts have become equated with civilizational advances. This biased approach excludes the whole universe of creative—moral, intellectual, and aesthetic—achievements which have no obvious connection with any particular levels or modes of energy use: there has been no obvious correlation between the modes and levels of energy use and any "refinement in cultural mechanisms." But such energetic determinism, like any other reductionist explanation, is highly misleading.

Georgescu-Roegen (1980, 264) suggested a fine analogy that also captures the challenge of historical explanations: geometry constrains the size of the diagonals in a square, not its color, and "how a square happens to be 'green', for instance, is a different and almost impossible question." And so every society's field of physical action and achievement is bound by the imperatives arising from the reliance on particular energy flows and prime movers—but even small fields can offer brilliant tapestries whose creation is not easy to explain. Mustering historical proofs for this conclusion is easy, in matters both grand and small.

Formulations of universal and lasting ethical precepts by ancient thinkers and moralists and founders of durable religions in the Middle East, India, and China were all done in low-energy societies where most of the population was preoccupied with basic physical survival. Christianity and

Islam, the two dominant monotheistic beliefs that continue to exercise immense influence on modern affairs, arose, respectively, about 20 and 13 centuries ago, in arid environments where agrarian societies had no technical means to convert abundant sunshine into useful energy. The Greeks of the classical era often referred to their slaves in terms clearly placing them on the level of working animals (calling them *andrapoda*, man-footed, as opposed to cattle, *tetrapoda*)—but they gave us the fundamental ideas of individual freedom and democracy. The concurrent advance of freedom and slavery is one of the most remarkable aspects of Greek history (Finley 1959), as is the affirmation of human equality and slavery at the beginning of the American republic.

The United States adopted its visionary constitution ("all men are created equal") while the society was energized primarily by wood and while its principal drafter and the country's fourth president, James Madison (1751–1836), was a slave owner, as were the first and the third presidents, George Washington (1732–1799) and Thomas Jefferson (1743–1826). Late nineteenth-century Germany embraced aggressive militarism and two generations later fascism just as it became continental Europe's leading consumer of energy—while Italy and Spain became dictatorships during, respectively, the 1920s and 1930s, when their per capita energy use was among the lowest on the continent, generations behind the German use.

And artistic accomplishments have had little to do with any specific level or energy use or with any particular kind of energy used at the time of their origin: the creation of timeless literature, painting, sculpture, architecture, or music shows no corresponding advances with the average level of a society's energy consumption. During the first decade of the sixteenth century an idler on Florence's Piazza della Signoria could have passed, in a matter of days, Leonardo da Vinci, Raphael, Michelangelo, and Botticelli, a concatenation of creative talent that is utterly inexplicable by the combustion of wood and the harnessing of draft animals, the common practices that could be seen in any other contemporary city in Italy, Europe, or Asia.

No energetic considerations can explain the presence of Gluck, Haydn, and Mozart in the same room in Joseph II's Vienna of the 1780s, or that in 1890s, in fin de siècle Paris, Émile Zola's latest novel could be read before one viewed the latest canvases by Claude Monet or Camille Pissarro on a day when Gustave Doret was conducting Claude Debussy's *L'Apres-midi d'un faun* (fig. 7.11). Moreover, art shows no progress commensurate with energy eras: paintings of animals from Neolithic caves of southern France, the proportions of the classical temples of Greece and southern Italy, and the sound of medieval chants from French cloisters are no less pleasing and captivating, no less modern, than the colorful compositions of Joan Miró, the sweeping curves of Kenzo Tange's buildings, or the drive and melancholy of Rachmaninov's music.

And during the twentieth century, level of energy use had little to do with the enjoyment of political and personal freedoms: they were expanded in the energy-rich United States as well as in energy-poor India; they were curtailed in the energy-rich USSR as they still are in energy-scarce Pakistan. After World War II the Stalinist and post-Stalinist USSR and some of the countries of the former Soviet empire used more energy than the West European democracies—yet they could not offer their people a comparable quality of life, a key factor that led eventually to communism's demise. And today's energy-rich Saudi Arabia has a much lower freedom rating than energy-poor India (Freedom House 2015).

Nor there have been any strong links between per capita energy use and subjective feelings of satisfaction with life or personal happiness (Diener,

Figure 7.11
Camille Pissarro, *Le Boulevard de Montmartre, Matinée de Printemps*. Oil on canvas, painted in 1897 (Google Art Project).

Suh, and Oishi 1997; Layard 2005; Bruni and Porta 2005). The top 20 nations with the highest satisfaction index included not only energy-rich Switzerland and Sweden but also such relatively low energy users as Bhutan, Costa Rice, and Malaysia, while Japan (in 90th place) ranks behind Uzbekistan and the Philippines (White 2007). And the 2015 World Happiness Report (Helliwell, Layard, and Sachs 2015) ranks among the top 25 countries such relatively moderate energy users as Mexico, Brazil, Venezuela, and Panama, all of them ahead of Germany, France, Japan, and Saudi Arabia.

Satisfying basic human needs obviously requires a moderate level of energy inputs, but international comparisons clearly show that further quality-of-life gains level off with rising energy consumption. Societies focusing more on human welfare than on frivolous consumption can achieve a higher quality of life while consuming a fraction of the fuels and electricity used by more wasteful nations. Contrasts between Japan and Russia, Costa Rica and Mexico, or Israel and Saudi Arabia make this obvious. In all of these cases the external realities of energy flows have obviously been of secondary importance to internal motivations and decisions. Very similar per capita energy use (for example, that of Russia and New Zealand) can produce fundamentally different outcomes, while highly disparate energy consumption rates have resulted in surprisingly similar levels of physical quality of life: South Korea and Israel have nearly identical human development index while the Korean per capita energy use is about 80% higher.

The image of spirit behind the façade of physical reality is equally fitting when looking at the virtual worldwide identity of high-energy structures and processes. The universal imperatives of their energy and material inputs and operating requirements make the blast furnaces of the U.S. Midwest, Germany's Ruhrgebiet, Ukraine's Donets region, China's Hebei province, Japan's Kyushu, and India's Bihar outwardly nearly identical—but they are not the same when seen in an all-encompassing context. Their distinctiveness relates to the amalgam of cultural, political, social, economic, and strategic settings in which they have originated and continue to operate, as well as to the eventual destination and quality of the final products made from the metal they smelt.

Another critical link where energy explanations are of limited usefulness is the effect of energy supply on population growth. Relatively most reliable long-term demographic reconstructions, those for Europe and China, show very long periods of slow growth composed of successive waves of expansions and crises caused by epidemics and wars (Livi-Bacci 2000, 2012). The

European total during the first half of the eighteenth century was about three times that at the beginning of the Common Era—but by 1900 it had more than tripled. Improved nutrition had to be a major reason for this rise, but making it virtually the only underlying factor (McKeown 1976) cannot be reconciled with careful reconstructions of average food energy intakes (Livi-Bacci 1991).

And if we were to ascribe Europe's post-1750 population rise to higher energy consumption (translated into better housing, hygiene, and health care), then how should we explain the concurrent rise of China's population during the Qing dynasty? In 1700 China's population total was only about three times the Han dynasty peak in the year 145—but by 1900 it nearly matched European growth by tripling to about 475 million people. Yet during that period there were no major shifts to new energy sources or prime movers, hardly any improvement in average per capita use of biomass fuels and coal, and no major gains in average per capita supply of food: indeed, the period had seen one of China's worst famines in 1876–1879.

Not surprisingly, energy considerations are also of limited help in trying to explain some of the greatest recurrent puzzles of history, the collapse of complex societies. Inquiries into this fascinating challenge (Tainter 1988; Ponting 2007; Diamond 2011; Faulseit 2015) proffer simple answers only when their authors are willing to ignore inconvenient complexities. The most notable energy-related explanations have included the effects of widespread ecosystemic degradation brought about by unsustainable methods of farming and by excessive deforestation and resulting in reduced food production. The impossibility of effectively integrating large empires because of poor land transportation and the rising burden of resources to protect distant territories (the imperial overstretch syndrome) are other common explanations.

But, as attested by scores of different reasons offered to explain the fall of the Roman Empire—by far the most studied "collapse" in history (Rollins 1983; Smil 2010c)—explanations favoring social dysfunction, internal conflict, invasions, epidemics, or climatic change are much more common. An indisputable fact is that many instances of sociopolitical collapse came about without any persuasive evidence of weakened energy bases. Neither the slow disintegration of the Western Roman Empire, nor Teotihuacan's sudden demise can be persuasively tied to badly degraded food production capacity, to any notable shifts in dominant prime movers, or to any dramatic changes in the use of biomass fuels. Conversely, a number of historically far-reaching consolidations and expansions—

including the gradual rise of Egypt's Old Kingdom, the emergence of the Roman republic as the dominant power in Italy, the swift spread of Islam during the seventh century, and the Mongolian invasions during the thirteenth century—cannot be connected to any major changes in using prime movers and fuels.

Extreme futures are easy to outline. On the one hand, it is conceivable that the understanding now mastered by Western civilization may take it beyond others in its basic behavioral patterns. The diffusion of this knowledge could create a true world civilization that will learn to live within the biospheric limits and prosper for millennia to come. Directly opposed to this is the argument that the biosphere is already subject to human actions that interfere with many fundamental life-supporting processes and even encroach on the planetary boundaries that define the safe operating space for the humanity (Stockholm Resilience Center 2015). Consequently, it is equally conceivable—leaving aside the possibility of a full-scale nuclear war—that the global, high-energy civilization may collapse long before approaching its resource limits. The vast space between these two extremes can be filled with many scenarios ranging from temporary continuation, or even deepening, of global inequality to slow but important progress toward more rational national and global policies.

Leaving aside the possibilities of an asteroid impact, megavolcanic eruptions, or unprecedented viral pandemics (for their appraisals, see Smil 2008b), the gradual dissipation brought on by degrading the biosphere beyond sustainable habitability would seem to have a higher probability than a sudden, Teotihuacan-like demise. I will offer no predictions regarding the chances of destructive social dysfunction, worldwide wars or epidemics but merely note the coexistence of two contradictory expectations concerning the energy basis of modern society: chronic conservatism (lack of imagination?) regarding the power of technical innovation, set against repeatedly exaggerated claims made on behalf of new energy sources.

The list of failed technical predictions is long (Gamarra 1969; Pogue 2012), and some of the choicest items concern the development and use of energy conversions (Smil 2003). Expert opinion of the day dismissed the possibility of gas lighting, steamships, incandescent light bulbs, telephones, gasoline engines, powered flight, alternating current, radio, rocket propulsion, nuclear energy communication satellites, and mass computing. This conservatism often persevered even after successful introduction of innovations. Transatlantic steamship voyages did not seem possible because it was thought that the vessels could not carry enough fuel for such long trips. In 1896 Lord Kelvin refused to join the Royal Aeronautical Society: his handwritten note to Baden F.S. Baden-Powell, an enthusiastic military aviation proponent, said that he had "not the smallest molecule of faith in aerial navigation other than ballooning" (Thomson 1896) At a time when scores of automakers were turning out more efficient and more reliable cars, Byrn (1900, 271) thought that "it is not probable that man will ever be able to get along without the horse."

The persistence of new energy myths is at least as remarkable. New energies are initially seen to carry few, if any, problems. They promise abundant and cheap supply, opening up the possibility of near-utopian social change (Basalla 1982; Smil 2003, 2010a). After millennia of reliance on biomass fuels, many nineteenth-century writers saw coal as an ideal energy source and the steam engine as a nearly miraculous prime mover. Heavy air pollution, land destruction, health hazards, mining accidents, and the need to turn to progressively poorer or deeper coal reserves soon swept that myth away. Electricity was the next carrier of unbounded possibilities, its powers eventually so far-reaching that they would cure poverty and disease (box 7.5).

What can be foreseen with great certainty is that much more energy will be needed during the coming generations to extend decent life to the majority of a still growing global population whose access to energy is well below the minima compatible with a decent quality of life. This may seem to be an overwhelming, even impossible, task. The global high-energy civilization already suffers economically and socially from its precipitous expansion, and its further growth threatens the biospheric integrity on which its very survival depends (Smil 2013a; Rockström et al. 2009).

Yet another great uncertainty is the long-term viability of urban living. Social cohesion and family nurturing so characteristic of rural life clearly do not prevail in modern cities. The strains of urban living on populations that have been for so long rural and cohesive are manifest in both rich and poor nations. Overall crime rates may have fallen in many nations, but large parts of many of the world's largest cities remain epitomes of violence, drug addiction, homelessness, child abandonment, prostitution, and squalid living. And yet, perhaps more than ever, the imperatives of modern economies demand social stability and continuity of effective cooperation. Cities have been always renewed by migration from villages—but what will happen to the already mostly urban civilization once the villages virtually disappear while the social structure of cities continues to disintegrate?

But there are hopeful signs. Precisely because the gross energy use does not determine the course of history, our commitment and our

Box 7.5
Electricity's promise that never ends

Electricity is the most versatile form of energy, and its multifaceted promise inspired innovators—Edison, Westinghouse, Steinmetz, Ford—and politicians alike, the latter often on opposite ends of the value spectrum, such as Lenin and Roosevelt. Even before the end of the Russian Civil War, Lenin (1920, 1) concluded that economic success "can be assured only when the Russian proletarian state effectively controls a huge industrial machine built on up-to-day technology; that means electrification." The "white coal" of hydroelectricity held a particularly special appeal among the Western technocrats until the 1950s, when it was surpassed by the unprecedented promise of nuclear power.

In 1954 Lewis L. Strauss (1896–1974), the chairman of the U.S. Atomic Energy Commission (he held the office between 1953 and 1958), told the National Association of Science Writers in New York:

> Our children will enjoy in their homes electrical energy too cheap to meter. It is not too much to expect that our children will know of great periodic regional famines in the world only as matters of history, will travel effortlessly over the seas and under them and through the air with a minimum of danger and at great speeds, and will experience a lifespan far longer than ours, as disease yields and man comes to understand what causes him to age. This is the forecast of an age of peace. (Strauss 1954, 5)

In 1971 Glen Seaborg, chairman of the U.S. Atomic Energy Commission, forecasted that by the year 2000, half of America's electricity generating capacity would come from nonpolluting and safe nuclear reactors, and that nuclear-powered spaceships would be ferrying men to Mars (Seaborg 1972). In reality, the 1980s saw an almost complete end of orders for new nuclear plants in the West, and fission's dismal prospects were further damaged in 1986 by the Chornobyl accident and in 2011 by multiple explosions at Fukushima. But wind turbines and photovoltaics, stepped into the mythical void created by fission's Western retreat with the promise to generate electricity so easily and so cheaply that decentralized power (eliminating all central stations) is to descend, manna-like on the modern world (a vision that requires ignoring the fact that most of the world's population will soon live in megacities, hardly the best location for decentralized power). And then there is, as there has been ever since 1945, the ultimate promise of electricity from nuclear fusion (although in practical terms we are no closer to that goal now than we were a generation ago).

inventiveness can go a long way toward first weakening and then even reversing the evolutionary link between civilization's advances and energy. We now realize that growing energy use cannot be equated with effective adaptations and that we should be able to stop and even to reverse that trend, to break the dictum of Lotka's (1925) law of maximum energy. This should be easier given the clear indications that it is counterproductive to maximize power outputs.

Indeed, higher energy use by itself does not guarantee anything except greater environmental burdens (Smil 1991). The historical evidence is clear. Higher energy use will not ensure a reliable food supply (wood-burning czarist Russia was a grain exporter; the USSR, the hydrocarbon superpower, had to import grain); it will not confer strategic security (the United States was surely more secure in 1915 than in 2015); it will not safely underpin political stability (whether in Brazil, Italy, or Egypt); it will not necessarily lead to a more enlightened governance (it surely has not in North Korea or Iran); and it will not bring widely shared increases in a nation's standard of living (it has not done so in Guatemala or Nigeria).

Opportunities for a grand transition to less energy-intensive society can be found primarily among the world's preeminent abusers of energy and materials in Western Europe, North America, and Japan. Many of these savings could be surprisingly easy to realize. I agree with Basalla (1980, 40) that

> if the energy-civilization equation is worthless and potentially dangerous it should be exposed and discarded because it supplies a supposedly scientific argument against efforts to adopt a style of living based upon lower levels of energy consumption. If it is a generalization of great truth and intellectual wealth, then it deserves a more sophisticated and rigorous handling then it has received from its supports to date.

Knowing about the enormous inefficiencies of resource use, whether that resource is energy, food, water, or metals, by modern civilization, I have always argued for more rational ways of consumption. Such a course would have profound consequences for assessing the prospects of a high-energy civilization—but any suggestions of deliberately reducing certain resource uses are rejected by those who believe that endless technical advances can satisfy steadily growing demand. In any case, the probability of adopting rationality, moderation, and restraint in resource consumption in general and energy use in particular, and even more so the likelihood of persevering on such a course, is impossible to quantify.

Life's two cardinal characteristics have been expansion and increasing complexity. Can we reverse these trends by adopting the technically feasible and environmentally desirable shift to moderated energy uses? Can we continue human evolution by concentrating only on those aspects that do not require maximization of energy flows, can we create an energetically invariable civilization that would be living strictly within its solar/biospheric limits? Could such a shift be accomplished without eventually converting to a no-growth economy and reducing the current global population? For individuals, this would mean a no less revolutionary delinking of social status from material consumption. Setting up such societies would be especially burdensome for the first generations making the transition. In the longer run, these new arrangements would also eliminate one of the mainsprings of Western progress, the quest for social and economic mobility. Or could new technical breakthroughs allow us to harness, directly and efficiently, a large share of incoming radiation and safeguard our addiction to a myriad of extrasomatic comforts?

Our current energy system is self-limiting: even on a historical time scale our high-energy civilization, exploiting the accumulated store of ancient radiation transformed into fuels, is just an interlude because even if the combustion of those fuels had no environmental impacts it could not, unlike its predecessors, based on harvesting near-instant solar energy flows, last for millennia. But the eventual exhaustion of fossil energies is most unlikely because the burning of coal and hydrocarbons is the principal source of anthropogenic CO_2 and the combustion of available fossil fuel resources would raise the tropospheric temperature high enough to eliminate the entire Antarctic ice sheet and cause a sea-level rise of about 58 m (Winkelmann et al. 2015).

With a majority of the world's population living in coastal regions, such a rise would have profound consequences for the civilization's survival. The available flows of renewable energies are large enough to avoid such a fate—but in order to maintain, and, for the billions living in low-income economies to expand, the prevailing levels of energy uses we would have to capture them, convert them, and store them on scales orders of magnitude higher than we have done so far. That epochal transition from the fossil fuel–dominated global energy system to a new arrangement based solely on renewable energy flows presents an enormous (and generally insufficiently appreciated) challenge: the ubiquity and the magnitude of our dependence on fossil fuels, and the need for further increases of global energy use, mean that even the most vigorously pursued transition could be accomplished only in the course of several generations.

And the complete transition would require the replacement of fossil fuels not only as the dominant providers of different kinds of energies but also as critical sources of raw materials: feedstocks for the synthesis of ammonia (about 175 Mt/year in 2015, mostly to supply nitrogen for crops) and other fertilizers and agrochemicals (herbicides and pesticides); feedstocks for now ubiquitous plastics (whose total output is about 300 Mt/year); metallurgical coke (now requiring every year about 1 Gt of coking coal and used not just as the source of energy for reducing iron oxides but for its structural role in supporting charged iron ore and flux in blast furnaces producing annually more than 1 Gt of iron); lubricants (essential for functioning of both stationary and transportation machines); and paving materials (inexpensive asphalt).

Our inability to comprehend the behavior of complex and interdependent wholes—the interactions of biospheric processes, energy production use, economic activities, technical advances, social changes, political developments, armed aggression—makes any specific (and now so commonly proffered) scenarios of distant futures mere speculation. In contrast, outlining the extremes is easy, as the visions of future range from dismal to ecstatic. Georgescu-Roegen's (1975, 379) was not hopeful: "Perhaps, the destiny of man is to have a short, but fiery, exciting and extravagant life rather than a long, uneventful and vegetative existence. Let other species—the amoebas, for example—which have no spiritual ambitions inherit the earth still bathed in plenty of sunshine." In contrast, techno-optimists see a future of unlimited energy, whether from superefficient PV cells or from nuclear fusion, and of humanity colonizing other planets suitably terraformed to the Earth's image. For the foreseeable future (two-four generations, 50–100 years) I see such expansive visions as nothing but fairy tales.

The only certainty is that the chances of succeeding in the unprecedented quest to create a new energy system compatible with the long-term survival of high-energy civilization remain uncertain. Given our degree of understanding, the challenge may not be relatively more forbidding than overcoming a number of barriers we have surmounted in the past. But understanding, no matter how impressive, will not be enough. What is needed is a commitment to change, so we could say with Senancour (1770–1846),

Man perisheth. That may be, but let us struggle even though we perish; and if the nothing is to be our portion, let it not come to us as a just reward. (Senancour 1901 [1804], 2:187)

Addenda

Basic Measures

Length, mass, time, and temperature are the basic units of scientific accounts. The meter (m) is the basic unit of length. For average-sized people it is roughly the distance between their waist and the ground. Most people are between 1.5 and 1.8 m tall; ceilings of American houses are about 2.5 m high; an Olympic track runs 400 m, a jet runway around 3,000 m. Standard Greek prefixes are used to express multiples of scientific units. Kilo is 1,000, and hence 3,000 m are 3 kilometers (km). A marathon run is 42.195 km, a coast-to-coast flight is 4,000 km, the equatorial circumference is about 40,000 km, light travels 300,000 km every second, and 150 million km separate Earth from the Sun. Standard Latin prefixes are used for fractional units. A centimeter is a hundredth part of one meter. A fist resting on a table with the thumb alongside the bent fingers will be about 10 cm (0.1 m). New pencils measure about 20 cm (0.2 m), newborn babies about 50 cm (0.5 m).

Commonly encountered area units range from cm^2 to km^2. A coaster covers about 10 cm^2, a bed about 2 m^2, the foundations of a small American bungalow about 100 m^2. This area (10 × 10 m) is called an *are*, and 100 (*hecto*) such squares add up to a *hectare* (ha), the basic metric unit used for measuring agricultural land. Chinese or Bangladeshis must feed themselves from less than 0.1 ha/capita, while Americans cultivate nearly 1 ha/capita. Outside agriculture, larger areas are usually expressed in square kilometers (km^2). North American cities of about one million people usually cover less than 500 km^2; small European countries have well below 100,000 km^2; and the United States takes up nearly 10 million km^2.

Basic mass units can be easily derived by filling cubes with water. A tiny cube enclosing one cubic centimeter (cm^3)—its side will be only as long as the width of a small fingernail—will weigh (or, more precisely, it will have a mass of) one gram (g) when filled with water. A fist-sized cube will enclose 1,000 cm³ (10 × 10 × 10 cm), or one liter (L) of volume. When filled with water it will have a mass of 1,000 grams (g) or one kilogram (kg). The kilogram is the basic unit of mass. Soft drinks weigh about a third of a kilogram (350 g), newborn babies between 3 and 4 kg, and most non-American adults between 50 and 90 kg. Compact cars have a mass around 1,000 kg, or one tonne (also called metric tons, t; the U.S. short ton has only 907 kg). A big horse will weigh as much as 1 t; railway cars range from 30 to 100 t, ships (fully loaded) from a few thousand tonnes to 500,000 t.

The second (s), a time span slightly longer than an average heartbeat, is the basic unit of time. When at rest, we take a breath every four seconds, and it takes about 10 seconds to drink a glass of water. Larger time units are exceptions within the metric system of scientific units. They do not go up with multiples of 10 but follow the ancient Sumerian-Babylonian sexegesimal (base 60) counting. A red light at a busy intersection lasts 60 seconds, or one minute. Hard boiling an egg takes eight minutes; an average work for classical symphony lasts 40 minutes. A normal pregnancy lasts 280 days; every nonleap year has 365 days or 31.536 million seconds; the average life span for females in Western countries has now surpassed 80 years; agriculture started to spread about 10,000 years ago; dinosaurs were plentiful 80 million years ago; and Earth is roughly 4.5 billion years old.

The scientific scale for temperature, degrees Kelvin, starts at absolute zero. The Celsius (C) scale is more common: it divides the span between the freezing and boiling point of water into 100 degrees (°C). On that scale absolute zero is –273.15°C, water freezes at 0°C, a fine spring day is around 20°C, and normal human temperature is 37°C. Water boils at 100°C, paper ignites at 230°C, iron melts at 1535°C, and the Sun's thermonuclear reactions proceed at 15 million °C.

Nearly all other scientific units can be derived from length, mass, time, and temperature. For energy and power the derivations are as follows. Force acting on a mass of 1 kg having an acceleration of 1 m/s² is equal to one newton (N). The force of 1 N applied over a distance of 1 m equals one joule (J), the basic unit of energy. Calorie, an energy unit often used in nutritional writings, is equal to 4.184 J. Both of these are small quantities: the daily food consumption of an active adult woman will be only 2,000 calories (2 Mcal) or 8.36 MJ. Power is energy per time; hence 1 J/s is equal to one watt (W). The section "Power in History," below, lists a wide variety of actions in ascending watt order.

Many units, such as those for speed—meters per second (m/s) or kilometers per hour (km/h)—and productivity—kilograms or tonnes per hour (kg/h, t/h) or tonnes per year (t/year)—have no special name. Working horses move at about 1 m/s; most highway speed limits are around 100 km/h. A slave milling grain with a stone quern produced flour at a rate no higher than 4 kg/h; an excellent crop of late medieval wheat was 1 t/ha.

Only a few measures described here account for most of the appearances in the text. Besides the energy and power units, they are the two basic physical units of length and mass (m and kg), two areal measures (ha and km^2), and the four markers of time (second, hour, day, and year). The full list of prefixes follows, but only few of them (going up: hecto, kilo, mega, giga; going down: milli, micro) are used frequently.

Scientific Units and Their Multiples and Submultiples

Basic SI units

Quantity	Name	Symbol
Length	meter	m
Mass	kilogram	kg
Time	second	s
Electric current	Ampere	A
Temperature	Kelvin	K
Amount of substance	mole	mol
Luminous intensity	candela	cd

Other units used in the text

Quantity	Name	Symbol
Area	hectare	ha
	square meter	m^2
Electric potential	Volt	V
Energy	Joule	J
Force	Newton	N
Mass	gram	g
	tonne	t
Power	Watt	W
Pressure	Pascal	Pa
Temperature	degree Celsius	°C
Volume	cubic meter	m^3

Multiples used in the International System of Units

Prefix	Abbreviation	Scientific notation
deka	da	10^1
hecto	h	10^2
kilo	k	10^3
mega	M	10^6
giga	G	10^9
tera	T	10^{12}
peta	P	10^{15}
exa	E	10^{18}
zeta	Z	10^{21}
yota	Y	10^{24}

Submultiples used in the International System of Units

Prefix	Abbreviation	Scientific notation
deci	d	10^{-1}
centi	c	10^{-2}
milli	m	10^{-3}
micro	μ	10^{-6}
nano	n	10^{-9}
pico	p	10^{-12}
femto	f	10^{-15}
atto	a	10^{-18}
zepto	z	10^{-21}
yocto	y	10^{-24}

Chronology of Energy-Related Developments

This list is compiled from a wide variety of sources cited in the text and in the bibliography. More extensive chronologies of technical advances can be found in Mumford (1934), Gille (1978), Taylor (1982), Williams (1987), and Bunch and Hellemans (1993), and readers wishing to see the most extensive chronologies of energy-related developments (listed by energy sources, applications, and impacts) should consult a nearly 1,000-page volume by Cleveland and Morris (2014). Space consideration restricts this list mostly to practical advances (and some notable failures): it excludes the underlying intellectual, scientific, political, and economic contributions. All early dates are inevitable approximations, and different sources may list different timings. Discrepancies exist even with modern advances: dates may refer to the original idea, to patenting, to the first practical application, or to successful commercialization. For problems in dating inventions, see Petroski (1993).

BCE

1,700,000 +	Oldowan stone tools (<0.5 m of edge/kg of stone)
250,000 +	Acheulean stone tools
150,000 +	Mousterian stone flake tools
50,000 +	Bone objects
30,000 +	Aurignacian stone tools
	Bow and stone arrows
15,000 +	Magdalenian stone tools (12 m of edge/kg of stone)
9,000 +	Sheep domesticated in the Middle East
7,400 +	Corn in Oaxaca Valley
7,000 +	Wheat in Mesopotamia
	Pigs domesticated in the Middle East
6,500 +	Cattle domesticated in the Middle East
6,000 +	Copper artifacts more common in the Middle East
5,000 +	Barley in Egypt
	Corn in the basin of Mexico
4,400 +	Potatoes in highland Peru and Bolivia
4,000 +	Light wooden plows in Mesopotamia
3,500 +	Pack asses in the Middle East
	Wooden ships in the Mediterranean
	Pottery and bricks fired in kilns in Mesopotamia
	Irrigation in Mesopotamia
3,200 +	Wheeled vehicles in Uruk
3,000 +	Square sail in Egypt
	Draft oxen in Mesopotamia
	Camel domesticated
	Potter's wheel in Mesopotamia
2,800 +	Pyramid construction in Egypt
2,500 +	Bronze in Mesopotamia
	Small glass objects in Egypt

2,000 +	Spoked wheel in Mesopotamia
	Horse-drawn vehicles in Egypt
	Shaduf in Mesopotamia
1,700 +	Horse riding
1,500 +	Copper in China
	Paddy rice in China
	Axle lubricants in the Middle East
1,400 +	Iron in Mesopotamia
1,300 +	Seed drill in Mesopotamia
	Horse-drawn chariots in China
1,200 +	Iron more common in India, Middle East, Europe
800 +	Mounted archers on Asian steppes
	Candles in the Middle East
600 +	Tin in Greece
	Penteconter ships common in Greece
	Archimedean screw in Egyptian irrigation
500 +	North Arabian camel saddle
	Trireme in Greece
400 +	Crossbow in China
432	Parthenon completed
300 +	Stirrups in China
	Gears in Egypt and Greece
312	Roman Via Appia and Aqua Appia completed
200 +	Breastband harness in China
	Sailings to windward advances in China
	Batten-strengthened sails in China
	Percussion drilling in Sichuan
	Crank handle in China
150 +	Iron moldboard plows in China
100 +	Beginnings of collar harness in China
	House heating by coal in China
	Waterwheels in Greece and Rome
	Wheelbarrow in China
	Norias in the Middle East
80 +	Hypocaust heating in Rome

CE

300	Roman cursus publicus surpasses 80,000 km
600 +	Windmills (Iran)
850 +	Triangular sail in the Mediterranean
900 +	Collar harness and horseshoes common in Europe
	Bamboo fire-lances in China
980 +	Canal pound lock in China
1000 +	Widespread adoption of waterwheels in Western Europe
1040	Clear directions for gunpowder preparation in China
1100 +	Long bow in England
1150 +	Windmills spreading in Western Europe
1200 +	Inca road construction
1280 +	Cannons in China
1300 +	Gunpowder and cannons in Europe
1327	Beijing-Hangzhou Grand Canal (1,800 km long) completed
1350 +	Handheld guns in Europe
1400 +	Heavy draft horses in Europe
	Drainage windmills in the Netherlands
	Blast furnaces in the Rhine region
1420 +	Portuguese caravels make longer sailings
1492	Columbus sails across the Atlantic
1497	Vasco da Gama sails to India
1519	Magellan's *Victoria* circumnavigates the Earth
1550 +	Large full-rigged sail ships with guns in Western Europe
1600 +	Ball bearings in Western Europe
1640 +	English coal mining expands
1690	Experiments with atmospheric steam engine (Denis Papin)
1698	Simple, small steam engine (Thomas Savery)
1709	Coke from bituminous coal (Abraham Darby)
1712	Atmospheric steam engine (Thomas Newcomen)
1745	Fantail for automatic turning of windmills
1750 +	Intensive canal construction in Western Europe
	Use of coke spreads in English iron making
	Newcomen's engine more common in English coal mines
1757	Precision-cutting lathe (Henry Maudslay)
1769	James Watt patents a separate condenser for steam engine

1770s	Factories powered by waterwheels
1775	Watt's patent extended to 1800
1782	Hot air balloon (Joseph and Etienne Montgolfier)
1794	Lamps with wick holders and glass chimneys (Aimé Argand)
1800	Electric battery (Alessandro Volta)
1800s	Steamboats (*Charlotte Dundas, Clermont*)
	High-pressure steam engines (R. Trevithick, O. Evans)
1805	Steam-powered crane (John Rennie)
	Coal (town) gas in England
1808	Arc lamp (Humphrey Davy)
1809	Chilean nitrates discovered
1816	Mine safety lamp (Humphrey Davy)
1820s	Designs of mechanical calculators (Charles Babbage)
	Iron ship hulls
1820	Electromagnetism (Hans C. Oersted)
1823	Silicon isolated (J. J. Berzelius)
1824	Portland cement (Joseph Aspdin)
	Aluminum isolated (Hans C. Oersted)
1825	Stockton-Darlington railway
1828	Hot blast in ironmaking (James Neilson)
1829	*Rocket* locomotive (Robert Stephenson)
1830s	Railway construction takes off in England
	Steamship cross the Atlantic
	Mechanical grain reaper (Cyrus McCormick, Obed Hussey)
1830	Thermostat (Andrew Ure)
	Liverpool-Manchester railway
1832	Water turbine (Benoît Fourneyron)
1833	Steel plow (John Lane)
	Steamship *Royal William* crosses from Quebec to London
1834	Free-standing kitchen range (Philo P. Stewart)
1837	Electric telegraph patented (William F. Cooke and Charles Wheatstone)
1838	Screw propulsion for steamships (John Ericsson)
	Telegraph code (Samuel Morse)
1840s	Peak decade of U.S. whaling
1841	Steam-driven threshing machine
	Thomas Cook offers holiday trips

1847	Inward-flow water turbine (James B. Francis)
1850s	Paraffin from oil for lighting
	Fast clipper ships on long voyages
1852	Hydrogen-filled airship (Henri Giffard)
1854	*Great Eastern* steamship (Isambard K. Brunel)
1856	Steel converter (Henry Bessemer)
1858	Grain harvester (C. W. and W. W. Marsh)
1859	Pennsylvania oil drilling (E. L. Drake)
1860s	Steam plowing of large American fields
1860	Horizontal internal combustion engine (J. J. E. Lenoir)
	Milking machine (L. O. Colvin)
1864	Open-hearth steelmaking process (W. and F. Siemens)
1865	Nitrocellulose (J. F. E. Schultze)
1866	Carbon-zinc battery (Georges Leclanche)
	Transatlantic cable in permanent operation
	Torpedo (Robert Whitehead)
1867	Refrigerated railway wagons in service
1869	Suez Canal completed
	U.S. transcontinental railroad completed
1870s	Refrigerated transport of meat by ocean ships
	Phosphate fertilizer industry begins
1871	Ring-wound armature dynamo (Z. T. Gramme)
1875	Dynamite (Alfred Nobel)
1874	Photographic film (George Eastman)
1876	Four-stroke internal combustion engine (N. A. Otto)
	Telephone patented (Alexander Graham Bell, Elisha Gray)
1877	Phonograph (Thomas A. Edison)
1878	Two-stroke internal combustion engine (Dugald Clerk)
	Filament light bulb (Joseph Swan)
	Twine knotter for grain harvester (John Appleby)
1879	Carbon filament light bulb (Thomas A. Edison)
1880s	Horse-drawn grain combines (California)
	Modern bicycles (J. K. Starley, William Sutton)
	Crude oil tankers
	Military high explosives formulated
1882	Edison's first electricity-generating plants

1883	Impulse steam turbine (Carl Gustaf de Laval)
	Four-stroke liquid-fueled engine (Gottlieb Daimler)
	Machine gun (Hiram S. Maxim)
1884	Steam turbine (Charles Parsons)
1885	Transformer (William Stanley)
	Karl Benz builds the first practical car
1886	Prestressed concrete (C. E. Dochring)
	Aluminum production (C. M. Hall and P. L. T. Héroult)
1887	Crude oil discovered in Texas
	Generation of electromagnetic waves (Heinrich Hertz)
1888	Induction electric motor (Nikola Tesla)
	Gramophone (Emile Berliner)
	Air-filled rubber tire (John B. Dunlop)
1889	Phonograph playing wax cylinders (Thomas A. Edison)
	Jet-driven water turbine (Lester A. Pelton)
1890s	Horses reach peak numbers in Western cities
	Electric household appliances introduced
1892	Diesel engine (Rudolf Diesel)
1894	Offshore oil drilling from jetties (California)
1895	Moving pictures (Louis and August Lumiére)
	X-rays (Wilhelm K. Roentgen)
1897	Cathode-ray tube (Ferdinand Braun)
1898	Tape recorder (Valdemar Poulsen)
1899	Radio signals transmitted across the English Channel (Guglielmo Marconi)
1900s	Electricity consumption takes off in United States and U.K.
	Volume production of cars begins
1900	Dirigible powered airship (Ferdinand von Zeppelin)
1901	Industrial air conditioning (Willis H. Carrier)
	Rotary drilling (Spindletop, Texas)
	Radio signals transmitted across the Atlantic (Guglielmo Marconi)
1903	Sustained controlled powered flight (Oliver and Wilbur Wright)
1904	Geothermal electricity generation (Lardarello, Italy)
	Vacuum diode (John A. Fleming)
1905	Photoelectric cell (Arthur Korn)
	Commercial tractor production in the United States
1906	British *Dreadnought* battleship launched
	Vacuum triode (Lee De Forest)

1908	Tungsten light bulb
	Ford Model T (manufactured until 1927)
1909	Rolling cutter rock-drilling bit (Howard Hughes)
	Louis Bleriot flies across the English Channel
	Bakelite, the first major plastic (Leo Baekeland)
1910	Neon light (Georges Claude)
	Synthetic gas from coal (Fischer-Tropsch, Germany)
1913	Moving production line (Ford Company)
	Panama Canal completed
	Ammonia synthesis (Fritz Haber and Carl Bosch)
	High-pressure crude oil cracking (W. M. Burton)
1914	World War I (until 1918): trench and gas warfare, airplanes, tanks
1919	Nonstop transatlantic flight (J. Alcock and A. W. Brown)
	Scheduled airline service (Paris-London)
1920s	Boilers burning pulverized coal
	Streamlined metal plane bodies
	Electric record players
	Radio broadcasting spreads in North America and Europe
	Liquefaction of coal (Friedrich Bergius)
1920	Axial-flow water turbine (Viktor Kaplan)
1922	Aircraft carrier *Hosho* launched in Japan
1923	Electronic camera tube (Vladimir Zworykin)
	Electric refrigerators by Electrolux
1927	Synthetic rubber (Buna)
	Nonstop solo transatlantic flight (Charles A. Lindbergh)
	Crude oil discovered in Kirkuk, Iraq
1928	Plexiglass (W. Bauer)
1929	Experimental TV broadcasts (UK)
1930s	Catalytical crude oil cracking (Eugene Houdry)
	Large hydrostations (United States and USSR)
	Long-distance bombers
	Chlorofluorocarbons in refrigeration
1933	Polyethylene (Imperial Chemical Industries)
1935	Fluorescent light (General Electric)
	Plastic magnetic tape (AEG Telefunken, I. G. Farben)
	Nylon (Wallace Carothers)
1936	Regular TV broadcasting (BBC)
	Gas turbine (Brown-Boveri)

1937	Fully pressurized aircraft (Lockheed XC-35)
1938	Prototype jet fighter (Hans Pabst von Ohain)
1939	Radar (UK)
	World War II (until 1945): Blitzkrieg
1940s	Military jet aircraft
	Electronic computers
1940	Helicopter (Igor Sikorsky)
1942	V-1 rockets (Wernher von Braun)
	Industrial production of silicone
	Controlled chain reaction (Enrico Fermi, Chicago)
1944	V-2 rockets
	DDT marketed
1945	Nuclear bombs (Trinity test, Hiroshima and Nagasaki)
	Electronic computer (ENIAC, United States)
	First herbicide (2,4-D) marketed
1947	Transistor (J. Bardeen, W. H. Brattan and W. B. Shockley)
	Offshore drilling out of sight of land (Louisiana)
	Piloted flight faster than speed of sound (Bell X-1)
1948	Basic-oxygen steelmaking furnace (Linz-Donawitz)
	The world's largest oil field (Saudi al-Ghawar) discovered
1949	First passenger jet aircraft (De Havilland Comet)
1950s	Rapidly growing worldwide crude oil consumption
	Continuous steel casting
	Electrostatic precipitators spread
	Commercial computers
	Stereo recordings
	Videotape recorders
1951	Automatic engine assembly (Ford Company)
	Transmission of color TV images
	Hydrogen (fusion) bomb
1952	British Comet jetliner in commercial service
1953	Microwave oven (Raytheon Manufacturing Company)
1954	U.S. Navy nuclear submarine *Nautilus* launched
1955	Fully transistorized Sony radio
1956	First commercial nuclear power plant (Calder Hall, UK)
	Transatlantic telephone cable
	Interstate highway construction begins in United States

1957	*Sputnik 1*, the first artificial Earthsatellite (USSR)
	First U.S. nuclear power plant (Shippingport, Pennsylvania)
1958	Integrated circuit (Texas Instruments)
	U.S. Boeing 707 jetliner in service
1960s	Semisubmersible platforms for offshore drilling
	Meteorological and communication satellites
	Very large crude oil tankers
	Large-scale deployment of intercontinental ballistic missiles (ICBMs)
	Largest Soviet fusion bombs tested in the atmosphere
	Spreading use of synthetic fertilizers and pesticides
	High-yielding crop varieties
1960	U.S. Minuteman weapon system ICBM tested
1961	U.S. nuclear-powered aircraft carrier *Enterprise* launched
	Manned space flight (Yuri Gagarin)
1962	Transatlantic television relay (Telstar)
1964	Shinkansen (Japan National Railways) starts operation
1966	U.S. jumbo jet aircraft Boeing 747 ordered
1969	British-French supersonic aircraft Concorde takes off
	Boeing 747 in commercial service
	U.S. *Apollo 11* spacecraft lands on Moon
1970s	Radio and television satellite broadcasting
	Concerns about fossil fuel supplies
	Acid rain over Europe and North America
	Japanese car exports soar
1971	First microprocessors (Intel, Texas Instruments)
1973	OPEC's first round of crude oil price increases (until 1974)
1975	Brazil starts producing automotive ethanol from sugar cane
1976	Concorde in commercial service
	Unmanned U.S. *Viking* spacecraft lands on the Mars
1977	Human-powered flight of *Gossamer Condor*
1979	OPEC's second round of crude oil price increases (until 1981)
1980s	Ownership of personal computers takes off
	More efficient appliances and cars
	Concerns about global environmental change
	Genetic engineering takes off
1982	CD player (Philips, Sony)
1983	French TGV starts operation (Paris-Lyon)

1985	Antarctic ozone hole identified
1986	Chornobyl nuclear reactor disaster
1989	World Wide Web introduced (Tim Berners-Lee)
1990	Global population surpasses 5 billion
1994	Netscape launched
1999	Mass adoption of smart phones begins
2000s	Widespread installation of wind turbines and PV cells
2000	German *Energiewende* begins
2003	Three Gorges Dam completed (Yangzi river, China)
2007	Hydraulic fracturing takes off in the United States
2009	China becomes world's largest consumer of energy
2011	Tsunami and mismanagement cause Fukushima nuclear disaster
	Global population reaches 7 billion
2014	United States is once again the world's largest producer of natural gas
2015	Average concentration of atmospheric CO_2 reaches 400 ppm

Power in History

Power ratings: From a candle to global civilization

Actions, prime movers, converters	Power (W)
Small wax candle burning (800 BCE)	5
Egyptian boy turning Archimedean screw (500 BCE)	25
Small U.S. windmill rotating (1880)	30
Chinese woman cranking a winnowing machine (100 BCE)	50
Steadily working French glass polishers (1700)	75
Strong man treading rapidly a wooden wheel (1400)	200
Donkey turning a Roman hourglass mill (100 BCE)	300
Weak pair of Chinese oxen plowing (1900)	600
Good English horse turning a whim (1770)	750
Dutch treadwheel powered by eight men (1500)	800
Very strong American horse pulling a wagon (1890)	1,000
Long-distance runner at the Olympic Games (600 BCE)	1,400
Roman vertical waterwheel turning a millstone (100 CE)	1,800
Newcomen's atmospheric engine pumping water (1712)	3,750

Actions, prime movers, converters	Power (W)
Engine of Ransom Olds's Curved Dash automobile (1904)	5,200
Greek penteconter with 50 oarsmen at full speed (600 BC)	6,000
Large German post windmill crushing oilseeds (1500)	6,500
Roman messenger horse galloping (200 CE)	7,200
Large Dutch windmill draining a polder (1750)	12,000
Engine of Ford Model T at full speed (1908)	14,900
Greek trireme with 170 oarsmen at full speed (500 BCE)	20,000
Watt's steam engine winding coal (1795)	20,000
Team of 40 horses pulling a California combine (1885)	28,000
Cascade of 16 Roman water mills at Barbegal (350 CE)	30,000
Benoît Fourneyron's first water turbine (1832)	38,000
Water pumps for Versailles at Marly (1685)	60,000
Engine of Honda Civic GL (1985)	63,000
Charles Parsons's steam turbine (1888)	75,000
Steam engine at Edison's Pearl Street Station (1882)	93,200
Watt's largest steam engine (1800)	100,000
Electricity use by a U.S. supermarket (1980)	200,000
Diesel engine of a German submarine (1916)	400,000
Lady Isabella, the world's largest waterwheel (1854)	427,000
Large steam locomotive at full speed (1890)	850,000
Parsons' steam turbine at Elberfeld Station (1900)	1,000,000
Shaw's water works at Greenock, Scotland (1840)	1,500,000
Large wind turbine (2015)	4,000,000
Rocket engine launching V-2 missile (1944)	6,200,000
Gas turbine powering a pipeline compressor (1970)	10,000,000
Japanese merchant ship's diesel engine (1960)	30,000,000
Four jet engines of Boeing 747 (1969)	60,000,000
Calder Hall nuclear reactor (1956)	202,000,000
Turbogenerator at Chooz nuclear power plant (1990)	1,457,000,000
Rocket engines launching Saturn C 5 (1969)	2,600,000,000
Kashiwazaki-kariwa nuclear power station (1997)	8,212,000,000
Japan's primary energy consumption (2015)	63,200,000,000
U.S. coal and biomass energy consumption (1850)	79,000,000,000
U.S. commercial energy consumption (2010)	3,050,000,000,000
Global commercial energy consumption (2015)	17,530,000,000,000

Maximum power of prime movers in field work, 1700–2015

Year	Actions, prime movers	Power (W)
1700	Chinese peasant hoeing a cabbage field	50
1750	Italian peasant harrowing with an old weak ox	200
1800	English farmer plowing with two small horses	1,000
1870	North Dakota farmer plowing with six powerful horses	4,000
1900	California farmer using 32 horses to pull a combine	22,000
1950	French farmer harvesting with a small tractor	50,000
2015	Manitoba farmer plowing with a large Diesel tractor	298,000

Maximum power of prime movers in land transportation, 1700–2015

Year	Prime movers	Power (W)
1700	Two oxen pulling a cart	700
1750	Four horses pulling a coach	2,500
1850	English steam locomotive	200,000
1900	The fastest American steam locomotive	1,000,000
1950	Powerful German diesel locomotive	2,000,000
2006	French TGV train by Alstom	9,600,000
2015	N700 series high-velocity *shinkansen* train	17,080,000

Average annual consumption (GJ/capita) of primary energy

	1750	1800	1850	1900	1950	2000
China	10	10	10	<15	<20	40
UK	30	60	80	115	100	150
France	<20	20	25	55	65	180
Japan	10	10	10	10	25	170
USA	<80	<100	105	135	245	345
World	<20	20	25	35	40	65

Note: All rates are rounded to the nearest 5 and include all phytomass (traditional and modern biofuels), fossil fuels, and primary electricity.

Bibliographical Notes

Advances in energy use are described systematically in multivolume histories of technical progress by Singer et al. (1954–1958), Forbes (1964–1972), and Needham et al. (1954–2015). Energy matters are covered with varying degrees of detail in many writings tracing the history of inventions and engineering practices. Their basic list should include books by Byrn (1900), Abbott (1932), Mumford (1934), Usher (1954), , Derry and Williams (1960), Burstall (1968), Kranzberg and Pursell (1967), Daumas (1969), Lindsay (1975), Gille (1978), L. White (1978), Landels (1980), Taylor (1982), Hill (1984), K. D. White (1984), Williams (1987), Basalla (1988), Pacey (1990), Finniston et al. (1992), Constable and Somerville (2003), Cleveland (2004), Smil (2005, 2006), McNeill et al. (2005), Billington and Billington (2006), Oleson (2008), Burke (2009), Weissenbacher (2009), Coopersmith (2010), Sørensen (2011), and Wei (2012).

The contributions of horses to civilization can be appreciated by consulting Lefebvre des Noëttes (1924), Smythe (1967), Dent (1974), Silver (1976), Villiers (1976), Telleen (1977), Langdon (1986), Hyland (1990), Anthony (2007), McShane and Tarr (2007), and Oleson (2008). The long history of waterwheels, and their importance during the time of early industrialization can be traced in volumes by Bresse (1876), Forbes (1965), Reynolds (1970), Hindle (1975), Reynolds (1983), Wikander (1983), Lewis (1997), Walton (2006), Malone (2009), and Mays (2010). The history of windmills and their economic importance are well reviewed in Wolff (1900), Skilton (1947), Freese (1957), Stockhuyzen (1963), Needham et al. (1965), Husslage (1965), Reynolds (1970), Wailes (1975), Torrey (1976), Harverson (1991), and Righter (2008). The development of sail ships is traced comprehensively in Chatterton (1914), Torr (1964), Armstrong (1969), and Chapelle (1988). Volumes on oared ships include Morrison and Gardiner (1995) and Morrison, Coates, and Rankov (2000).

Indispensable sources for the history of steam engines and their uses are Farey (1827), Fry (1896), Croil (1898), Dalby (1920), Dickinson (1939), Watkins (1967), Jones (1973), von Tunzelmann (1978), Hunter (1979), Ellis (1981), O'Brien (1983), Hills (1989), and Garrett and Wade-Matthews (2015). The development of internal combustion engines and gas turbines is reviewed in Diesel (1913), Constant (1981), Taylor (1984), Gunston (1986 and 1999), Cumpsty (2006), and Smil (2010b). The automobile age is chronicled by Beaumont (1906), Kennedy (1941), Sittauer (1972), May (1975), Flower and Jones (1981), Flink (1988), Cummins (1989), Ling (1990), Womack, Jones, and Roos (1990), and Maxton and Wormald (2004). The history of flying can be followed in Wright (1953), Constant (1981), Taylor (1989), Jakab (1990), Heppenheimer (1995), U.S. Centennial of Flight Commission (2003), Blériot (2015), and McCullough (2015).

The properties and uses of biomass energies are covered in Earl (1973), Smil (1983), Sieferle (2001), and Perlin (2005). Histories of the coal industry are presented by Bald (1812), Jevons (1865), Nef (1932), Eavenson (1942), Flinn et al. (1984–1993), Church, Hall, and Kanefsky (1986), and Thomson (2003). The development of the oil and gas industry is covered in Brantly (1971), Perrodon (1985), Yergin (2008), and Smil (2015a). The pioneering decades of the electrical industry and its subsequent expansion are traced by Jehl (1937), MacLaren (1943), Lilienthal (1944), Josephson (1959), Dunsheath (1962), Electricity Council (1973), Hughes (1983), Cheney (1981), Friedel and Israel (1986), Schurr et al. (1990), Cantelon, Hewlett, and Williams (1991), Nye (1992), Beauchamp (1997), Bowers (1998), and Hausman, Hertner, and Wilkins (2008).

The literature on the history of productive human activities is very rich. Perspectives on agricultural development, from farming's origins to the twentieth century, can be found in Bailey (1908), King (1927), Seebohm (1927), Buck (1930, 1937), Leser (1931), Lizerand (1942), Haudricourt and Delamarre (1955), Geertz (1963), Slicher van Bath (1963), Allan (1965), Boserup (1965, 1976), Perkins (1969), Titow (1969), Clark and Haswell (1970), White (1970), Fussell (1972), Ho (1975), Schlebecker (1975), Cohen (1977), Abel (1962), Xu and Dull (1980), Bray (1984), Rindos (1984), Mazoyer and Roudart (2006), Federico (2008), and Tauger (2010). Details on water lifting and irrigation are contained in Ewbank (1870), Molenaar (1956), Needham et al. (1965), Butzer (1976), Oleson (1984, 2008), and Mays (2010). The energy costs of modern agriculture are reviewed in Pimentel (1980), Fluck (1992), and Smil (2008a).

Interdisciplinary insights into the origins, process, and consequences of industrialization can be found in Kay (1832), Clapham (1926), Ashton

(1948), Landes (1969), Falkus (1972), Mokyr (1976, 2002), Clarkson (1985), Rosenberg and Birdzell (1986), Blumer (1990), and Stearns (2012). Many aspects of construction activities are chronicled and explained by Ashby (1935), Fitchen (1961), Bandaranayke (1974), Baldwin (1977), Hodges (1989), Lepre (1990), Waldron (1990), Wilson (1990), Gies and Gies (1995), Lehner (1997), and Ching, Jarzombek, and Prakash (2011). Contributions to the history of transportation include books by Savage (1959), Hadfield (1969), Sitwell (1981), Piggott (1983), Ratcliffe (1985), Ville (1990), Gerhold (1993), Herlihy (2004), Levinson (2006), and Smil (2010b).

Metallurgical progress can be traced in books by Biringuccio (1959 [1540]), Agricola (1912 [1556]), Bell (1884), Greenwood (1907), King (1948), Needham (1964), Straker (1969), Hogan (1971), Hyde (1977), Gold et al. (1984), Haaland and Shinnie (1985), Harris (1988), Geerdes, Toxopeus, and van der Vliet (2009), and Smil (2016). Weapons from ancient to modern times, and their effects on societies are reviewed in Mitchell (1932), Kloss (1963), Cipolla (1965), Ziemke (1968), Egg (1971), Singer and Small (1972), Kesaris (1977), McNeill (1989), Keegan (1994), Chase (2003), Parker (2005), Buchanan (2006), and Archer et al. (2008).

Writings on broad social implications of energy use include books by Ostwald (1912), Ellul (1964), Jones (1971), Odum (1971), Adams (1975, 1982), Smil (1991, 2008,), and Schobert (2014). Finally, anybody who wants to study history through the evolution of tools and machines must consult books that are appropriately illustrated. The two unsurpassed classical works are Ramelli (1976 [1588]) and Diderot and d'Alembert (1769–1772). Ardrey (1894), Abbott (1932), Hommel (1937), Burstall (1968), Hopfen (1969), Williams (1987), Basalla (1988), Finniston et al. (1992), Smil (2005, 2006), and DK Publishing (2012) are among the many modern contributions.

References

Abbate, J. 1999. *Inventing the Internet*. Cambridge, MA: MIT Press.

Abbott, C. G. 1932. *Great Inventions*. Washington, DC: Smithsonian Institution.

Abel, W. 1962. *Geschichte der deutschen Landwirtschaft von frühen Mittelalter bis zum 19 Jahrhundert*. Stuttgart: Ulmer.

Adam, J.-P. 1994. *Roman Building: Materials and Techniques*. London: Routledge.

Adams, R. N. 1975. *Energy and Structure: A Theory of Social Power*. Austin: University of Texas Press.

Adams, R. N. 1982. *Paradoxical Harvest: Energy and Explanation in British History, 1870–1914*. Cambridge: Cambridge University Press.

Adler, D. 2006. *Daimler & Benz: The Complete History: The Birth and Evolution of the Mercedes-Benz*. New York: Harper.

Adshead, S. A. M. 1992. *Salt and Civilization*. New York: St. Martin's Press.

Agricola, G. 1912 (1556). *De re metallica*. Trans. H. C. Hoover and L. H. Hoover. London: The Mining Magazine.

Aiello, L. C. 1996. Terrestriality, bipedalism and the origin of language. *Proceedings of the British Academy* 88:269–289.

Aiello, L. C., and J. C. K. Wells. 2002. Energetics and the evolution of the genus *Homo*. *Annual Review of Anthropology* 31:323–338.

Aiello, L. C., and P. Wheeler. 1995. The expensive-tissue hypothesis. *Current Anthropology* 36:199–221.

Alberici, S., et al. 2014. *Subsidies and Costs of EU Energy*. Brussels: EU Commission. https://ec.europa.eu/energy/sites/ener/files/documents/ECOFYS%202014%20 Subsidies%20and%20costs%20of%20EU%20energy_11_Nov.pdf.

Aldrich, L. J. 2002. *Cyrus McCormick and the Mechanical Reaper*. Greensboro, NC: Morgan Reynolds.

Allan, W. 1965. *The African Husbandman*. Edinburgh: Oliver & Boyd.

Allen, R. 2003. *Farm to Factory: A Reinterpretation of the Soviet Industrial Revolution*. Princeton, NJ: Princeton University Press.

Allen, R. C. 2007. *How Prosperous Were the Romans? Evidence from Diocletian's Price Edict (301 AD)*. Oxford: Oxford University, Department of Economics.

Allen, R. C., et al. 2011. Wages, prices, and living standards in China, 1738–1925: In comparison with Europe, Japan, and India. *Economic History Review* 64 (S1): 8–38.

Allianz. 2010. *The Sixth Kondratieff: Long Waves of Prosperity*. Frankfurt am Main: Allianz. https://www.allianz.com/v_1339501901000/media/press/document/other/kondratieff_en.pdf.

Alvard, M. S., and L. Kuznar. 2001. Deferred harvests: The transition from hunting to animal husbandry. *American Anthropologist* 103:295–311.

Amitai, R., and M. Biran, eds. 2005. *Mongols, Turks, and Others: Eurasian Nomads and the Sedentary World*. Leiden: Brill.

Amontons, G. 1699. Moyen de substituer commodement l'action du feu, à la force des hommes et des chevaux pour mouvoir les machines. *Mémoires de l'Académie Royale* 1699:112–126.

Andersen, S. O., and K. M. Sarma. 2002. *Protecting the Ozone Layer*. London: Earthscan.

Anderson, B. D. 2003. *The Physics of Sailing Explained*. Dobbs Ferry, NY: Sheridan House.

Anderson, E. N. 1988. *The Food of China*. New Haven, CT: Yale University Press.

Anderson, M. S. 1988. *War and Society in Europe of the Old regime, 1618–1789*. New York: St. Martin's Press.

Anderson, R. 1926. *The Sailing Ship: Six Thousands Years of History*. London: George Harrap.

Anderson, R. C. 1962. *Oared Fighting Ships: From Classical Times to the Coming of Steam*. London: Percival Marshall.

Angelo, J. E. 2003. *Space Technology*. Westport, CT: Greenwood Press.

Anthony, D. W. 2007. *The Horse, the Wheel, and Language: How Bronze-Age Riders from the Eurasian Steppes Shaped the Modern World*. Princeton, NJ: Princeton University Press.

Anthony, D., D. Y. Telegin, and D. Brown. 1991. The origin of horseback riding. *Scientific American* 265 (6): 94–100.

Apt, J., and P. Jaramillo. 2014. *Variable Renewable Energy and the Electricity Grid*. Washington, DC: Resources for the Future.

Archer, C. I., et al. 2008. *World History of Warfare*. Lincoln: University of Nebraska Press.

Ardrey, L. R. 1894. *American Agricultural Implements*. Chicago: L. R. Ardrey.

Arellano, C. J., and R. Kram. 2014. Partitioning the metabolic cost of human running: A task-by-task approach. *Integrative and Comparative Biology* 54:1084–1098.

Armelagos, G. J., and K. N. Harper. 2005. Genomics at the origins of agriculture, part one. *Evolutionary Anthropology* 14:68–77.

Armstrong, R. 1969. *The Merchantmen*. London: Ernest Benn.

Army Air Forces. 1945. *Army Air Forces Statistical Digest, World War II*. http://www .afhra.af.mil/shared/media/document/AFD-090608-039.pdf.

Army Technology. 2015. M1A1/2 Abrams Main Battle Tank, United States of America. http://www.army-technology.com/projects/abrams.

Ashby, T. 1935. *The Aqueducts of Ancient Rome*. Oxford: Oxford University Press.

Ashton, Thomas S. 1948. *The Industrial Revolution, 1760–1830*. Oxford: Oxford University Press.

Astill, G., and J. Langdon, eds. 1997. *Medieval Farming and Technology: The Impact of Agricultural Change in Northwest Europe*. Leiden: Brill.

Åström, K. J., and R. M. Murray. 2009. *Feedback Systems: An Introduction for Scientists and Engineers*. Princeton, NJ: Princeton University Press; http://www.cds.caltech .edu/~murray/books/AM05/pdf/am08-complete_22Feb09.pdf.

Atalay, S., and C. A. Hastorf. 2006. Food, meals, and daily activities: Food *habitus* at Neolithic Çatalhöyük. *American Antiquity* 71:283–319.

Atkins, S. E. 2000. *Historical Encyclopedia of Atomic Energy*. Westport, CT: Greenwood Press.

Atsmon, Y., and V. Dixit. 2009. Understanding China's wealthy. *McKinsey Quarterly* . http://www.mckinsey.com/insights/marketing_sales/understanding_chinas _wealthy.

Atwater, W. O., and C. F. Langworthy. 1897. *A Digest of Metabolism Experiments in Which the Balance of Income and Outgo Was Determined*. Washington, DC: U.S. GPO.

Atwood, C. P. 2004. *Encyclopedia of Mongolia and the Mongol Empire*. New York: Facts on File.

Atwood, R. 2009. Maya roots. *Archaeology* 62:18–66.

Augarten, S. 1984. *Bit by Bit*. Boston: Ticknor & Fields.

Axelsson, E., et al. 2013. The genomic signature of dog domestication reveals adaptation to a starch-rich diet. *Nature* 495:360–364.

Ayres, R. U. 2014. *The Bubble Economy: Is Sustainable Growth Possible?* Cambridge, MA: MIT Press.

Ayres, R. U., L. W. Ayres, and B. Warr. 2003. Exergy, power and work in the UA economy, 1900–1998. *Energy* 28:219–273.

Baars, C. 1973. *De Geschiedenis van de Landbouw in de Bayerlanden.* Wageningen: PUDOC (Centrum voor Landbouwpublicaties en Landbouwdocumentatie).

Bailey, L. H., ed. 1908. *Cyclopedia of American Agriculture.* New York: Macmillan.

Bailey, R. C., G. Head, M. Jenike, et al. 1989. Hunting and gathering in tropical rain forest: Is it possible? *American Anthropologist* 91:59–82.

Bailey, R. C., and T. N. Headland. 1991. The tropical rain forest: Is it a productive environment for human foragers? *Human Ecology* 19:261285.

Baines, D. 1991. *Emigration from Europe 1815–1930.* London: Macmillan.

Bairoch, P. 1988. *Cities and Economic Development: From the Dawn of History to the Present.* Chicago: University of Chicago Press.

Baker, T. L. 2006. *A Field Guide to America Windmills.* Tempe, AZ: ACMRS (Arizona Center for Medieval and Renaissance Studies), University of Arizona.

Bald, R. 1812. *A General View of the Coal Trade of Scotland, Chiefly that of the River Forth and Mid-Lothian. To Which is Added An Inquiry Into the Condition of the Women Who Carry Coals Under Ground in Scotland. Known by the Name of Bearers.* Edinburgh: Oliphant, Waugh and Innes.

Baldwin, G. C. 1977. *Pyramids of the New World.* New York: G. P. Putnam's Sons.

Bamford, P. W. 1974. *Fighting Ships and Prisons: The Mediterranean Galleys of France in the Age of Louis XIV.* Cambridge: Cambridge University Press.

Bandaranayke, S. 1974. *Sinhalese Monastic Architecture.* Leiden: E. J. Brill.

Bank of Nova Scotia. 2015. Global Auto Report. http://www.gbm.scotiabank.com/English/bns_econ/bns_auto.pdf.

Bapat, N. 2012. How Indians defied gravity and achieved success in Silicon Valley. http://www.forbes.com/sites/singularity/2012/10/15/how-indians-defied-gravity-and-achieved-success-in-silicon-valley.

Bar-Yosef, O. 2002. The Upper Paleolithic revolution. *Annual Review of Anthropology* 31:363–393.

Bardeen, J., and W. H. Brattain. 1950. *Three-electron Circuit Element Utilizing Semiconductive Materials.* US Patent 2,524,035, October 3. Washington, DC: USPTO. http://www.uspto.gov.

Barjot, D. 1991. *L'énergie aux XIXe et XXe siècles.* Paris: Presses de l'E.N.S.

Barker, A. V., and D. J. Pilbeam. 2007. *Handbook of Plant Nutrition*. Boca Raton, FL: CRC Press.

Barles, S. 2007. Feeding the city: Food consumption and flow of nitrogen, Paris, 1801–1914. *Science of the Total Environment* 375:48–58.

Barles, S., and L. Lestel. 2007. The nitrogen question: Urbanization, industrialization, and river quality in Paris 1830–1939. *Journal of Urban History* 33:794–812.

Barnes, B. R. 2014. Behavioural change, indoor air pollution and child respiratory health in developing countries: A review. *International Journal of Environmental Research and Public Health* 11:4607–4618.

Barro, R. J. 1997. *Determinants of Economic Growth: A Cross-Country Empirical Study*. Cambridge, MA: MIT Press.

Bartosiewicz, L. et al. 1997. *Draught Cattle: Their Osteological Identification and History*. Tervuren: Musée royal de l'Afrique central.

Basalla, G. 1980. Energy and civilization. In *Science, Technology and the Human Prospect*, ed. C. Starr and P. C. Ritterbusch, 39–52. Oxford: Pergamon Press.

Basalla, G. 1982. Some persistent energy myths. In *Energy and Transport*, ed. G. H. Daniels and M. H. Rose, 27–38. Beverley Hills, CA: Sage.

Basalla, G. 1988. *The Evolution of Technology*. Cambridge: Cambridge University Press.

Basile, S. 2014. *Cool: How Air Conditioning Changed Everything*. New York: Fordham University Press.

Basso, L. C., T. O. Basso, and S. N. Rocha. 2011. *Ethanol Production in Brazil: The Industrial Process and Its Impact on Yeast Fermentation, Biofuel Production: Recent Developments and Prospects*. http://cdn.intechopen.com/pdfs/20058/InTech -Ethanol_production_in_brazil_the_industrial_process_and_its_impact_on_yeast _fermentation.pdf.

Bayley, J., D. Dungworth, and S. Paynter. 2001. *Archaeometallurgy*. London: English Heritage.

Beauchamp, K. G. 1997. *Exhibiting Electricity*. London: Institution of Electrical Engineers.

Beaumont, W. W. 1902. *Motor Vehicles and Motors: Their Design, Construction and Working by Steam, Oil and Electricity*. Westminster: Archibald Constable and Company.

Beaumont, W. W. 1906. *Motor Vehicles and Motors: Their Design, Construction and Working by Steam, Oil and Electricity*. Westminster: Archibald Constable and Co.

Beevor, A. 1998. *Stalingrad*. London: Viking.

Behera, B., et al. 2015. Household collection and use of biomass energy sources in South Asia. *Energy* 85:468–480.

Bell, L. 1884. *Principles of the Manufacture of Iron and Steel*. London: George Routledge & Sons.

Bell System Memorial. 2011. Who really invented the transistor? http://www.porticus.org/bell/belllabs_transistor1.html.

Bennett, M. K. 1935. British wheat yield per acre for seven centuries. *Economy and History* 3:12–29.

Benoît, C. 1996. Le Canon de 75: Une gloire centenaire. Vincennes, France: Service Historique de l'Armée de Terre.

Benoit, F. 1940. L'usine de meunerie hydraulique de Barbegal (Arles). *Review of Archaeology* 15:19–80.

Beresford, M. W., and J. G. Hurst. 1971. *Deserted Medieval Villages*. London: Littleworth.

Berklian, Y. U., ed. 2008. *Crop Rotation*. New York: Nova Science Publishers.

Bernard, L., A. V. Gevorkyan, T. Palley, and W. Semmler. 2013. Time scales and mechanisms of economic cycles: A review of theories of long waves. Political Economy Research Institute Working Paper, no.337, 1–21. Amherst, MA: University of Massachusetts.

Bessemer, H. 1905. *Sir Henry Bessemer, F.R.S.: An Autobiography* . London: Offices of Engineering.

Bettencourt, L., and G. West. 2010. A unified theory of urban living. *Nature* 467:912–913.

Bettinger, R. L. 1991. *Hunter-Gatherers: Archaeological and Evolutionary Theory*. New York: Plenum Press.

Betz, A. 1926. *Wind-Energie und ihre Ausnutzung durch Windmühlen*. Göttingen: Bandenhoeck & Ruprecht.

Billington, D. P., and D. P. Billington, Jr. 2006. *Power, Speed, and Form: Engineers and the Making of the Twentieth Century*. Princeton, NJ: Princeton University Press.

bin Laden, U. 2004. Message to the American people. http://english.aljazeera.net/NR/exeres/79C6AF22-98FB-4A1C-B21F-2BC36E87F61F.htm.

Bird-David, N. 1992. Beyond "The Original Affluent Society." *Current Anthropology* 33:25–47.

Biringuccio, V. 1959 (1540). *De la pirotechnia [The pirotechnia]*. Trans. C. S. Smith and M. T. Gnudi. New York: Basic Books.

Bishop, C. 2014. *The Illustrated Encyclopedia of Weapons of World War I: The Comprehensive Guide to Weapons Systems, Including Tanks, Small Arms, Warplanes, Artillery.* London: Amber.

Blériot, L. 2015. *Blériot: Flight into the XXth Century.* London: Austin Macauley.

Blumenschine, R. J., and J. A. Cavallo. 1992. Scavenging and human evolution. *Scientific American* 267 (4): 90–95.

Blumer, H. 1990. *Industrialization as an Agent of Social Change.* New York: Aldine de Gruyter.

Blyth, R. J., A. Lambert, and J. Ruger, eds. 2011. *The Dreadnought and the Edwardian Age.* Farnham: Ashgate.

Boden, T., and B. Andres. 2015. *Global CO_2 Emissions from Fossil-Fuel Burning, Cement Manufacture, and Gas Flaring: 1751–2011.* Oak Ridge, TN: CDIAC (Carbon Dioxide Information Analysis Center), Oak Ridge National Laboratory. http://cdiac.ornl.gov/trends/emis/tre_glob_2011.html.

Boden. T., B. Andres, and G. Marland. 2016. Global CO_2 emissions from fossil fuel burning, cement manufacture, and gas flaring: 1751–2013. http://cdiac.ornl.gov/ftp/ndp030/global.1751_2013.ems.

Boeing. 2015. Boeing history. http://www.boeing.com/history.

Bogin, B. 2011. Kung nutritional status and the original "affluent society": A new analysis. *Anthropologischer Anzeiger* 68:349–366.

Bono, P., and C. Boni. 1996. Water supply of Rome in antiquity and today. *Environmental Geology* 27:126–134.

Boonenburg, K. 1952. *Windmills in Holland.* The Hague: Netherlands Government Information Service.

Borghese, A., ed. 2005. *Buffalo Production and Research.* Rome: FAO.

Bos, M. G. 2009. *Water Requirements for Irrigation and the Environment.* Dordrecht: Springer.

Bose, S., ed. 1991. *Shifting Agriculture in India.* Calcutta: Anthropological Survey of India.

Boserup, E. 1965. *The Conditions of Agricultural Growth: The Economics of Agrarian Change under Population Pressure.* Chicago: Aldine.

Boserup, E. 1976. Environment, population, and technology in primitive societies. *Population and Development Review* 2:21–36.

Bott, R. D. 2004. *Evolution of Canada's Oil and Gas Industry.* Calgary, AB: Canadian Centre for Energy Information.

Boulding, K. E. 1974. The social system and the energy crisis. *Science* 184:255–257.

Bowers, B. 1998. *Lengthening the Day: A History of Lighting Technology.* Oxford: Oxford University Press.

Bowers, B. 2001. *Sir Charles Wheatstone: 1802–1875,* 2nd ed. London: Institution of Engineering and Technology.

Boxer, C. R. 1969. *The Portuguese Seaborne Empire 1415–1825.* London: Hutchinson.

BP (British Petroleum). 2016. *Statistical Review of World Energy 2016.* https://www .bp.com/content/dam/bp/pdf/energy-economics/statistical-review-2015/bp -statistical-review-of-world-energy-2015-full-report.pdf.

Bramanti, B., et al. 2009. Genetic discontinuity between local hunter-gatherers and Central Europe's first farmers. *Science* 326:137–140.

Bramble, D. M., and D. E. Lieberman. 2004. Endurance running and the evolution of *Homo. Nature* 432:345–352.

Brandstetter, T. 2005. "The most wonderful piece of machinery the world can boast of": The water-works at Marly, 1680–1830. *History and Technology* 21:205–220.

Brantly, J. E. 1971. *History of Oil Well Drilling.* Houston, TX: Gulf Publishing.

Braudel, F. 1982. *On History.* Chicago: University of Chicago Press.

Braun, D. R., et al. 2010. Early hominin diet included diverse terrestrial and aquatic animals 1.95 Ma in East Turkana, Kenya. *Proceedings of the National Academy of Sciences of the United States of America* 107:10002–10007.

Braun, G. W., and D. R. Smith. 1992. Commercial wind power: Recent experience in the United States. *Annual Review of Energy and the Environment* 17:97–121.

Bray, F. 1984. *Science and Civilisation in China.* Vol. 6, Part II. *Agriculture.* Cambridge: Cambridge University Press.

Bresse, M. 1876. *Water-Wheels or Hydraulic Motors.* New York: John Wiley.

Brodhead, M. J. 2012. *The Panama Canal: Writings of the U. S. Army Corps of Engineers Officers Who Conceived and Built It.* Alexandria, VA: U.S. Army Corps of Engineers History Office.

Brody, S. 1945. *Bioenergetics and Growth.* New York: Reinhold.

Bronson, B. 1977. The earliest farming: Demography as cause and consequence. In *Origins of Agriculture ,* ed. C. Reed, 23–48. The Hague: Mouton.

Brooks, D. R., and E. O. Wiley. 1986. *Evolution as Entropy.* Chicago: University of Chicago Press.

Brown, G. I. 1999. *Count Rumford: The Extraordinary Life of a Scientific Genius.* Stroud: Sutton Publishing.

Brown, K. S., et al. 2009. Fire as an engineering tool of early modern humans. *Science* 325:859–862.

Brown, K. S., et al. 2012. An early and enduring advanced technology originating 71,000 years ago in South Africa. *Nature* 491:590–593.

Brown, S., P. Schroeder, and R. Birdsey. 1997. Aboveground biomass distribution of US eastern hardwood forests and the use of large trees as an indicator of forest development. *Forest Ecology and Management* 96:31–47.

Bruce, A. W. 1952. *The Steam Locomotive in America.* New York: Norton.

Brunck, R. F. P. 1776. *Analecta Veterum Poetarum Graecorum.* Strasbourg: I. G. Bauer & Socium.

Bruni, L., and P. L. Porta. 2006. *Economics and Happiness.* New York: Oxford University Press.

Brunner, K. 1995. Continuity and discontinuity of Roman agricultural knowledge in the early Middle Ages. In *Agriculture in the Middle Ages,* ed. D. Sweeney, 21–39. Philadelphia: University of Pennsylvania Press.

Brunt, L. 1999. *Estimating English Wheat Production in the Industrial Revolution.* Oxford: University of Oxford. http://www.nuffield.ox.ac.uk/economics/history/paper35/dp35a4.pdf.

Buchanan, B. J., ed. 2006. *Gunpowder, Explosives and the State: A Technological History.* Aldershot: Ashgate.

Buck, J. L. 1930. *Chinese Farm Economy.* Nanking: University of Nanking.

Buck, J. L. 1937. *Land Utilization in China.* Nanking: University of Nanking.

Buckley, T. A. 1855. *The Works of Horace.* New York: Harper & Brothers.

Budge, E. A. W. 1920. *An Egyptian Hieroglyphic Dictionary.* London: John Murray.

Bulliet, R. W. 1975. *The Camel and the Wheel.* Cambridge, MA: Harvard University Press.

Bulliet, R. W. 2016. *The Wheel: Inventions and Reinventions.* New York: Columbia University Press.

Bunch, B. H., and A. Hellemans. 1993. *The Timetables of Technology: A Chronology of the Most Important People and Events in the History of Technology.* New York: Simon & Schuster.

Burke, E., III. 2009. Human history, energy regimes and the environment. In *The Environment and World History,* ed. E. Burke III and K. Pomeranz, 33–53. Berkeley: University of California Press.

Burstall, A. F. 1968. *Simple Working Models of Historic Machines.* Cambridge, MA: MIT Press.

Burton, R. F. 1880. *The Lusiads*. London: Tinsley Brothers.

Butler, J. H., and S. A. Montzka. 2015. The NOAA Annual Greenhouse Gas Index. Boulder, CO: NOAA. http://www.esrl.noaa.gov/gmd/aggi/aggi.html.

Butzer, K. W. 1976. *Early Hydraulic Civilization in Egypt*. Chicago: University of Chicago Press.

Butzer, K. W. 1984. Long-term Nile flood variation and political discontinuities in Pharaonic Egypt. In *From Hunters to Farmers*, ed. J. D. Clark and S. A. Brandt, 102–112. Berkeley: University of California Press.

Byrn, E. W. 1900. *The Progress of Invention in the Nineteenth Century*. New York: Munn & Co.

Caidin, M. 1960. *A Torch to the Enemy: The Fire Raid on Tokyo*. New York: Balantine Books.

Cairns, M. F., ed. 2015. *Shifting Cultivation and Environmental Change: Indigenous People, Agriculture and Forest Conservation*. London: Earthscan Routledge.

Cameron, R. 1982. The Industrial Revolution: A misnomer. *History Teacher* 15 (3): 377–384.

Cameron, R. 1985. A new view of European industrialization. *Economic History Review* 3:1–23.

Campbell, B. M. S., and M. Overton. 1993. A new perspective on medieval and early modern agriculture: Six centuries of Norfolk farming, *c.* 1250-*c.* 1850. *Past & Present* 141 (1): 38–105.

Campbell, H. R. 1907. *The Manufacture and Properties of Iron and Steel*. New York: Hill Publishing.

Cantelon, P. L., R. G. Hewlett, and R. C. Williams, eds. 1991. *The American Atom: A Documentary History of Nuclear Policies from the Discovery of Fission to the Present*. Philadelphia: University of Pennsylvania Press.

Capulli, M. 2003. *Le Navi della Serenissima: La Galea Veneziana di Lazise*. Venezia: Marsilio Editore.

Cardwell, D. S. L. 1971. *From Watt to Clausius: The Rise of Thermodynamics in the Early Industrial Age*. Ithaca, NY: Cornell University Press.

Caro, R. A. 1982. *The Years of Lyndon Johnson: The Path to Power*. New York: Knopf.

Caron, F. 2013. *Dynamics of Innovation: The Expansion of Technology in Modern Times*. New York: Berghahn.

Carrier, D. R. 1984. The energetic paradox of human running and hominid evolution. *Current Anthropology* 25:483–495.

Carter, R. A. 2000. *Buffalo Bill Cody: The Man behind the Legend.* New York: John Wiley.

Carter, W. E. 1969. *New Lands and Old Traditions: Kekchi Cultivators in the Guatemala Lowlands.* Gainesville: University of Florida Press.

Casson, L. 1994. *Ships and Seafaring in Ancient Times.* Austin: University of Texas Press.

CDC (Centers for Disease Control and Prevention). 2015. Overweight & Obesity. http://www.cdc.gov/nchs/fastats/obesity-overweight.htm.

CDFA (Clean Diesel Fuel Alliance). 2015. Ultra Low Sulfur Diesel (ULSD). http://www.clean-diesel.org/index.htm.

Centre des Recherches Historiques. 1965. *Villages Desertes et Histoire Economique.* Paris: SEVPEN.

Ceruzzi, P. E. 2003. *A History of Modern Computing.* Cambridge, MA: MIT Press.

CFM International. 2015. Discover CFM. http://www.cfmaeroengines.com/files/brochures/Brochure_CFM_2015.pdf.

Chandler, T. 1987. *Four Thousand Years of Urban Growth: An Historical Census.* Lewiston, NY: Edwin Mellen Press.

Chapelle, H. I. 1988. *The History of American Sailing Ships.* Modesto, CA: Bonanza Books.

Charette, R. N. 2009. This car runs on code. *IEEE Spectrum 2009* (February). http://spectrum.ieee.org/green-tech/advanced-cars/this-car-runs-on-code/0.

Charles, C., and P. Wooders. 2011. *Subsidies to Liquid Transport Fuels: A comparative review of estimates.* Geneva: IISD.

Chartrand, R. 2003. *Napoleon's Guns 1792–1815. Botley.* Osprey Publishing.

Chase, K. 2003. *Firearms: A Global History to 1700.* Cambridge: Cambridge University Press.

Chatterton, E. K. 1914. *Sailing Ships: The Story of Their Development from the Earliest Times to the Present Day.* London: Sidgwick & Jackson.

Chatterton, E. K. 1926. *The Ship Under Sail.* London: Fisher Unwin.

Chauvois, L. 1967. *Histoire merveilleuse de Zénobe Gramme.* Paris: Albert Blanchard.

Cheney, Margaret. 1981. *Tesla: Man out of Time.* New York: Dorset Press.

Chevedden, P. E., et al. 1995. The trebuchet. *Scientific American* 273 (1): 66–71.

China Energy Group. 2014. *Key China Energy Statistics 2014.* Berkeley, CA: Lawrence Berkeley National Laboratory.

Chincold. 2015. Three Gorges Project. http://www.chincold.org.cn/dams/rootfiles/2010/07/20/1279253974143251-1279253974145520.pdf.

Ching, F. D. K., M. Jarzombek, and V. Prakash. 2011. *A Global History of Architecture.* Hoboken, NJ: John Wiley & Sons.

Chorley, G. P. H. 1981. The agricultural revolution in Northern Europe, 1750–1880: Nitrogen, legumes, and crop productivity. *Economic History Review* 34 (1):71–93.

Choudhury, P. C. 1976. *Hastividyarnava.* Gauhati: Publication Board of Assam.

Christ, K. 1984. *The Romans.* Berkeley: University of California Press.

Church, R., Hall, A. and J. Kanefsky. 1986. *History of the British Coal Industry.* Vol. 3, *Victorian Pre-Eminence.* Oxford: Oxford University Press.

Cipolla, C. M. 1965. *Guns, Sails and Empires: Technological Innovation and the Early Phases of European Expansion, 1400–1700.* New York: Pantheon Books.

City Population. 2015. Major agglomerations of the world. http://www.citypopulation.de/world/Agglomerations.html.

Clapham, J. H. 1926. *An Economic History of Modern Britain.* Cambridge: Cambridge University Press.

Clark, C., and M. Haswell. 1970. *The Economics of Subsistence Agriculture.* London: Macmillan.

Clark, G. 1987. Productivity growth without technical change in European agriculture before 1850. *Journal of Economic History* 47:419–432.

Clark, G. 1991. Yields per acre in English agriculture, 1250–1850: Evidence from labour inputs. *Economic History Review* 44:445–460.

Clark, G., M. Huberman, and P. H. Lindert. 1995. A British food puzzle, 1770–1850. *Economic History Review* 48:215–237.

Clarke, R., and M. Dubravko. 1983. *Soviet Economic Facts, 1917–1981.* London: Palgrave Macmillan.

Clarkson, L. A. 1985. *Proto-Industrialization: The First Phase of Industrialization?* London: Macmillan.

Clavering, E. 1995. The coal mills of Northeast England: The use of waterwheels for draining coal mines, 1600–1750. *Technology and Culture* 36:211–241.

Clerk, D. 1909. *The Gas, Petrol, and Oil Engine.* London: Longmans, Green and Co.

Cleveland, C. J., ed. 2004. *Encyclopedia of Energy,* 6 vols. Amsterdam: Elsevier.

Cleveland, C. J., and C. Morris. 2014. *Handbook of Energy.* Vol. 2, *Chronologies, Top Ten Lists, and World Clouds.* Amsterdam: Elsevier.

CMI (Center for Military History). 2010. *War in the Persian Gulf: Operations Desert Shield and Desert Storm*, August 1990–March 1991. http://www.history.army.mil/html/books/070/70-117-1/cmh_70-117-1.pdf.

Coates, J. F. 1989. The trireme sails again. *Scientific American* 261 (4): 68–75.

Cobbett, J. P. 1824. *A Ride of Eight Hundred Miles in France*. London: Charles Clement.

Cochrane, W. W. 1993. *The Development of American Agriculture: A Historical Analysis*. Minneapolis: University of Minnesota Press.

Cockrill, W. R., ed. 1974. *The Husbandry and Health of the Domestic Buffalo*. Rome: FAO.

Cohen, B. 1990. *Benjamin Franklin's Science*. Cambridge, MA: Harvard University Press.

Cohen, N. M. 1977. *The Food Crisis in Prehistory*. New Haven, CT: Yale University Press.

Collier, B. 1962. *The Battle of Britain*. London: Batsford.

Collins, E. V., and A. B. Caine. 1926. *Testing Draft Horses. Iowa Experimental Station Bulletin* 240.

Coltman, J. W. 1988. The transformer. *Scientific American* 258 (1): 86–95.

Committee for the Compilation of Materials on Damage Caused by the Atomic bombs in Hiroshima and Nagasaki. 1991. *Hiroshima and Nagasaki: The Physical, Medical and Social Effects of the Atomic Bombing*. New York: Basic Books.

Conklin, H. C. 1957. *Hanunoo Agriculture*. Rome: FAO.

Conquest, Robert. 2007. *The Great Terror: A Reassessment*. 40th Anniversary Edition. Oxford: Oxford University Press.

Constable, G., and B. Somerville. 2003. *A Century of Innovation*. Washington, DC: Joseph Henry Press.

Constant, E. W. 1981. *The Origins of Turbojet Revolution*. Baltimore, MD: Johns Hopkins University Press.

Coomes, O. T., F. Grimard, and G. J. Burt. 2000. Tropical forests and shifting cultivation: Secondary forest fallow dynamics among traditional farmers of the Peruvian Amazon. *Ecological Economics* 32:109–124.

Coopersmith, J. 2010. *Energy, the Subtle Concept: The Discovery of Feynman's Blocks from Leibniz to Einstein*. Oxford: Oxford University Press.

Copley, Frank B. 1923. *Frederick W. Taylor: Father of Scientific Management*. New York: Harper & Brothers.

Cornways. 2015. Combine. http://www.cornways.de/hi_combine.html.

Cotterell, B., and J. Kamminga. 1990. *Machines of Pre-industrial Technology*. Cambridge: Cambridge University Press.

Coulomb, C. A. 1799. Résultat de plusieurs expériences destinées à déterminer la quantité d'action que les hommes peuvent fournir par leur travail journalier. ... *Mémoires de l'Institut national des sciences et arts—Sciences mathématiques et physique* 2:380–428.

Coulton, J. J. 1977. *Ancient Greek Architects at Work*. Ithaca, NY: Cornell University Press.

Cowan, R. 1990. Nuclear power reactors: A study in technological lock-in. *Journal of Economic History* 50:541–567.

Craddock, P. T. 1995. *Early Metal Mining and Production*. Edinburgh: Edinburgh University Press.

Crafts, N. F. R., and C. K. Harley. 1992. Output growth and the British Industrial Revolution. *Economic History Review* 45:703–730.

Crafts, N., and T. Mills. 2004. Was 19th century British growth steam-powered? The climacteric revisited. *Explorations in Economic History* 41:156–171.

Croil, J. 1898. *Steam Navigation*. Toronto: William Briggs.

Crossley, D. 1990. *Post-medieval Archaeology in Britain*. Leicester: Leicester University Press.

Cummins, C. L. 1989. *Internal Fire*. Warrendale, PA: Society of Automotive Engineers.

Cumpsty, N. 2006. *Jet Propulsion*. Cambridge: Cambridge University Press.

Cuomo, S. 2004. The sinews of war: Ancient catapults. *Science* 303:771–772.

Curtis, W. H. 1919. *Wood Ship Construction*. New York: McGraw-Hill.

Daggett, S. 2010. *Costs of Major U.S. Wars*. Washington, DC: Congressional Research Service. http://cironline.org/sites/default/files/legacy/files/June2010CRScostofuswars .pdf.

Dalby, W. E. 1920. *Steam Power*. London: Edward Arnold.

Darby, H. C. 1956. The clearing of the woodland of Europe. In *Man's Role in Changing the Face of the Earth*, ed. W. L. Thomas, 183–216. Chicago: University of Chicago Press.

Darling, K. 2004. *Concorde*. Marlborough: Crowood Press.

Daugherty, C. R. 1927. The development of horse-power equipment in the United States. In *Power Capacity and Production in the United States*, ed. C. R. Daugherty, A. H. Horton and R. W. Davenport, 5–112. Washington, DC: U.S. Geological Survey.

Daumas, M., ed. 1969. *A History of Technology and Invention*. New York: Crown Publishers.

David, P. 1985. Clio and the economics of QWERTY. *American Economic Review* 75:332–337.

David, P. A. 1991. The hero and the herd in technological history: Reflections on Thomas Edison and the Battle of the Systems. In *Favorites of Fortune: Technology, Growth and Economic Development since the Industrial Revolution*, ed. P. Higonett, D. S. Landes and H. Rosovsky, 72–119. Cambridge, MA: Harvard University Press.

Davids, K. 2006. River control and the evolution of knowledge: A comparison between regions in China and Europe, c. 1400–1850. *Journal of Global History* 1:59–79.

Davies, N. 1987. *The Aztec Empire: The Toltec Resurgence*. Norman: University of Oklahoma Press.

Davis, M. 2001. *Late Victorian Holocausts*. New York: Verso.

de Beaune, S. A., and R. White. 1993. Ice age lamps. *Scientific American* 266 (3): 108–113.

de la Torre, I. 2011. The origins of stone tool technology in Africa: A historical perspective. *Philosophical Transactions of the Royal Society of London. Series B, Biological Sciences* 366 (1567): 1028–1037.

De Zeeuw, J. W. 1978. Peat and the Dutch Golden Age: The historical meaning of energy-attainability. *A.A.G. Bijdragen* 21:3–31.

Deffeyes, K. S. 2001. *Hubbert's Peak: The Impending World Oil Shortage*. Princeton, NJ: Princeton University Press.

Demarest, A. 2004. *Ancient Maya: The Rise and Fall of a Rainforest Civilization*. Cambridge: Cambridge University Press.

Dempsey, P. 2015. Notes on the Liberty aircraft engine. http://www.enginehistory.org/Before1925/Liberty/LibertyNotes.shtml.

Denevan, W. H. 1982. Hydraulic agriculture in the American tropics: Forms, measures, and recent research. In *Maya Subsistence*, ed. K. V. Flannery, 181–203. New York: Academic Press.

Denny, M. 2004. The efficiency of overshot and undershot waterwheels. *European Journal of Physics* 25:193–202.

Denny, M. 2007. *Ingenium: Five Machines That Changed the World*. Baltimore, MD: Johns Hopkins University Press.

Dent, A. 1974. *The Horse*. New York: Holt, Rinehart and Winston.

Department of Energy & Climate Change, UK Government. 2015. Historical coal data: Coal production, availability and consumption 1853 to 2014. https://www

.gov.uk/government/statistical-data-sets/historical-coal-data-coal-production
-availability-and-consumption-1853-to-2011.

Derry, T. K., and T. I. Williams. 1960. *A Short History of Technology*. Oxford: Oxford University Press.

Diamond, J. 2011. *Collapse: How Societies Choose to Fail or Succeed*. New York: Penguin Books.

Dickens, C. 1854. *Hard Times*. London: Bradbury & Evans.

Dickey, P. A. 1959. The first oil well. *Journal of Petroleum Technology* 59:14–25.

Dickinson, H. W. 1939. *A Short History of the Steam Engine*. Cambridge: Cambridge University Press.

Dickinson, H. W., and R. Jenkins. 1927. *James Watt and the Steam Engine*. Oxford: Oxford University Press.

Diderot, D., and J.L.R. D'Alembert. 1769–1772. *L'Encyclopedie ou dictionnaire raisonne des sciences des arts et des métiers*. Paris: Avec approbation et privilege du roy.

Dieffenbach, E. M., and R. B. Gray. 1960. The development of the tractor. In *Power to Produce: 1960 Yearbook of Agriculture* , 24–45. Washington, DC: U.S. Department of Agriculture.

Dien, A. 2000. The stirrup and its effect on Chinese military history. http://www .silk-road.com/artl/stirrup.shtml.

Diener, E., E. Suh, and S. Oishi. 1997. Recent findings on subjective well-being. *Indian Journal of Clinical Psychology* 24:25–41.

Diesel, E. 1937. *Diesel: Der Mensch, das Werk, das Schicksal*. Hamburg: Hanseatische Verlagsanstalt.

Diesel, R. 1893a. Arbeitsverfahren und Ausführungsart für Verbrennungskraftmaschinen. https://www.dhm.de/lemo/bestand/objekt/patentschrift-von-rudolf-diesel -1893.html.

Diesel, R. 1893b. *Theorie und Konstruktion eines rationellen Wärmemotors zum Ersatz der Dampfmaschinen und der heute bekannten Verbrennungsmotoren*. Berlin: Julius Springer.

Diesel, R. 1903. *Solidarismus: Natürliche wirtschaftliche Erlösung des Menschen*. Munich (repr., Augsburg: Maro Verlag, 2007).

Diesel, R. 1913. *Die Entstehung des Dieselmotors*. Berlin: Julius Springer.

Dikötter, F. 2010. *Mao's Great Famine: The History of China's Most Devastating Catastrophe, 1958–1962*. London: Walker Books.

DK Publishing. 2012. *Military History: The Definitive Visual Guide to the Objects of Warfare*. New York: DK Publishing.

Domínguez-Rodrigo, M. 2002. Hunting and scavenging by early humans: The state of the debate. *Journal of World Prehistory* 16:1–54.

Donnelly, J. S. 2005. *The Great Irish Potato Famine*. Stroud: Sutton Publishing.

Doorenbos, J., et al. 1979. *Yield Response to Water*. Rome: FAO.

Dowson, D. 1973. Tribology before Columbus. *Mechanical Engineering* 95 (4): 12–20.

Doyle, J., B. Francis, and A. Tannenbaum. 1990. *Feedback Control Theory*. London: Macmillan.

Drews, R. 2004. *Early Riders: The Beginnings of Mounted Warfare in Asia and Europe*. New York: Routledge.

Duby, G. 1968. *Rural Economy and Country Life in the Medieval West*. London: Edward Arnold.

Duby, G. 1998. *Rural Economy and Country Life in the Medieval West*. Philadelphia: University of Pennsylvania Press.

Dukes, J. S. 2003. Burning buried sunshine: Human consumption of ancient solar energy. *Climatic Change* 61:31–44.

Duncan-Jones, R. 1990. *Structure and Scale in the Roman Economy*. Cambridge: Cambridge University Press.

Dunsheath, P. 1962. *A History of Electrical Industry*. London: Faber and Faber.

Dupont, B., D. Keeling, and T. Weiss. 2012. Passenger fares for overseas travel in the 19th and 20th centuries. Paper presented at the Annual Meeting of the Economic History Association, Vancouver, BC, September 21–23. http://eh.net/eha/wp -content/uploads/2013/11/Weissetal.pdf.

Dyer, Frank L., and Thomas C. Martin. 1929. *Edison: His Life and Inventions*. New York: Harper & Brothers.

Eagar, T. W., and C. Musso. 2001. Why did the World Trade Center collapse? Science, engineering, and speculation. *JOM* 53:8–11. http://www.tms.org/pubs/ journals/JOM/0112/Eagar/Eagar-0112.html.

Earl, D. 1973. *Charcoal and Forest Management*. Oxford: Oxford University Press.

Eavenson, H. N. 1942. *The First Century and a Quarter of American Coal Industry*. Pittsburgh, PA: Privately printed.

Eckermann, E. 2001. *World History of the Automobile*. Warrendale, PA: SAE Press.

ECRI (Economic Cycle Research Institute). 2015. Economic cycles. https://www .businesscycle.com.

Eden, F. M. 1797. *The State of the Poor*. London: J. Davis.

Edison, T. A. 1880. Electric Light. Specification forming part of Letters Patent No. 227,229, dated May 4, 1880. Washington, DC: U.S. Patent Office. http://www.uspto .gov.

Edison, T. A. 1889. The dangers of electric lighting. *North American Review* 149:625–634.

Edgerton, D. 2007. *The Shock of the Old: Technology and Global History since 1900.* Oxford: Oxford University Press.

Edgerton, S. Y. 1961. Heat and style: Eighteenth-century house warming by stoves. *The Journal of the Society of Architectural Historians* 20:20–26.

Edwards, J. F. 2003. Building the Great Pyramid: Probable construction methods employed at Giza. *Technology and Culture* 44:340–354.

Egerton, W. 1896. *Indian and Oriental Armour.* London: W. H. Allen.

Egg, E., et al. 1971. *Guns.* Greenwich, CT: New York Graphic Society.

Electricity Council. 1973. *Electricity Supply in Great Britain: A Chronology—From the Beginnings of the Industry to 31 December 1972.* London: Electricity Council.

Elliott, D. 2013. *Fukushima: Impacts and Implications.* Houndmills: Palgrave Macmillan.

Ellis, C. H. 1983. *The Lore of the Train.* New York: Crescent Books.

Ellison, R. 1981. Diet in Mesopotamia: The evidence of the barley ration texts. *Iraq* 45:35–45.

Ellul, J. 1954. *La Technique ou l'enjeu du siècle.* Paris: Armand Colin.

Elphick, P. 2001. *Liberty: The Ships That Won the War.* Annapolis, MD: Naval Institute Press.

Elton, A. 1958. Gas for light and heat. In *A History of Technology*, vol. 4, ed. C. Singer et al., 258–275. Oxford: Oxford University Press.

Engels, F. 1845. *Die Lage der arbeitenden Klasse in England.* Leipzig: Otto Wigand.

Erdkamp, P. 2005. *The Grain Market in the Roman Empire: A Social, Political and Economic Study.* Cambridge: Cambridge University Press.

Erickson, C. L. 1988. Raised field agriculture in the Lake Titicaca Basin. *Expedition* 30 (1): 8–16.

Erlande-Brandenburg, A. 1994. *The Cathedral: The Social and Architectural Dynamics of Construction.* Cambridge: Cambridge University Press.

Esmay, M. L., and C. W. Hall, eds. 1968. *Agricultural Mechanization in Developing Countries.* Tokyo: Shin-Norinsha.

Evangelou, P. 1984. *Livestock Development in Kenya's Maasailand*. Boulder, CO: Westview Press.

Evans, O. 1795. *The Young Millwright and Miller's Guide*. Philadelphia: O. Evans.

Evelyn, J. 1607. *Silva*. London: R. Scott.

Ewbank, T. 1870. *A Descriptive and Historical Account of Hydraulic and Other Machines for Raising Water*. New York: Scribner.

Executive Office of the President. 2013. *Economic Benefits of Increasing Electric Grid Resilience to Weather Outages*. Washington, DC: The White House.

Fairlie, S. 2011. Notes on the history of the scythe and its manufacture. http://scytheassociation.org/history.

Faith, J. T. 2007. Eland, buffalo, and wild pigs: Were Middle Stone Age humans ineffective hunters? *Journal of Human Evolution* 55:24–36.

Falkenstein, A. 1939. *Zehnter vorläufiger Bericht über die von der Notgemeinschaft der deutschen Wissenschaft in Uruk-Warka unternommen Ausgrabungen*. Berlin: Verlag Akademie der Wissenschaften.

Falkus, M. E. 1972. *The Industrialization of Russia, 1700–1914*. London: Macmillan.

Fant, K. 2014. *Alfred Nobel: A Biography*. New York: Arcade Publishing.

FAO (Food and Agriculture Organization). 2004. *Human Energy Requirements. Report of a Joint FAO/WHO/UNU Consultation*. Rome: FAO.

FAO. 2015a. FAOSTAT. http://faostat3.fao.org/home/E.

FAO. 2015b. The state of food insecurity in the world 2015. http://www.fao.org/hunger/key-messages/en.

Faraday, M. 1832. Experimental researches in electricity. *Philosophical Transactions of the Royal Society of London* 122:125–162.

Farey, J. 1827. *A Treatise on the Steam Engine*. London: Longman, Rees, Orme, Brown and Green.

Faulseit, R. K., ed. 2015. *Beyond Collapse: Archaeological Perspectives on Resilience, Revitalization, and Transformation in Complex Societies*. Carbondale, IL: Southern Illinois University Press.

Federico, G. 2008. *Feeding the World: An Economic History of Agriculture, 1800–2000*. Princeton, NJ: Princeton University Press.

Ferguson, E. F. 1971. The measurement of the "man-day." *Scientific American* 225 (4): 96–103.

Fernández-Armesto, F. 1988. *The Spanish Armada: The Experience of War in 1588*. New York: Oxford University Press.

Feuerbach, A. 2006. Crucible Damascus steel: A fascination for almost 2,000 years. *Journal of Metals* (May): 48–50.

Feugang, J. M., P. Konarski, D. Zou, F. C. Stintzing, and C. Zou. 2006. Nutritional and medicinal use of cactus pear (*Opuntia* spp.) cladodes and fruits. *Frontiers in Bioscience* 11:2574–2589.

Feynman, R. 1988. *The Feynman Lectures on Physics*. Redwood City, CA: Addison-Wesley.

Fiedel, S., and G. Haynes. 2004. A premature burial: Comments on Grayson and Meltzer's "Requiem for overkill." *Journal of Archaeological Science* 31:121–131.

Figuier, L. 1888. *Les nouvelles conquêtes de la science: L'électricité*. Paris: Manpir Flammarion.

Finley, M. I. 1959. Was Greek civilization based on slave labour? *Historia. Einzelschriften* 1959:145–164.

Finley, M. I. 1965. Technical innovation and economic progress in the ancient world. *Economic History Review* 18:29–45.

Finniston, M. et al. 1992. *Oxford Illustrated Encyclopedia of Invention and Technology*. Oxford: Oxford University Press.

Fish, J. L., and C. A. Lockwood. 2003. Dietary constraints on encephalization in primates. *American Journal of Physical Anthropology* 120:171–181.

Fitchen, J. 1961. *The Construction of Gothic Cathedrals: A Study of Medieval Vault Erection*. Chicago: University of Chicago Press.

Fitzhugh, B., and J. Habu, eds. 2002. *Beyond Foraging and Collecting: Evolutionary Change in Hunter-Gatherer Settlement Systems*. Berlin: Springer.

Flannery. K.V., ed. 1982. *Maya Subsistence*. New York: Academic Press.

Flink, J. J. 1988. *The Automobile Age*. Cambridge, MA: MIT Press.

Flinn, M. W. et al. 1984–1993. *History of the British Coal Industry*, 5 vols. Oxford: Oxford University Press.

Flower, R., and M. W. Jones. 1981. *100 Years of Motoring: An RAC Social History of Car*. Maidenhead: McGraw-Hill.

Fluck, R. C., ed. 1992. *Energy in Farm Production*. Amsterdam: Elsevier.

Fogel, R. W. 1991. The conquest of high mortality and hunger in Europe and America: Timing and mechanisms. In *Favorites of Fortune*, ed. P. Higgonet et al., 33–71. Cambridge, MA: Harvard University Press.

Foley, R. A., and P. C. Lee. 1991. Ecology and energetics of encephalization in hominid evolution. *Philosophical Transactions of the Royal Society of London* 334:223–232.

Fontana, D. 1590. Della trasportatione dell'obelisco Vaticano et delle fabriche di nostro signore Papa Sisto V. Roma: Domenico Basa. http://www.rarebookroom.org/Control/ftaobc/index.html.

Forbes, R. J. 1958. Power to 1850. In *A History of Technology*, vol. 4, ed. C. Singer et al., 148–167. Oxford: Oxford University Press.

Forbes, R. J. 1964–1972. *Studies in Ancient Technology*. 9 volumes. Leiden: E. J. Brill.

Forbes, R. J. 1964. Bitumen and petroleum in antiquity. In *Studies in Ancient Technology*. vol. 1, 1–124. Leiden: E. J. Brill.

Forbes, R. J. 1965. *Studies in Ancient Technology*, vol. 2. Leiden: E. J. Brill.

Forbes, R. J. 1966. Heat and heating. In *Studies in Ancient Technology*, vol. 6, 1–103. Leiden: E. J. Brill.

Forbes, R. 1972. Copper. In *Studies in Ancient Technology*, vol. 6, 1–133. Leiden: E. J. Brill.

Forbes. 2015. The world's biggest public companies. http://www.forbes.com/global2000/list/#tab:overall.

Fores, M. 1981. The Myth of a British Industrial Revolution. *History* 66:181–198.

Foster, D. R., and J. D. Aber. 2004. *Forests in Time: The Environmental Consequences of 1,000 Years of Change in New England*. New Haven, CT: Yale University Press.

Foster, N., and L. D. Cordell. 1992. *Chilies to Chocolate: Food the Americas Gave the World*. Tucson: University of Arizona Press.

Fouquet, R. 2008. *Heat, Power and Light: Revolutions in Energy Services*. London: Edward Elgar.

Fouquet, R. 2010. The slow search for solutions: Lessons from historical energy transitions by sector and service. *Energy Policy* 38:6586–6596.

Fouquet, R., and P. J. G. Pearson. 2006. Seven centuries of energy services: The price and use of light in the United Kingdom (1300–2000). *Energy Journal* 27:139–177.

Fox, R. F. 1988. *Energy and the Evolution of Life*. San Francisco: W. H. Freeman.

Francis, D. 1990. *The Great Chase: A History of World Whaling*. Toronto: Penguin Books.

Frankenfield, D. C., E. R. Muth, and W. A. Rowe. 1998. The Harris-Benedict studies of human basal metabolism: History and limitations. *Journal of the American Dietetic Association* 98:439–445.

FRED (Federal Reserve Economic Data). 2015. Real gross domestic product per capita. https://research.stlouisfed.org/fred2/series/A939RX0Q048SBEA.

Freedman, B. 2014. *Global Environmental Change*. Amsterdam: Springer Netherlands.

Freedom House. 2015. Freedom in the world 2015. https://freedomhouse.org/report/ freedom-world/freedom-world-2015#.Vfcs74dRGM8.

Freese, S. 1957. *Windmills and Millwrighting*. Cambridge: Cambridge University Press.

French, J. C., and C. Collins. 2015. Upper Palaeolithic population histories of southwestern France: A comparison of the demographic signatures of ^{14}C date distributions and archaeological site counts. *Journal of Archaeological Science* 55:122–134.

Friedel, R., and P. Israel. 1986. *Edison's Electric Light*. New Brunswick, NJ: Rutgers University Press.

Friedman, H. B. 1992. DDT (dichlorodiphenyltrichloroethane): A chemist's tale. *Journal of Chemical Education* 69:362–365.

Frison, G. C. 1987. Prehistoric hunting strategies. In *The Evolution of Human Hunting*, ed. M. H. Nitecki and D. V. Nitecki, 177–223. New York: Plenum Press.

Froment, A. 2001. Evolutionary biology and health of hunter-gatherer populations. In *Hunter-gatherers: An Interdisciplinary Perspective*, ed. C. Panter-Brick, R. Layton and P. Rowley-Conwy, 239–266. Cambridge: Cambridge University Press.

Fry, H. 1896. *History of North Atlantic Steam Navigation*. London: Sampson, Low, Marston & Company.

Fujimoto, T. 1999. *The Evolution of a Manufacturing System at Toyota*. New York: Oxford University Press.

Fussell, G. E. 1952. *The Farmer's Tools, 1500–1900*. London: A. Melrose.

Fussell, G. E. 1972. *The Classical Tradition in West European Farming*. Rutherford: Fairleigh Dickinson University Press.

Gaastra, F. S. 2007. *The Dutch East India Company*. Zutpen: Walburg Press.

Gaier, C. 1967. The origin of Mons Meg. *Journal of the Arms and Armour Society London* 5:425–431.

Galaty, J. G., and P. C. Salzman, eds. 1981. *Change and Development in Nomadic and Pastoral Societies*. Leiden: E. J. Brill.

Gales, B., et al. 2007. North versus South: Energy transition and energy intensity in Europe over 200 years. *European Review of Economic History* 2:219–253.

Galloway, J. A., D. Keene, and M. Murphy. 1996. Fuelling the city: Production and distribution of firewood and fuel in London's region, 1290–1400. *Economic History Review* 49:447–472.

Galor, O. 2005. *From Stagnation to Growth: Unified Growth Theory*. Amsterdam: Elsevier.

Gamarra, N. T. 1969. *Erroneous Predictions and Negative Comments*. Washington, DC: Library of Congress.

Gans, P. J. 2004. The medieval horse harness: Revolution or evolution? A case study in technological change. In *Villard's Legacy: Studies in Medieval Technology, Science and Art in Memory of Jean Gimpel*, ed. M.-T. Zenner, 175–187. London: Routledge.

Garcke, E. 1911. Electric lighting. In *Encyclopaedia Britannica*, 11th ed., vol. 9., 651–673. Cambridge: Cambridge University Press.

Gardiner, R. 2000. *The Heyday of Sail: The Merchant Sailing Ship 1650–1830*. New York: Chartwell Books.

Gardner, J., ed. 2011. *Gilgamesh*. New York: Knopf Doubleday.

Garrett, C., and M. Wade-Matthews. 2015. *The Ultimate Encyclopedia of Steam and Rail*. London: Southwater Publishing.

Gartner. 2015. Gartner says Smartphone sales surpassed one billion units in 2014. http://www.gartner.com/newsroom/id/2996817.

Gaskell, E. 1855. *North and South*. London: Chapman & Hall.

Gates, D. 2011. *The Napoleonic Wars 1803–1815*. New York: Random House.

Geerdes, M., H. Toxopeus, and C. van der Vliet. 2009. *Modern Blast Furnace Ironmaking*. Amsterdam: IOS Press.

Geertz, C. 1963. *Agricultural Involution*. Berkeley: University of California Press.

Gehlsen, D. 2009. *Social Complexity and the Origins of Agriculture*. Saarbrücken: VDM Verlag.

Georgescu-Roegen, N. 1975. Energy and economic myths. *Ecologist* 5:164–174, 242–252.

Georgescu-Roegen, N. 1980. Afterword. In *Entropy: A New World View*, ed. J. Rifkin, 261–269. New York: Viking Press.

Geothermal Energy Association. 2014. *2014 Annual U.S. & Global Geothermal Power Production Report*. http://geo-energy.org/events/2014%20Annual%20US%20&%20Global%20Geothermal%20Power%20Production%20Report%20Final.pdf.

Gerhold, D. 1993. *Road Transport before the Railways*. Cambridge: Cambridge University Press.

Gesner, J. M., ed. 1735. *Scriptores rei rusticae*. Leipzig: Fritsch.

Giampietro, M., and K. Mayumi. 2009. *The Biofuel Delusion*. London: Earthscan.

Gies, F., and J. Gies. 1995. *Cathedral Forge and Waterwheel: Technology and Invention in the Middle Ages*. New York: Harper.

Gill, R. B. 2000. *The Great Maya Droughts: Water, Life, and Death*. Albuquerque: University of New Mexico Press.

Gille, B. 1978. *Histoire des techniques*. Paris: Gallimard.

Gimpel, J. 1997. *The Medieval Machine*. New York: Penguin Books.

Ginouvès, R. 1962. *Balaneutikè: Recherches sur le bain dans l'antiquité grecque*. Paris: de Boccard.

Glaser, B. 2007. Prehistorically modified soils of central Amazonia: A model for sustainable agriculture in the twenty-first century. *Philosophical Transactions of the Royal Society of London. Series B, Biological Sciences* 362:187–196.

Global Wind Energy Council. 2015. Global wind statistics 2014. http://www.gwec .net/wp-content/uploads/2015/02/GWEC_GlobalWindStats2014_FINAL_10.2.2015 .pdf.

Godfrey, F. P. 1982. *An International History of the Sewing Machine*. London: R. Hale.

Goe, M. R., and R. E. Dowell. 1980. *Animal Traction: Guidelines for Utilization*. Ithaca, NY: Cornell University, Department of Animal Science.

Gold, B., et al. 1984. *Technological Progress and Industrial Leadership: The Growth of the U.S. Steel Industry, 1900–1970*. Lexington, MA: D. C. Heath and Co.

Goldsmith. R. W. 1946. The power of Victory: Munitions output in World War II. *Military Affairs* 10:69–80.

Goldstein, D. B., S. Martinez, and R. Roy. 2011. Are there rebound effects from energy efficiency? An analysis of empirical data, internal consistency, and solutions. *Electricity Policy* 2011:1–18.

Gómez, J. J. H., V. Marquina, and R. W. Gómez. 2013. On the performance of Usain Bolt in the 100 m sprint. *European Journal of Phycology* 34:1227–1233.

Goren-Inbar, N., et al. 2004. Evidence of hominin control of fire at Gesher Benot Ya'aqov, Israel. *Science* 304:725–727.

Goudsblom, J. 1992. *Fire and Civilization*. London: Allen Lane.

Grayson, D. K., and F. Delpech. 2002. Specialized early Upper Paleolithic hunters in southwestern France? *Journal of Archaeological Science* 29:1439–1449.

Greene, A. N. 2008. *Horses at Work*. Cambridge, MA: Harvard University Press.

Greene, K. 2000. Technological innovation and economic progress in the ancient world: M. I. Finley re-considered. *Economic History Review* 53:29–59.

Greeno, F. L., ed. 1912. *Obed Hussey Who, of All Inventors, Made Bread Cheap*. Rochester, NY: Rochester Herald Publishing Co.

Greenwood, W. H. 1907. *Iron*. London: Cassell.

Griffiths, J. 1992. *The Third Man: The Life and Times of William Murdoch 1754–1839*. London: Andre Deutsch.

Grigg, D. B. 1974. *The Agricultural Systems of the World*. Cambridge: Cambridge University Press.

Grigg, D. B. 1992. *The Transformation of Agriculture in the West*. Oxford: Blackwell.

Grimal, N. 1992. *A History of Ancient Egypt*. Oxford: Blackwell.

Gronow, P., and I. Saunio. 1999. *International History of the Recording Industry*. London: Bloomsbury Academic.

Grousset, R. 1938. *L'empire des steppes*. Paris: Payot.

Grousset, R. 1970. *The Epic of the Crusades*. New York: Orion Press.

GSI (Global Subsidies Initiative). 2015. Global Subsidies Initiative. https://www.iisd.org/gsi/fossil-fuel-subsidies.

Gulflink. 1991. Fast facts about operations Desert Shield/Desert Storm. http://www.gulflink.osd.mil/timeline/fast_facts.htm.

Gunston, B. 1986. *World Encyclopedia of Aero Engines*. Wellingborough: Patrick Stephens.

Gunston, B. 1999. *The Development of Piston Aero Engines*. Yeovil: Patrick Stephens.

Gunston, B. 2002. *Aviation: The First 100 Years*. Hauppauge, NY: Barron's Educational Series.

Haaland, R., and P. Shinnie, eds. 1985. *African Iron Working: Ancient and Traditional*. Oslo: Norwegian University Press.

Hadfield, C. 1969. *The Canal Age*. New York: Praeger.

Hadland, T., and H.-E. Lessing. 2014. *Bicycle Design: An Illustrated History*. Cambridge, MA: MIT Press.

Haile-Selassie, Y., et al. 2015. New species from Ethiopia further expands Middle Pliocene hominin diversity. *Nature* 521:483–488.

Hair, T. H. 1844. *Sketches of the Coal Mines in Northumberland and Durham*. London: J. Madden & Co.

Hammel, E. M. 1985. *The Root: The Marines in Beirut, August 1982–February 1984*. New York: Harcourt Brace Jovanovich.

Hansell, M. H. 2005. *Animal Architecture*. Oxford: Oxford University Press.

Hansen, P. V. 1992. Experimental reconstruction of the medieval trebuchet. *Acta Archaeologica* 63:189–208.

Hanson, N. 2011. *The Confident Hope of a Miracle: The True History of the Spanish Armada*. New York: Random House.

Harlan, J. R. 1975. *Crops and Man*. Madison, WI: American Society of Agronomy.

Harlow, J. H. 2012. *Electric Power Transformer Engineering*. Boca Raton, FL: CRC Press.

Harmand, S., et al. 2015. 3.3-Million-year-old stone tools from Lomekwi 3, West Turkana, Kenya. *Nature* 521:310–315.

Harris, J. R. 1988. *The British Iron Industry 1700–1850*. London: Macmillan.

Harris, M. 1966. The cultural ecology of India's sacred cattle. *Current Anthropology* 7:51–66.

Harrison, P. D., and B. L. Turner, eds. 1978. *Pre-Hispanic Maya Agriculture*. Albuquerque: University of New Mexico Press.

Harris, J. A., and F. G. Benedict. 1919. *A Biometric Study of Basal Metabolism in Man*. Washington, DC: Carnegie Institution.

Hart, J. F. 2004. *The Changing Scale of American Agriculture*. Charlottesville: University of Virginia Press.

Hartmann, F. 1923. *L'agriculture dans l'ancienne Egypte*. Paris: Libraire-Imprimerie Réunies.

Harverson, M. 1991. *Persian Windmills*. The Hague: International Molinological Society.

Hashimoto, T., et al. 2013. Hand before foot? Cortical somatotopy suggests manual dexterity is primitive and evolved independently of bipedalism. *Philosophical Transactions B* 368 (1630): 1–12.

Hassan, F. A. 1984. Environment and subsistence in Predynastic Egypt. In *From Hunters to Farmers*, ed. J. D. Clark and S. A. Brandt, 57–64. Berkeley: University of California Press.

Haudricourt, A. G., and M. J. B. Delamarre. 1955. *L'Homme et la Charrue à travers le Monde*. Paris: Gallimard.

Haug, G. H., et al. 2003. Climate and collapse of Maya civilization. *Science* 299:1731–1735.

Haugaasen, J. M. T., et al. 2010. Seed dispersal of the Brazil nut tree (*Bertholletia excelsa*) by scatter-hoarding rodents in a central Amazonian forest. *Journal of Tropical Ecology* 26:251–262.

Hausman, W. J., P. Hertner, and M. Wilkins. 2008. *Global Electrification: Multinational Enterprise and International Finance in the History of Light and Power, 1878–2007*. Cambridge: Cambridge University Press.

Hawkes, K., J. F. O'Connell, and N. G. Blurton Jones. 2001. Hadza meat sharing. *Evolution and Human Behavior* 22:113–142.

Hayden, B. 1981. Subsistence and ecological adaptations of modern hunter/gatherers. In *Omnivorous Primates*, ed. R. S. O. Harding and G. Teleki, 344–421. New York: Columbia University Press.

Haynie, D. 2001. *Biological Thermodynamics.* Cambridge: Cambridge University Press.

Headland, T. N., and L. A. Reid. 1989. Hunter-gatherers and their neighbors from prehistory to the present. *Current Anthropology* 30:43–66.

Heidenreich, C. 1971. *Huronia: A History and Geography of the Huron Indians.* Toronto: McClelland and Stewart.

Heinrich, B. 2001. *Racing the Antelope: What Animals Can Teach Us about Running and Life.* New York: HarperCollins.

Heizer, R. F. 1966. Ancient heavy transport, methods and achievements. *Science* 153:821–830.

Helland, J. 1980. *Five Essays on the Study of Pastoralists and the Development of Pastoralism.* Bergen: Universitet i Bergen.

Helliwell, J. F., R. Layard, and J. Sachs eds. 2015. *World Happiness Report 2015.* http://worldhappiness.report/wp-content/uploads/sites/2/2015/04/WHR15-Apr29-update.pdf.

Hemphill, R. 1990. Le transport de l'obélisque du Vatican. *Etudes Francaises* 26 (3): 111–116.

Henry, A. G., A. S. Brooks, and D. R. Piperno. 2014. Plant foods and the dietary ecology of Neanderthals and early modern humans. *Journal of Human Evolution* 69:44–54.

Heppenheimer, T. A. 1995. *Turbulent Skies: The History of Commercial Aviation.* New York: John Wiley.

Herlihy, D. V. 2004. *Bicycle: The History.* New Haven, CT: Yale University Press.

Herodotus. n.d. *Book of Histories.* Excerpt at http://www.cheops-pyramide.ch/khufu-pyramid/herodotus.html.

Herring, H. 2004. Rebound effect in energy conservation. In *Encyclopedia of Energy*, ed. C. Cleveland et al., vol. 5, pp. 411–423. Amsterdam: Elsevier.

Herring, H. 2006. Energy efficiency: A critical view. *Energy* 31:10–20.

Heston, A. 1971. An approach to the sacred cow of India. *Current Anthropology* 12:191–209.

Heyne, E. G., ed. 1987. *Wheat and Wheat Improvement.* Madison, WI: American Society of Agronomy.

Hildinger, E. 1997. *Warriors of the Steppe: A Military History of Central Asia, 500 B.C. to A.D. 1700*. New York: Sarpedon Publishers.

Hill, A. V. 1922. The maximum work and mechanical efficiency of human muscles and their most economical speed. *Journal of Physiology* 56:19–41.

Hill, D. 1984. *A History of Engineering in Classical and Medieval Times*. La Salle, IL: Open Court Publishing.

Hills, R. 1989. *Power from Steam: A History of the Stationary Steam Engine*. Cambridge: Cambridge University Press.

Hindle, B., ed. 1975. *America's Wooden Age: Aspects of Its Early Technology*. Tarrytown, NY: Sleepy Hollow Restorations.

Hippisley, J. C. 1823. *Prison Treadmills*. London: W Nicol.

Hitchcock, R. K., and J. I. Ebert. 1984. Foraging and food production among Kalahari hunter/gatherers. In *From Hunters to Farmers*, ed. J. D. Clark and S. A. Brandt, 328–348. Berkeley: University of California Press.

Ho, P. 1975. *The Cradle of the East*. Hong Kong: Chinese University of Hong Kong Press.

Hodge, A. T. 1990. A Roman factory. *Scientific American* 263 (5): 106–111.

Hodge, A. T. 2001. *Roman Aqueducts & Water Supply*. London: Duckworth.

Hodges, P. 1989. *How the Pyramids Were Built*. Longmead: Element Books.

Hoffmann, H. 1953. *Die chemische Veredlung der Steinkohle durch Verkokung*. http://epic.awi.de/23532/1/Hof1953a.pdf.

Hogan, W. T. 1971. *Economic History of the Iron and Steel Industry in the United States*. 5 vols. Lexington, MA: Lexington Books.

Hogg, I. V. 1997. *German Artillery of World War Two*. Mechanicsville, PA: Stackpole Books.

Holley, I. B. 1964. *Buying Aircraft: Matériel Procurement for the Army Air Forces*. Washington, DC: Department of the Army.

Holliday, M. A. 1986. Body composition and energy needs during growth. In *Human Growth: A Comprehensive Treatise*, ed. F. Falkner and J. M. Tanner, vol. 2, 101–117. New York: Plenum Press.

Holt, P. M. 2014. *The Age of the Crusades: The Near East from the Eleventh Century to 1517*. London: Routledge.

Holt, R. 1988. *The Mills of Medieval England*. Oxford: Oxford University Press.

Homewood, K. 2008. *Ecology of African Pastoralist Societies*. Oxford: James Curry.

Hommel, R. P. 1937. *China at Work*. Doylestown, PA: Bucks County Historical Society.

Hong, S. 2001. *Wireless: From Marconi's Black-Box to the Audio*. Cambridge, MA: MIT Press.

Hopfen, H. J. 1969. *Farm Implements for Arid and Tropical Regions*. Rome: FAO.

Hough, R. and D. Richards. 2007. *Battle of Britain*. Barnsley: Pen & Sword Aviation.

Hounshell, D. A. 1981. Two paths to the telephone. *Scientific American* 244 (1): 157–163.

Howell, J. M. 1987. Early farming in Northwestern Europe. *Scientific American* 257 (5): 118–126.

Howell, J. W., and H. Schroeder. 1927. *The History of the Incandescent Lamp*. Schenectady, NY: Maqua Co.

Hoyt, E. P. 2000. *Inferno: The Fire Bombing of Japan, March 9–August 15, 1945*. New York: Madison Books.

Hua, J. 1983. The mass production of iron castings in ancient China. *Scientific American* 248:120–128.

Huang, N. 1958. *China Will Overtake Britain*. Beijing: Foreign Languages Press.

Hubbard, F. H. 1981. *Encyclopedia of North American railroading: 150 years of railroading in the United States and Canada*. New York: McGraw-Hill.

Hublin, J.-J., and M. P. Richards, eds. 2009. *The Evolution of Hominin Diets: Integrating Approaches to the Study of Palaeolithic Subsistence*. Berlin: Springer.

Hudson, P. 1990. Proto-industrialisation. *Recent Findings of Research in Economics and Social History* 10:1–4.

Hughes, Thomas P. 1983. *Networks of Power*. Baltimore, MD: Johns Hopkins University Press.

Hugill, P. J. 1993. *World Trade Since 1431*. Baltimore, MD: Johns Hopkins University Press.

Humphrey, W. S., and J. Stanislaw. 1979. Economic growth and energy consumption in the UK, 1700–1975. *Energy Policy* 7:29–42.

Hunley, J. D. 1995. The Enigma of Robert H. Goddard. *Technology and Culture* 36:327–350.

Hunter, L. C. 1975. Water power in the century of steam. In *America's Wooden Age: Aspects of Its Early Technology*, ed. B. Hindle, 160–192. Tarrytown, PA: Sleepy Hollow Restorations.

Hunter, L. 1979. *A History of Industrial Power in the US, 1780–1930*, vol. 1. Charlottesville: University of Virginia Press.

Hunter, L. C., and L. Bryant. 1991. *A History of Industrial Power in the United States, 1780–1930*. Vol. 3, *The Transmission of Power*. Cambridge, MA: MIT Press.

Husslage, G. 1965. *Windmolens: Een overzicht van de verschillende molensoorten en hun werkwijze*. Amsterdam: Heijnis.

Huurdeman, A. A. 2003. *The Worldwide History of Telecommunications*. New York: John Wiley & Sons.

Hyde, C. K. 1977. *Technological Change and the British Iron Industry 1700–1870*. Princeton, NJ: Princeton University Press.

Hyland, A. 1990. *Equus: The Horse in the Roman World*. New Haven, CT: Yale University Press.

IBIS World. 2015. Bicycle manufacturing in China. http://www.ibisworld.com/industry/china/bicycle-manufacturing.html.

IEA (International Energy Agency). 2015a. *Energy Balances of Non-OECD Countries*. Paris: IEA.

IEA. 2015b. World balance. http://www.iea.org/sankey.

Ienaga, S. 1978. *The Pacific War, 1931–1945*. New York: Pantheon Books.

ICCT (International Council on Clean Transportation). 2014. *European Vehicle Market Statistics. Pocketbook 2014*. http://www.theicct.org/sites/default/files/publications/EU_pocketbook_2014.pdf.

IFIA (International Fertilizer Industry Association). 2015. Market outlook reports. http://www.fertilizer.org/MarketOutlooks.

Illich, I. 1974. *Energy and Equity*. New York: Harper and Row.

IMF (International Monetary Fund). 2015. Counting the cost of energy subsidies. http://www.imf.org/external/pubs/ft/survey/so/2015/new070215a.htm.

Intel. 2015. Moore's law and Intel innovation. http://www.intel.com/content/www/us/en/history/museum-gordon-moore-law.html.

International Labour Organization. 2015. Forced labour, human trafficking and slavery. http://www.ilo.org/global/topics/forced-labour/lang--en/index.htm.

IPCC (Intergovernmental Panel on Climate Change). 2015. [*Synthesis Report Summary for Policymakers*. Geneva: IPCC.] *Climatic Change:*2014.

Irons, W., and N. Dyson-Hudson, eds. 1972. *Perspective on Nomadism*. Leiden: E. J. Brill.

IRRI (International Rice Research Institute). 2015. Rice milling. http://www .knowledgebank.irri.org/ericeproduction/PDF_&_Docs/Teaching_Manual_Rice _Milling.pdf.

Jakab, P. L. 1990. *Visions of a Flying Machine: The Wright Brothers and the Process of Invention*. Washington, DC: Smithsonian Institution Press.

Jamasmie, C. 2015. End of an era for UK coal mining: Last mines close up shop. http://www.mining.com/end-of-an-era-for-uk-coal-mining-last-mines-close -up-shop.

James, A. 2015. *Global PV Demand Outlook 2015–2020: Exploring Risk in Downstream Solar Markets*. GTM Research, June. http://www.greentechmedia.com/research/ report/global-pv-demand-outlook-2015-2020.

Janick, J. 2002. Ancient Egyptian agriculture and the origins of horticulture. *Acta Horticulturae* 582:23–39.

Jansen, M. B. 2000. *The Making of Modern Japan*. Cambridge, MA: Belknap Press of Harvard University Press.

Jehl, F. 1937. *Menlo Park Reminiscences*. Dearborn, MI: Edison Institute.

Jenkins, B. 1993. *Properties of Biomass, Appendix to Biomass Energy Fundamentals*. Palo Alto, CA: EPRI.

Jenkins, R. 1936. *Links in the History of Engineering and Technology from Tudor Times*. Cambridge: Cambridge University Press.

Jensen, H. 1969. *Sign, Symbol and Script*. New York: G. P. Putnam's Sons.

Jevons, W. S. 1865. *The Coal Question: An Inquiry Concerning the Progress of the Nation, and the Probable Exhaustion of our Coal Mines*. London: Macmillan.

Jing, Y., and R. K. Flad. 2002. Pig domestication in ancient China. *Antiquity* 76:724–732.

Johannsen, O. 1953. *Geschichte des Eisens*. Dusseldorf: Verlag Stahleisen.

Johanson, D. 2006. How bipedalism arose. PBS, *Nova*, October 1. http://www.pbs .org/wgbh/nova/evolution/what-evidence-suggests.html.

Johnson, E. D. 1973. *Communication: An Introduction to the History of the Alphabet, Writing, Printing, Books, and Libraries*. Metuchen, NJ: Scarecrow Press.

Jones, C. F. 2014. *Routes of Power*. Cambridge, MA: Harvard University Press.

Jones, H. M. 1971. *The Age of Energy*. New York: Viking Press.

Jones, H. 1973. *Steam Engines*. London: Ernest Benn.

Josephson, M. 1959. *Edison: A Biography*. New York: McGraw-Hill.

J.P. Morgan. 2015. *A Brave New World: Deep Decarbonization of Electricity Grids*. New York: J. P. Morgan.

Juleff, G. 2009. Technology and evolution: A root and branch view of Asian iron from first-millennium BC Sri Lanka to Japanese steel. *World Archaeology* 41:557–577.

Junqueira, A. B, G. H. Shepard, and C. R. Clement. 2010. Secondary forests on anthropogenic soils in Brazilian Amazonia conserve agrobiodiversity. *Biodiversity and Conservation* 19:1933–1961.

Kander, A. 2013. The second and third industrial revolutions. In *Power to the People: Energy in Europe Over the Last Five Centuries*, by A. Kander, P. Malanima, and P. Warde, 249–386. Princeton, NJ: Princeton University Press.

Kander, A., P. Malanima, and P. Warde. 2013. *Power to the People: Energy in Europe over the Last Five Centuries*. Princeton, NJ: Princeton University Press.

Kander, A., and P. Warde. 2011. Energy availability from livestock and agricultural productivity in Europe, 1815–1913: A new comparison. *The Economic History Review* 64:1–29.

Kanigel, R. 1997. *The One Best Way: Frederick Winslow Taylor and the Enigma of Efficiency*. New York: Viking.

Kaplan, D. 2000. The darker side of the "Original Affluent Society." *Journal of Anthropological Research* 56:301–324.

Karim, M. R., and M. S. H. Fatt. 2005. Impact of the Boeing 767 aircraft into the World Trade Center. *Journal of Engineering Mechanics* 131:1066–1072.

Karkanas, P., et al. 2007. Evidence for habitual use of fire at the end of the Lower Paleolithic: Site-formation processes at Qesem Cave, Israel. *Journal of Human Evolution* 53:197–212.

Kaufer, D. S., and K. M. Carley. 1993. *Communication at a Distance: The Influence of Print on Sociocultural Organization and Change*. Hillsdale, NJ: Lawrence Erlbaum Associates.

Kaufmann, R. K. 1992. A biophysical analysis of the energy/real GDP ratio: Implications for substitution and technical change. *Ecological Economics* 6:35–56.

Kay, J. P. 1832. *The Moral and Physical Condition of the Working Classes Employed in the Cotton Manufacture in Manchester*. London: Ridgway.

Keay, J. 2010. *The Honourable Company: A History of the English East India Company*. London: HarperCollins UK.

Keegan, J. 1994. *A History of Warfare*. New York: Vintage.

Keeling, C. D. 1998. Rewards and penalties of monitoring the Earth. *Annual Review of Energy and the Environment* 23: 25–82.

Kelly, R. L. 1983. Hunter-gatherer mobility strategies. *Journal of Anthropological Research* 39:277–306.

Kendall, A. 1973. *Everyday Life of Incas*. London: B. T. Batsford.

Kennedy, C. A., et al. 2015. Energy and material flows of megacities. *Proceedings of the National Academy of Sciences of the United States of America* 112:5985–5990.

Kennedy, E. 1941. *The Automobile Industry: The Coming of Age of Capitalism's Favorite Child*. New York: Reynal & Hitchcock.

Kesaris, P. 1977. *Manhattan Project: Official History and Documents*. Washington, DC: University Publications of America.

Khaira, G. 2009. Coal transportation logistics. Annual Community Coal Forum, Tumbler Ridge, BC.

Khalturin, V. I., et al. 2005. A review of nuclear testing by the Soviet Union at Novaya Zemlya, 1955–1990. *Science & Global Security* 13 (1): 1–42.

Khazanov, A. M. 1984. *Nomads and the Outside World*. Cambridge: Cambridge University Press.

Khazanov, A. M. 2001. *Nomads in the Sedentary World*. London: Curzon.

Kilby, Jack S. 1964. *Miniaturized Electronic Circuits*. U.S. Patent 3,138,743, June 23, 1964. Washington, DC: USPTO.

King, C. D. 1948. *Seventy-five Years of Progress in Iron and Steel*. New York: American Institute of Mining and Metallurgical Engineers.

King, F. H. 1927. *Farmers of Forty Centuries*. New York: Harcourt, Brace & Co.

King, P. 2011. The choice of fuel in the eighteenth century iron industry: The Coalbrookdale accounts reconsidered. *Economic History Review* 64:132–156.

King, R. 2000. *Brunelleschi's Dome: How a Renaissance Genius Reinvented Architecture*. London: Chatto & Windus.

King, P. 2005. The production and consumption of bar iron in early modern England and Wales. *Economic History Review* 58:1–33.

Kingdon, J. 2003. *Lowly Origin: Where, When, and Why Our Ancestors First Stood Up*. Princeton, NJ: Princeton University Press.

Klein, H. A. 1978. Pieter Bruegel the Elder as a guide to 16th-century technology. *Scientific American* 238 (3): 134–140.

Klima, B. 1954. Paleolithic huts at Dolni Vestonice, Czechoslovakia. *Antiquity* 28:4–14.

Kloss, E. 1963. *Der Luftkrieg über Deutschland, 1939–1945*. Munich: DTV.

Komlos, J. 1988. Agricultural productivity in America and Eastern Europe: A comment. *Journal of Economic History* 48:664–665.

Konrad, T. 2010. MV Mont, Knock Nevis, Jahre Viking—World's largest supertanker. *gCaptain* July 18,2020. http://gcaptain.com/mont-knock-nevis-jahre-viking-worlds -largest-tanker-ship/#.Vc3zB4dRGM8.

Kongshaug, G. 1998. *Energy Consumption and Greenhouse Gas Emissions in Fertilizer Production*. Paris: International Fertilizer Association.

Kopparapu, R. K., et al. 2014. Habitable zones around main sequence stars: Depen- dence on planetary mass. *Astrophysical Journal. Letters* 787:L29.

Kranzberg, M., and C. W. Pursell, eds. 1967. *Technology in Western Civilization*, vol. 1. New York: Oxford University Press.

Krausmann, F., and H. Haberl. 2002. The process of Industrialization from an ener- getic metabolism point of view: Socio-economic energy flows in Austria 1830–1995. *Ecological Economics* 41:177–201.

Kumar, S. N. 2004. Tanker transportation. In *Encyclopedia of Energy*, vol. 6, ed. C. Cleveland et al., 1–12. Amsterdam: Elsevier.

Kushnirs, I. 2015. Gross Domestic Product (GDP) in USSR. http://kushnirs.org/ macroeconomics/gdp/gdp_ussr.html#leader1.

Kuthan, J. and J. Royt. 2011. *Katedrála sv. Víta, Václava a Vojtěcha: Svatyně českých patronů a králů*. Praha: Nakladatelství Lidové noviny.

Kuthan, M., et al. 2003. Domestication of wild *Saccharomyces cerevisiae* is accompa- nied by changes in gene expression and colony morphology. *Molecular Microbiology* 47:745–754.

Kuznets, S. S. 1971. *Economic Growth of Nations: Total Output and Production Structure*. Cambridge, MA: Belknap Press of Harvard University Press.

Lacey, J. M. 1935. *A Comprehensive Treatise on Practical Mechanics*. London: Technical Press.

Laloux, R., et al. 1980. Nutrition and fertilization of wheat. In *Wheat*, 19–24. Basel: CIBA-Geigy.

Lancaster, L. C. 2005. *Concrete Vaulted Construction in Imperial Rome: Innovations in Context*. Cambridge: Cambridge University Press.

Landels, J. G. 1980. *Engineering in the Ancient World*. London: Chatto & Windus.

Landes, David. 1969. *The Unbound Prometheus: Technological Change and Industrial Development in Western Europe from 1750 to the Present*. Cambridge: Cambridge University Press.

Langdon, J. 1986. *Horses, Oxen, and Technological Innovation*. Cambridge: Cambridge University Press.

Lannoo, B. 2013. Energy consumption of ICT networks. Brussels: TREND Final Workshop. http://www.fp7-trend.eu/.../energyconsumptionincentives-energy-efficient-net.

Lardy, N. 1983. *Agriculture in China's Modern Economic Development*. Cambridge: Cambridge University Press.

Latimer, B. 2005. The perils of being bipedal. *Annals of Biomedical Engineering* 33:3–6.

Lawler, A. 2016. Megaproject asks: What drove the Vikings? *Science* 352:280–281.

Layard, A. H. 1853. *Discoveries among the Ruins of Nineveh and Babylon*. New York: G.P. Putnam & Company.

Layard, R. 2005. *Happiness: Lessons from a New Science*. New York: Penguin Press.

Layton, E. T. 1979. Scientific technology, 1845–1900: The hydraulic turbine and the origins of American industrial research. *Technology and Culture* 20:64–89.

Leach, E. R. 1959. Hydraulic society in Ceylon. *Past & Present* 15:2–26.

Lécuyer, C., and D. C. Brock. 2010. *Makers of the Microchip*. Cambridge, MA: MIT Press.

Lee, R. B., and R. Daly, eds. 1999. *The Cambridge Encyclopaedia of Hunters and Gatherers*. Cambridge: Cambridge University Press.

Lee, R. B., and I. DeVore, eds. 1968. *Man the Hunter*. New York: Aldine de Gruyter.

Lefebvre des Noëttes, R. 1924. *La Force Motrice animale à travers les Âges*. Paris: Berger-Levrault.

Legge, A. J., and P. A. Rowley-Conwy. 1987. Gazelle killing in Stone Age Syria. *Scientific American* 257 (2): 88–95.

Lehner, M. 1997. *The Complete Pyramids*. London: Thames and Hudson.

Lenin, V. I. 1920. Speech delivered to the Moscow Gubernia Conference of the R.C.P. (B.), November 21, 1920. https://www.marxists.org/archive/lenin/works/1920/nov/21.htm.

Lenstra, J. A., and D. G. Bradley. 1999. Systematics and phylogeny of cattle. In *The Genetics of Cattle*, ed. R. Fries and A. Ruvinsky, 1–14. Wallingford: CABI.

Leon, P. 1998. *The Discovery and Conquest of Peru, Chronicles of the New World Encounter*, ed. and trans. A. P. Cook and N. D. Cook. Durham, NC: Duke University Press.

Leonard, W. R., J. J. Snodgrass, and M. L. Robertson. 2007. Effects of brain evolution on human nutrition and metabolism. *Annual Review of Nutrition* 27:311–327.

Leonard, W. R., et al. 2003. Metabolic correlates of hominid brain evolution. *Comparative Biochemistry and Physiology Part A* 136:5–15.

Lepre, J. P. 1990. *The Egyptian Pyramids*. Jefferson, NC: McFarland & Co.

Lerche, G. 1994. *Ploughing Implements and Tillage Practices in Denmark from the Viking Period to about 1800: Experimentally Substantiated*. Herning: P. Kristensen.

Leser, P. 1931. *Entstehung und Verbreitung des Pfluges*. Münster: Aschendorff.

Lesser, I. O. 1991. *Oil, the Persian Gulf, and Grand Strategy*. Santa Monica, CA: Rand Corp.

Leveau, P. 2006. *Les moulins de Barbegal (1986–2006)* . http://traianus.rediris.es.

Levine, A. J. 1992. *The Strategic Bombing of Germany, 1940–1945*. London: Greenwood.

Levinson, M. 2006. *The Box: How the Shipping Container Made the World Smaller and the World Economy Bigger*. Princeton, NJ: Princeton University Press.

Levinson, M. 2012. *U.S. Manufacturing in International Perspective*. Washington, DC: Congressional Research Service; http://www.fas.org/sgp/crs/misc/R42135.pdf.

Lewin, R. 2004. *Human Evolution: An Illustrated Introduction*. Oxford: Wiley.

Lewis, M. J. T. 1993. The Greeks and the early windmill. *History and Technology* 15:141–189.

Lewis, M. J. T. 1994. The origins of the wheelbarrow. *Technology and Culture* 35:453–475.

Lewis, M. J. T. 1997. *Millstone and Hammer: The Origins of Water-Power*. Hull: University of Hull Press.

Li, L. 2007. *Fighting Famine in North China: State, Market, and Environmental Decline, 1690s-1990s*. Stanford, CA: Stanford University Press.

Liebenberg, L. 2006. Persistence hunting by modern hunter-gatherers. *Current Anthropology* 47:1017–1025.

Lighting Industry Association. 2009. Lamp history. http://www.thelia.org.uk/lighting-guides/lamp-guide/lamp-history.

Lilienfeld, E. J. 1930. *Method and apparatus for controlling electric currents*. US Patent 1,745,175, January 28, 1930. Washington, DC: USPTO.

Lilienthal, D. E. 1944. *TVA: Democracy on the March*. New York: Harper and Brothers.

Lindgren, M. 1990. *Glory and Failure*. Cambridge, MA: MIT Press.

Lindsay, R. B. 1975. *Energy: Historical Development of the Concept*. Stroudsburg, PA: Dowden, Hutchinson & Ross.

Ling, P. J. 1990. *America and the Automobile: Technology, Reform and Social Change*. Manchester: Manchester University Press.

Linsley, J. W., E. W. Rienstra, and J. A. Stiles. 2002. *Giant under the Hill: History of the Spindletop Oil Discovery at Beaumont, Texas, in 1901*. Austin: Texas State Historical Association.

Livi-Bacci, M. 1991. *Population and Nutrition*. Cambridge: Cambridge University Press.

Livi-Bacci, M. 2000. *The Population of Europe*. Oxford: Blackwell.

Livi-Bacci, M. 2012. *A Concise History of World Population*. Oxford: Wiley-Blackwell.

Lizerand, G. 1942. *Le régime rural de l'ancienne France*. Paris: Presses Universitaires.

Lizot, J. 1977. Population, resources and warfare among the Yanomami. *Man* 12:497–517.

Lockwood, A. H. 2012. *The Silent Epidemic: Coal and the Hidden Threat to Health*. Cambridge, MA: MIT Press.

Looney, R. 2002. *Economic Costs to the United States Stemming from the 9/11 Attacks*. Monterey, CA: Center for Contemporary Conflict.

López, A. E. 2014. *La conquista de América*. Barcelona: RBA Libros.

Lotka, A. J. 1922. Contribution to the energetics of evolution. *Proceedings of the National Academy of Sciences of the United States of America* 8:147–151.

Lotka, A. 1925. *Elements of Physical Biology*. Baltimore, MD: Williams and Wilkins.

Lovejoy, C. O. 1988. Evolution of human walking. *Scientific American* 259 (5): 82–89.

Lowrance, R., et al., eds. 1984. *Agricultural Ecosystems*. New York: John Wiley.

Lubar, S. 1992. "Do not fold, spindle or mutilate": A cultural history of the punch card. *Journal of American Culture* 15 (4): 43–55.

Lucas, A. R. 2005. Industrial milling in the ancient and medieval Worlds. A survey of the evidence for an industrial revolution in medieval Europe. *Technology and Culture* 4: 1–30.

Lucassen, J., and R. W. Unger. 2011. Shipping, productivity and economic growth. In *Shipping Efficiency and Economic Growth 1350–1850*, ed. R. W. Unger, 3–44. Leiden: Brill.

Lucchini, F. 1996. *Pantheon*. Roma: Nova Italia Scientifica.

Luknatskii, N.N. 1936. Podnyatie Aleksandrovskoi kolonny v 1832. *Stroitel'naya Promyshlennost'* 1936 (13) :31–34.

Lüngen, H. B. 2013. Trends for reducing agents in blast furnace operation. http://www.dkg.de/akk-vortraege/2013-_-2rd_polnisch_deutsches_symposium/abstract-luengen_reducing-agents.pdf.

MacDonald, W. L. 1976. *The Pantheon Design, Meaning, and Progeny*. Cambridge, MA: Harvard University Press.

Macedo, I. C., M. R. L. V. Leal, and J. E. A. R. da Silva. 2004. *Assessment of Greenhouse Gas Emissions in the Production and Use of Fuel Ethanol in Brazil*. São Paulo: Government of the State of São Paulo; http://unica.com.br/i_pages/files/pdf_ingles.pdf.

Machiavello, C. M. 1991. *La construcción del sistema agrario en la civilización andina*. Lima: Editorial Econgraf.

MacLaren, M. 1943. *The Rise of the Electrical Industry During the Nineteenth Century*. Princeton, NJ: Princeton University Press.

Madden, J. 2015. How much software is in your car? From the 1977 Toronado to the Tesla P85D. http://www.qsm.com/blog/2015/how-much-software-your-car-1977-toronado-tesla-p85d.

Maddison Project. 2013. Maddison Project. http://www.ggdc.net/maddison/maddison-project/home.htm.

Madureira, N. L. 2012. The iron industry energy transition. *Energy Policy* 50:24–34.

Magee, D. 2005. *The John Deere Way: Performance That Endures*. New York: Wiley.

Mak, S. 2010. *Rice Cultivation—The Traditional Way*. Solo, Java: CRBOM (Center for River Basin Organizations and Management).

Malanima, P. 2006. Energy crisis and growth 1650–1850: The European deviation in a comparative perspective. *Journal of Global History* 1:101–121.

Malanima, P. 2013a. Energy consumption in the Roman world. In *The Ancient Mediterranean Environment between Science and History*, ed. W. V. Harris, 13–36. Leiden: Brill.

Malanima, P. 2013b. Pre-industrial economies. In *Power to the People: Energy in Europe Over the Last Five Centuries*, ed. A. Kander, P. Malanima, and P. Warde, 35–127. Princeton, NJ: Princeton University Press.

Malik. J. 1985. *The Yields of Hiroshima and Nagasaki Explosions*. Los Alamos, NM: Los Alamos National Laboratory. http://atomicarchive.com/Docs/pdfs/00313791.pdf.

Malone, P. M. 2009. *Waterpower in Lowell: Engineering and Industry in Nineteenth-Century America*. Baltimore, MD: Johns Hopkins University Press.

Manx National Heritage. 2015. The Great Laxey Wheel. http://www .manxnationalheritage.im/attractions/laxey-wheel.

Marchetti, C. 1986. Fifty-year pulsation in human affairs. *Futures* 18:376–388.

Marder, T. A., and M. W. Jones. 2015. *The Pantheon: From Antiquity to the Present.* Cambridge: Cambridge University Press.

Mark, J. 1985. Changes in the British brewing industry in the twentieth century. In *Diet and Health in Modern Britain*, ed. D. J. Oddy and D. P. Miller, 81–101. London: Croom Helm.

Marlowe, F. W. 2005. Hunter-gatherers and human evolution. *Evolutionary Anthropology* 14:54–67.

Marshall, R. 1993. *Storm from the East: From Genghis Khan to Khublai Khan.* Berkeley: University of California Press.

Martin, C., and G. Parker. 1988. *The Spanish Armada.* London: Hamish Hamilton.

Martin, P. S. 1958. Pleistocene ecology and biogeography of North America. *Zoogeography* 151:375–420.

Martin, P. S. 2005. *Twilight of the Mammoths.* Berkeley: University of California Press.

Martin, T. C. 1922. *Forty Years of Edison Service, 1882–1922: Outlining the Growth and Development of the Edison System in New York City.* New York: New York Edison Company.

Mason, S. L. R. 2000. Fire and Mesolithic subsistence: Managing oaks for acorns in northwest Europe? *Palaeogeography, Palaeoclimatology, Palaeoecology* 164:139–150.

Mauthner, F., and W. Weiss. 2014. *Solar Heat Worldwide 2012.* Paris: IEA.

Maxton, G. P., and J. Wormald. 2004. *Time for a Model Change: Re-engineering the Global Automotive Industry.* Cambridge: Cambridge University Press.

Maxwell, J. C. 1865. A dynamical theory of the electromagnetic field. *Philosophical Transactions of the Royal Society of London* 155:459–512.

May, G. S. 1975. *A Most Unique Machine: The Michigan Origins of the American Automobile Industry.* Grand Rapids, MI: William B. Eerdmans Publishing.

May, T. 2013. *The Mongol Conquests in World History.* London: Reaktion Books.

Mayhew, H., and J. Binny. 1862. *The Criminal Prisons of London: And Scenes of Prison Life.* London: Griffin, Bohn, and Co.

Mays, L. W., ed. 2010. *Ancient Water Technologies.* Berlin: Springer.

Mays, L. W., and Y. Gorokhovich. 2010. Water technology in the ancient American Societies. In *Ancient Water Technologies*, ed. L. W. Mays, 171–200. Berlin: Springer.

Mazoyer, M., and L. Roudart. 2006. *A History of World Agriculture: From the Neolithic Age to the Current Crisis*. New York: Monthly Review Press.

McCalley, B. 1994. *Model T Ford: The Car That Changed the World*. Iola, WI: Krause Publications.

McCartney, A. P., ed. 1995. *Hunting the Largest Animals: Native Whaling in the Western Arctic and Subarctic*. Studies in Whaling 3. Edmonton, AB: Canadian Circumpolar Institute.

McCloy, S. T. 1952. *French Inventions of the Eighteenth Century*. Lexington: University of Kentucky Press.

McCullough, D. 2015. *The Wright Brothers*. New York: Simon & Schuster.

McDougall, I., F. H. Brown, and J. G. Fleagle. 2005. Stratigraphic placement and age of modern humans from Kibish, Ethiopia. *Nature* 433:733–736.

McGranahan, G., and F. Murray, eds. 2003. *Air Pollution and Health in Rapidly Developing Countries*. London: Routledge.

McHenry, H. M., and K. Coffing. 2000. *Australopithecus* to *Homo*: Transformations in body and mind. *Annual Review of Anthropology* 29:125–146.

McKeown, T. 1976. *The Modern Rise of Population*. London: Arnold.

McNeill, J. R. 2001. *Something New Under the Sun: An Environmental History of the Twentieth-Century*. New York: W. W. Norton.

McNeill, W. H. 1980. *The Human Condition*. Princeton, NJ: Princeton University Press.

McNeill, W. H. 1989. *The Age of Gunpowder Empires, 1450–1800*. Washington, DC: American Historical Association.

McNeill, W. H. 2005. *Berkshire Encyclopedia of World History 5 Volumes*. Great Barrington, MA: Berkshire Publishing.

McShane, C., and J. A. Tarr. 2007. *The Horse in the City*. Baltimore, MD: Johns Hopkins University Press.

Medeiros, L. C., et al. 2001. *Nutritional Content of Game Meat*. Laramie: University of Wyoming. http://www.wyomingextension.org/agpubs/pubs/B920R.pdf.

Meldrum, R. A., and C. E. Hilton, eds. 2004. *From Biped to Strider: The Emergence of Modern Human Walking, Running, and Resource Transport*. Berlin: Springer.

Mellars, P. A. 1985. The ecological basis of social complexity in the Upper Paleolithic of Southwestern France. In *Prehistoric Hunter-Gatherers*, ed. T. D. Price and J. A. Brown, 271–297. Orlando, FL: Academic Press.

Mellars, P. 2006. Why did modern human populations disperse from Africa ca. 60000 years ago? A new model. *Proceedings of the National Academy of Sciences of the United States of America* 103:9381–9386.

Melosi, M. V. 1982. Energy transition in the nineteenth-century economy. In *Energy and Transport*, ed. G. H. Daniels and M. H. Rose, 55–67. Beverly Hills, CA: Sage Publications.

Melville, H. 1851. *Moby-Dick or the Whale*. New York: Harper & Brothers.

Mendels, F. F. 1972. Proto-industrialization: The first phase of the industrialization process. *Journal of Economic History* 32:241–261.

Mendelssohn, K. 1974. *The Riddle of the Pyramids*. London: Thames and Hudson.

Mensch, Gerhard. 1979. *Stalemate in Technology*. Cambridge, MA: Ballinger.

Mercer, D. 2006. *The Telephone: The Life Story of a Technology*. New York: Greenwood Publishing Group.

Merrill, A. L., and B. K. Watt. 1973. *Energy Value of Foods: Basis and Derivation*. Washington, DC: United States Department of Agriculture.

Meyer, J. H. 1975. *Kraft aus Wasser: Vom Wasserrad zur Pumpturbine*. Innertkirchen: Kraftwerke Oberhasli.

Mill, J. S. 1913. *The Panama Canal. A History and Description of the Enterprise*. New York: Sully & Kleinteich.

Minchinton, W. 1980. Wind power. *History Today* 30 (3): 31–36.

Minchinton, W., and P. Meigs. 1980. Power from the sea. *History Today* 30 (3): 42–46.

Minetti, A. E. 2003. Efficiency of equine express postal systems. *Nature* 426: 785–786.

Minetti, A. E., et al. 2002. Energy cost of walking and running at extreme uphill and downhill slopes. *Journal of Applied Physiology* 93:1039–1046.

Mir-Babaev, M. F. 2004. *Kratkaia khronologiia istorii azerbaidzhanskogo neftiianogo dela*. Baku: Sabakh.

Mitchell, W. A. 1931. *Outlines of the World's Military History*. Harrisburg, PA: Military Service Publishing.

mobiForge. 2015. Global mobile statistics 2014. https://mobiforge.com/research -analysis/global-mobile-statistics-2014-part-a-mobile-subscribers-handset-market -share-mobile-operators.

Mokyr, J. 1976. *Industrialization in the Low Countries, 1795–1850*. New Haven, CT: Yale University Press.

504

References

Mokyr, J. 2002. *The Gifts of Athena: Historical Origins of the Knowledge Economy.* Princeton, NJ: Princeton University Prss.

Mokyr, J. 2009. *The Enlightened Economy: An Economic History of Britain 1700–1850.* New Haven, CT: Yale University Press.

Molenaar, A. 1956. *Water Lifting Devices for Irrigation.* Rome: FAO.

Moore, G. 1965. Cramming more components onto integrated circuits. *Electronics* 38 (8): 114–117.

Moore, G. E. 1975. Progress in digital integrated electronics. *Technical Digest, IEEE International Electron Devices Meeting,* 11–13.

Morgan, R. 1984. *Farm Tools, Implements, and Machines in Britain: Pre-history to 1945.* Reading: University of Reading and the British Agricultural History Society.

Moritz, L. A. 1958. *Grain-Mills and Flour in Classical Antiquity.* Oxford: Clarendon Press.

Moritz, M. 1984. *The Little Kingdom: The Private Story of Apple Computer.* New York: W. Morrow.

Morrison, J. S., and J. F. Coates. 1986. *The Athenian Trireme.* Cambridge: Cambridge University Press.

Morrison, J. S., J. F. Coates, and B. Rankov. 2000. *The Athenian Trireme: The History and Reconstruction of an Ancient Greek Warship.* Cambridge: Cambridge University Press.

Morrison, J. S., and R. Gardiner, eds. 1995. *The Age of the Galley: Mediterranean Oared Vessels since Pre-Classical Times.* London: Conway Maritime.

Morton, H. 1975. *The Wind Commands: Sailors and Sailing Ships in the Pacific.* Vancouver: University of British Columbia Press.

Mozley, J. H. 1928. *Statius. Silvae: Thebaid I–IV.* London: William Heinemann.

Mukerji, C. 1981. *From Graven Images: Patterns of Modern Materialism.* New York: Columbia University Press.

Muldrew, C. 2011. *Food, Energy and the Creation of Industriousness: Work and Material Culture in Agrarian England, 1550–1780.* Cambridge: Cambridge University Press.

Muller, G., and K. Kauppert. 2004. Performance characteristics of water wheels. *Journal of Hydraulic Research* 42:451–460.

Müller, I. 2007. *A History of Thermodynamics: The Doctrine of Energy and Entropy.* Berlin: Springer.

Müller, W. 1939. *Die Wasserräder.* Detmold: Moritz Schäfer.

Mumford, L. 1934. *Technics and Civilization*. New York: Harcourt, Brace & Company.

Mumford, L. 1961. *The City in History: Its Origins, Its Transformations, and Its Prospects*. New York: Harcourt, Brace & World.

Mumford, L. 1967. *Technics and Human Development*. New York: Harcourt, Brace & World.

Mundlak, Y. 2005. Economic growth: Lessons from two centuries of American agriculture. *Journal of Economic Literature* 43:989–1024.

Murdock, G. P. 1967. Ethnographic atlas. *Ethnology* 6:109–236.

Murphy, D. J. 2007. *People, Plants, and Genes: The Story of Crops and Humanity*. Oxford: Oxford University Press.

Murphy, D. J., and C. A. S. Hall. 2010. EROI or energy return on (energy) invested. *Annals of the New York Academy of Sciences* 1185:102–118.

Murra, J. V. 1980. *The Economic Organization of the Inka State*. Greenwood, CT: JAO Press.

Mushet, D. 1804. Experiments on wootz or Indian steel. *Philosophical Transactions of the Royal Society of London. Series A, Mathematical and Physical Sciences* 95:175.

Mushrush, G. W., et al. 2000. Use of surplus napalm as an energy source. *Energy Sources* 22:147–155.

Mussatti, D. C. 1998. *Coke Ovens: Industry Profile*. Research Triangle Park, NC: U.S. Environmental Protection Agency.

Musson, A. E. 1978. *The Growth of British Industry*. New York: Holmes & Meier.

Nagata, T. 2014. *Japan's Policy on Energy Conservation*. Tokyo: Ministry of Economy, Trade and Industry. http://www.meti.go.jp/english/policy/energy_environment/.

Napier, J. R. 1970. *The Roots of Mankind*. Washington, DC: Smithsonian Institution Press.

National Coal Mining Museum. 2015. *National Coal Mining Museum for England*. https://www.ncm.org.uk.

National Geographic Society. 2001. Pearl Harbor ships and planes. http://www.nationalgeographic.com/pearlharbor/history/pearlharbor_facts.html.

Naville, E. 1908. *The Temple of Deir el Bahari. Part VI*. London: The Egyptian Exploration Fund.

Needham, J. 1964. *The Development of Iron and Steel in China*. London: The Newcomen Society.

Needham, J. 1965. *Science and Civilisation in China*. Vol. 4, Part II. *Physics and Physical Technology*. Cambridge: Cambridge University Press.

Needham, J. et al. 1954–2015. *Science and Civilisation in China*. 7 volumes. Cambridge: Cambridge University Press.

Needham, J., et al. 1971. *Science and Civilisation in China*. Vol. 4, Part III. *Civil Engineering and Nautics*. Cambridge: Cambridge University Press.

Needham, J., et al. 1986. *Science and Civilisation in China*. Vol. 5, Part VII. *Military Technology: The Gunpowder Epic*. Cambridge: Cambridge University Press.

Nef, J. U. 1932. *The Rise of the British Coal Industry*. London: G. Routledge.

Nelson, W. H. 1998. *Small Wonder: The Amazing Story of the Volkswagen Beetle*. Cambridge, MA: Robert Bentley.

Nesbitt, M., and G. Prance. 2005. *The Cultural History of Plants*. London: Taylor & Francis.

Newhall, B. 1982. *The History of Photography: From 1839 to the Present*. New York: Museum of Modern Art.

Newitt, M. 2005. *A History of Portuguese Overseas Expansion, 1400–1668*. London: Routledge.

Nicholson, J. 1825. *Operative Mechanic, and British Machinist*. London: Knight and Lacey.

Niel, F. 1961. *Dolmens et menhirs*. Paris: Presses Universitaires de France.

Nishiyama, M., and G. Groemer. 1997. *Edo Culture: Daily Life and Diversions in Urban Japan, 1600–1868*. Honolulu: University of Hawaii Press.

NOAA. 2015. Trends in atmospheric carbon dioxide. ftp://aftp.cmdl.noaa.gov/products/trends/co2/co2_annmean_mlo.txt.

Noelker, K., and J. Ruether. 2011. Low energy consumption ammonia production: Baseline energy consumption, options for energy optimization. Nitrogen + Syngas Conference 2011, Düsseldorf. http://www.thyssenkrupp-industrial-solutions.com/fileadmin/documents/publications/Nitrogen-Syngas-2011/Low_Energy_Consumption _Ammonia_Production_2011_paper.pdf.

Noguchi, Tatsuo, and Toshishige Fujii. 2000. Minimizing the effect of natural disasters. *Japan Railway & Transport Review* 23:52–59.

Nordhaus, W. D. 1998. *Do Real-Output and Real-Wage Measures Capture Reality? The History of Lighting Suggests Not*. New Haven, CT: Cowles Foundation for Research in Economics at Yale University.

Norenzayan, A. 2013. *Big Gods: How Religion Transformed Cooperation and Conflict*. Princeton, NJ: Princeton University Press.

Norgan, N. G., et al. 1974. The energy and nutrient intake and the energy expenditure of 204 New Guinean adults. *Philosophical Transactions of the Royal Society of London. Series B, Biological Sciences* 268:309–348.

Norris, J. 2003. *Early Gunpowder Artillery: 1300–1600*. Marlborough: Crowood Press.

North American Electric Reliability Corporation. 2015. *State of Reliability 2015*. http://www.nerc.com/pa/RAPA/PA/Performance%20Analysis%20DL/2015%20State%20of%20Reliability.pdf.

Noyce, Robert N. 1961. *Semiconductor Device-and-Lead Structure*. U.S. Patent 2,981,877, April 25, 1961. Washington, DC: USPTO.

Nutrition Value. 2015. Nutrition value. http://www.nutritionvalue.org.

Nye, D. E. 1992. *Electrifying America: Social Meaning of a New Technology*. Cambridge, MA: MIT Press.

Nye, D. E. 2013. *America's Assembly Line*. Cambridge, MA: MIT Press.

Oberg, E., et al. 2012. *Machinery's Handbook*, 29th ed. South Norwalk, CT: Industrial Press.

O'Brien, P., ed. 1983. *Railways and the Economic Development of Western Europe, 1830–1914*. New York: St. Martin's Press.

Odend'hal, S. 1972. Energetics of Indian cattle in their environment. *Human Ecology* 1:3–22.

Odum, H. T. 1971. *Environment, Power, and Society*. New York: Wiley-Interscience.

Okigbo, B. N. 1984. *Improved Production Systems as an Alternative to Shifting Cultivation*. Rome: FAO.

Oklahoma State University. 2015. Horses. http://www.ansi.okstate.edu/breeds/horses.

Oleson, J. P. 1984. *Greek and Roman Mechanical Water-Lifting Devices: The History of a Technology*. Toronto: University of Toronto Press.

Oleson, J. P., ed. 2008. *The Oxford Handbook of Engineering and Technology in the Classical World*. Oxford: Oxford University Press.

Oliveira, A. R. E. 2014. *A History of the Work Concept: From Physics to Economics*. Dordrecht: Springer.

Olivier, J. G. J. 2014. *Trends in Global CO_2 Emissions: 2014 Report*. The Hague: Netherlands Environmental Assessment Agency. http://edgar.jrc.ec.europa.eu/news_docs/jrc-2014-trends-in-global-co2-emissions-2014-report-93171.pdf.

Olson, M. 1982. *The Rise and Fall of Nations*. New Haven, CT: Yale University Press.

Olsson, F. 2007. *Järnhanteringens dynamic: Produktion, lokalisering och agglomerationer i Bergslagen och Mellansverige 1368–1910*. Umeå: Umeå Studies in Economic History.

Olsson, M., and P. Svensson, eds. 2011. *Growth and Stagnation in European Historical Agriculture*. Turnhout: Brepols.

Ohno, T. 1988. *Toyota Production System: Beyond Large-Scale Production*. Cambridge, MA: Productivity Press.

OPEC (Organization of Petroleum Exporting Countries). 2015. Who gets what from imported oil? http://www.opec.org/opec_web/en/publications/341.htm.

Orme, B. 1977. The advantages of agriculture. In *Hunters, Gatherers and First Farmers beyond Europe*, ed. J. V. S. Megaw, 41–49. Leicester: Leicester University Press.

Orwell, G. 1937. *The Road to Wigan Pier*. London: Victor Gollancz.

Osirisnet. 2015. Djehutyhotep. http://www.osirisnet.net/tombes/el_bersheh/djehoutyhotep/e_djehoutyhotep_02.htm.

Ostwald, W. 1912. *Der energetische Imperativ* . Leipzing: Akademische Verlagsgesselschaft.

Outram, A. K., et al. 2009. The earliest horse harnessing and milking. *Science* 323:1332–1335.

Ovitt, G. 1987. *The Restoration of Perfection: Labor and Technology in Medieval Culture*. New Brunswick, NJ: Rutgers University Press.

Owen, D. 2004. *Copies in Seconds*. New York: Simon and Schuster.

Pacey, A. 1990. *Technology in World Civilization*. Cambridge, MA: MIT Press.

Palgrave Macmillan, ed. 2013. *International Historical Statistics*. London: Palgrave Macmillan; http://www.palgraveconnect.com/pc/connect/archives/ihs.html.

Pan, W., et al. 2013. Urban characteristics attributable to density-driven tie formation. *Nature Communications* . http://hdl.handle.net/1721.1/92362.

Park, J., and T. Rehren. 2011. Large-scale 2nd and 3rd century AD bloomery iron smelting in Korea. *Journal of Archaeological Science* 38:1180–1190.

Parker, G. 1996. *The Military Revolution: Military Innovation and the Rise of the West, 1500–1800*. Cambridge: Cambridge University Press.

Parker, G., ed. 2005. *The Cambridge History of Warfare*. Cambridge: Cambridge University Press.

Parris, H. S., M.-C. Daunay, and J. Janick. 2012. Occidental diffusion of cucumber (*Cucumis sativus*) 500–1300 CE: Two routes to Europe. *Annals of Botany* 109: 117–126.

Parrott, A. 1955. *The Tower of Babel*. London: SCM Press.

Parsons, J. T. 1976. The role of chinampa agriculture in the food supply of Aztec Tenochtitlan. In *Cultural Change and Continuity*, ed. C. Clelland, 233–257. New York: Academic Press.

Parsons, R. H. 1936. *The Development of Parsons Steam Turbine*. London: Constable & Co.

Patton, P. 2004. *Bug: The Strange Mutations of the World's Most Famous Automobile*. Cambridge, MA: Da Capo Press.

Patwhardan, S. 1973. *Change among India's Harijans*. New Delhi: Orient Longman.

Pearson, P. J. G., and T. J. Foxon. 2012. A low carbon industrial revolution? Insights and challenges from past technological and economic transformations. *Energy Policy* 50:117–127.

Pentzer, W. T. 1966. The giant job of refrigeration. In *USDA Yearbook*, 123–138. Washington, DC: USDA.

Perdue, P. C. 1987. *Exhausting the Earth: State and Peasant in Hunan, 1500–1850*. Cambridge, MA: Harvard University Press.

Perdue, P. C. 2005. *China Marches West: The Qing Conquest of Central Asia*. Cambridge, MA: Belknap Press of Harvard University Press.

Perkins, D. S. 1969. *Agricultural Development in China, 1368–1968*. Chicago: University of Chicago Press.

Perkins, S. 2013. Earth is only just within the Sun's habitable zone. *Nature*. doi:10.1038/nature.2013.14353.

Perlin, J. 2005. *Forest Journey: The Story of Wood and Civilization*. Woodstock, VT: Countryman Press.

Perrodon, A. 1985. *Histoire des Grandes Decouvertes Petrolieres*. Paris: Elf Aquitaine.

Pessaroff, N. 2002. An electric idea. … Edison's electric pen. *Pen World International* 15 (5): 1–4.

Pétillon, J.-M., et al. 2011. Hard core and cutting edge: Experimental manufacture and use of Magdalenian composite projectile tips. *Journal of Archaeological Science* 38:1266–1283.

Petroski, H. 1993. On dating inventions. *American Scientist* 81:314–318.

Petroski, H. 2011. Moving obelisks. *American Scientist* 99:448–451.

Pfau, T., et al. 2009. Modern riding style improves horse racing times. *Science* 325:289–291.

Phocaides, A. 2007. *Handbook on Pressurized Irrigation Techniques.* Rome: FAO.

Piggott, S. 1983. *The Earliest Wheeled Transport.* Ithaca, NY: Cornell University Press.

Pimentel, D., ed. 1980. *Handbook of Energy Utilization in Agriculture.* Boca Raton, FL: CRC Press.

Pinhasi, R., J. Fort, and A. J. Ammerman. 2005. Tracing the origin and spread of agriculture in Europe. *PLoS Biology* 3:2220–2228.

PISA. 2015. PISA 2012 Results. http://www.oecd.org/pisa/keyfindings/pisa-2012 -results.htm.

Plutarch. 1961. *Plutarch's Lives.* Trans. B. Perrin. Cambridge, MA: Harvard University Press.

Pobiner, B. L. 2015. New actualistic data on the ecology and energetics of hominin scavenging opportunities. *Journal of Human Evolution* 80:1–16.

Pogue, S. 2012. Use it better: The worst tech predictions of all time. *Scientific American* http://www.scientificamerican.com/article/pogue-all-time-worst-tech -predictions.

Polimeni, J. M., et al. 2008. *The Jevons Paradox and the Myth of Resource Efficiency Improvements.* London: Earthscan.

Polmar, N. 2006. *Aircraft Carriers: A History of Carrier Aviation and Its Influence on World Events.* Vol. 1., *1909–1945.* Lincoln, NB: Potomac Press.

Polmar, N., and T. B. Allen. 1982. *Rickover: Controversy and Genius.* New York: Simon and Schuster.

Pomeranz, K. 2002. Political economy and ecology on the eve of industrialization: Europe, China, and the global conjuncture. *American Historical Review* 107:425–446.

Ponting, C. 2007. *A New Green History of the World: The Environment and the Collapse of Great Civilizations.* New York: Penguin Books.

Pope, F. L. 1894. *Evolution of the Electric Incandescent Lamp.* New York: Boschen & Wefer.

Pope, S. T. 1923. A study of bows and arrows. *University of California Publications in American Archaeology and Ethnology* 13:329–414.

Prager, F. D., and G. Scaglia. 1970. *Brunelleschi: Studies of His Technology and Inventions.* Cambridge, MA: MIT Press.

Pratap, A., and J. Kumar. 2011. *Biology and Breeding of Food Legumes.* Wallingford: CAB.

Price, T. 1991. The Mesolithic of Northern Europe. *Annual Review of Anthropology* 20:211–233.

Price, T. D., and O. Bar-Yosef. 2011. The origins of agriculture: New data, new ideas. *Current Anthropology* 52 (Supplement): S163–S174.

Prigogine, I. 1947. *Étude thermodynamique des phenomenes irreversibles*. Paris: Dunod.

Prigogine, I. 1961. *Introduction to Thermodynamics of Irreversible Processes*. New York: Interscience.

Prost, Antoine. 1991. Public and private spheres in France. In *A History of Private Life*, vol. 5, ed. Antoine Prost and Gérard Vincent., 1–103. Cambridge, MA: Belknap Press of Harvard University Press.

Protzen, J.-P. 1993. *Inca Architecture and Construction at Ollantaytambo*. Oxford: Oxford University Press.

Pryor, F. L. 1983. Causal theories about the origin of agriculture. *Research in Economic History* 8:93–124.

Pryor, A. J. E., et al. 2013. Plant foods in the Upper Palaeolithic at Dolní Vestonice? Parenchyma redux. *Antiquity* 87 (338): 971–984.

Quick, D. 2012. World record 1,626 miles on one tank of diesel. http://www.gizmag .com/tank-diesel-distance-world-record/22488.

Raepsaet, G. 2008. Land transport, part 2: Riding, harnesses, and vehicles. In *The Oxford Handbook of Engineering and Technology in the Classical World*, ed. J. P. Oleson, 580–605. Oxford: Oxford University Press.

Rafiqul, I., et al. 2005. Energy efficiency improvements in ammonia production: Perspectives and uncertainties. *Energy* 30:2487–2504.

Raghavan, B., and J. Ma. 2011. The energy and emergy of the Internet. *Hotnets '11*: 1–6. http://www1.icsi.berkeley.edu/~barath/papers/emergy-hotnets11.pdf.

Ramelli, A. 1976 (1588). *Le diverse et artificiose machine*. Trans. M. Teach Gnudi. Baltimore, MD: Johns Hopkins University Press.

Ranaweera, M. P. 2004. Ancient stupas in Sri Lanka: Largest brick structure sin the world. *Construction History Society Newsletter* 70:1–19.

Rankine, W. J. M. 1866. *Useful Rules and Tables Relating to Mensuration, Engineering Structures and Machines*. London: G. Griffin & Co.

Rapoport, B. I. 2010. Metabolic factors limiting performance in marathon runners. *PLoS Computational Biology* 6:1–13.

Rappaport, R. A. 1968. *Pigs for the Ancestors*. New Haven, CT: Yale University Press.

Ratcliffe, M. 1985. *Liquid Gold Ships: A History of the Tanker, 1859–1984*. London: Lloyd's of London Press.

Rea, M. S., ed. 2000. *IESNA Handbook*. New York: Illuminating Engineering Society of North America.

Reader, J. 2008. *Propitious Esculent: The Potato in World History*. New York: Random House.

Recht, R. 2008. *Believing and Seeing: The Art of Gothic Cathedrals*. Chicago: University of Chicago Press.

Reid, T. R. 2001. *The Chip: How Two Americans Invented the Microchip and Launched a Revolution*. New York: Simon and Schuster.

REN21. 2016. *Renewables 2016 Global Status Report*. Paris: REN21. http://www.ren21 .net/wp-content/uploads/2016/06/GSR_2016_KeyFindings1.pdf.

Revel, J. 1979. Capital city's privileges: Food supply in early-modern Rome. In *Food and Drink in History*, ed. R. Foster and O. Ranum, 37–49. Baltimore, MD: Johns Hopkins University Press.

Revelle, R., and H. E. Suess. 1957. Carbon dioxide exchange between atmosphere and ocean and the question of an increase of atmospheric CO_2 during the past decades. *Tellus* 9:18–27.

Reynolds, J. 1970. *Windmills and Watermills*. London: Hugh Evelyn.

Reynolds, S. C., and A. Gallagher, eds. 2012. *African Genesis: Perspectives on Hominin Evolution*. Cambridge: Cambridge University Press.

Reynolds, T. S. 1979. Scientific influences on technology: The case of the overshot waterwheel, 1752–1754. *Technology and Culture* 20:270–295.

Reynolds, T. S. 1983. *Stronger Than a Hundred Men: A History of the Vertical Water Wheel*. Baltimore, MD: Johns Hopkins University Press.

Rhodes, J. A., and S. E. Churchill. 2009. Throwing in the Middle and Upper Paleolithic: inferences from an analysis of humeral retroversion. *Journal of Human Evolution* 56:1–10.

Ricci, M. 2014. *Il genio di Brunelleschi e la costruzione della Cupola di Santa Maria del Fiore*. Livorno: Casa Editrice Sillabe.

Richerson, P.J., R. Boyd, and R. L. Bettinger. 2001. Was agriculture impossible during the Pleistocene but mandatory during the Holocene? A climate change hypothesis. *American Antiquity* 66:387–411.

Richmond, B. G., et al. 2001. Origin of human bipedalism: The knuckle-walking hypothesis revisited. *Yearbook of Physical Anthropology* 44:71–105.

Rickman, G. E. 1980. The grain trade under the Roman Empire. *Memoirs from the American Academy in Rome* 36:261–276.

Riehl, S., M. Zeidi, and N. J. Conard. 2013. Emergence of agriculture in the foothills of the Zagros Mountains of Iran. *Science* 341:65–67.

Righter, R. W. 2008. *Wind Energy in America: A History*. Norman: University of Oklahoma Press.

Rindos, D. 1984. *The Origins of Agriculture: An Evolutionary Perspective*. Orlando, FL: Academic Press.

Robson, G. 1983. *Magnificent Mercedes: The History of the Marque*. New York: Bonanza Books.

Roche, D. 2000. *A History of Everyday Things: The Birth of Consumption in France, 1600–1800*. Cambridge: Cambridge University Press.

Rockström, J., et al. 2009. A safe operating space for humanity. *Nature* 461:472–475.

Rogin, L. 1931. *The Introduction of Farm Machinery*. Berkeley: University of California Press.

Rollins, A. 1983. *The Fall of Rome: A Reference Guide*. Jefferson, NC: McFarland & Co.

Rolt, L.T.C. 1963. *Thomas Newcomen: The Prehistory of the Steam Engine*. Dawlish: David and Charles.

Rose, D. J. 1974. Nuclear eclectic power. *Science* 184:351–359.

Rosen, W. 2012. *The Most Powerful Idea in the World: The Story of Steam, Industry, and Invention*. Chicago: University of Chicago Press.

Rosenberg, N. 1975. America's rise to woodworking leadership. In *America's Wooden Age: Aspects of Its Early Technology*, ed. B. Hindle, 37–62. Tarrytown, PA: Sleepy Hollow Restorations.

Rosenberg, N., and L. E. Birdzell. 1986. *How the West Grew Rich: The Economic Transformation of the Industrial World*. New York: Basic Books.

Rosenblum, N. 1997. *A World History of Photography*. New York: Abbeville Press.

Rostow, W. W. 1965. *The Stages of Economic Growth*. Cambridge: Cambridge University Press.

Rostow, W. W. 1971. *The Stages of Economic Growth: A Non-Communist Manifesto*. Cambridge: Cambridge University Press.

Rothenberg, B., and F. G. Palomero. 1986. The Rio Tinto enigma—no more. *IAMS* 8:1–6. https://www.ucl.ac.uk/iams/newsletter/accordion/journals/iams_08/iams_8_1986_rothenberg_palomero.

Rouse, J. E. 1970. *World Cattle*. Norman: University of Oklahoma Press.

Rousmaniere, P., and N. Raj. 2007. Shipbreaking in the developing world: Problems and prospects. *International Journal of Occupational and Environmental Health* 13:359–368.

Rowland, F. S. 1989. Chlorofluorocarbons and the depletion of stratospheric ozone. *American Scientist* 77:36–45.

Rubio, M., and M. Folchi. 2012. Will small energy consumers be faster in transition? Evidence form early shift from coal to oil in Latin America. *Energy Policy* 50:50–61.

Ruddle, K., and G. Zhong. 1988. *Integrated Agriculture-Aquaculture in South China*. Cambridge: Cambridge University Press.

Ruff, C. B., et al. 2015. Gradual decline in mobility with the adoption of food production in Europe. *Proceedings of the National Academy of Sciences of the United States of America* 112:7147–7152.

RWEDP (Regional Wood Energy Development Programme in Asia). 1997. *Regional Study of Wood Energy Today and Tomorrow*. Rome: FAO-RWEDP. http://www.rwedp.org/fd50.html.

Ryder, H. W., H. J. Carr, and P. Herget. 1976. Future performance in footracing. *Scientific American* 224 (6): 109–119.

Sagui, C. L. 1948. Le meunerie de Barbegal (France) et les roués hydrauliques les ancients et au moyen âge. *Isis* 38:225–231.

Sahlins, M. 1972. *Stone Age Economics*. Chicago: Aldine.

Salkield, L. U. 1970. Ancient slags in the wouth west of the Iberian Peninsula. Paper presented at the Sixth International Mining Congress, Madrid, June 1970.

Salzman, P. C. 2004. *Pastoralists: Equality, Hierarchy, and the State*. Boulder, CO: Westview Press.

Samedov, V.A. 1988. *Neft' i ekonomika Rossii: 80–90e gody XIX veka*. Baku: Elm.

Sanders, W. T., J. R. Parsons, and R. S. Santley. 1979. *The Basin of Mexico: Ecological Processes in the Evolution of a Civilization*. New York: Academic Press.

Sanz, M., J. Call, and C. Boesch, eds. 2013. *Tool Use in Animals: Cognition and Ecology*. Cambridge: Cambridge University Press.

Sarkar, D. 2015. *Thermal Power Plant: Design and Operation*. Amsterdam: Elsevier.

Sasada, T., and A. Chunag. 2014. Irom smelting in the nomadic empire of Xiongnu in ancient Mongolia. *ISIJ International* 54:1017–1023.

Savage, C. I. 1959. *An Economic History of Transport*. London: Hutchinson.

Schlebecker, J. T. 1975. *Whereby We Thrive*. Ames: Iowa State University Press.

Schmidt, M. J. 1996. Working elephants. *Scientific American* 274 (1): 82–87.

Schmidt, P., and D. H. Avery. 1978. Complex iron smelting and prehistoric culture in Tanzania. *Science* 201:1085–1089.

Schobert, H. H. 2014. *Energy and Society: An Introduction*. Boca Raton, FL: CRC Press.

Schram, W. D. 2014. Greek and Roman Siphons. http://www.romanaqueducts.info/siphons/siphons.htm.

Schumpeter, J. A. 1939. *Business Cycle: A Theoretical and Statistical Analysis of the Capitalist Processes*. New York: McGraw-Hill.

Schurr, S. H., and B. C. Netschert. 1960. *Energy in the American Economy 1850–1975*. Baltimore, MD: Johns Hopkins University Press.

Schurr, S. H., et al. 1990. *Electricity in the American Economy: Agent of Technological Progress*. New York: Greenwood Press.

Schurz, W. L. 1939. *The Manila Galleon*. New York: E. P. Dutton.

Scott, D. A. 2002. *Copper and Bronze in Art: Corrosion, Colorants, Conservation*. Los Angeles: Getty Conservation Institute.

Scott, R. A. 2011. *Gothic Enterprise A Guide to Understanding the Medieval Cathedral*. Berkeley: University of California Press.

Seaborg, G. T. 1972. Opening Address. In *Peaceful Uses of Atomic Energy: Proceedings of the Fourth International Conference on the Peaceful Uses of Atomic Energy*, 29–35. New York: United Nations.

Seavoy, R. E. 1986. *Famine in Peasant Societies*. New York: Greenwood Press.

Seebohm, M. E. 1927. *The Evolution of the English Farm*. London: Allen & Unwin.

comte de Ségur, P.-P. 1825. *History of the Expedition to Russia, Undertaken by Emperor Napoleon, in the Year 1812*. London: Treuttel and Würtz.

Self. 2015. Nuts, brazilnuts. Dried, unblanched. Self.com. http://nutritiondata.self.com/facts/nut-and-seed-products/3091/2.

Sellin, H. J. 1983. The large Roman water mill at Barbégal (France). *History and Technology* 8:91–109.

Senancour, E. P. 1901 (1804). *Obermann*. Trans. J. D. Frothingham. Cambridge: Riverside Press.

Sexton, A. H. 1897. *Fuel and Refractory Materials*. London: Vlackie and Son.

Sharma, R. 2012. *Wheat Cultivation Practices: With Special Reference to Nitrogen and Weed Management*. Saarbrücken: LAP Lambert Academic Publishing.

Shannon, C. E. 1948. A mathematical theory of communication. *Bell System Technical Journal* 27:379–423, 623–656.

Sheehan, G. W. 1985. Whaling as an organizing focus in Northwestern Eskimo society. In *Prehistoric Hunter-Gatherers*, ed. T. D. Price and J. A. Brown, 123–154. Orlando, FL: Academic Press.

Sheldon, C. D. 1958. *The Rise of the Merchant Class in Tokugawa Japan, 1600–1868: An Introductory Survey.* New York: J. J. Augustin.

Shen, T. H. 1951. *Agricultural Resources of China.* Ithaca, NY: Cornell University Press.

Shift Project. 2015. Redesigning Economy to Achieve Carbon Transition. http://www.theshiftproject.org.

Shockley, W. 1964. Transistor technology evokes new physics. In *Nobel Lectures: Physics 1942–1962*, 344–374. Amsterdam: Elsevier.

Shulman, P. A. 2015. *Coal and Empire: The Birth of Energy Security in Industrial America.* Baltimore, MD: Johns Hopkins University Press.

Sieferle, R. P. 2001. *The Subterranean Forest.* Cambridge: White Horse Press.

Siemens, C. W. 1882. Electric lighting, the transmission of force by electricity. *Nature* 27:67–71.

Sierra-Macías, M., et al. 2010. Caracterización agronómica, calidad industrial y nutricional de maíz para el trópico mexicano. *Agronomía Mesoamericana* 21:21–29.

Sillitoe, P. 2002. Always been farmer-foragers? Hunting and gathering in the Papua New Guinea Highlands. *Anthropological Forum* 12:45–76.

Silver, C. 1976. *Guide to the Horses of the World.* Oxford: Elsevier Phaidon.

Simons, G. 2014. *Comet! The World's First Jet Airliner.* Barnsley: Pen and Sword Books.

Singer, C. et al., eds. 1954–1958. *A History of Technology.* 5 volumes. Oxford: Oxford University Press.

Singer, J. D., and M. Small. 1972. *The Wages of War 1816–1965: A Statistical Handbook.* New York: John Wiley.

Sinor, D. 1999. The Mongols in the West. *Journal of Asian History* 33:1–44.

Sittauer, H. L. 1972. *Gebändigte Explosionen.* Berlin: Transpress Verlag für Verkehrswesen.

Sitwell, N. H. 1981. *Roman Roads of Europe.* New York: St. Martin's Press.

Siuru, B. 1989. Horsepower to the people. *Mechanical Engineering (New York)* 111 (2): 42–46.

Skilton, C. P. 1947. *British Windmills and Watermills.* London: Collins.

Slicher van Bath, B. H. 1963. *The Agrarian History of Western Europe, A.D. 500–1850*. London: Arnold.

Smeaton, J. 1759. An experimental enquiry concerning the natural power of water and wind to turn mills, and other machines, depending on a circular motion. *Philosophical Transactions of the Royal Society of London* 51:100–174.

Smil, V. 1976. *China's Energy*. New York: Praeger.

Smil, V. 1981. China's food. *Food Policy* 6:67–77.

Smil, V. 1983. *Biomass Energies*. New York: Plenum Press.

Smil, V. 1985. *Carbon Nitrogen Sulfur: Human Interference in Grand Biospheric Cycles*. New York: Plenum Press.

Smil, V. 1987. *Energy Food Environment*. Oxford: Oxford University Press.

Smil, V. 1988. *Energy in China's Modernization*. Armonk, NY: M. E. Sharpe.

Smil, V. 1991. *General Energetics*. New York: John Wiley.

Smil, V. 1994. *Energy in World History*. Boulder, CO: Westview.

Smil, V. 1997. *Cycles of Life*. New York: Scientific American Library.

Smil, V. 2000a. Energy in the twentieth century: Resources, conversions, costs, uses, and consequences. *Annual Review of Energy and the Environment* 25:21–51.

Smil, V. 2000b. *Feeding the World*. Cambridge, MA: MIT Press.

Smil, V. 2000c. Jumbo. *Nature* 406:239.

Smil, V. 2001. *Enriching the Earth: Fritz Haber, Carl Bosch and the Transformation of World Food Production*. Cambridge, MA: MIT Press.

Smil, V. 2003. *Energy at the Crossroads: Global Perspectives and Uncertainties*. Cambridge, MA: MIT Press.

Smil, V. 2004. War and energy. In *Encyclopedia of Energy*, ed. C. Cleveland et al., vol. 6, 363–371. Amsterdam: Elsevier.

Smil, V. 2005. *Creating the Twentieth Century: Technical Innovations of 1867–1914 and Their Lasting Impact*. New York: Oxford University Press.

Smil, V. 2006. *Transforming the Twentieth Century: Technical Innovations and Their Consequences*. New York: Oxford University Press.

Smil, V. 2008a. *Energy in Nature and Society: General Energetics of Complex Systems*. Cambridge, MA: MIT Press.

Smil, V. 2008b. *Global Catastrophes and Trends*. Cambridge, MA: MIT Press.

Smil, V. 2008c. *Oil*. Oxford: Oneworld Press.

Smil, V. 2010a. *Energy Transitions: History, Requirements, Prospects.* Santa Barbara, CA: Praeger.

Smil, V. 2010b. *Prime Movers of Globalization: The History and Impact of Diesel Engines and Gas Turbines.* Cambridge: MIT Press.

Smil, V. 2010c. *Why America Is Not a New Rome.* Cambridge, MA: MIT Press.

Smil, V. 2013a. *Harvesting the Biosphere: What We Have Taken from Nature.* Cambridge, MA: MIT Press.

Smil, V. 2013b. Just how polluted is China, anyway? *The American,* January 31, 2013. http://www.vaclavsmil.com/wp-content/uploads/smail-article-20130131.pdf.

Smil, V. 2013c. *Made in the USA: The Rise and Retreat of American Manufacturing.* Cambridge, MA: MIT Press.

Smil, V. 2013d. *Should We Eat Meat?* Chichester: Wiley Blackwell.

Smil, V. 2014a. Fifty years of the *Shinkansen. Asia-Pacific Journal: Japan Focus,* December 1, 2014. http://www.vaclavsmil.com/wp-content/uploads/shinkansen.pdf.

Smil, V. 2014b. *Making the Modern World: Materials and Dematerialization.* Chichester: Wiley.

Smil, V. 2015a. *Natural Gas: Fuel for the 21st Century.* Chichester: Wiley.

Smil, V. 2015b. *Power Density: A Key to Understanding Energy Sources and Uses.* Cambridge, MA: MIT Press.

Smil, V. 2015c. Real price of oil. *IEEE Spectrum* 26 (October). http://www.vaclavsmil.com/wp-content/uploads/10.OIL_.pdf.

Smil, V. 2016. *Still the Iron Age: Iron and Steel in the Modern World.* Amsterdam: Elsevier.

Smith, K. 2013. *Biofuels, Air Pollution, and Health: A Global Review.* Berlin: Springer.

Smith, K. P., and A. Anilkumar. 2012. *Rice Farming.* Saarbrücken: Lambert Academic Publishing.

Smith, N. 1980. The origins of the water turbine. *Scientific American* 242 (1): 138–148.

Smith, P. C. 2015. *Mitsubishi Zero: Japan's Legendary Fighter.* Barnsley: Pen & Sword Books.

Smith, N. 1978. Roman hydraulic technology. *Scientific American* 238:154–161.

Smythe, R. H. 1967. *The Structure of the Horse.* London: J. A. Allen & Co.

Sobel, D. 1995. *Longitude: The True Story of a Lone Genius Who Solved the Greatest Scientific Problem of His Time.* New York: Penguin.

Sockol, M. D., D. A. Raichlen, and H. Pontzer. 2007. Chimpanzee locomotor energetics and the origin of human bipedalism. *Proceedings of the National Academy of Sciences of the United States of America* 104:12265–12269.

Soddy, F. 1933. *Money versus Man: A Statement of the World Problem from the Standpoint of the New Economics.* New York: E. P. Dutton.

Soedel, W., and V. Foley. 1979. Ancient catapults. *Scientific American* 240 (3): 150–160.

Solomon, B. D., J. R. Barnes, and K. E. Halvorsen. 2007. Grain and cellulosic ethanol: History, economics, and energy policy. *Biomass and Bioenergy* 31:416–425.

Solomon, F., and R. Q. Marston, eds. 1986. *The Medical Implications of Nuclear War.* Washington, DC: National Academies Press.

Sørensen, B. 2011. *History of Energy: Northern Europe from the Stone Age to the Present Day.* London: Routledge.

Speer, A. 1970. *Inside the Third Reich: Memoirs.* New York: Macmillan.

Spence, K. 2000. Ancient Egyptian chronology and the astronomical orientation of pyramids. *Nature* 408:320–324.

Spencer, J. E. 1966. *Shifting Cultivation in Southeastern Asia.* Berkeley: University of California Press.

Spinardi, G. 2008. *From Polaris to Trident: The Development of US Fleet Ballistic Missile Technology.* Cambridge: Cambridge University Press.

Sponheimer, M., et al. 2013. Isotopic evidence of early hominin diets. *Proceedings of the National Academy of Sciences of the United States of America* 110:10513–10518.

Sprague, G. F., and J. W. Dudley, eds. 1988. *Corn and Corn Improvement.* Madison, WI: American Society of Agronomy.

Spring-Rice, M. 1939. *Working-Class Wives.* Hardmonsworth: Penguin.

Spruytte, J. 19837. *Études expérimentales sur l'attelage: Contribution à l'histoire du cheval.* Paris: Crépin-Lebond.

Stanhill, G. 1976. Trends and deviations in the yield of the English wheat crop during the last 750 years. *Agro-ecosystems* 3:1–10.

Stanley, W. 1912. Alternating-current development in America. *Journal of the Franklin Institute* 173:561–580.

Starbuck, A. 1878. *History of the American Whale Fishery.* Waltham, MA: A. Starbuck.

Stearns, P. N. 2012. *The Industrial Revolution in World History.* Boulder, CO: Westview Press.

Stern, D. I. 2004. Economic growth and energy. In *Encyclopedia of Energy*, ed. C. Cleveland et al., vol. 2, 35–51. Amsterdam: Elsevier.

Stern, D. I. 2010. *The Role of Energy in Economic Growth*. Canberra: Australian National University.

Stewart, I., D. De, and A. Cole. 2015. Technology and people: The great job-creating machine. Deloitte. http://www2.deloitte.com/uk/en/pages/finance/articles/technology-and-people.html.

Stockholm Resilience Center. 2015. The nine planetary boundaries. http://www.stockholmresilience.org/research/planetary-boundaries/planetary-boundaries/about-the-research/the-nine-planetary-boundaries.html.

Stockhuyzen, F. 1963. *The Dutch Windmill*. New York: Universe Books.

Stoltzenberg, D. 2004. *Fritz Haber: Chemist, Nobel Laureate, German, Jew*. Philadelphia: Chemical Heritage Press.

Stopford, M. 2009. *Maritime Economics*. London: Routledge.

Straker, E. 1969. *Wealden Iron*. New York: Augustus M. Kelley.

Strauss, L. L. 1954. Speech to the National Association of Science Writers, New York City, September 16. Cited in *New York Times*, September 17, 5.

Stross, R. E. 1996. *The Microsoft Way: The Real Story of How the Company Outsmarts its Competition*. Reading, MA: Addison-Wesley.

Subcommittee on Horse Nutrition. 1978. *Nutrient Requirements of Horses*. Washington, DC: NAS.

Sullivan, R. J. 1990. The revolution of ideas: Widespread patenting and invention during the English Industrial Revolution. *Journal of Economic History* 50:349–362.

Swade, D. 1991. *Charles Babbage and His Calculating Engines*. London: Science Museum.

Taeuber, I. B. 1958. *The Population of Japan*. Princeton, NJ: Princeton University Press.

Tainter, J. A. 1988. *The Collapse of Complex Societies*. New York: Cambridge University Press.

Takamatsu, N., et al. 2014. Steel recycling circuit in the world. *Tetsu To Hagane* 100:740–749.

Tanaka, Y. 1998. The cyclical sensibility of Edo-period Japan. *Japan Echo* 25 (2): 12–16.

Tata Steel. 2011. Tata Steel announces completion of 100 years of its A-F Blast Furnace's existence. http://www.tatasteel.com/UserNewsRoom/usershowcontent.asp?id

=785&type=PressRelease&REFERER=http://www.tatasteel.com/media/press-release
.asp.

Tate, K. 2009. America's Moon Rocket Saturn V. http://www.space.com/18422
-apollo-saturn-v-moon-rocket-nasa-infographic.html.

Tauger, M. B. 2010. *Agriculture in World History*. London: Routledge.

Taylor, A. 2013. A luxury car club is stirring up class conflict in China. http://www
.businessinsider.com/chinas-sports-car-club-envy-2013-4.

Taylor, C. F. 1984. *The Internal-Combustion Engine in Theory and Practice*. Cambridge,
MA: MIT Press.

Taylor, G. R., ed. 1982. *The Inventions That Changed the World*. London: Reader's
Digest Association.

Taylor, M. J. H., ed. 1989. *Jane's Encyclopedia of Aviation*. New York: Portland House.

Taylor, F. S. 1972. *A History of Industrial Chemistry*. New York: Arno Press.

Taylor, F. W. 1911. *Principles of Scientific Management*. New York: Harper & Brothers.

Taylor, N. A. S. 2006. Ethnic differences in thermoregulation: Genotypic versus
phenotypic heat adaptation. *Journal of Thermal Biology* 31:90–104.

Taylor, N. A. S., and C. A. Machado-Moreira. 2013. Regional variations in transepi-
dermal water loss, eccrine sweat gland density, sweat secretion rates and electrolyte
composition in resting and exercising humans. *Extreme Physiology & Medicine*
2:1–29.

Taylor, R. 2007. The polemics of eating fish in Tasmania: The historical evidence
revisited. *Aboriginal History* 31:1–26.

Taylor, T. S. 2009. *Introduction to Rocket Science and Engineering*. Boca Raton, FL: CRC
Press.

Telleen, M. 1977. *The Draft Horse Primer*. Emmaus, PA: Rodale Press.

Termuehlen, H. 2001. *100 Years of Power Plant Development*. New York: ASME Press.

Tesla, N. 1888. *Electro-magnetic Motor. Specification forming part of Letters Patent No.
391,968, dated May 1, 1888*. Washington, DC: U.S. Patent Office. http://www.uspto
.gov.

Testart, A. 1982. The significance of food storage among hunter-gatherers: Residence
patterns, population densities, and social inequalities. *Current Anthropology*
23:523–537.

Thieme, H. 1997. Lower Paleolithic hunting spears from Germany. *Nature*
385:807–810.

Thomas, B. 1986. Was there an energy crisis in Great Britain in the 17th century? *Explorations in Economic History* 23:124–152.

Thomas Edison Papers. 2015. Edison's patents. http://edison.rutgers.edu/patents .htm.

Thompson, W. C. 2002. *Thompson Releases Report on Fiscal Impact of 9/11 on New York City*. New York: NYC Comptroller.

Thomsen, C. J. 1836. *Ledetraad til nordisk oldkyndighed*. Copenhagen: L. Mellers.

Thomson, K. S. 1987. How to sit on a horse. *American Scientist* 75:69–71.

Thomson, E. 2003. *The Chinese Coal Industry: An Economic History*. London: Routledge.

Thomson, W. 1896. Letter to Major Baden Baden-Powell, December 8, 1896. *Correspondence of Lord Kelvin*. http://zapatopi.net/kelvin/papers/letters.html#baden -powell.

Thoreau, H. D. 1906. *The Journal of Henry David Thoreau, 1837–1861*. Boston: Houghton-Mifflin.

Thrupp, L. A., et al. 1997. *The Diversity and Dynamics of Shifting Cultivation: Myths, Realities, and Policy Implications*. Washington, DC: World Resources Institute.

Thurston, R. H. 1878. *A History of the Growth of the Steam-Engine*. New York: D. Appleton Co.

Titow, J. Z. 1969. *English Rural Society, 1200–1350*. London: George Allen and Unwin.

Tomaselli, I. 2007. *Forests and Energy in Developing Countries*. Rome: FAO.

Tomlinson, R. 2002. The invention of e-mail just seemed like a neat idea. SAP INFO. http://www.sap.info.

Tompkins, P. 1971. *Secrets of the Great Pyramid*. New York: Harper & Row.

Tompkins, P. 1976. *Mysteries of Mexican Pyramids*. New York: Harper & Row.

Torii, M. 1995. Maximal sweating rate in humans. *Journal of Human Ergology* 24:137–152.

Torr, G. 1964. *Ancient Ships*. Chicago: Argonaut Publishers.

Torrey, V. 1976. *Wind-Catchers: American Windmills of Yesterday and Tomorrow*. Brattleboro, VT: Stephen Greene Press.

Tresemer, D. 1996. *The Scythe Book*. Chambersburg, PA: Alan C. Hood.

Trinkaus, E. 1987. Bodies, brawn, brains and noses: Human ancestors and human predation. In *The Evolution of Human Hunting*, ed. M. Nitecki and D. V. Nitecki, 107–145. New York: Plenum.

Trinkaus, E. 2005. Early modern humans. *Annual Review of Anthropology* 34:207–230.

TsSU (Tsentral'noie statisticheskoie upravlenie). 1977. *Narodnoie khoziaistvo SSSR za 60 let*. Moscow: Statistika.

Turner, B. L. 1990. The rise and fall of population and agriculture in the Central Maya Lowlands 300 B.C. to present. In *Hunger in History*, ed. L. F. Newman, 78–211. Oxford: Blackwell.

Tvengsberg, P. M. 1995. Rye and swidden cultivation. *Tools and Tillage* 7:131–146.

Tyne Built Ships. 2015. *Glückauf*. http://www.tynebuiltships.co.uk/G-Ships/gluckauf1886.html.

UNDP (United Nations Development Programme). 2015. *Human Development Report 2015*. New York: UNDP.

UNESCO. 2015a. Head-Smashed-In Buffalo Jump. http://whc.unesco.org/en/list/158.

UNESCO. 2015b. Mount Qingcheng and Dujianyang Irrigation System. http://whc.unesco.org/en/list/1001.

Unger, R. 1984. Energy sources for the Dutch Golden Age. *Research in Economic History* 9:221–253.

United Nations Organization. 1956. World energy requirements in 1975 and 2000. In *Proceedings of the International Conference on the Peaceful Uses of Atomic Energy*, vol. 1, 3–33. New York: UNO.

Upham, C. W., ed. 1851. *The life of General Washington: First President of the United States*. vol. 2. London: National Illustrated Library.

Urbanski, T. 1967. *Chemistry and Technology of Explosives*. New York: Pergamon Press.

U.S. Strategic Bombing Survey. 1947. *Effects of Air Attack on Urban Complex Tokyo-Kawasaki-Yokohama*. Washington, DC: U.S. Strategic Bombing Survey.

USBC (U.S. Bureau of the Census). 1954. *U.S. Census of Manufacturers: 1954*. Washington, DC: U.S. GPO.

USBC. 1975. *Historical Statistics of the United States: Colonial Times to 1970*. Washington, DC: USBC.

U.S. Centennial of Flight Commission. 2003. *History of Flight*. Washington, DC.

U.S. Centennial of Flight Commission. http://www.centennialofflight.gov/hof/index.htm.

USDA (U.S. Department of Agriculture). 1959. *Changes in Farm Production and Efficiency*. Washington, DC: USDA.

USDA. 2011. *National Nutrient Database for Standard Reference* . http://ndb.nal.usda.gov.

USDA. 2014. *Multi-Cropping Practices: Recent Trends in Double Cropping.* Washington, DC: USDA.

USDOE (U.S. Department of Energy). 2011. *Biodiesel Basics.* http://www.afdc.energy.gov/pdfs/47504.pdf.

USDOE. 2013. Energy efficiency of LEDs. http://apps1.eere.energy.gov/buildings/publications/pdfs/ssl/led_energy_efficiency.pdf.

USDOL (U.S. Department of Labor). 2015. Employment by major industry sector. http://www.bls.gov/emp/ep_table_201.htm.

USEIA (U.S. Energy Information Agency). 2014. Consumer energy expenditures are roughly 5% of disposable income, below long-term average. http://www.eia.gov/todayinenergy/detail.cfm?id=18471.

USEIA. 2015a. *Annual Coal Report.* http://www.eia.gov/coal/annual.

USEIA. 2015b. China. http://www.eia.gov/beta/international/analysis.cfm?iso=CHN.

USEIA. 2015c. *Direct Federal Financial Interventions and Subsidies in Energy in Fiscal Year 2013.* Washington, DC: USEIA. http://www.eia.gov/analysis/requests/subsidy.

USEIA. 2015d. Energy intensity. http://www.eia.gov/cfapps/ipdbproject/iedindex3.cfm?tid=92&pid=46&aid=2.

USEIA. 2016a. Coal. http://www.eia.gov/coal.

USEIA. 2016b. U.S. imports from Iraq of crude oil and petroleum products. https://www.eia.gov/dnav/pet/hist/LeafHandler.ashx?n=pet&s=mttimiz1&f=a.

USEPA (U.S. Environmental Protection Agency). 2004. Photochemical smog. http://www.epa.sa.gov.au/files/8238_info_photosmog.pd.

USEPA. 2015. *Light-Duty Automotive Technology, Carbon Dioxide Emissions, and Fuel Economy Trends: 1975 Through 2015.* https://www3.epa.gov/fueleconomy/fetrends/1975-2015/420r15016.pdf.

USGS (U.S. Geological Survey). 2015. Commodity statistics and information. http://minerals.usgs.gov/minerals/pubs/commodity.

Usher, A. P. 1954. *A History of Mechanical Inventions.* Cambridge, MA: Harvard University Press.

Utley, F. 1925. *Trade Guilds of the Later Roman Empire.* London: London School of Economics.

Van Beek, G. W. 1987. Arches and vaults in the ancient Near East. *Scientific American* 257 (2): 96–103.

van Duijn, J. J. 1983. *The Long Wave in Economic Life*. London: George Allen & Unwin.

Van Noten, F., and J. Raymaekers. 1988. Early iron smelting in Central Africa. *Scientific American* 258:104–111.

Varvoglis, H. 2014. *History and Evolution of Concepts in Physics*. Berlin: Springer.

Vasko, T., R. Ayres, and L. Fontvieille, eds. 1990. *Life Cycles and Long Waves*. Berlin: Springer-Verlag.

Vavilov, N. I. 1951. *Origin, Variation, Immunity and Breeding of Cultivated Plants*. Waltham, MA: Chronica Botanica.

Veraverbeke, W. S., and J. A. Delcour. 2002. Wheat protein composition and properties of wheat glutenin in relation to breadmaking functionality. *Critical Reviews in Food Science and Nutrition* 42:179–208.

Versatile. 2015. Versatile. http://www.versatile-ag.ca.

Vikingeskibs Museet. 2016. Wool sailcloth. http://www.vikingeskibsmuseet.dk/en/professions/boatyard/experimental-archaeological-research/maritime-crafts/maritime-technology/woollen-sailcloth.

Ville, S. P. 1990. *Transport and the Development of European Economy, 1750–1918*. London: Macmillan.

Villiers, G. 1976. *The British Heavy Horse*. London: Barrie and Jenkins.

Vogel, H. U. 1993. The Great Wall of China. *Scientific American* 268 (6): 116–121.

Volkswagen, A. G. 2013. Ivan Hirst. http://www.volkswagenag.com/content/vwcorp/info_center/en/publications/2013/11/ivan_hirst.bin.html/binarystorageitem/file/VWAG_HN_4_Ivan-Hirst-eng_2013_10_18.pdf.

von Bertalanffy, L. 1968. *General System Theory*. New York: George Braziller.

von Braun, W., and F. I. Ordway. 1975. *History of Rocketry and Space Travel*. New York: Thomas Y. Crowell.

von Hippel, Frank, et al. 1988. Civilian casualties from counterforce attacks. *Scientific American* 259 (3): 36–42.

von Tunzelmann, G. N. 1978. *Steam Power and British Industrialization to 1860*. Oxford: Clarendon Press.

Wailes, R. 1975. *Windmills in England: A Study of Their Origin, Development and Future*. London: Architectural Press.

Waldron, A. 1990. *The Great Wall of China*. Cambridge: Cambridge University Press.

Walther, R. 2007. *Pechelbronn: A la source du pétrole, 1735–1970*. Strasbourg: Hirlé.

Walton, S. A., ed. 2006. *Wind and Water in the Middle Ages: Fluid Technologies from Antiquity to the Renaissance*. Tempe: Arizona Center for Medieval and Renaissance Studies.

Walz, W., and H. Niemann. 1997. *Daimler-Benz: Wo das Auto Anfing*. Konstanz: Verlag Stadler.

Wang, Z. 1991. *A History of Chinese Firearms*. Beijing: Military Science Press.

War Chronicle. 2015. Estimated war dead World War II. http://warchronicle.com/numbers/WWII/deaths.htm.

Warburton, M. 2001. Barefoot running. *Sportscience* 5 (3): 1–4.

Warde, P. 2007. *Energy Consumption in England and Wales, 1560–2004*. Naples: Consiglio Nazionale della Ricerche.

Warde, P. 2013. The first industrial revolution. In *Power to the People: Energy in Europe Over the Last Five Centuries*, ed. A. Kander, P. Malanima, and P. Warde, 129–247. Princeton, NJ: Princeton University Press.

Washlaski, R. A. 2008. Manufacture of Coke at Salem No. 1 Mine Coke Works. http://patheoldminer.rootsweb.ancestry.com/coke2.html.

Waterbury, J. 1979. *Hydropolitics of the Nile Valley*. Syracuse, NY: Syracuse University Press.

Watkins, G. 1967. Steam power—an illustrated guide. *Industrial Archaeology* 4 (2): 81–110.

Watt, J. 1855 (1769). *Steam Engines, &c. 29 April 1769*. Patent reprint by G. E. Eyre and W. Spottiswoode. https://upload.wikimedia.org/wikipedia/commons/0/0d/James_Watt_Patent_1769_No_913.pdf.

Watters, R. F. 1971. *Shifting Cultivation in Latin America*. Rome: FAO.

Watts, P. 1905. *The Ships of the Royal Navy as They Existed at the Time of Trafalgar*. London: Institution of Naval Architects.

Wei, J. 2012. *Great Inventions that Changed the World*. Hoboken, NJ: Wiley.

Weissenbacher, M. 2009. *Sources of Power: How Energy Forges Human History*. Santa Barbara, CA: Praeger.

Weisskopf, V. F. 1983. Los Alamos anniversary: "We meant so well." *Bulletin of the Atomic Scientists*, August–September, 24–26.

Weller, J. A. 1999. Roman traction systems. http://www.humanist.de/rome/rts.

Welsch, R. L. 1980. No fuel like an old fuel. *Natural History* 89 (11): 76–81.

Wendel, J. F., et al. 1999. Genes, jeans, and genomes: Reconstructing the history of cotton. In *Seventh International Symposium of the International-Organization-of-Plant-Biosystematists*, ed. L. W. D. VanRaamsdonk and J. C. M. DenNijs, 133–159.

Wesley, J. P. 1974. *Ecophysics*. Springfield, IL: Charles C. Thomas.

Whaples, R. 2005. Child Labor in the United States. *EH.Net Encyclopedia* http://eh.net/encyclopedia/child-labor-in-the-united-states.

Wheat Foods Council. 2015. Wheat facts. http://www.wheatfoods.org/resources/72.

Whipp, B. J., and K. Wasserman. 1969. Efficiency of muscular work. *Journal of Applied Physiology* 26:644–648.

White, A. 2007. A global projection of subjective well-being: A challenge to positive psychology? *Psychtalk* 56:17–20.

White, K. D. 1967. *Agricultural Implements of the Roman World*. Cambridge: Cambridge University Press.

White, K. D. 1970. *Roman Farming*. London: Thames & Hudson.

White, K. D. 1984. *Greek and Roman Technology*. Ithaca, NY: Cornell University Press.

White, L. A. 1943. Energy and the evolution of culture. *American Anthropologist* 45:335–356.

White, L. 1978. *Medieval Religion and Technology*. Berkeley: University of California Press.

White, P., and T. Denham, eds. 2006. *The Emergence of Agriculture: A Global View*. London: Routledge.

Whitmore, T. M., et al. 1990. Long-term population change. In *The Earth as Transformed by Human Action*, ed. B. L. Turner II et al., 25–39. Cambridge: Cambridge University Press.

WHO (World Health Organization). 2002. *Protein and Amino Acid Requirements in Human Nutrition*. Geneva: WHO.

WHO. 2015a. Life expectancy. http://apps.who.int/gho/data/node.main.688.

WHO. 2015b. Road traffic injuries. http://www.who.int/mediacentre/factsheets/fs358/en.

Wier, S. K. 1996. Insight from geometry and physics into the construction of Egyptian Old Kingdom pyramids. *Cambridge Archaeological Journal* 6:150–163.

Wikander, Ö. 1983. *Exploitation of Water-Power or Technological Stagnation?* Lund: CWK Gleerup.

Wilkins, J., et al. 2012. Evidence for early hafted hunting technology. *Science* 338:942–946.

Williams, M. 2006. *Deforesting the Earth :From Prehistory to Global Crisis*. Chicago: Chicago University Press.

Williams, M. R. 1997. *History of Computing Technology*. Los Alamitos, CA: IEEE Computer Society.

Williams, T. 1987. *The History of Invention: From Stone Axes to Silicon Chips*. New York: Facts on File.

Wilson, A. M. 1999. Windmills, cattle and railroad: The settlement of the Llano Estacado. *Journal of the West* 38 (1): 62–67.

Wilson, A. M. et al. 2001. Horses damp the spring in their step. *Nature* 414:895–899.

Wilson, C. 1990. *The Gothic Cathedral: The Architecture of the Great Church 1130–1530*. London: Thames and Hudson.

Wilson, C. 2012. Up-scaling, formative phases, and learning in the historical diffusion of energy technologies. *Energy Policy* 50:81–94.

Wilson, D. G. 2004. *Bicycling Science*. Cambridge, MA: MIT Press.

Winkelmann, R., et al. 2015. Combustion of available fossil fuel resources sufficient to eliminate the Antarctic Ice Sheet. *Science Advances* 1:e1500589.

Winter, T. N. 2007. *The Mechanical Problems in the Corpus Aristotle*. Lincoln:: University of Nebraska, Classics and Religious Studies Department.

Winterhalder, B., R. Larsen, and R. B. Thomas. 1974. Dung as an essential resource in a highland Peruvian community. *Human Ecology* 2:89–104.

Wirfs-Brock, J. 2014. Explore 15 years of power outages. http://insideenergy.org/2014/08/18/data-explore-15-years-of-power-outages/.

WNA (World Nuclear Association). 2014. Decommissioning nuclear facilities. http://www.world-nuclear.org/info/nuclear-fuel-cycle/nuclear-wastes/decommissioning-nuclear-facilities.

WNA. 2015a. Uranium enrichment. http://www.world-nuclear.org/info/Nuclear-Fuel-Cycle/Conversion-Enrichment-and-Fabrication/Uranium-Enrichment.

WNA. 2015b. World nuclear power reactors & uranium requirements. http://www.world-nuclear.org/info/Facts-and-Figures/World-Nuclear-Power-Reactors-and-Uranium-Requirements.

Wolfe, D. A., and A. Bramwell. 2008. Innovation, creativity and governance: Social dynamics of economic performance in city-regions. *Innovation: Management. Policy & Practice* 10:170–182.

Wolff, A. R. 1900. *The Windmill as Prime Mover*. New York: John Wiley.

Wölfel, W. 1987. *Das Wasserrad: Technik und Kulturgeschichte*. Wiesbaden: U. Pfriemer.

Womack, J. P., D. T. Jones, and D. Roos. 1990. *The Machine that Changed the World: The Story of Lean Production.* New York: Simon and Schuster.

Wood, W. 1922. *All Afloat.* Toronto: Glasgow, Brook & Company.

Woodall, F. P. 1982. Water wheels for winding. *Industrial Archaeology* 16:333–338.

Woolfe, J. A. 1987. *The Potato in the Human Diet.* Cambridge: Cambridge University Press.

World Bank. 2015a. Energy use. http://data.worldbank.org/indicator/EG.USE.PCAP .KG.OE.

World Bank. 2015b. Motor vehicles (per 1,000 people). http://data.worldbank.org/ indicator/IS.VEH.NVEH.P3.

World Bank. 2015c. Trade. http://data.worldbank.org/indicator/NE.TRD.GNFS.ZS.

World Bank. 2015d. Urban population (% total). http://data.worldbank.org/ indicator/SP.URB.TOTL.IN.ZS.

World Coal Association. 2015. Coal mining. http://www.worldcoal.org/coal/coal -mining.

World Digital Library. 2014. Telegram from Orville Wright in Kitty Hawk, North Carolina, to his father announcing four successful flights, 1903 December 17. http:// www.wdl.org/en/item/11372.

World Economic Forum. 2012. Energy for economic growth. http://www3.weforum. org/docs/WEF_EN_EnergyEconomicGrowth_IndustryAgenda_2012.pdf.

Wrangham, R. 2009. *Catching Fire: How Cooking Made Us Human.* New York: Basic Books.

Wright, O. 1953. *How We Invented the Airplane.* New York: David McKay.

Wrigley, E. A. 2002. The transition to an advanced organic economy: Half a millennium of English agriculture. *Economic History Review* 59:435–480.

Wrigley, E. A. 2006. The transition to an advanced organic economy: Half a millennium of English agriculture. *Economic History Review* 59:435–480.

Wrigley, E. A. 2010. *Energy and the English Industrial Revolution.* Cambridge: Cambridge University Press.

Wrigley, E. A. 2013. Energy and the English Industrial Revolution. *Philosophical Transactions of the Royal Society A* 371. doi:10.1098/rsta.2011.0568.

Wu, K. C. 1982. *The Chinese Heritage.* New York: Crown Publishers.

Wulff, H. E. 1966. *The Traditional Crafts of Persia.* Cambridge, MA: MIT Press.

Xie, Y., and Y. Jin. 2015. Household wealth in China. *China Sociological Review* 47: 203–229.

Xinhua. 2015. China boasts world's largest highspeed railway network. http://news.xinhuanet.com/english/photo/2015-01/30/c_133959250.htm.

Xu, Z., and J. L. Dull. 1980. *Han Agriculture: The Formation of Early Chinese Agrarian Economy, 206 B.C.–A.D. 220.* Seattle: University of Washington Press.

Yang, J. 2012 *Tombstone: The Great Chinese Famine, 1958–1962.* New York: Farrar, Straus and Giroux.

Yates, P. 2012. *Evaluation and Model of the Chinese Kang System.* Fort Collins, CO: University of Colorado.

Yates, R. S. 1990. War, food shortages, and relief measures in early China. In *Hunger in History,* ed. L. F. Newman, 147–177. Oxford: Basil Blackwell.

Yenne, B. 2006. *The American Aircraft Factory in World War II.* Minneapolis, MN: Zenith Press.

Yergin, D. 2008. *The Prize: The Epic Quest for Oil, Money, and Power.* New York: Simon and Schuster.

Yesner, D. R. 1980. Maritime hunter-gatherers: Ecology and prehistory. *Current Anthropology* 21:727–750.

Yonekura, S. 1994. *The Japanese Iron and Steel Industry, 1850–1990: Continuity and Discontinuity.* New York: St. Martin's Press.

Zaanse Schans. 2015. Zaanse Schans. http://www.dezaanseschans.nl/en.

Zeder, M. 2011. The origins of agriculture in the Near East. *Current Anthropology* 52 (Supplement): S221–S235.

Ziemke, E. F. 1968. *The Battle for Berlin: End of the Third Reich.* New York: Ballantine Books.

Name Index

Subject Index

Irrigation, 51–52, 63–65, 74, 76–82, 86,
    88, 90–93, 98–99, 116, 118, 124,
    254, 306–307, 311, 388, 391, 408,
    420, 423
  devices, 76–81
    bucket lift, 76–77, 79–80
    noria, 79–80
    shaduf, 77, 80
    water ladder, 77, 79–81, 93
  energy cost of, 81
  energy return of, 93, 95
  power requirements of, 80–81

Japan, 60, 207, 314–315, 324, 327–329,
    331, 343, 364, 376–377, 379, 387,
    416, 429, 434, 439
  agriculture in, 95, 111, 115–116, 169
  energy use in, 272, 277–278, 280–281,
    284, 290, 305, 321, 348–350, 358,
    363–364, 366
Java, 50, 95, 115

Kenya, 24, 356
Kuwait, 279, 367, 380

Labor, 18–19, 36–37, 45, 50, 52, 58–61,
    94, 101, 108–110, 127, 130–132,
    171, 320, 324, 355, 360, 364,
    387–392, 394, 397, 405, 407–408,
    410, 419–420, 423
  in agriculture, 64–65, 74, 94, 101, 104,
    108–110, 112–115, 122, 125–126,
    307
  child, 131–132
  energy cost of, 18–19, 58–60
  massed, 127, 181, 390–391
Legumes, 45, 49, 58–59, 84–87, 95, 100,
    111, 121
Levers, 6, 23, 26, 77, 128, 130, 133–35,
    180, 202, 218–219, 245
Life, quality of, 1, 259, 295, 351,
    355–364, 405, 409–410, 413–414,
    421, 433–434, 437

Lights, 14, 176–177, 248, 257–262, 268,
    297, 402, 411, 436
  arc, 257–258
  cost of, 406
  efficiency of, 402
  fluorescent, 14, 262, 402, 411
  incandescent, 232, 258–262, 268, 297,
    402, 436
Loads, 131, 133–134, 136, 145, 187,
    199, 221–222
  carrying of, 18, 132, 180–181,
    231–232
  pulling of, 71, 128, 181–182, 185,
    190
Locomotives
  Diesel, 290
  steam, 139, 161–162, 187, 240–242,
    244, 255, 272, 298, 302–303, 316,
    403
London, 152, 168, 229, 232, 239, 241,
    258, 261, 265, 325, 333, 353–354,
    428
  transportation in, 146, 186, 244
Longitude
  determination of, 425–426

Maasai, 46–47
Machines, 138, 151, 154, 157, 160, 187,
    199, 202, 235–239, 241–243, 249,
    250–255, 258, 263, 266–267, 290,
    314–315, 337, 360, 371, 375, 379,
    438. *See also* Airplanes; Cars;
    Engines; Motors
Macronutrients, 10, 40, 51, 82, 307
Malnutrition, 111, 120–121, 413
Mammoths, 27, 33–35, 37
Manufacturing, 83, 127, 131, 155, 157,
    168–169, 236, 238, 242–243, 255,
    267, 314, 318–321, 324, 340,
    348–349, 376, 388, 390, 392, 413,
    416, 420, 422
  assembly line in, 153, 320
  labor in, 320–321